AGGREGATION OF THERAPEUTIC PROTEINS

AGGREGATION OF THERAPEUTIC PROTEINS

Edited by

Wei Wang

Christopher J. Roberts

WILEY

A John Wiley & Sons, Inc., Publication

Published by John Wiley & Sons, Inc., Hoboken, New Jersey
Published simultaneously in Canada

Library of Congress Cataloging-in-Publication Data
Aggregation of therapeutic proteins / edited by Wei Wang, Christopher J. Roberts.
p. ; cm.
Includes bibliographical references.
ISBN 978-0-470-41196-4 (hardback)
1. Protein drugs. 2. Aggregation (Chemistry) I. Wang, Wei, 1957 Mar. 10– II. Roberts, Christopher John, 1972–
[DNLM: 1. Recombinant Proteins—therapeutic use. 2. Cell Aggregation. 3. Protein Conformation. 4. Protein Folding. 5. Recombinant Proteins—metabolism. QU 55 A266 2010]
RS431.P75A34 2010
615.5′8—dc22

2010008428

Printed in the United States of America

10 9 8 7 6 5 4 3 2 1

To my parents, Jiamei Wang and Yuzhen Ma, for their unlimited love and great appreciation of higher education.

WW

To my wife, for her unending patience, support, and love.

CJR

CONTENTS

Preface xvii

Contributors xxi

1. Fundamental Structures and Behaviors of Proteins 1
Jennifer S. Laurence and C. Russell Middaugh

1.1 The Problem of Protein Aggregation 1
1.1.1 Structural Features of Proteins 4
1.1.2 Structural Features of Protein Aggregates 5

1.2 Parallels to Protein Folding 11

1.3 Views of Protein Stability and Aggregation 12
1.3.1 Physicochemical Properties of Proteins 12
1.3.2 Surface Properties and Packing Arrangements 18
1.3.3 Solvent Interactions 21

1.4 Models of Aggregation 22
1.4.1 Monomer Conversion 23
1.4.2 Oligomeric Intermediates 24
1.4.3 Nucleation in Protein Folding 25
1.4.4 Domain Swapping 26

1.5 Models of Protein Folding 29
1.5.1 Classical Models Based on Chemical Equilibrium 29
1.5.2 Statistical Mechanical Models 31
1.5.3 Computational Models 33
1.5.4 Application of Coarse-Grained Models 35
1.5.5 Information Theory (IT) 37

1.6 Influences of Chemical Alteration on Aggregation 40
1.6.1 Length of the Polypeptide Chain 41
1.6.2 Methionine Oxidation 41
1.6.3 Covalent Cross-Linking of Strands 42

1.6.4 Deamidation 44
1.6.5 Proline Isomerization 45
1.7 Approaches to Predicting Aggregation 46
1.8 Conclusions 49
References 50

2. Protein Aggregation Pathways, Kinetics, and Thermodynamics 63
Yi Li and Christopher J. Roberts

2.1 Introduction 63
2.2 Native and Nonnative Aggregation Pathways 66
2.3 Thermodynamics of Reversible Self-Association 69
2.3.1 Self-Association of Folded/Native Proteins 70
2.3.2 Nonnative Self-Association 73
2.4 Aggregation Kinetics and Distinguishing Kinetic Pathways 75
2.4.1 Aggregation Kinetics 75
2.4.2 Distinguishing Aggregation Pathways and Rate-Limiting Steps 78
2.4.3 Influence of Adsorption to Macroscopic Surfaces 80
2.5 Chemical Modifications 80
2.6 Effects of Cosolvents or Cosolutes 82
2.6.1 Transfer Free Energies and Preferential Interaction Parameters 82
2.6.2 Relating $\Gamma_{32}^{(\mu_3)}$ to Molecular Interactions 85
2.6.3 An Additive Approach to $\Delta\mu_2^{tr}$, $\partial\mu_2^{ex}/\partial m_3$, and $\Gamma_{32}^{(\mu_3)}$ 89
2.6.4 Predicting Cosolvent/Cosolute Effects on Unfolding, Association, and Aggregation 93
Appendix—Derivation of Γ_{32} for van der Waals (vdW) Mixture 94
Acknowledgments 97
References 97

3. Identification and Impact of Aggregation-Prone Regions in Proteins and Therapeutic Monoclonal Antibodies 103
Sandeep Kumar, Xiaoling Wang, and Satish K. Singh

3.1 Introduction 103
3.2 Energy Landscapes, Protein Folding, and Aggregation 105
3.3 Prediction of APRs in Proteins and Biotherapeutics 106

3.3.1 Computational Tools 106
3.3.2 Experimentally Studied Aggregation-Prone Sequences and Sequence Patterns 109
3.3.3 Prediction of APRs in Therapeutic mAbs 110
3.3.4 Other Useful Studies 113

3.4 Conclusions and Future Directions 114

Acknowledgments 115

References 115

4. External Factors Affecting Protein Aggregation 119
Wei Wang, Ning Li, and Stan Speaker

4.1 Introduction 119

4.2 Protein Aggregation Pathways 120
4.2.1 Physical Aggregation through Formation of Unfolding Intermediates 120
4.2.2 Direct Aggregation through Self-Association or Chemical Linkages 127
4.2.3 Indirect Aggregation through Chemical Degradation 129
4.2.4 Protein Aggregation from the Unfolded State 129
4.2.5 Aggregation Nucleation 130
4.2.6 Multiplicity in Protein Aggregation 131

4.3 Effects of Temperature 132
4.3.1 Relationship between Temperature and Protein Stability 132
4.3.2 Effect of High Temperature on Protein Aggregation 133
4.3.3 Melting Temperature (T_m) and Protein Aggregation 135
4.3.4 Effects of Low Temperature 136

4.4 Effects of Solution Conditions and Composition on Protein Aggregation 136
4.4.1 Solution pH 137
4.4.2 Type and Concentration of Buffering Agents 140
4.4.3 Ionic Strength 141
4.4.4 Excipients/Additives 143
4.4.5 Protein Concentration 154
4.4.6 Metal Ions 157

4.4.7 Denaturant and Reducing Agents 159
4.4.8 Impurities 160
4.4.9 Organic Solvents 161
4.4.10 Containers/Closures 162
4.4.11 Sources of Proteins 163
4.4.12 Light 163

4.5 Effects of Processing Steps on Protein Aggregation 164
4.5.1 Fermentation/Expression 164
4.5.2 Unfolding/Refolding 165
4.5.3 Purification 168
4.5.4 Freeze-Thaw 168
4.5.5 Agitation 169
4.5.6 High Hydrostatic Pressure 170
4.5.7 Drying 171
4.5.8 Preparation of Protein Delivery Systems 173
4.5.9 Analytical Methodologies 173
4.5.10 Miscellaneous Processes 173

4.6 Effects of Solid-State Condition and Composition on Protein Aggregation 174
4.6.1 Solid-State "pH" 174
4.6.2 Excipients and Excipient Levels 174
4.6.3 Physical State of the Solid 176
4.6.4 Moisture Content 177

4.7 Summary 177

Acknowledgment 178

References 178

5. Experimental Detection and Characterization of Protein Aggregates **205**
Vikas K. Sharma and Devendra S. Kalonia

5.1 Introduction 205

5.2 Aggregate Classification 206
5.2.1 Aggregate Nomenclature 209

5.3 Analytical Tools for the Characterization of Aggregates 212
5.3.1 SEC 221
5.3.2 Light Scattering 223
5.3.3 Analytical Ultracentrifugation (AUC) 228

5.3.4 FFF 231
5.3.5 Electrophoresis 233
5.3.6 Turbidimetry and Nephelometry 235
5.3.7 Analytical Techniques for Subvisible Particles 237
5.3.8 Visible Particles 240
5.3.9 Miscellaneous Technologies for Characterization of Protein Aggregates 241
5.3.10 Protein Structural Characterization in Aggregates: Spectroscopy-Based Techniques 245

5.4 Summary 246

References 247

6. Approaches to Control Protein Aggregation during Bulk Production 257

Linda O. Narhi, Yijia Jiang, Rohini Deshpande, Sohye Kang, and Joseph Shultz

6.1 Introduction 257

6.2 Candidate Selection 257
6.2.1 Sequence Analysis 258
6.2.2 Stability to Process Conditions 259
6.2.3 Focus on Stability to Formulation and Long-Term Storage Conditions 265
6.2.4 Specific Considerations for Mammalian Cell Culture-Derived Protein 268
6.2.5 Specific Considerations for Bacteria-Derived Protein 268

6.3 Protein Aggregation and Cell Culture 269
6.3.1 Expression Strategies 269
6.3.2 Bioreactor Conditions 270

6.4 Protein Aggregation and Purification 271
6.4.1 Preventing Aggregation with Appropriate Solvent Conditions 274
6.4.2 Removal of Aggregates during Processing 281
6.4.3 Specific Considerations for Mammalian Cell-Derived Protein 285
6.4.4 Specific Considerations for Bacterial-Derived Proteins 285
6.4.5 Bulk DS Stability and Storage 292
6.4.6 Scale-Up 293

6.5 Summary 295

References 295

7. Protein Aggregation and Particle Formation: Effects of Formulation, Interfaces, and Drug Product Manufacturing Operations **301**

Hanns-Christian Mahler, Stefan Fischer, Theodore W. Randolph, and John F. Carpenter

7.1 Introduction 301

7.2 Roles of Conformational and Colloidal Stability in Reducing Rates of Aggregation 302

7.2.1 Conformational Stability 303

7.2.2 Colloidal Stability 305

7.3 Effects of Interfaces on Protein Aggregation 305

7.3.1 Effects of Air–Water Interface 305

7.3.2 Protein Aggregation at Solid–Liquid Interfaces 306

7.3.3 Effects of Excipients on Structural Perturbation and Aggregation at Surfaces 307

7.3.4 Interactions of Proteins with Microparticle Surfaces 309

7.3.5 Protein Aggregation at Silicone Oil/Water Interfaces 309

7.3.6 Shear Effects on Protein Aggregation 309

7.4 Critical Processing Steps during Drug Product Manufacturing of Biopharmaceuticals 310

7.4.1 Freezing and Thawing Operations 310

7.4.2 Mixing of Bulk Drug Solution after Thawing or Pooling 313

7.4.3 Preparation of Final Formulation with Excipients 313

7.4.4 Filtration 314

7.4.5 Filling 314

7.5 Particles in Parenteral Products and Visible Inspection 316

7.5.1 Introduction: Understanding of Particles from the Parenteral Product Viewpoint is Important for Protein Biopharmaceutical Developers 316

7.5.2 The History of Visible Inspection and Particle Contamination Control in Parenteral Products 316

7.5.3 Current Regulatory Requirements on Particulate Contamination and Control of Subvisible and Visible Particles in Parenteral Products in the EU, United States, and Japan 317
7.5.4 Subvisible Particle Measurements 317
7.5.5 Subvisible Particle Acceptance Criteria 318
7.5.6 Visual Inspection Requirements 320
7.5.7 Visual Inspection Methods 321
7.5.8 What Does "Visible" Mean? 322
7.5.9 Potential Causes of Particulate Contamination in Parenteral Products 323

7.6 Summary and Outlook 324

References 325

8. Approaches to Managing Protein Aggregation in Product Development **333**
Wei Wang and Nicholas W. Warne

8.1 Introduction 333

8.2 Approaches in Formulation Development 334
8.2.1 Traditional Formulation Development 334
8.2.2 High-Throughput Formulation Development (HTFD) 344
8.2.3 Computer-Assisted Design of Formulations 344

8.3 Protection of Proteins in Various Processing Steps 345
8.3.1 Shaking 346
8.3.2 Freezing/Thawing 346
8.3.3 Manufacturing Processes for Liquid Drug Product 347
8.3.4 Drying 348
8.3.5 Reconstitution 350
8.3.6 Preparation of Controlled Protein Delivery Systems 350
8.3.7 Miscellaneous Processes 351

8.4 Aggregation Control by Structural Modifications 351
8.4.1 Mutagenesis 351
8.4.2 PEGylation 352
8.4.3 Glycosylation 353

8.5 Summary 353

References 354

9. Case Studies Involving Protein Aggregation **367**

Rahul S. Rajan, Tiansheng Li, and Tsutomu Arakawa

9.1 Introduction 367

9.2 Case Study 1: Aggregation in the Liquid State: The Role of Osmolytes in Stabilizing KGF toward Aggregation 368

9.2.1 Background 368

9.2.2 Model Protein and Conditions 369

9.2.3 Results and Interpretation 369

9.2.4 Discussion and Lessons Learned 372

9.3 Case Study 2: Aggregation in the Liquid State: Heterogeneity and Non-Linearity in IgG2 Aggregation during Long-Term Storage 376

9.3.1 Background 376

9.3.2 Model Protein and Conditions 378

9.3.3 Results and Interpretation 378

9.3.4 Discussion and Lessons Learned 380

9.4 Case Study 3: Aggregation in the Frozen State: The Role of Excipient Crystallization 381

9.4.1 Background 381

9.4.2 Model Protein and Conditions 382

9.4.3 Results and Interpretation 382

9.4.4 Discussion and Lessons Learned 383

9.5 Case Studies 4 and 5: Aggregation in the Lyophilized State: Role of Residual Moisture and Mechanisms of Excipient Stabilization 385

9.5.1 Background 385

9.5.2 Model Protein and Conditions (Case Study 4) 386

9.5.3 Results and Interpretation (Case Study 4) 387

9.5.4 Model Protein and Conditions (Case Study 5) 388

9.5.5 Results and Interpretation (Case Study 5) 389

9.5.6 Discussion and Lessons Learned 390

9.6 Case Study 6: Protein Particulation Due to Nucleation by Foreign Material in Fill/Finish Manufacturing Operations 391

9.6.1 Background 391

9.6.2 Model Protein and Conditions 391

9.6.3 Results and Interpretation 392
9.6.4 Discussion and Lessons Learned 393
9.7 Overall Discussion 394
9.7.1 Are Aggregates Created Equal? 394
9.7.2 Future Directions 396
Acknowledgments 396
References 396

10. Aggregation and Immunogenicity of Therapeutic Proteins 403
Vasco Filipe, Andrea Hawe, Huub Schellekens, and Wim Jiskoot

10.1 Introduction 403
10.2 Immunogenicity of Therapeutic Proteins 404
10.2.1 Incidence of Immunogenicity during Therapy 404
10.2.2 Consequences of Protein Immunogenicity for Patients 405
10.2.3 Factors Playing a Role in Protein Immunogenicity 406
10.3 Immune Mechanisms Related to Protein Immunogenicity 409
10.3.1 Immune Responses to Therapeutic Proteins 410
10.3.2 Immune Tolerance 411
10.3.3 B-Cell Activation Mechanisms 411
10.3.4 Breaking of B-Cell Immune Tolerance 415
10.4 Aggregates and Immunogenicity 415
10.4.1 Evidence of Aggregates as a Risk Factor for Immunogenicity 417
10.4.2 Proposed Mechanisms of Antibody Induction by Protein Aggregates 420
10.4.3 Possible Factors Responsible for the Immunogenicity of Protein Aggregates 423
10.4.4 Risk Analysis of Protein Aggregates 426
10.4.5 Predictive Models to Access the Immunogenicity of Aggregates 426
10.5 Conclusions 427
References 428

11. Regulatory Perspective on Aggregates as a Product Quality Attribute 435
Wendy C. Weinberg, Linan Ha, Susan L. Kirshner, and Daniela I. Verthelyi

11.1 Introduction 435

11.2 An Overview of the Regulatory Process 436

11.2.1 Drug Development to Licensure 436

11.2.2 Pre- and Post-Marketing Comparability 438

11.3 Product Aggregates and Safety Concerns 438

11.3.1 Aggregates and Immunogenicity 438

11.4 The Assessment of Aggregates: Regulatory Approaches to Controlling Product Aggregation 440

11.4.1 Factors Contributing to the Formation of Aggregates 440

11.4.2 Selection and Validation of Detection Methods 440

11.4.3 Setting Specifications for Aggregates as Part of Lot Release and Stability Testing 444

11.4.4 Commonly Used Guidance Documents Relating to Control of Aggregates 445

11.5 Future Challenges 446

11.5.1 Control of Subvisible Particulates: A Gap in Regulatory Science 446

11.5.2 Assessment of Biosimilars 446

11.6 Summary 447

Acknowledgments 447

Disclaimer 447

References 448

Index **453**

PREFACE

Protein therapeutics have played an indispensable and increasing role in human health for the past three decades. Protein drugs, or biopharmaceuticals, are used in the treatment of a number of debilitating or even deadly human diseases, including diabetes and a number of forms of cancer. The growth of therapeutic proteins as candidates in drug development pipelines is at a record speed, outpacing significantly that for small molecule drugs. Such a pace is fueled by the record investment in biotechnology. According to a recent survey, the global biotech industry raised a total of $53 billion in 2007, a 13% growth compared to the previous year. It is estimated that the high annual growth rate may continue at levels above 15% over the next 10 years.

However, development of commercial protein drugs has not been straightforward in most cases. In comparison to small molecule drugs, production of recombinant protein drugs requires sophisticated and careful control of long fermentation/cell culture and purification processes, characterization with a much more complex suite of analytical chemistry tools, and protection of proteins against a variety of sources for instability and degradation. One such instability is the high tendency of protein molecules to aggregate under a wide range of processing and storage conditions. This propensity for aggregation is arguably the most common and troubling manifestation of protein instability during development of protein biotherapeutics. This is because protein aggregates usually exhibit either reduced or no biological activity and, more importantly, might have stronger immunogenicity and/or cellular toxicities. As such, protein aggregation has to be controlled to a satisfactory level before such a product can be commercialized. Although significant progress has been made in recent years, control or even moderate inhibition of protein aggregation has been largely semiempirical in practice.

Much of the published fundamental work in protein and polypeptide aggregation over the past three decades has focused on two main areas: aggregation as an off-pathway reaction during recombinant *in vivo* expression and aggregates as potentially causative agents in so-called protein deposition diseases such as Alzheimer's disease, Parkinson's disease, prion diseases (bovine spongiform encephalopathy and Creutzfeldt–Jakob disease), Huntington's disease, Down's syndrome, as well as a variety of other inherited or contracted amyloidoses. At least some of the basic physics and chemistry of many of the well-studied systems are expected to hold for aggregation of therapeutic proteins, although the latter are often larger and more structurally complex than many of their model counterparts. In addition, therapeutic proteins often encounter a much more diverse set of environments than those typically explored in biological contexts, for example, extremes of pH, solvent composition, temperature, and exposure to air–water and other bulk interfaces.

Therefore, understanding and overcoming protein aggregation remains a key area of intensive research, both in academia and biopharmaceutical companies. It is inherently interdisciplinary, spanning fields such as structural biology, chemistry and biochemistry, biophysics, pharmaceutics, and biochemical engineering. Although it is impossible to be exhaustive in any text covering so broad a field as protein aggregation, the intention of this text is to summarize current understanding and recent progress regarding protein aggregation in the context of biopharmaceutical products. This will hopefully facilitate and stimulate new and continued investigations into the principles of protein aggregation, and the application of those principles to more rapid and predictable commercialization of protein drugs and effective protection of human lives in the future.

This book is composed of 11 chapters. The first two chapters offer the audience overviews of the principles and importance of protein structure, folding and misfolding, native versus non-native interactions, self-association, competing aggregation pathways, how these manifest thermodynamically and kinetically, and how they change as a function of canonical experimental parameters such as solvent composition and temperature.

Chapters 3 and 4, respectively, discuss the internal or intrinsic (protein sequence-related) and external or extrinsic (environmental) factors that affect/control protein aggregation. Chapter 5 discusses and compares analytical methodologies available for monitoring and/or characterizing protein aggregation in research and in commercial development settings, and proposes a systematic nomenclature to help alleviate confusion and ambiguities that often arise when comparing reports from different laboratories.

Chapters 6–8 focus on control, inhibition, and monitoring of aggregation: during expression and purification (Chapter 6), during downstream processing (Chapters 6 and 7), from the context of product formulation (Chapters 6–8), as it relates to bulk interfaces (Chapter 7), when trying to optimize the final drug product (Chapter 8), and how this is coupled to current and historical questions of protein and non-protein particulates in biopharmaceutical products (Chapter 7).

Illustrative case studies on protein aggregation are offered in Chapter 9, providing more detailed discussions to complement the more general presentations in previous chapters. Chapter 10 presents the current understanding as well as key outstanding questions regarding immunogenicity issues for therapeutic proteins and protein aggregates. The last chapter provides an overview of the regulatory process and considerations in developing protein drugs, with a focus on protein aggregation and stability issues.

We are indebted to all of the authors who contributed to this text and offer our sincere thanks for their efforts and dedication, as well as their patience with us during the long process that building a text such as this becomes. We also thank Dr. Erinc Sahin for critical reading of the full manuscript and for composing the cover art.

We hope the resulting volume serves as a valuable and comprehensive resource for scientists and engineers in the industry and academia and for both experienced and new researchers in the field.

Wei Wang
Chris Roberts
February 2010

CONTRIBUTORS

Tsutomu Arakawa, Alliance Protein Laboratories, Thousand Oaks, CA

John F. Carpenter, Center for Pharmaceutical Biotechnology, Department of Pharmaceutical Sciences, University of Colorado Denver, Aurora, CO

Rohini Deshpande, Amgen, Inc., Thousand Oaks, CA

Vasco Filipe, Leiden/Amsterdam Center for Drug Research (LACDR), Division of Drug Delivery Technology, Biologics Formulation Group, Gorlaeus Laboratories, Leiden, The Netherlands

Stefan Fischer, Formulation R&D Biologics and Parenterals, Pharmaceutical and Analytical R&D, F. Hoffmann-La Roche Ltd., Basel, Switzerland

Linan Ha, Division of Monoclonal Antibodies, Office of Biotechnology Products, Office of Pharmaceutical Science, Center for Drug Evaluation and Research, US Food and Drug Administration, Bethesda, MD

Andrea Hawe, Leiden/Amsterdam Center for Drug Research (LACDR), Division of Drug Delivery Technology, Biologics Formulation Group, Gorlaeus Laboratories, Leiden, The Netherlands

Yijia Jiang, Amgen, Inc., Thousand Oaks, CA

Wim Jiskoot, Leiden/Amsterdam Center for Drug Research (LACDR), Division of Drug Delivery Technology, Biologics Formulation Group, Gorlaeus Laboratories, Leiden, The Netherlands

Devendra S. Kalonia, Department of Pharmaceutical Sciences, University of Connecticut, Storrs, CT

Sohye Kang, Amgen, Inc., Thousand Oaks, CA

Susan L. Kirshner, Division of Therapeutic Proteins, Office of Biotechnology Products, Office of Pharmaceutical Science, Center for Drug Evaluation and Research, US Food and Drug Administration, Bethesda, MD

Sandeep Kumar, BioTherapeutics Pharmaceutical Sciences, Pfizer Inc., Chesterfield, MO

Jennifer S. Laurence, Department of Pharmaceutical Chemistry, University of Kansas, Lawrence, KS

Ning Li, BioTherapeutics Pharmaceutical Sciences, Pfizer Inc., Chesterfield, MO

Tiansheng Li, HTL Biosolutions Inc., Newbury Park, CA

Yi Li, Merck Research Laboratories, West Point, PA

Hanns-Christian Mahler, Pharmaceutical and Device Development, Pharma Technical Development Biologics Europe, F. Hoffmann-La Roche Ltd., Basel, Switzerland

C. Russell Middaugh, Department of Pharmaceutical Chemistry, University of Kansas, Lawrence, KS

Linda O. Narhi, Amgen, Inc., Thousand Oaks, CA

Rahul S. Rajan, Process and Product Development, Amgen Inc., Thousand Oaks, CA

Theodore W. Randolph, Center for Pharmaceutical Biotechnology, Department of Chemical and Biological Engineering, University of Colorado, Boulder, CO

Christopher J. Roberts, Department of Chemical Engineering and Center for Molecular and Engineering Thermodynamics, University of Delaware, Newark, DE

Huub Schellekens, Department of Pharmaceutics, Utrecht Institute for Pharmaceutical Sciences (UIPS), Utrecht University, Utrecht, The Netherlands

Vikas K. Sharma, Early Stage Pharmaceutical Development, Genentech, Inc., South San Francisco, CA

Joseph Shultz, Amgen, Inc., Thousand Oaks, CA

Satish K. Singh, BioTherapeutics Pharmaceutical Sciences, Pfizer Inc., Chesterfield, MO

Stan Speaker, BioTherapeutics Pharmaceutical Sciences, Pfizer Inc., Chesterfield, MO

Daniela I. Verthelyi, Division of Therapeutic Proteins, Office of Biotechnology Products, Office of Pharmaceutical Science, Center for Drug Evaluation and Research, US Food and Drug Administration, Bethesda, MD

Wei Wang, BioTherapeutics Pharmaceutical Sciences, Pfizer Inc., Chesterfield, MO

Xiaoling Wang, BioTherapeutics Pharmaceutical Sciences, Pfizer Inc., Chesterfield, MO

Nicholas W. Warne, BioTherapeutics Pharmaceutical Sciences, Pfizer Inc., Andover, MA

Wendy C. Weinberg, Division of Monoclonal Antibodies, Office of Biotechnology Products, Office of Pharmaceutical Science, Center for Drug Evaluation and Research, US Food and Drug Administration, Bethesda, MD

CHAPTER 1

Fundamental Structures and Behaviors of Proteins

JENNIFER S. LAURENCE and C. RUSSELL MIDDAUGH

Protein aggregation has been increasingly recognized as a problem limiting the efficacy and shelf life of protein therapeutics and as an indicator and cause of numerous disease states. Elucidating the molecular mechanisms behind aggregation has become a central focus of investigation in order to improve therapeutics and to understand the relationship between aggregate formation and cellular toxicity in protein misfolding diseases. Innovations in analysis techniques, particularly of solid-state materials, and computational molecular modeling approaches have provided higher resolution information about the structure of aggregates as well as key insights into the mechanisms of aggregate formation. These breakthroughs, coupled with understanding gained from solution experiments and biological systems, have just begun to enable strategies to combat aggregation, including the design and evaluation of peptides and small molecules that inhibit the growth or that facilitate the dissociation of aggregates. This chapter describes the fundamental properties of proteins and the current understanding of underlying mechanisms that influence native folding and the formation of aggregates.

1.1 THE PROBLEM OF PROTEIN AGGREGATION

Protein aggregation has significant influence in the pathology, onset, and progression of most, if not all, misfolding diseases. Over 40 human diseases have been linked to aggregation of a specific protein, including hemoglobin in sickle cell anemia, the widely recognized Aβ peptides in Alzheimer's disease, the PrP prion protein in Creutzfeldt–Jakob's and related diseases, expanded polyglutamine tracts in Huntington's disease, amylin-induced β-cell death in

Aggregation of Therapeutic Proteins, Edited by Wei Wang and Christopher J. Roberts

diabetes, and α-synuclein in Parkinson's disease.[1] Moreover, studies of non-disease-associated proteins *in vitro* show that aggregates and amyloid fibers can be induced to occur from almost any protein, suggesting it is a ubiquitous phenomenon reflecting a common mechanism.[2] Therapeutic proteins used to treat various diseases can also produce ill effects when aggregates are present, in some cases contributing to amyloid plaque formation *in vivo*.[3,4] Aggregates have been observed to form in therapeutic proteins during purification and storage, and the administration of proteins containing aggregates has been shown to stimulate immune responses, causing effects ranging from mild skin irritation to anaphylaxis.[5,6] As such, major efforts are underway to stabilize therapeutic proteins against aggregation. Thus, the goal of understanding the fundamental properties of proteins that contribute to aggregation and the mechanisms by which they aggregate is of critical importance for determining how to prevent and treat numerous diseases.

In vivo protein aggregation appears to be an ever-present problem caused by thermal fluctuations and chemical changes that disrupt the physical structure of these delicate molecules. Consequently, cells have evolved several mechanisms by which they prevent aggregates from interfering with normal function.[7] Improperly folded proteins are removed from cells before they can initiate aggregation by being degraded into smaller peptides via the proteosome or lysosomal enzymes. Alternatively, intracellular proteins can be refolded to their native conformation by interaction with chaperone proteins, which are often expressed at elevated levels in response to thermal (heat shock) or chemical stress. Chaperones bind to hydrophobic patches on misfolded proteins and use an energy-dependent process to alter their conformation, therein providing the protein with additional attempts to find its native fold.[8] When the capacity of the aforementioned machinery is exceeded, aggregates may form,[9] as is often observed in recombinant expression systems. As one might expect, coexpression with chaperones can reduce the formation of aggregates in some cases. Chaperones have been demonstrated to affect aggregation not only by improving recovery of soluble protein but also conversely to promote aggregation when present at high levels. When aggregates form *in vivo*, sequestration mechanisms exist that recognize aggregated species and shuttle them to designated storage locations within the cell, such as the bacterial inclusion body and aggresome or newly discovered IPOD and JUNQ sites in eukaryotic cells.[10,11] When the cellular machinery is overwhelmed by excessive damage to normal proteins or by mutations that generate a less stable form of a protein that accelerates aggregation, disease or death may result. Evidence for this is found in that increased amounts of proteosomal and chaperone proteins are found colocalized with aggregates in these inclusions.

Recombinant expression has become an increasingly important method for producing large amounts of protein for therapeutic and biotechnology applications. Production of recombinant protein is often frustrated by aggregation in the host. Yield can sometimes be improved by decreasing the temperature at which the protein is made or by coexpression with chaperone proteins (e.g., GroEL) to aid folding *in vivo* and to reduce sequestration to inclusion bodies.[8] Nonetheless, proteins are often shuttled to inclusion bodies. Aggregated proteins, however, may be folded *in vitro* from the insoluble state. Single-domain proteins less than 150 residues, which are directed to inclusion bodies, can sometimes be extracted from the solid aggregate and refolded. Denaturing conditions are used to disrupt associations between chains, and the denatured material is diluted into a non-denaturing solution so that it may refold into its native form. When the native form of the protein contains disulfide bonds, folding is carried out under defined redox conditions to facilitate proper disulfide formation. This approach is not very efficient, typically resulting in a substantial fraction of the protein returning to an insoluble state. This observation suggests that proteins may follow different pathways during the course of folding, of which only some are productive. Very limited success has been had using this approach with large, multidomain proteins or those with numerous or more complex posttranslational modifications. The difficulty in refolding these proteins probably derives from increased competition between alternative interactions with those of the native state. These incorrect associations may lead to misfolding when the rate of protein production or the context in which the protein is produced is altered. Addition of chaperones at the dilution step has been used to enhance refolding of proteins that otherwise aggregate. The strategy is also being applied to stabilize purified proteins during storage. Once the active form is purified, proteins are commonly maintained at cold temperatures to restrict their conformational flexibility and to preserve their structural integrity.

Protein folding and unfolding may not follow the same pathway.[12] Some proteins fold and unfold reversibly, yet may accomplish each event using a different approach to overcome transition state barriers. Other proteins require assistance to attain their native conformation but subsequently are quite resistant to unfolding. These observations suggest that a protein's ability to arrive at its native conformation and to maintain it is not necessarily ruled to the same degree by the same parameters. Proteins are not static and undergo a variety of different types of conformational fluctuations. The range of states sampled is dictated by noncovalent interactions that stabilize the native fold and the effect of external influences like temperature and solution conditions on their interactions. Factors that drive folding influence the stability of the folded form, but coincident interactions that develop as a consequence of the folded state also impact retention of the native fold and help determine the frequency of transitions to partially unfolded conformations and subsequent progression to aggregated states. Once the folded protein is obtained, stabilization against transitions to aggregated states becomes a critical issue. Understanding how to prevent aggregation has primarily been based on empirical studies. The next objective is to elucidate the mechanisms that determine how transitions from the native ensemble promote aggregation.

1.1.1 Structural Features of Proteins

Proteins are linear polymers. Their primary structure is composed of 20 naturally occurring amino acids having diverse chemical properties. The amino acids are typically alpha amino acids and have L chirality. Each is joined by a peptide bond, which has a planar character that restricts the conformational freedom of the backbone of the polypeptide chain. As such, common structural features are observed among folded proteins, most broadly falling into the categories of alpha helix, beta sheet, turns, and disordered regions. These first three secondary structural elements are developed as a result of hydrogen bonding interactions that involve atoms from within the polypeptide backbone. Disordered regions lack such hydrogen bonding patterns. Alpha helices are almost always right handed and have a register in which the carbonyl oxygen from residue i forms a hydrogen bond with the amide proton four residues to its C-terminus (i + 4). The alpha helix contains 3.6 residues per turn, and due to the slight offset in vertical alignment, a secondary twist develops with elongation. Rarely, a short 3_{10}-helix has been observed to form, in which i to i + 3 bonding occurs. The even rarer π-helix utilizes i to i + 5 bonding. Due to favored dihedral backbone angles and steric constraints, alpha is the most favorable helical organization. It often occurs in isolation, whereas the other two forms are found only in small segments in folded proteins in which the structural context provides stabilization for these less favorable structural elements. Beta-sheet structure is also favorable and can arise from a parallel or antiparallel alignment of the strands. The pattern of hydrogen bonding differs between these two sheet organizations. Antiparallel strands may form from contiguous or discontinuous primary structure, but parallel association necessarily occurs between sequences that have intervening secondary structural elements.

Contiguous stretches of repeated H-bonds stabilize each structural element and help compact the polymer within local regions of the sequence. This limits the number of possible arrangements between distal segments, which also facilitates the establishment of a preferred three-dimensional conformation (tertiary structure). Three-dimensional coalescence into a compact state generally relies on interactions between amino acid side chains. The diverse chemical composition of the side chains produces both attractive and repulsive forces, and the native configuration derives from the formation of the most energetically favorable associations between distal moieties that stabilize the packed arrangement within a folded domain. The majority of interactions that contribute to protein folding are noncovalent, but covalent bonding between the thiol-containing moieties of cysteine residues may occur to generate a disulfide bond under oxidizing conditions. Disulfide bonds are common among secreted proteins, where they often greatly enhance the ability of the protein to resist unfolding.

Globular domain folds are classified into families that range from all helical to mixed alpha-beta to all beta composition. Regardless of the domain architecture, separate polypeptide chains can further associate into homotypic or heterotypic oligomers to yield a quaternary structure. Individual subunits in an oligomeric complex can simply physically associate based on surface complementation, but they may also be covalently tethered. Covalent attachment ensures close proximity and is most often accomplished through intermolecular disulfide bonds.

Noncovalent association between subunits vary in affinity based on the same principles that dictate protein folding, and several modes of interaction have been described, including lock and key, induced fit, and preexisting equilibrium/conformational selection mechanisms. Lock-and-key binding implies the structure is unaltered by the binding event. Induced fit models suggest that the protein adopts a new state in response to binding to its partner, whereas conformational selection indicates that in the associated complex, an existing state is stabilized. Analogous modes of interaction may also apply to protein–protein associations that pertain to physical aggregation. A relatively new area of investigation has demonstrated the diversity of conformations that can result from the same sequence. For example, natively disordered proteins have been suggested to adopt distinct conformations in different contexts to perform discrete functions. Recent studies also reveal that globular proteins can maintain more than one unique stable conformation. Moreover, crystal structures are often reported for the same protein in distinct oligomeric states depending on solution conditions. These findings suggest that preferences in the conformation of a protein, even those below the current limit of detection, are influenced by the context in which the protein resides.

1.1.2 Structural Features of Protein Aggregates

Macroscopic attributes of aggregates have been described from data acquired using a variety of microscopy techniques. In the most general terms, the morphological features commonly observed are usually categorized as amorphous or fibrillar. Amorphous aggregates are present in inclusion bodies *in vivo* and often emerge during the course of processing and storing protein samples. Amorphous aggregates lack long-range order and are often opaque if they are not soluble. They were originally thought to contain completely unfolded material held together by random associations between hydrophobic residues. The hypothesis that amorphous aggregates lack discernable structure was derived from (1) early observation that harsh denaturants like sodium dodecyl sulfate (SDS), urea, and guanidine-HCl (Gdn) are required to resolubilize proteins from inclusion bodies; and (2) a lack of data concerning structural features owing to the fact that amorphous materials often scatter light, interfering with the spectroscopic analyses typically used to characterize structure. In contrast, some aggregates retain native activity, and the active form of some proteins can only be recovered from inclusion bodies when mild denaturing conditions are used, whereas aggressive denaturation leads to an inability to refold the protein (e.g., hGH),[13] suggesting that the species present in the inclusion body contain elements of native structure that facilitate refolding. Additionally, staining methods have been used to reveal the presence of regular structure within amorphous aggregates. Staining of some aggregates and not others by dyes like Congo red (CR) and thioflavin T (ThT) suggests that proteins directed to inclusion bodies contain at least some ordered structural elements common to fibrillar aggregates.[14,15] Many aggregates have increased beta-sheet content and diminished alpha helicity compared to the native state, which is presumably developed through intermolecular contacts. Although the molecular organization of amorphous aggregates remains rather coarsely described overall, the Weliky lab recently provided evidence for native-like structure within amorphous aggregates. Their examination of influenza virus hemagglutinin within an inclusion body using solid-state nuclear magnetic resonance (NMR) indicates the presence of a substantial retention of helical structure. The residues that form helices in the aggregate correspond well to those present in the native conformation observed in the available crystal structure.[16] This result, combined with the fact that amorphous aggregates from inclusion bodies bind CR and ThT, suggests that aggregates are composed of a combination of native- and fibrillar-like structures.

Fibrillar aggregates are commonly, but not always, associated with the formation of amyloid plaques and are named for their long, thin, fibrous shape. Great diversity in the diameter, length, and interconnection between amyloid strands has been observed, but each have hallmark characteristics of being birefringent when stained with dyes such as CR and of being detergent insoluble. CR and ThT bind to fibrils and were first used to identify the presence of a regular structure in fibrillar aggregates derived from proteins of diverse composition. Fibril variation depends on the protein from which they are derived and the conditions under which they are made. Such macroscopic differences suggest that despite having common features, structural diversity exists.[17] Fibrillar aggregates contain long-range order and, as a result, have been described in greater detail than amorphous forms. Although fibrils are not amenable to atomic-level structure determination by X-ray crystallography or by solution NMR, analysis using medium to high-resolution techniques and reconstructive methodologies has begun to provide molecular details of their organization.[18–20] These aggregates have been shown to be composed of substantial amounts of beta-sheet structure and lacking in helical content. When helical structure is present, it is significantly diminished compared to the native state of the corresponding protein. Both parallel and antiparallel sheet orientations have been observed in fibrils using IR spectroscopy, electron microscopy, and NMR.[21] Aggregates composed of antiparallel organization are typically formed from short peptide strands. Longer fragments in which the same sequence is embedded arrange themselves preferentially into parallel strands, as was observed from studying Aβ peptides. Most fibrils are composed of parallel strands.

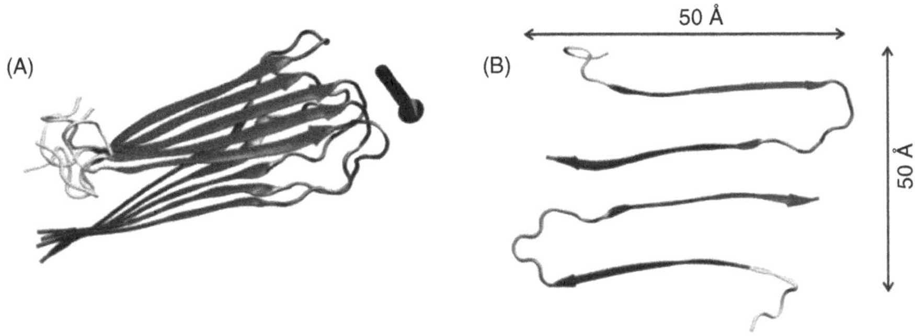

Figure 1.1. Ribbon diagram of cross-beta amyloid structure. The long axis is composed of two parallel sheets (A) stacked vertically (B). Reproduced with permission from Reference 27.

One particularly interesting feature of fibrillar aggregates is that they seem to depend largely on backbone interactions that are distinguishable from those of folded proteins. Antibodies have been generated that recognize only the fibrillar form of a peptide and not the monomer or other intermediate aggregation states.[22,23] Because these antibodies can also bind to fibrils composed of entirely unrelated peptide sequences, it was concluded that the antibodies must recognize a common backbone configuration that exists uniquely in fibrillar structures. Higher-resolution X-ray diffraction and NMR data lend credibility to this assertion by revealing that amyloid fibrils have a cross-β structural organization (Fig. 1.1) that has not been observed in globular proteins. The cross-β spine has been shown to occur through a parallel arrangement of beta strands[24] in which the side chains between two facing sheets interdigitate, forming a steric zipper (Fig. 1.2).[25–27] The zipper interface is tightly packed and is completely dehydrated.[26] The strands run perpendicular to the long axis of the fibril with 4.7- to 4.8-Å spacing between each strand.[28] The cross-β organization has been observed in aggregates of several unrelated amyloid peptides, suggesting this is a regular structural element in fibrillar aggregates. The same peptide sequence, however, has the ability to participate in different morphologies, and these are hypothesized to derive from unique packing arrangements of the spine. The Eisenberg lab proposed a structure for the fibril that is consistent with distances established by X-ray diffraction, electron paramagnetic resonance (EPR), and intrinsic fluorescence of multiple peptides.[26] Viewed down the long axis, it is a dimer of dimers with a left-handed twist of 3.4° per layer, which has a diameter of 64 Å when based on a 20-residue peptide (Fig. 1.3). Earlier models are similar in many regards. The *serpentine* model of the amylin fiber is analogous but is based on fewer sheets.[29] The Tycko model has four sheets but does not have as tightly interdigitated side-chain packing because it was based on distance restraints derived from solid-state NMR experiments that indicate closer interactions between certain side chains than are used in the Eisenberg model.[27] Recalculation of the Eisenberg

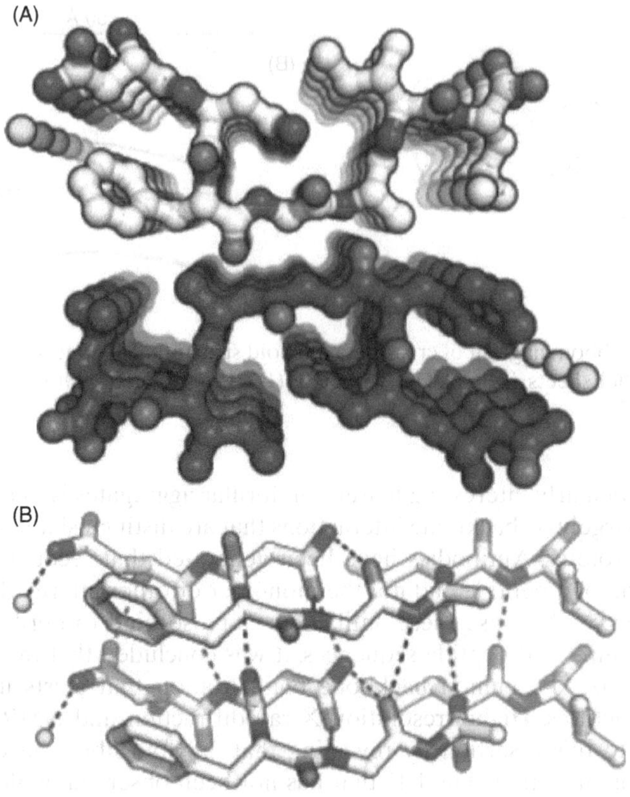

Figure 1.2. Molecular model of cross-beta structure showing interdigitation of side chains that form the tight packing arrangement in the islet amyloid polypeptide (IAPP) sequence NNFGAIL (A) and the backbone hydrogen bonding between strands (B). See color insert.

model to include the NMR-based side-chain interactions resulted in poorer alignment between the sheets. The common element of the cross-β arrangement among unrelated sequences indicates that a backbone conformation amenable to beta structure and compatible interaction between side chains in the strands stabilize fibrillar aggregates. The disparities between the models suggest that differences in sequence, which affect side-chain packing, may be responsible for the morphological differences observed in aggregates.

The phenomenon of aggregation is complex and seems to involve a series of intermediate structural states, some of which may be amorphous, along the path toward forming highly ordered fibrillar amyloid structures (Fig. 1.4).[9] Some of these states have been described in varying degrees of detail. Many plaque-forming proteins undergo an early transition from native monomers to oligomeric aggregates before fibrils are detected. The shape of many such species appears spherical or ellipsoidal (e.g., Aβ, α-synuclein, prion protein,

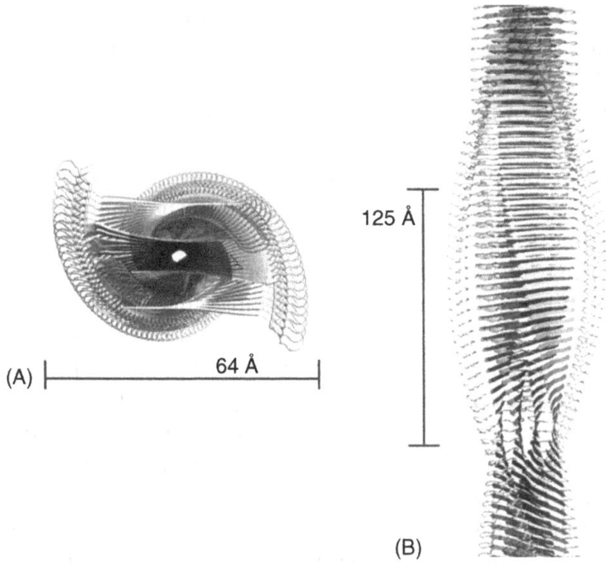

Figure 1.3. Model of macroscopic fibril structure composed of four sheets shown looking down the long axis (A) and from the side, showing the twist (B). From Reference 26.

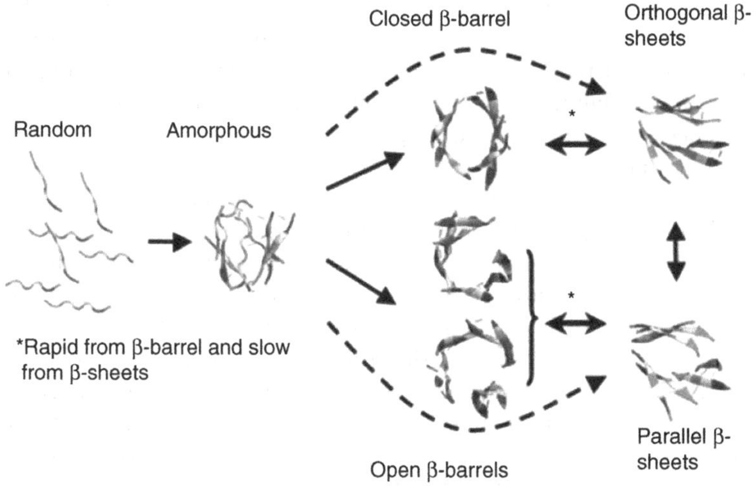

Figure 1.4. Peptide fragments arrange in beta barrel formations during simulations. Closed or open barrels result from different peptide sequences. Reproduced with permission from Reference 33.

cystatin C).[22] Their diameters vary, but often it has been shown that a discrete number of protein molecules are involved, suggesting that specific structural contacts are made between each polypeptide chain. These contacts are often so stable that they resist dissociation in the presence of SDS, earning the designation *stable aggregates*. Aβ forms several distinct stable oligomers depending on a variety of factors. In the presence of trimethylamine N-oxide (TMAO), Aβ (1–40) generates elliptical aggregates of $5.0 \times 1.5 \times 6.0$ nm, which must be composed of four or five strands.[30] SDS-insoluble Aβ oligomers composed of 3–24 strands have been observed, and metastable tetramers and dodecamers have been shown to convert into fibrillar aggregates. It is evident in some micrographs of protein aggregates that fibrils emerge from association between oligomers, as *beads on a string*-like structures are apparent in the images preceding fibrillation.[31] Because fibrillar forms are easily detected, it was originally thought that the various protein solubility diseases were actually directly caused by the fibers. More recent evidence indicates that fibers may be more a consequence of disease processes, which are instead caused by oligomeric aggregates. In support of this, Aβ species having globular structures of 4–5 nm have been shown specifically to be toxic,[22] whereas smaller oligomers and larger fibrils are less damaging. Experiments involving other amyloidogenic proteins, including transthyretin, islet amyloid protein, α-synuclein, immunoglobulin light chain, and β2-microglobulin (β2m), as well as non-disease-related proteins, reconfirm that in general, non-fibrillar, SDS-stable, low-molecular-weight aggregates are cytotoxic.[22] This common finding among so many unrelated sequences is suggestive that similar mechanisms and structures underlie protein pathogenesis.[32] Although the exact structure of toxic species is unknown, these soluble aggregates all possess the ability to interact with and cross membranes and to disrupt ionic gradients. As such, it was proposed that beta barrels might form in the membrane, mimicking the structure of ion channels (Fig. 1.4).[33] Amyloids have been induced to form in the presence of membranes, detergents, surfactants, polyanions, and compounds that change the dielectric of the solvent, which suggests that conformational changes mediate aggregation. Computational analyses of peptide aggregation lend support to the formation of β-barrels by aggregation-prone sequences.[33] Depending on the chain length and number of strands that participate in oligomerization, either open or closed barrels may form. Closed barrels convert to orthogonal sheets, while open barrels generate parallel sheets. This finding suggests that different proteins may undergo semi-common intermediate aggregation states that lead to distinctly different final morphologies. Pathogenesis may not be a consequence of a single universal mechanism since linear protofilbrils greater than 400 nm composed of *tau* protein are notably toxic as well.[30] Higher-resolution structural information about each of these systems is needed to investigate the β-barrel hypothesis and to better understand the relationship between aggregate structure and membrane permeability.

Most proteins unfold as the temperature is elevated, and at temperatures below their unfolding transition (T_m), where only partial unfolding is observed, aggregates begin to form. Typically, the aggregates are amorphous, as detected by light scattering and/or by the formation of opaque precipitates. For example, water-soluble aggregates of ovalbumin were induced to form by heat denaturation, and these soluble aggregates were observed to develop cross-β structure as recognized by ThT binding.[14] The finding suggests that the formation of amorphous aggregates may involve mechanisms common to fibrillar aggregates, but no high-resolution structural information is available to validate this inference. Because it is very difficult to ascertain atomic-level structural details from amorphous systems, identification of parallels between amorphous aggregates and those having a long-range order provides a basis for a better understanding how amorphous aggregates are stabilized.

1.2 PARALLELS TO PROTEIN FOLDING

Anfinsen's Nobel prize-winning studies on ribonuclease demonstrated the seminal importance of hydrophobic collapse in protein folding, generating the *thermodynamic hypothesis*.[34] This hypothesis states that the condensed form of the polypeptide is spontaneously achieved in aqueous solution because burial of hydrophobic moieties in the protein core is energetically favorable. Numerous experiments have been conducted on many proteins since then, reinforcing the general nature of this conclusion for proteins.[35] For practical reasons, in the case of protein folding, it has been presumed that the free energy thermodynamic minimum is met in the native state; nevertheless, conversion from the native conformation to more stable aggregated states regularly occurs from the normal range of fluctuations experienced by proteins.[36] The factors that drive proteins to the condensed state are determined by thermodynamics and kinetics. The native state is defined by the thermodynamic minimum of the active protein under physiological conditions. Nonnative states that are more thermodynamically stable may emerge, but they are not active and/or kinetically accessible on the biologically relevant time frame (ms-sec) of protein folding.[37]

Although the thermodynamics of protein folding were clearly established by Anfinsen and colleagues, the mechanism or pathway for folding was not. Originally, protein folding was presumed to occur via a random process in which the polypeptide chain could adopt any possible conformation on the way to the native state. In this model, an astronomical number of conformations are possible and equally probable. Levinthal pointed out the paradox that the timescale on which proteins must fold is many, many orders of magnitude shorter than a completely unbiased random process would require.[38] Since experimental results show that proteins achieve native folds in ms-sec rather than the predicted expanses of years, the hypothesis emerged that folding must be directed in some way toward the native state. Since then, many have worked to identify and explain the parameters that permit rapid folding of proteins to their native state. Proteins having different amino acid sequences have different rates of folding and unfolding as well as differing stabilities. Single amino acid substitutions in a protein can dramatically affect these processes. These observations provided the foundation for the idea that the amino acid sequence encodes structural information that influences folding and stability. As such, some have sought to identify sequence motifs responsible for aggregation, but no well-defined common sequences have been found. Instead, it appears that general features of the polypeptide sequence direct cooperative folding.[39] The properties of the side chain influence both the local backbone conformation and packing interactions, imposing restrictions on the folding pathway.

Both protein sequence and environmental conditions have been shown to affect the structural stability and aggregation of proteins. The Gibbs free energy depends on the totality of interactions within a system, and as such, the thermodynamic minimum for a protein is determined not only by the amino acid sequence but also by the environment in which it is contained. Temperature, pH, ionic strength, viscosity, counterions, and other factors have been shown experimentally to influence protein stability, indicating that environmental conditions are critical determinants of the lowest energy conformation of a protein.[40] In many cases where misfolded proteins cause disease, mutation is responsible for destabilizing the native state under normal conditions, thereby lowering the barrier to structural transitions that generate non-native conformations and promote aggregation.[36]

1.3 VIEWS OF PROTEIN STABILITY AND AGGREGATION

Views of protein stability have largely emerged from studies of well-behaved, single-domain proteins that undergo reversible folding to unfolding transitions. Numerous experiments have been carried out that employ thermal or chemical denaturation to examine folding or unfolding by following fluorescence or absorbance changes, hydrogen/deuterium exchange, or structural transitions by circular dichroism (CD) or NMR. Reversible analyses are almost exclusively performed using chemical denaturants, because thermally induced unfolding typically leads to rapid aggregation as the structure begins to be altered. At high concentrations, chemical denaturants compete for binding to the moieties involved in self-association, thereby promoting retention of the unfolded state. The most frequently characterized and well-described systems display two-state kinetics, although a few examples of multistate kinetics have been reported. These simple systems have obvious utility for investigating fundamental influences on protein folding, and the findings are relevant to protein aggregation. The fundamental parameters that influence the behavior of proteins in solution and the models that have been developed to explain their behavior are discussed below.

1.3.1 Physicochemical Properties of Proteins

Although proteins appear complex, their behavior is ultimately dictated by basic chemical and physical principles.[35,41] From a thermodynamic perspective, collapse of the polypeptide chain is an organizing event and also requires that water be removed from parts of the protein that form the core of the folded species. The energetic cost of removing water and the entropic penalty (ΔS) must be compensated in order for folding to occur. This is generally achieved by a decrease in the enthalpy (ΔH) of the protein. The free energy (ΔG) in the native folded state is typically 5–20 kcal/mol more stable than the unfolded state.[35,40] This ΔG value represents only a fractional difference in the total energy of both the folded and unfolded state.[40] It is approximately equivalent to two to four hydrogen bonds, and even small proteins employ tens of intramolecular H-bonds in their fold,[7] thus indicating that the sum of many small energy differences contributes to achieving a folded state. The major factors that have been shown to play a role in the specificity and stability of protein folds are derived from the chemical attributes of the amino acid sequence, including hydrophobicity, aromatic stacking, electrostatic interactions, steric constraints, and hydrogen bonding (secondary structure), as well as the efficiency of packing in the core and surface properties of the folded molecule. The contributions of each component to protein folding and stability are discussed below.

$$\Delta G = \Delta H - T\Delta S. \tag{1.1}$$

Hydrophobicity and apolarity have been shown to be the primary factors that drive proteins to the condensed state because of the large heat capacity change associated with this process. The driving force for hydrophobic collapse is modulated by the temperature, pH, and dielectric constant (ε) of the solvent. At extremes of any condition, structural and dynamic changes are apparent. Low pH (<3) conditions often cause native folds to relax, transforming into molten globule (MG) states, especially at high ionic strengths.[42] The MG retains much of the secondary structure of the native state, but tertiary packing is disrupted such that the residues in the core are solvated efficiently. This has been shown by H/D exchange experiments and by heat capacity measurements, which indicate these properties are more akin to the unfolded state than the native form.[43] Titration analyses performed on numerous proteins reveal that the stability of the fold, as determined by changes in structure and aggregation, is also significantly affected by pH.[44] In this regard, it has been shown that benzyl alcohol, a hydrophobic additive used as an antimicrobial agent, destabilizes the tertiary structure of recombinant human interleukin-1 receptor antagonist (rhIL-1ra) and granulocyte colony-stimulating factor (GCSF), inducing aggregation.[45,46] Changes in solution conditions that diminish the driving force for hydrophobic association concomitantly interfere with core packing and disrupt the native fold. Increasing temperature causes a decrease in dielectric constant and, above approximately room temperature, an increase in the solubility of hydrophobic groups.[47] Because the difference in free energy between the folded and unfolded state is relatively small, changes in temperature and the dielectric properties of the solution

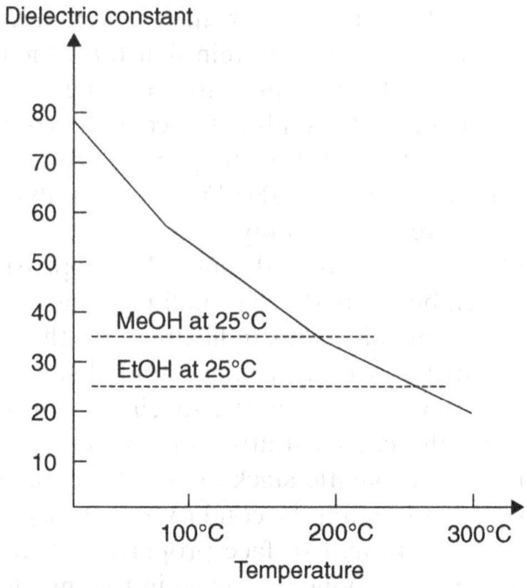

Figure 1.5. Graph showing the effect of temperature on the dielectric constants of water, methanol, and ethanol. The dielectric constants of the alcohols are virtually unchanged with increased temperature, whereas the value for water is dramatically altered. From http://www.greenfluids.org/SubcriticalWater.htm (accessed January 2009).

presumably lead to unfolding through alteration of the driving force for hydrophobic collapse. The dielectric constant for water at 20°C is approximately 80, while it is estimated that the ε value of the protein core is 5–20.[48] The dielectric of water is strongly affected by temperature such that its ε decreases with increasing temperature. The dielectric of hydrophobic moieties are largely unaffected by increasing temperature in the absence of water, but hydration due to increased thermal motion increases the dielectric (Fig. 1.5).[41,49] As such, raising the temperature lowers the barrier to unfolding, which is consistent with experimental observations.[41,50]

Aromatic residues (Phe, Trp, and Tyr) contribute to hydrophobic interactions that stabilize the core because they have relatively large apolar surface areas. Additionally, they provide enhanced stability when arranged in a stacked configuration, where they not only maximize van der Waals contacts but benefit from pi–pi interactions.[51] Moreover, the stacked orientation is an efficient packing arrangement. A well-packed core further contributes to protein stability.[52] Aromatic stacking has been proposed to stabilize amyloid fibrils,[53] and a parameter-free model indicates pi-stacking should facilitate aggregation.[54] Pi bonds produce a dipole, which also enables favorable interaction with cations. There are many examples of structures in which Lys and Arg side chains are found to interact with aromatic rings in folded proteins. These associations are believed to be primarily electrostatic, but regular distances and geometries also suggest hydrogen bonding may be a mode of interaction.[54–56] Aromatic–aromatic and aromatic–amine interaction also provide an enthalpic contribution to protein stability.[56]

Charged residues, including Glu and Asp (acidic) and Lys and Arg (basic) most often appear on the surface of folded proteins. Because they interact favorably with water, their solvent accessibility is a dominant factor in the overall solubility of proteins. The isoelectric point (pI) of a protein to some extent determines the solubility of a protein at a given pH, with the lowest solubility theoretically occurring at the pH equivalent to the pI. Titration of the pH away from the pI to either more basic or more acidic conditions often improves solubility within the limit of structure retention. Although His is a basic amino acid, it commonly exists in the neutral state and is located in the core of folded proteins. The pKa of its imidazole is 6.7 in solution but can be perturbed substantially in the context of a folded protein due to the local microenvironment. His participates in hydrogen bonding, where it can act as a donor or as an acceptor. It is paired with buried Tyr in many enzymes, where it lends specificity to the interaction between two structural regions that undergo a conformational change induced by substrate binding and turnover.[57–59] The uncharged species is easily accommodated in the hydrophobic core because it can add stability through hydrophobic and aromatic interactions as well. Electrostatic interactions involving Asp, Glu, Lys, and Arg are also observed in the core of folded proteins in the form of salt bridges. The pairing of a positive and negative charge permits the incorporation of these residues into the apolar core and provides structural specificity to the packing arrangement.[60,61] Uncompensated charge, however, disrupts hydrophobic packing, decreasing stability.[62,63] Mutations that result in protein aggregation often introduce uncompensated charge or polar moieties into an otherwise hydrophobic environment, thereby inducing partial unfolding in which significant amounts of apolar surface area become solvent accessible.[64] Interestingly, evolution seems to have selected for protection against aggregation at the edge strands of beta sheets using this strategy. Lys is often positioned in the center of terminal strands where it disfavors interaction between the outside edges of β-strands.[56] This idea has been applied to test the aggregation potential of amyloid peptides.[65] The substitution of a hydrophobic with a charged residue ameliorates aggregation and, when added at high concentrations, promotes dissociation of aggregated species. Modeling studies predict that an amyloid peptide is bound by the mutant peptide and this disrupts the pattern required to support fibrillation.

The most important steric constraint that limits the number of possible backbone conformations is the planarity of the peptide bond. The original proposed models for possible secondary structures did not account for this chemistry and consequently presented many more configurations than are possible.[66] It was the subsequent recognition of the significance of the planar bond in protein structures that earned Linus Pauling a Nobel Prize.[66,67] The consequence of this constraint is obvious in the now standard Ramachandran plots of globular proteins where ϕ/ψ dihedral angles largely cluster into two main regions corresponding to the α-helix and β-strand conformations.[68] Quite a range of deviation is observed around the optimal angles for alpha helix and beta structures, indicating substantial flexibility is permitted in the formation of these regular elements. Secondarily, dihedral angles are affected by steric restrictions that the side chain imposes on the backbone. The degree of solvation and preferred dihedral angles of the backbone are correlated with the degree of solvation of the side chain, such that amino acids and peptide sequences have an inherent propensity for specific secondary structures.[69–71] The degree to which this bias exists depends on the amino acid and on the sequence in which it is embedded. Polyalanine sequences have a strong tendency to form alpha helices. Leu is found approximately 60% of the time in helices and 25% of the time in beta strands, whereas beta-branched amino acids (Val, Ile) tend to adopt an extended beta structure, such that Ile is located in beta strands in about 75% of sequences. Many sequences have only a marginal bias such that the actual development of a secondary structure depends on context, which may be derived from neighboring sequences and structures in the protein or may be influenced by solution conditions. There are several pieces of evidence that support this claim. Peptides that form stable helices or strands in folded proteins often exist as random coils in isolation.[72,73] Alterations in the dielectric constant of the solvent, commonly by addition of alcohols such as TMAO or trifluoroethanol (TFE), induce secondary structure in random coil peptides.[46,74] At low concentrations of these alcohols, alpha helicity increases, but at concentrations where the cosolvent comprises a significant fraction of the volume, beta structure has been observed. It is important to note, however, that peptides from specific regions in proteins do not necessarily convert into the same structures in the presence of alcohols.[75] These additives change the dielectric of the solvent such that hydration of the backbone is altered by the differences in backbone and side-chain solubility compared to water.[70]

All secondary structures are stabilized by hydrogen bonds.[7] Alpha helices emerge from local, sequential interactions, with their bonding pattern involving repeating interactions between each CO at residue i and NH at residue $i + 4$. Sequences that contain hydrophobic residues every seventh and third or fourth position tend to form helices because the pattern of apolar side chains is in register with the 3.6 amino acids per turn inherent in the α-helical structure. This produces an amphipathic arrangement and supports stabilization of the helix by interaction between its hydrophobic face and another hydrophobic surface. Often the partner is another helix, as seen in coiled-coil structures or helical bundles, but interaction with beta sheets is common as well. When polar or charged residues occupy the remaining positions, the opposite face of the helix is amenable to interaction with water resulting in a soluble arrangement. Helices composed mostly of hydrophobic residues are poorly solvated and must be embedded in the core of a larger assembly. Beta sheets rely on the formation of hydrogen bonds between residues that are distant in the sequence. They have been shown to form from sequences in which hydrophobic (H) residues alternate with polar (P) residues, that is, HPHPHP.[65] This pattern generates compatible interactions between strands and produces the expanded hydrophobic surface necessary to sustain collapse as well as an apposed hydrophilic surface to support the solubility of the folded form. Consistent with this, a large number of proteins fold into beta sandwiches in which their hydrophobic surfaces face inward to stabilize the interaction between the two sheets composing the sandwich, and their outer surfaces are amenable to solvation by water. In the cross-beta structure formed in amyloids, the hydrophobic residues are positioned on the interior of the strand and the orientation of the strands is straight with respect to one another, such that the addition of strands to both ends is favorable.[19,20] Aggregation involves not only hydrophobic association but also depends strongly on backbone hydrogen bonding to form an intermolecular beta structure. Methylation of the backbone amide nitrogen inhibits aggregation of amyloid peptides despite the presence of an extensive hydrophobic surface area and strongly indicates that intermolecular hydrogen bonding is essential to the formation of stable aggregates.[76]

The propensity of individual amino acids to form alpha and beta structures has been characterized in a variety of ways. Statistical approaches have been used to assess the frequency of each amino acid in helices and sheets in known structures and in the NMR chemical shifts of backbone resonances.[77–80] Experiments were performed on helical peptides to characterize the helical propensity of each amino acid. Theoretical studies revealed that the basis for the observed helicity has to do with the dihedral angles of the side chain, because entropic losses are incurred upon folding.[81,82] Alanine most strongly favors the helical conformation because no loss occurs. Modeling of dipeptides has been performed to obtain information on preferred backbone torsion angles. The dominant contribution to beta-sheet propensity was found to be the avoidance of steric clashes between the backbone and the associated side chain.[71] There are contextual constraints imposed on β-strand association, but this study indicates that inherent β-propensity is strongly tied to local features. An interesting outcome of the study was that Asn displayed an exceptionally high propensity for β-structure, which was attributed to the unique hydrogen-bonding capabilities of its side chain.

An important additional consideration is the kinetics of folding as it pertains to individual secondary structure elements. The contacts involved in beta structure are more disbursed and cannot assemble as rapidly as adjacent interactions within individual helices because the segments forming β-strands are discontinuous. This is not to say that all helices form equally rapidly, as was noted above in the discussion of sequence effects. The antiparallel arrangement may form more rapidly when a short turn, particularly a well-defined turn, separates the strands, whereas parallel configurations are necessarily separated by long stretches of residues not involved in the sheet. Proteins with such arrangements often receive assistance from chaperones to fold *in vivo*. For example, rhodanese contains a parallel beta sheet in which the strand sequences are separated by helical segments. Refolding experiments show that rhodanese aggregates readily following dilution from denaturants, despite a significant retention of the secondary structure. Inclusion of several chaperones is necessary to prevent aggregation and facilitate its refolding.[59] The results of this study indicate that both extended structure and an MG-like state exist, which must be stabilized during the folding process to avoid aggregation. The inference then is that the sequences involved in sheet formation are protected from aggregation by chaperones until they are oriented properly with respect to each other. Antiparallel sheets composed of strands with long intervening sequences are akin to parallel configurations in this regard, indicating a longer time may be needed for their formation. This difference in kinetics may provide an explanation for why alpha to beta conversion is observed in aggregation. When alpha helices are destabilized, beta strands that exist may become amenable to rapid intermolecular association since they are no longer protected by intramolecular structural elements. A survey of protein structures revealed that beta sheets are often protected from intermolecular association in their native conformation because they are covered by helices. Alternatively, the newly uncoiled region may sample β-conformations more frequently and may directly participate in aggregation. If a competition exists between the two secondary structures in the same sequence, structural stabilization and conditions that favor helical organization would prevent aggregation. If the exposure of the existing beta structure is responsible for aggregation, then preservation of helicity in the absence of tertiary structure retention would be insufficient to prevent aggregation.

Although most aggregates have diminished helical content and/or increased beta structure, it is possible that association between helices could confer nonnative oligomerization via open domain-swapped conformations, which subsequently may promote further aggregation. Even when an equivalent amount of hydrophobic surface area is buried, helices lack the hydrogen bonds that stabilize intermolecular beta interactions. The upshot is that intermolecular beta sheets are more likely to produce more stable (irreversible) aggregates than alpha helical associations, which is consistent with available data.

1.3.2 Surface Properties and Packing Arrangements

The driving force for folding is primarily burial of apolar surface area, but the composition of the outside surface also has significant implications for stability once the native fold is attained. Polar and charged residues constitute the majority of the surface area on protein exteriors because they are more easily solvated by water than by hydrophobic side chains. On average, greater than 70% of the surface area of monomeric proteins is hydrophilic. Charged moieties are more easily hydrated than non-charged species and have a greater driving force for interaction with water. The number, density, and location of charged residues on the surface determine solubility and also exert a profound influence on the stability of proteins.[83,84] In globular proteins, the more charge per square angstrom, the greater the thermodynamic stabilizing force on the protein. There seems to be an optimal charge density, however, because proteins having extremely high percentages of charged residues are often natively disordered. Thermophilic organisms are adapted to function at greatly elevated temperatures, and their proteins can occasionally resist thermal unfolding above 100°C. A computational study comparing the stability of thermophilic and mesophilic proteins showed that the enhanced stability of the former was imparted by surface residues. The hydrophobic core of the proteins contributed equally to stability, but the increased density of packing among the surface residues of the thermophilic proteins correlated strongly with stabilization.[83] Proteins from thermophiles were noted to display higher percentages of Lys, Arg, and Glu and lower fractions of Ala, Asp, Asn, Gln, Thr, Ser, and His on their solvent-accessible surface than mesophiles. Additionally, increased enthalpy changes at the melting temperature have been observed for proteins with higher thermostability, suggesting that electrostatic moieties participate in preventing unfolding.[61] This principle may also underlie low pH induction of the MG state.[85] By protonating the acidic groups and by reducing the net surface charge, the differential solubility between surface and core residues is decreased, thus lowering the barrier to unfolding. Not only is the total number of charges important, but the density and location of charged moieties also influence stability. Because like charges repel each other, compensation is important for stability, such that stabilization is bolstered by attractive interactions outweighing repulsive forces.[84] Individual salt bridges on the surface have been shown to contribute significantly to the stabilization of many proteins.[61,86–90] Networks of charged residues and hydrogen bonds are found on the surface of proteins and have been shown to enhance stability synergistically. Salt bridge formation on the surface of a protein can improve structural stability by contributing a favorable tertiary interaction between nonsequential residues. Clustering of positive or negative charges on compatible surfaces can also result in intermolecular ion pairing to generate soluble or insoluble oligomers or fibrous structures.[91] Oligomers that retain their native structure can be solubilized reversibly by pH adjustment or high salt, as has been shown for the CC chemokines.[92] If the interaction alters the tertiary structure, aggregation may be irreversible as a result of more extensive structural rearrangement.

Although most of the protein core is composed of hydrophobic moieties, structures in the Protein Data Bank (PDB) show that acidic and basic side chains are often buried in the protein core. The proteins exist in stable folded structures as long as the charge is offset in a way that its presence is more stabilizing than destabilizing. This can be accomplished by the pairing of oppositely charged groups. Salt bridges, N–O pairings, and longer-range ion-pairings help decrease the energy cost associated with burial in a nonpolar environment.[61] The pairing is most effective when the center of charge and at least one N and O atom from each residue are within 4 Å, constituting a salt bridge. The energy is higher when only the second criterion is met but is still effectively stabilizing in many proteins. Longer-range pairings are stabilizing when a larger network of interactions, often involving hydrogen bonds, is present or alternative parings exist within ~5 Å. An intermolecular salt bridge has been proposed to help stabilize fibrillar aggregates of Aβ, in which Lys16 interacts with Glu22 in the antiparallel configuration.[21,93] Several modeling studies confirm charge complementarity and support interstrand and beta-sheet association.[94–96]

An alternative means of accommodation of charged moieties in protein cores is perturbation of side-chain pKa or stabilization through hydrogen bonding.[97] The pKa of the hydrated carboxylic acid moieties in Asp and Glu in solution is 3.5 and 4.1, respectively.[98] The presence of adjacent groups greatly affects the pKa of these residues, causing it to range from 0.5 to 8.8. For example, the decreased dielectric constant experienced by a moiety within the hydrophobic protein core can raise pKa values sufficiently to make the carboxyl group neutral at physiological pH. pKas are much more substantially altered by hydrogen bonding, which lowers the pKa. Each H-bond depresses the pKa by approximately 1.6 and contributes 2 kcal/mol toward stabilizing proteins.[97] Asp76 in RNase T1 makes three strong hydrogen bonds, and its pKa is 0.7. Disruption of such hydrogen bonds in RNase T1 and in many other proteins reduces their thermal stability.[97,99] Groupings of charged residues on the surface of a protein can also influence the net charge by perturbing the pKa of residues in close proximity, but this usually has only a small effect, inducing fractional changes in the measured pKa. Structures of Aβ fibrils align such that strings of Glu residues are positioned adjacent to each other along the long axis of the fiber. The Glu side chains are surrounded by a high density of hydrophobic moieties on both sides. The high degree of hydrophobicity and close proximity of like charges both should contribute to elevating the pKa of the carboxylate group, lowering the barrier to association between sheets.

It follows from earlier arguments that the amount of apolar surface area exposed on the outside of folded proteins should influence their solubility. One way that surface hydrophobicity is mitigated to preserve solubility is through protein oligomerization. Many proteins function as dimers, trimers, tetramers, or higher-order oligomers *in vivo*. Analysis of their structures reveals that the interfaces between subunits have an extensive hydrophobic contact area to support self-association. Heterodimerization also helps preserve the integrity of protein monomers with an exposed hydrophobic surface area, and isolation of one protein in the absence of its partner often leads to aggregation.[100] Protein–protein interaction involves the association of complementary surfaces, which are often approximately planar.[101,102] This is common among dimers that interact by abutting edge beta strands to form an extended sheet structure (e.g., λ-cro, chemokines).[92,103] Association also results from interactions between hydrophobic clusters found in loops or helices (e.g., hemoglobin, PRL-1).[104,105]

1.3.3 Solvent Interactions

It is apparent from countless studies that the solution environment surrounding a protein strongly influences its structural retention. The long-standing, dominant theory that explains this phenomenon argues that interactions between the solvent/cosolvents and protein are very weak and nonspecific, such that simple parameters such as osmotic pressure and surface tension dictate stabilization. The interaction between proteins and water is still poorly understood, and little is certain at this point about how stabilization is achieved at the molecular level. In aqueous solutions containing proteins, water molecules typically possess two diffusion rates: one associated with the bulk solution and another with the hydration sphere associated with the protein. Water in the hydration sphere is more static than in bulk solution, forming a layer approximately 7Å thick around the surface of the protein. The individual water molecules can be retained from tens of picoseconds to nanoseconds, differentiating them from bulk solutions, which have residence times of approximately 1–8ps.[106,107] Based on residence time, little associated water is detected in unfolded proteins. This data suggest that the water molecules may be organized into a more structured network around the protein. This hydration shell is thought to promote stability in part because the water binds more tightly to the folded protein than does bulk water to the unfolded form, a simple mass action effect. The basis for increased residence time has not been quantitatively established as of yet and does not appear to correlate with solubility. Although proteins with larger charged surface areas are often more soluble and it might be expected that water would be retained around charged moieties, it has already been noted that too high a charge density often produces disordered proteins, which have reduced retention of water at their surface. Because the arrangement of water around the folded protein is rather poorly understood at the molecular level, it is possible that the shape, specific interactions that occur between moieties in close proximity on the surface of the protein, and the local structure and chemical properties of the surface-exposed groups may be major factors that contribute to the transient organization of water molecules in the hydration sphere.

Ions are also retained at the surface of proteins when present at low concentration due to field potentials induced by charged residues.[108] The interaction of salt ions with charged residues on a protein's surface has been modeled, and they have been found to interact preferentially based on residence time. Attraction of ions is stabilizing to a limited extent, depending on the protein. For example, lysozyme binds a Cl^- ion that slows thermal denaturation. High concentrations of ions are destabilizing because they outcompete the protein in binding to water molecules, often inducing precipitation through the well-known salting out phenomenon.[109] This phenomenon may alternatively induce conformational rearrangement by shielding stabilizing interactions between charged species on the surface of the protein. Alternatively, very high concentrations of some larger ions such as amino acids can stabilize proteins through a preferential hydration effect.[110] Charge neutralization may also be accomplished through the interaction of polyanions with binding partners. Destabilization of protein structure has been reported in the presence of sulfated glycosaminoglycans (GAGs) and proteoglycans and has been found to accelerate nucleation and fibril polymerization of Aβ, albeit through an unknown mechanism.[111] Conversely, many native proteins are dramatically stabilized by the presence of polyanionic compounds,[44,112–115] suggesting that specific knowledge of the mechanisms by which each event occurs is needed to explain the differing results. One possible mechanism of aggregation is that binding may induce and stabilize conformational changes that expose the apolar surface area. Further rearrangement may then be permitted due to neutralization of charged moieties through electrostatic interactions.[63] When binding leads to stabilization of a protein, the conformation would be such that the accessible hydrophobic surface area is limited or it persists in a conformation that is not amenable to self-association.

1.4 MODELS OF AGGREGATION

Several mechanisms have been proposed to describe the formation of aggregates that are based on the observed kinetics of assembly. Each begins with a partial unfolding of the native state. Partial unfolding permits the association of monomers through exposure of previously inaccessible residues. Metastable monomeric intermediates are thought to be responsible for aggregate formation because partial unfolding of proteins with low to moderate amounts of denaturant or heat produces aggregation, whereas complete unfolding results in reduced or no aggregation at high denaturant concentrations or temperatures.[116] Natively unstructured proteins do not inherently aggregate rapidly and these too must be provoked to do so by partial ordering.[117] It seems reasonable, based on the understanding gained from protein folding studies, that a critical density of hydrophobic contacts would be required to stabilize any condensed state regardless of whether contacts stabilize the native state or are involved in aggregation.[63] In many proteins, it is apparent that the appropriate conditions for aggregation are met in partially unfolded states, but experimental requirements do not permit determination of whether aggregation proceeds solely from a partially unfolded state or whether the fully unfolded form also aggregates rapidly. In some cases, however, completely unfolded forms of proteins do not appear to aggregate. When high concentrations of denaturants are used, the unfolded state does not readily aggregate, but this appears to be due to competition between the chemical denaturant and protein molecules for interaction. Theoretically, if transition to the fully unfolded state could be accomplished instantaneously and enough thermal energy is imparted to completely disrupt the structure (which also seems to be required for aggregate formation), then the heat-denatured state should be less prone to self-association as well. The folding and unfolding processes of globular proteins have been studied extensively using proteins that unfold reversibly and obey two-state kinetics.[118,119] Aggregation has been considered with respect to the parameters derived from these models, but because aggregates are difficult to study directly and in most cases are irreversible or are only very slowly reversible, these models are unique and continue to be revised. Aggregation models are based on forward rates of association at varied concentrations and conditions to form large aggregates and intermediates that precede the final stable form. Because the morphology and size of fibrous aggregates are easier to correlate with specific changes in structure than large amorphous agglomerates, most models have been developed based on data from amyloidogenic peptides and proteins.

1.4.1 Monomer Conversion

Templated assembly (TA) was proposed soon after the discovery of prion proteins to explain their mechanism of infectivity. A dominant conformational change in one monomer (α) provokes rearrangement in native monomers (α') by acting as a template (α_2) from which the aggregate grows.[120] A favorable energy difference driving α' to α is proposed, but the barrier to conversion would be large for the wild-type protein under normal physiological conditions. This makes formation of the dimer highly improbable. The existence of α_2, however, would then lower the energy barrier for a subsequent conformational rearrangement of α', allowing α_2 to act as a template for propagation. The spontaneous onset of disease and conversion could be explained by any perturbation that lowers the initial activation barrier to permit the formation of α_2, such as a mutation or a change in the environment. This model predicts that the rate of assembly will be directly proportional to the concentration of α_2 and will not depend on the concentration of α' over a wide range unless α_2 is in excess. Based on the assumptions of this model, α' will always be in great excess over α_2 since the formation of the dimer is rate limiting. Because prions are infectious, they must remain soluble in aqueous solution and must cross lipid membranes. This duality suggests that the existence of a fluid equilibrium between multiple states is necessary for transmission. The existence of multiple equilibria in disease states should apply equally well to non-disease proteins in concept, although the structural transitions may not occur in such uniform ways to generate highly organized fibrils. The monomer-directed conversion (MDC) model is based on the same concept involving a conformational change in aggregate assembly.[121] In this model, however, the aggregate does not act as a template. Instead, the barrier to structural rearrangement is minimal and assembly occurs as a consequence of frequent conversion to the aggregation-prone conformation. The rate of formation of aggregates increases with increasing concentration of protein in this model because the probability of productive collisions increases.

1.4.2 Oligomeric Intermediates

The nucleation–polymerization (NP) model was developed to account for the lag phase preceding the growth of insoluble fibers from untainted protein samples. The NP model resembles crystallization processes in which nucleation is the rate-determining step and is followed by a more rapid growth phase.[122] The formation of nuclei depends primarily on the initial concentration of soluble protein, such that the rate of formation increases exponentially with increasing concentration.[123] The rate also depends on the equilibrium of conversion between the native and aggregation-prone states. If the kinetic barrier is easily surmounted, nucleation will proceed more rapidly than if the barrier is more difficult to cross. The upshot of this model is that there is a critical concentration at which stable nuclei are able to form and persist long enough for growth to occur. The oligomeric nucleus must remain associated longer than the effective rate of diffusional encounter with monomers having an aggregation-prone conformation. The growth phase is predicted to be constant at monomer concentrations in large excess of nuclei and to taper off as the concentration diminishes through incorporation of protein into the aggregate. If preformed nuclei are introduced into a solution of monomers, they act to seed growth, eliminating the lag phase. This model was developed based on *in vitro* studies involving the amyloid-forming peptide PrP96-111M, which complies with all aspects of this kinetic model of aggregation. Although the results of these studies support the NP model and full-length PrP forms amyloid plaques *in vivo*, it is evident from other studies that neither the TA nor the NP model explains fully the mechanism of PrP^{C} conversion to PrP^{SC} with regard to both infectivity and plaque formation.

The nucleation conformational conversion (NCC) mechanism employs elements of both TA and NP to explain the transformation of amorphous aggregates into highly ordered aggregated structures. Experiments conducted using the NM (glutamine-rich N-terminal and highly charged middle regions) of the yeast prion protein Sup35 showed that the kinetic rates of assembly were inconsistent with both the TP and NP models.[124] The rates did not vary much with protein concentration over a broad range and were limited to a maximum rate at very high ratios of monomer to nuclei. In NCC, like the NP model, the kinetic barrier and rate-limiting step in aggregation is oligomerization rather than conversion of the monomeric conformation. The oligomeric species and monomers exist in equilibrium, but the concentration of oligomer dictates the rate of fibril formation, as opposed to the monomer concentration directly determining the rate of further growth. The growth phase proceeds through the fusion of preformed nuclei, generating larger aggregates as a result of interaction between oligomers. This mechanism implies that the conformation of the monomer in the context of the oligomeric structure has a lower barrier to further aggregation than does the soluble monomer. In this model, the oligomer must be accessible from the normal conformational milieu of the native protein or a partially unfolded state in which a specific segment becomes exposed, which initiates aggregation.[36] Perturbations, such as mutation or changes in pH, which increase exposure of such nucleation sites, have been identified for several proteins.[125,126] PrP aggregation appears to depend on residues 96–111, which is buried in the native structure, and many peptides derived from the core regions of various normal and disease-causing proteins aggregate rapidly in solution. The increased rate of aggregation of the peptides would therefore be a consequence of removing the protection afforded by the native fold.

1.4.3 Nucleation in Protein Folding

Because proteins usually fold cooperatively, there is a need to understand the interplay between local and distal interactions that synergize to develop a condensed state. One idea that has been presented is that a local structural nucleation site (or sites) exists around which the rest of the protein coalesces. A foldon is a short, contiguous amino acid sequence with a preference for a particular conformation. Foldons were proposed to act as nucleation sites within protein sequences that aid in the specificity of folding. To achieve cooperative folding following two-state kinetics, the foldons would have to overlap. If the foldons are separated in space, then multistate kinetics will be observed as cooperativity breaks down and intermediates accumulate while the separate foldons search for preferential modes of interaction that lead to the native conformation.[127] The foldon concept fundamentally implies that sequence-encoded information dictates folding. Cooperativity would occur largely through predisposed backbone angles with concomitant compatible side-chain interactions both locally and distally. With respect to this idea, partial unfolding of a protein may lead to increased exposure of regions that retain such structural features, and these foldon-like segments may then nucleate intermolecular assembly.

The diffusion–collision (DC) model presented by Karplus and Weaver is based on the formation of microdomains within the protein structure, such as transiently stable helices. The kinetic process of folding is dictated by the diffusion and collision of microdomains, such that productive collisions result in the coalescence of higher structure.[128] In most applications of this model, only native collision–coalescence events are permitted. Nonnative interactions were allowed, however, in a study by Beck et al., when modeling a four-helix bundle to characterize transient nonnative collisions, potentially indicative of misfolding events.[129] DC and foldons share the common idea that preformed structural elements or tendencies are encoded in the sequence, but DC differs from the foldon concept in that folding need not be fully cooperative and is quantified using a molecular modeling approach. The existence of microdomains simply restricts the folding pathway options, which permits the formation of off-pathway associations. This view is consistent with observations from *in vivo* systems because it accommodates the possibility of off-pathway and slow-folding intermediates, which are formed often in proteins and require assistance from chaperones to achieve their native conformation. No particular kinetic mechanism is inherent in the DC model of protein folding, and as such, it is amenable to characterizing proteins that have great diversity in the rate of folding and in the number of paths available regardless of whether intermediates form. The microdomains arise from fundamental features of the sequence, and the collision between domains depends on the distances separating the domains. With this approach, mutations can be introduced to assess easily the effect on folding rate and the formation of intermediates. A similar concept may play a role in the aggregation of proteins, in which aggregation "hot spots" are unveiled as a protein unfolds. Peptide sequences derived from proteins that aggregate are prone to rapid assembly when examined in the absence of the full, folded sequence.

1.4.4 Domain Swapping

Domain swapping among dimers has been observed in a number of proteins. In some cases, the structure of the fold is virtually identical to the native monomer forming a *closed interface*, while in others, the interaction results in a distinctly different conformation for the participating residues.[130,131] Proteins that swap domains often aggregate further, and it has been proposed that domain swapping is an intermediate on the way to the formation of larger oligomers.[132] The phenomenon has been investigated in detail for human cystatin C (HCC), where mutation of L68 to Gln causes amyloid angiopathy.[64,133,134] This cysteine protease inhibitor may function as a monomer, but when destabilized by the L68Q mutation, it associates into dimers. The mutation occurs in the third beta strand (β3) and disrupts packing interactions by introducing a polar moiety into an otherwise hydrophobic space. Consequently, the energy barrier between the native and unfolded state is lowered, allowing the first two beta strands (β1 and β2) and alpha helix (α1) to cross over and interact with the sheet from the partner domain in the swapped configuration (Fig. 1.6).[64] The original architecture is retained on each half of the symmetric dimer, but a new "open interface" is created by the bridging interaction between the domains. The dimer suffers from the same packing problem as the monomer but overall is stabilized thermodynamically compared to the native state by the additional β-strand H-bonds in the linker region (βL).[64] It has been proposed based on an examination of other domain-swapped dimers that the open interface is responsible for further aggregation.[130] It is plausible that in HCC, conversion to larger oligomers may emerge at the open interface of the domain-swapped state because of the same packing defect, which concomitantly diminishes interaction between residue 68 and α1 via Y34 and V31 from the partner molecule. Movement of this helix would be a prerequisite to the formation of an amyloid cross-β structure, which HCC forms. Because α1 and β2 are more flexible in the L68Q mutant, disulfide bonds were engineered into this protein to anchor α1 and/or β2 to the main sheet where β3 and residue 68 are located.[133] Each set of cysteine mutations stabilized the monomer, pre-

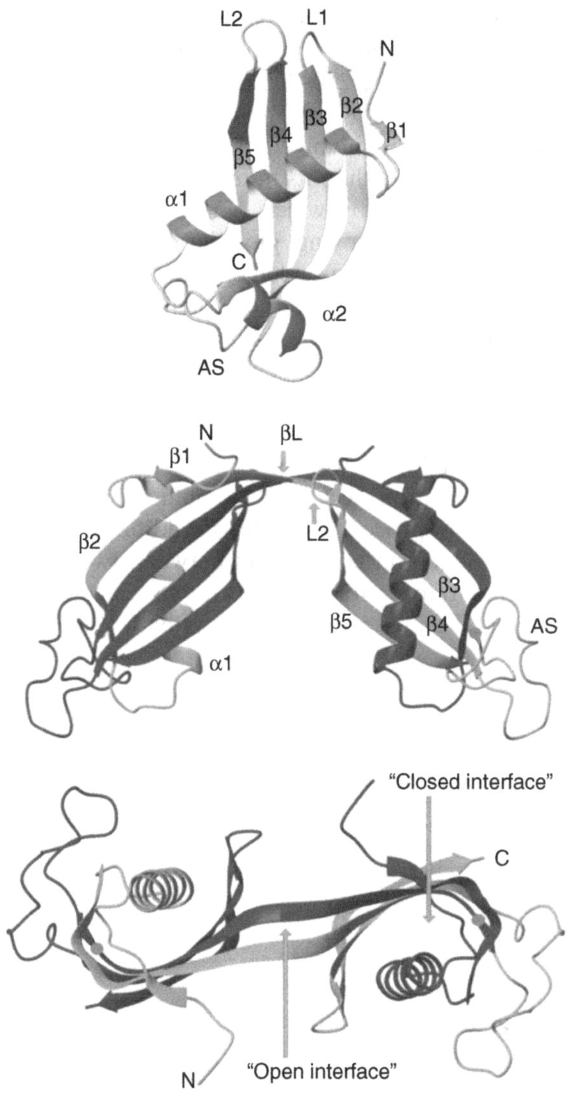

Figure 1.6. Domain-swapped dimer structure of cystatin C. The fold of chicken cystatin (PDB entry 1CEW), defining the topology of this class of proteins (top). Domain-swapped dimer of HCC in a view similar to (middle) and in a perpendicular orientation (bottom) emphasizing the beta sheet in the domain switch region (open interface) and the site of the L68Q mutation (red dot). From Reference 64. See color insert.

vented domain swapping, and greatly diminished amyloid fibril formation, indicating the domain-swapped dimer is an intermediate that propagates aggregation. Domain-swapped HCC subsequently forms metastable oligomers having a doughnut-like morphology with uniform outer and inner diameters of 13.4 and 2.7 nm, respectively.[134] The oligomeric state (or states) could not be accurately defined because they are not amenable to analysis by gel filtration, but they appear to be 200–300 kDa by sodium dodecylsulfate polyacrylamide gel electrophoresis (SDS-PAGE). The oligomeric species seems to be a metastable intermediate involved in fibril formation because it converts into fibrils much more rapidly than its monomeric and dimeric precursors.

An alternative mechanism by which fibrils form directly from domain-swapped dimers without occupying globular intermediates requires that domain swapping produce an open-ended oligomer in which additional monomers can assemble at both ends of the structure. This so-called runaway domain swapping inherently bears the requirement of high sequence specificity and was confirmed to exist using T7 endonuclease I (T7EI).[132] Cys residues were engineered into the T7EI sequence such that intermolecular disulfide bonds would specifically result when embedded in the proposed organization. When aggregation was allowed to proceed under reducing conditions, wild-type protein-like fibrillization occurred. Domain-swapped dimers pre-stabilized by oxidation did not proceed into fibrils. Addition of increasing amounts of reducing agent revealed an equilibrium between two dimeric forms, only one of which was able to form fibrils, suggesting that a conformational change in the domain-swapped dimer precedes fibril growth (Fig. 1.7).[132]

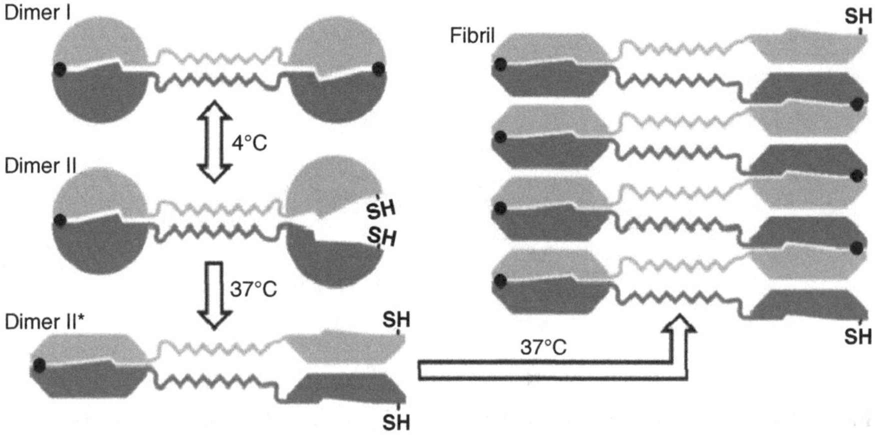

Figure 1.7. Schematic model for the fibril formation of T7EI L19C/S95C, illustrating the need for a structural transition from the dimeric state to generate the fibrillar arrangement. Each subunit is colored either in light or dark gray. Black dots represent disulfide bonds, and SH represents free cysteine. In dimer I, both disulfide bonds are intact and the protein is locked in close-ended dimers and is unable to form fibrils. In dimer II, half of the domain-swapped molecule is unlatched by the reduction of one of the two disulfide bonds. Upon incubation at 37°C, dimer II changes to an open form (dimer II*), in which half of the dimer opens up, exposing the interface that remains protected in dimer I. Open-ended dimer II* readily fibrillizes via runaway domain swapping. The hinge–loop region of the domain-swapped protein forms a zipper spine in the fibrils. Reproduced with permission from Reference 132.

From the studies described in this section as well as numerous others, it is clear that aggregates can have well-defined structures and can participate in specific interactions involving hydrogen bonding between beta strands. What remains to be explained is the selective basis for and the mechanisms by which a polypeptide achieves the native fold or an aggregate-prone intermediate conformation. While coordinated unfolding at specific sequences and association is required to generate fibrillar aggregates, amorphous aggregation may arise from heterogeneity among the unfolded population in which associations between several states have approximately equal free energies. This could permit an irregular macroscopic organization to develop as a result of intermolecular interaction at multiple sites.

1.5 MODELS OF PROTEIN FOLDING

1.5.1 Classical Models Based on Chemical Equilibrium

The earliest models developed to explain protein folding were based on chemical reactions at equilibrium, as defined by kinetics. The simplest conception envisioned a direct conversion between the unfolded (U) and native (N) state: U $\leftrightarrow$ N. Inherent in the chemical equilibrium view is the idea that the protein has a single path by which U and N interconvert. In this case, conversion involves a single step with a free energy barrier separating them. Folding must be completely cooperative to fit this model. Indeed, many early observations made concerning protein folding and unfolding obeyed such two-state kinetics. These early studies were almost all performed on small globular proteins that fold cooperatively, and the results show that this model describes folding well for this type of protein, at least to a first approximation. Because the timescale for folding is often on the order of milliseconds, only the unfolded and native states were observed in equilibrium experiments, validating the model. With this line of thought, it follows that aggregation must proceed from the unfolded state, since the native form is uniquely folded. As such, aggregates were believed to be composed of random coils held together through nonspecific hydrophobic interactions. This model is vastly oversimplified based on the knowledge available today, which indicates aggregation usually arises from partially structured species that form organized interfaces.

During early folding experiments, a few proteins were observed to fold in a manner that did not obey first-order kinetics, suggesting that folding involved multiple steps and the formation of intermediates. Such intermediates had to persist sufficiently long and to accumulate in enough quantity to be detected on the experimental timescale. Some proteins were hypothesized and were later shown to undergo isomerization at Pro residues or to be altered by the incorporation of a metal or heme group, slowing the approach toward the native state and allowing the observation of a specific intermediate.[118] As a result, the model was amended to include a limited number of intermediates in series along a single pathway: U ↔ U′ ↔ (U″ ↔ U^n) ↔ N. This multistate model permits the formation of on-pathway intermediates along the reaction coordinate toward achieving the native conformation. More extensive analyses have shown that these intermediates largely resemble the native form with minor deviations, implying that they are late-folding or early-unfolding intermediates. Because tools were not available to detect states that did not accumulate, the formation of transient intermediates could not be unambiguously established until the advent of stopped-flow and rapid-quench methods. Disulfide bonds added additional complexity to this scheme by providing evidence for off-pathway intermediates. This was evident even in Anfinsen's early studies in which refolding performed in the absence of a reducing agent led to a heterogeneous population of covalently cross-linked inter- and intramolecular species. These species remained soluble and were easily eliminated by refolding from the denaturant before allowing oxidation to occur. The mass action view of folding has limited options with which to describe off-pathway intermediates because it is based on a series of only a few sequential steps. Aggregates would then result from structures that are part of the native folding pathway. This model also does not lend itself to providing satisfactory explanations for the phenomena of nucleation and growth of aggregates.

$$\begin{array}{ccccccccc} U & \leftrightarrow & U' & \leftrightarrow & (U'' & \leftrightarrow & U^n) & \leftrightarrow & N \\ & & & & \updownarrow & & \updownarrow & & \\ & & & & A & & A' & & \end{array}$$

Scheme 1.1.

Applications of classical models to protein folding often result in a poor fit to experimental data. Hydrogen–deuterium exchange (HX) experiments have revealed the coexistence of subpopulations of folding intermediates, indicating that multiple, diverse pathways exist for arriving at the final native structure from the unfolded state or vice versa.[135] It was in part because of HX data that a new model for protein folding and unfolding emerged based in statistical mechanics.[136] This is discussed further in the next section.

The ideal gas law has also been used as a limiting law tool for describing protein folding. This model is of value because it incorporates the concept that a continuous range of states exists during protein folding, opening up many options for how a protein transitions from the unfolded to the folded state. In this model, the boundary is set by the volume the protein is permitted to sample, and entropy is a function of volume (V) and depends on the number of residues in the chain (N) according to the following equation: $S(V) = \mathrm{N}k_B\ln(V)$.[137] The entropy of the unfolded chain increases with larger numbers of residues and is presumed to be ~$2.3k_B\mathrm{N}$.[138] The residues are permitted to diffuse within the bounded space according to a potential function that drives them away from the edge of the spherical boundary and toward the center of the sphere, promoting the condensed, folded state. Although this simple model was useful for beginning to explore the effect of entropy on the unrestricted and bounded states, the function used is often a poor mimic of reality, because different proteins of the same length do not necessarily condense to the same architecture, indicating that other factors exert major influence on how a protein folds. This does not invalidate the thermodynamic model but simply implies that details of the protein sequence are critical to understanding the mechanism and kinetics of folding as well. The poor fit of data to any model stimulates testing of additional parameters. The problem with using the ideal gas model is that there is no phase transition. Condensation requires attraction between atoms or molecules, as exist with nonideal gas systems. As confirmed by Anfinsen's and others' studies, hydrophobic interaction provides an attractive force to drive condensation during protein folding. The simple gas law-based approach to examining entropic effects was an essential first step in understanding protein folding and was the basis for the development of more sophisticated models that introduced attractive forces, such as hydrophobic interaction, and other restraints into the calculations. Another outcome of applying the ideal gas model to protein folding was the hypothesis that similar to phase transitions between the gaseous and condensed states, the formation of a nucleus may be required to seed the transition. The nucleation hypothesis applied to protein folding provides a possible mechanism by which Levinthal's paradox is resolved.[137]

1.5.2 Statistical Mechanical Models

A particularly important innovation in understanding protein folding was the idea that proteins do not follow a single path to the native state but rather have manifold options by which to arrive at the native conformation. Along with this idea came the consideration that proteins are unlikely to populate a single static state but instead exist as an ensemble of conformers. Such models are based on statistical mechanics, in which the minimum energy state corresponds to a collection of native structures and folding proceeds along any trajectory that is not energetically forbidden.[136] The result of this type of analysis produces an enormous number of different viable conformations that could be adopted along the way from unfolded states to the native structures. Moreover, a Boltzmann-like ensemble is obtained for the folded protein; it is made up of similar but not identical structures that define the native state. A major difficulty in the interpretation of such data emanates from the inability to observe individual conformers experimentally. The most detailed view that can be obtained currently is with solution NMR, which is useful in probing the conformational dynamics of a protein and for observing even small subpopulations (>1%) within the sample.[139] In practice, it can be quite expensive and time-consuming to acquire and interpret this type of NMR data. As such, computational analysis provides an alternative, efficient way to explore the

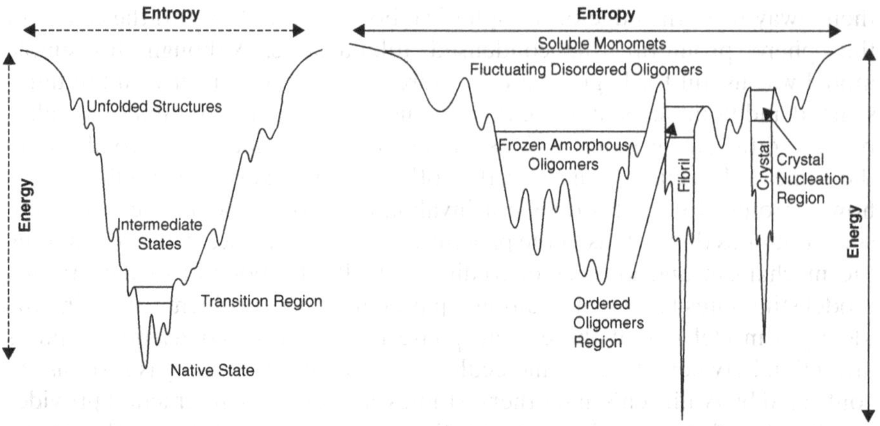

Figure 1.8. Two-dimensional energy landscape diagram of entropy versus energy. A semi-smooth, single funnel represents a typical protein folding landscape (left), whereas the landscape depicting aggregation is more complex and has multiple, well-separated minima corresponding to distinct stable structures (right). From Reference 141.

mechanisms behind protein folding and interactions, including aggregation.[140] *In silico* experiments are easily modified to assess the effect of mutations and already are providing very useful insights concerning the mechanisms of folding and misfolding. The value is maximized when the results of simulations are used to inform experiments and vice versa.

The computational approach involves calculations reliant upon several physicochemical parameters, corresponding to a highly multidimensional space of the function $E(\Theta_1, \Theta_2, \Theta_3 \ldots, \Theta_n)$, where E is the energy derived from an equation involving one or more parameters (Θ). The output is often simplified for visualization in the form of an energy landscape. The axes of the two-dimensional landscape are entropy on the x-axis and internal free energy on the y-axis. The profile for a protein that undergoes fast folding looks much like a funnel with rough edges (Fig. 1.8).[141] The better the protein complies with the classical two-state cooperative folding model, the smoother the funnel appears. The landscape is often rendered as a three-dimensional plot in which the z-axis is the internal free energy of a chain conformation and the axes in the transverse plane reflect specific parameters such as torsion angle energies, hydrogen bonding, and van der Waals contact, defined by $E(\Theta_1, \Theta_2)$.[142] Each point on the energy surface corresponds to a conformation of the protein. In general, the appearance of the landscape reflects kinetic differences between the conformational states, such that points at similar latitudes correspond to states that are equally probable. The slope of the surface is indicative of the rate of folding, in which steep edges indicate fast rates. Importantly, the results of these models can be evaluated by comparison to measured energy values obtained from various experiments. There are many approaches used to calculate the energy, and the variations are derived from observations made using a wide range of experimental tools that help further refine our understanding of the parameters that influence folding and aggregation.

While protein folding landscapes typically are funnel shaped, theoretical studies of protein aggregation produce much more complex landscapes with many traps.[141] The degree of complexity reflects the kinetic and equilibrium behavior of proteins at high concentration, in which native contacts compete with alternative intermolecular interactions.

1.5.3 Computational Models

Technological advances that enhance computer processing speed and data storage capacity have enabled the ability to model protein folding and other interactions. Structure-based approaches have provided the ability to test the influence of individual parameters on protein folding, such as the strength of hydrophobic and electrostatic interactions. Results can be validated and the accuracy of the result assessed by comparison to experiments. The validation of the approach and the role specific parameters exert on folding has made ab initio calculations possible to perform to assess the folding pathways of proteins and to identify intermediate states based on first principles without the need for a priori knowledge of the structure. Computations are based on parameters observed through experiment to affect protein folding rates guided using an energy function and restraints derived from experimental values. Ab initio calculations are not practical for use on systems with a large number of atoms, and for this reason, models have been simplified in a number of ways to make computation feasible.

One of the earliest models used to investigate protein folding using a computational approach was the HP model in which residues were designated as either hydrophobic (H) or polar (P).[143] The protein is placed into a two-dimensional lattice such that each residue occupies a position on the lattice. Residues are arranged randomly in an unfolded state with their connectivity preserved. Folding is dictated solely by interaction between H residues (HH) due to the assignment of a favorable energy term, while HP and PP interactions are ignored because no energy term is assigned. Energy is introduced into the system, prompting the residues to rearrange through movement to open lattice positions. An exhaustive search is conducted to identify all possible conformations and their associated energies. Folding using this approach shows multiphase kinetics in which collapse occurs rapidly followed by slower steps corresponding to local rearrangements. Rearrangements are slower because this requires energy barriers to be surmounted. Based on comparisons to experiments, this model reproduces well general features of protein folding, likely because it provides a complete characterization of the energy landscape and reveals interactions that are preferred during the folding process. The landscape is rough due to the presence of numerous kinetic traps caused by use of a single parameter. The HP+ model incorporates an energy penalty for nonnative interactions, speeding the process of folding and minimizing the depth and number of kinetic traps, making the landscape smoother. This method cannot be applied to a protein with unknown structure.[142] Proteins display non-Arrhenius dependencies on temperature during folding and unfolding experiments. By incorporating the temperature dependence of hydrophobic interactions derived from oil/water partitioning experiments, these simple models were able to reproduce qualitatively non-Arrhenius behavior and to attribute it to the temperature dependence of the hydrophobic interaction. This model was also able to explain chevron plots and observed deviations from ideal behavior (linear dependence of rate on denaturant concentration). The results underscore the central role of hydrophobicity in forming the condensed state.

The lattice method has been modified for use in three dimensions and with expanded alphabets to attain higher specificity. Additionally, more sophisticated terms are incorporated to account for polar interactions. When full alphabets with all 20 amino acids are involved, the Miyazawa–Jernigan (MJ) amino acid interaction matrix is typically used to define the interaction potentials between residues.[144] The MJ potential is derived from empirical data, which presumably makes it more accurate than simpler potential functions like HP when used in lattice model calculations.[145,146] Off-lattice models have also been developed, and this approach is more realistic and more accurately mimics protein folding.[147] Specifically, it is better suited to predict the structures and assembly process of aggregates.[148] It was used to analyze a 46-residue fragment of staphylococcal protein A that forms a three-helix bundle.[52] The inclusion of all side-chain atoms in the model produced dynamics closely resembling experimental results, and this positive outcome was attributed to realistic side-chain packing. Constant temperature, simulated annealing (SA), or replica exchange protocols may be implemented using molecular dynamics (MD), discrete molecular dynamics (DMD), or Monte Carlo simulations, depending on the time frame of interest in the study and the parameters used. DMD is a good choice for aggregation analyses because it is able to explore significantly longer time frames, which is necessary for examining the formation and rearrangement of intermolecular interactions in assembly processes.[91]

Go models were developed for use in the early stages of computational modeling of protein folding and are still useful. They rely on pairwise summation of contact energies derived from the known native fold. The method has been used extensively to characterize and test how individual parameters affect folding of proteins to the native conformation. This model cannot be applied to proteins or aggregates where the high-resolution structure has not been determined. Distance cutoffs are applied to define contacts and a binary (–1 or 0) approach is used to calculate the energy. It can be used with any bead model to further reduce the cost of the computation. The primary advantage of Go is that there are large energy gaps between native and other states such that folding proceeds rapidly in a cooperative manner. Because Go relies on known native contacts, aggregates that are based on nonnative contacts cannot be observed using Go models.[149] They are, however, amenable to producing aggregates in which native contacts are preserved. An expanded version of this model was used to identify early stable local elementary structures (LESs) involved in the folding process. Pairwise potentials were added based on $\Delta\Delta G$ measurements from mutants that affect the native conformation.[150] The LESs were found to provide the strongest interactions at the early stages of folding. These are needed to surmount the transition state barrier and to limit sufficiently the number of possible conformations to accomplish rapid, cooperative folding. This finding is consistent with the microdomain and foldon concepts of modular assembly and nucleation theories of folding.[127,151]

1.5.4 Application of Coarse-Grained Models

Protein folding and aggregation are analyzed and predicted based on knowledge of parameters that affect these phenomena *in vitro*. Terms that describe intramolecular forces affecting the protein and solvent effects on the protein are included. Specifically, energy terms related to hydrophobicity, backbone (ϕ, ψ) and side-chain torsion angles (χ_1, χ_2), steric strain, electrostatic attraction and repulsion, dielectric constants, hydrogen bond angles and lengths, aromatic interactions, and van der Waals contact and overlap may be included. Each is applied based on an energy function, often a continuous (Lennard–Jones) or a square well potential. The more constraints used in performing the calculation, the more time-consuming the computations become. Likewise, the computational time increases as more atoms are involved, such that all atom simulations are very computationally expensive and intractable for systems with greater than a few tens of atoms. Consequently, simplified *coarse-grain* models have been used extensively to examine proteins because they greatly increase the speed with which computations are performed.[152,153] Despite being greatly simplified, the results typically agree with experimental data, indicating the value of the simplified approach. Also, for this reason, solvent is not explicitly defined in many calculations, and instead, an implicit value for solvent is applied uniformly to the nonprotein space that reflects the average properties of water. It is more feasible to include explicit water in computations involving short peptides, where the number of atoms is relatively small. This offers the advantage of providing information about interactions between water molecules in the hydration sphere and the peptide that may influence the peptide's conformation. In the past few years, coarse-grained models have been used increasingly for the examination of protein–protein interactions and aggregation of peptides because they reduce the cost of calculations sufficiently to permit the study of molecular systems with very large numbers of atoms. These systems are difficult to study experimentally and the information provided by computational methods has enabled the design of tractable experiments to test and refine hypotheses about the structure and mechanisms of protein aggregation phenomena.

Coarse-grained models are simplified because they replace individual atoms with a single *bead* that reflects the composite properties of the group it represents. The group can be defined in a number of ways, and typically, bead models employ one to six beads per amino acid instead of using all atoms. The coarser the model, the more important the accuracy of the parameter-based force field is in interpreting the result, and the more general the conclusions must be about the system. One-bead models reflect the average properties of an entire amino acid and place the bead at the centroid of the residue or Cα position. A small number of parameters are included in one-bead calculations that reflect the general size, geometry, and properties of the amino acids. The force field parameters are simplified to correspond with the one-bead approximation. This model is useful for comparing differences in the mechanism of folding between similar proteins where the topology is known, but not for de novo folding. The simplicity of the approach requires that the angle and dihedral terms be known, which means a high-resolution structure must be available. The simulation is then guided by terms reflecting the side-chain contributions to folding. The method has provided increased understanding about how specific side-chain interactions influence protein folding. These models lack the molecular detail needed to assess the contribution of specific backbone interactions and, as such, do not provide sufficient detail to explain the mechanism or structural differences underlying the formation of aggregates.

Two-bead models have the advantage of providing information about backbone geometry and hydrogen bonding. One bead corresponds to the backbone atoms and a second to the centroid of the side chain. Separating the two groups increases the accuracy of the prediction by making it reasonable to incorporate additional terms into the force field, including angle–dihedral angle correlations or secondary structure propensity.[154] The addition of a third bead to longer side chains further improves accuracy by allowing more flexibility and distinct features to behave independently in the simulation.[155] The additional beads greatly improve folding studies, but the model still has limited utility for examining aggregation because hydrogen bonding and backbone angles are crucial determinants, and these parameters remain too generalized in this approach.

Models with a minimum of four beads are needed to define explicitly the backbone atoms and to examine hydrogen bonding. Three of the beads are assigned to backbone atoms (N, Cα, and CO), and side chains are represented by one (commonly Cβ) or more beads. The array of four- to six-bead modeling approaches that can be performed is expansive, and it is beyond the scope of this chapter to cover them in detail so this description is limited to examples pertaining to aggregation. A four-bead model was used to examine dimerization of Aβ(1-40) and (1-42) peptides. Nine structures of planar beta-sheet dimers having equal free energies were observed, and no significant difference in energy was found between the two peptide sequences.[156] The result demonstrates the ability of this model to generate the expected secondary structure characteristics of fibrous aggregates but, as the authors point out, also obviates the need for improved parameters to help differentiate between these structural isoforms since all atom simulations are intractable. Polyalanine peptides have been modeled under different conditions, and the results demonstrate the critical role of solvent, temperature, and peptide concentration on the distribution of monomer and aggregate structures obtained during the modeling process.[157,158] Refinement of the parameters appears to be a significant challenge, which is just beginning to be addressed.

Only a few computational studies have been published concerning protein aggregation. The aggregates studied involve small peptides known to form fibrillar structures. The first employed mid- to high-resolution modeling methods and obtained flat beta-sheet structures.[147] A lower resolution approach using one degree of freedom for the backbone dihedral and two intramolecular dipoles also was shown to generate fibrils.[159] The major deficit of these approaches is that neither was able to reproduce the twist observed in fibrillar sheets. A mid-resolution approach using a simplified alphabet was developed to rapidly model a peptide aggregation that reproduces the twist associated with fibrils as well as the known interstrand and intersheet associations.[96] This approach retains most of the backbone degrees of freedom, which are necessary for hydrogen bonding and chirality, and includes a single bead for the side chain to account for intersheet associations. The peptide contained alternating HP known to encourage beta structure and included oppositely charged residues at the termini. This model revealed the formation of a single hydrogen-bonded sheet (tape) early in the simulation, in which the strands formed an antiparallel arrangement. Double tapes emerged from the single tape only after a "critical nucleus" was obtained. In this study, the strength of the hydrophobic interaction and the polarity of the charged side chains were modulated. Double tapes formed solely from association of hydrophobic surfaces, which is consistent with the experimentally observed cross-β structure of fibrils. Higher-order tape formation requires the association of polar surfaces, and these species were only observed to form when the polarity of the charged residues was reduced in the simulation. The kinetics of fibril formation in this model are consistent with a nucleation–growth mechanism, in which a single tape of critical length serves as the nucleus and electrostatic effects influence nucleation. This model clearly provides a good platform for further investigation of factors that contribute to nucleation, growth and the structure of aggregates.

1.5.5 Information Theory (IT)

IT is a network analysis model focused on quantification of information and relationships based on the probability of an event's occurrence. Originally, IT was developed in 1924 at Bell Labs by Harry Nyquist for the compression and reliable transmission of data by electronic communication. This fundamental approach relies on simple concepts and has been expanded since then for use in a wide variety of applications, including molecular networks. The goal is robustness or to maximize efficiency of information transfer without compromising the quality of the output on the receiving end. With electronic data, this means determination of the minimum number of bits necessary to achieve an accurate and correct interpretation of the information during transmission over noisy, lossy lines. Line noise introduces uncertainty into the transmission and is considered to be equivalent to the entropy of the system. The efficiency of information transfer emerges from the use of probability functions in this model, because communicating the mean and distribution that describes a relationship more efficiently summarizes the system compared to transmitting the entire set of individual measured values that compose the system separately. Both approaches generate a similar interpretation following transmission, but one is more compact than the other. The two main features of quantification are *entropy* and *mutual information*. Entropy reflects the randomness of a variable, and mutual information describes the relationship between variables. In thermodynamics, the Boltzmann distribution function and Gibbs free energy equation encompass the base mathematical components of IT.

IT-based approaches have been applied to predict protein folding and aggregation. They parallel the simple thermodynamic model that based entropy on chain length such that molecular details of the proteins are not included in the model. Instead, general principles and macroscopic observables that contribute to association are used for the calculations. Early on, an approach that accurately describes the primitive hydrophobic effect of solutes in water was developed by Pratt and coworkers.[160] The model predicts the probability of observing a solute-sized cavity in water based on the density of water and on the radial distribution of the central oxygen atoms in the solvent molecules. The two-moment model emphasizes the thermodynamics of hydration of the more hydrophobic solute and the association between hydrophobic entities, simply modeled as hard spheres. The model quantifies the hydrophobic effect using the excess chemical potential ($\Delta\mu_{ex}$) of the solute as the variable, because it reflects the probability (p_o) of finding a cavity of a given size/shape in the solvent according to the function $\Delta\mu_{ex} = -k_B T \ln(p_o)$. This simple approach reproduces accurate plots of cavity-formation thermodynamics and the hydrophobic effect when applied to two simple, small-molecule model systems. A study of methane in water was performed, yielding the same profile as the more complex Smith and Haymet model. Application of the approach to *n*-butane generated lower free energies for the more compact *cis* and *gauche* conformations compared with *trans*, paralleling the results obtained in a more complex analysis by Begelov and Roux.

In one application to protein aggregation, Kumar et al. used a similar approach to characterize the interactions between neurofilaments (NFs) in axons.[161] NFs are composed of monomers that form a central core along the long axis and a brush-like structure on the outside that is composed of a largely disordered sequence having an extremely high percentage of charged residues. Association of NF was modeled using IT based on experimentally derived values. The cross-sectional radial-spatial distribution function (RDF) of NF in axons describes the density of NFs around a single central NF, providing the probability distribution ($g(r)$). The potential mean force (u_{MF}) can then be determined for the average interaction potential over all configurations and orientations of NFs in the system, according to $u_{MF} = -kT\ln g(r)$. A central NF was randomly selected and the number of particles counted within a defined spaced was used to determine the occupancy probability distribution (OPD). Fitting to the IT model was then accomplished using the RDF and OPD data. Other modeling approaches require that interparticle distances be known or estimated. Because the measurement is not possible in the NF system, the IT model is advantageous because this data is not needed for the calculation.

Several IT-based models have been used to assess aggregation propensity for proteins using the amino acid sequence. The Thornton group used an approach based on the propensity of each individual amino acid to form a β structure to predict the preference of a peptide to then form pairings between strands that would lead to parallel or antiparallel beta-sheet structures.[162] The method was able to identify the correct register less than 50% of the time but was able to distinguish aggregation-prone from random sequences. Since it has been established that hydrophobicity is directly related to the amount of solvent-accessible surface area for polar and nonpolar groups alike, another approach to understanding aggregation was based on prediction of the solvent accessibility of amino acids in the context of different sequences.[163] Calculations were performed using values for each individual amino acid alone and for each in the context of a 17-residue peptide. The results confirmed that the side-chain solubility is strongly influenced by local context, and a revised hydropathy scale was presented to account for neighbor effects on solubility. Like the two aforementioned approaches, most studies have relied on small, hand-selected data sets in order to focus on analyzing a specific type of interaction. A systematic, high-throughput method was developed by Ofran and Rost to sort the very large structural data set available in the PDB to identify differences in composition correlated with distinct types of protein interfaces.[164] The approach confirmed findings from the previous studies on smaller data sets and also revealed the presence of six types of interfaces. The distinctions reflect the probability of interaction, which is derived from amino acid contact preference and interatomic distance, as determined by proximity within the polypeptide sequence and through covalent (obligate) attachment between chains. Homo- and heterochain interactions were found to have different bases for association, as was also noted among transient and obligate partners. The composition of amino acids within each type of interface was unique among the six classifications. In all types of interfaces, however, lysine, serine, alanine, and glycine were underrepresented, while arginine was overrepresented. Hydrophobic residues as a class were found with approximately equal frequency in all types of interfaces, but the specific apolar residues present varied between interface types. For example, on average, leucine and valine are abundant in protein interfaces, but the percentage of each varies dramatically between unique types of interfaces. Hot spots for association were previously reported to have an abundance of tryptophan, tyrosine, and arginine and to lack serine, threonine, valine, and isoleucine.[165] Separating the interface into two regions, the core and rim, leads to further distinction in the involvement of individual amino acids.[166] While these general observations were affirmed in this study, more specificity among residues was gleaned from each type of interface. For example, on average, tryptophan is overrepresented in protein interfaces, but it is extremely underrepresented in homocomplexes.

Mintseris and Weng further simplify the IT modeling approach by employing a five-letter alphabet.[167] They grouped the residues by type to more rapidly examine the influence of specific properties of the side chain on folding and interfacial association. Grouping reduces the amount of mutual information, making the calculation more efficient. Analogous to coarse-grained bead models, the groupings encompass contributions from the main chain and from hydrophobic, positively charged, negatively charged, and polar side chains. This study predicts that hydrophobicity within a sequence has a greater influence on monomer folding than on aggregation and also that polar and electrostatic interactions are important factors in interface formation.

The major advantage of an IT-based approach is the simplicity of the model, which remains capable of producing results analogous to more complex approaches. Alternative or additional physical insights can be incorporated directly into the model, allowing it to be applied to the investigation of more sophisticated systems like proteins. Some additional applications of its use for the study of proteins not discussed here include interactions of ligands at binding sites, the effects of mutations on protein stability, and examination of conformational equilibria of protein side chains.

1.6 INFLUENCES OF CHEMICAL ALTERATION ON AGGREGATION

Proteins are subject to both enzymatic and nonenzymatic changes to their chemical composition, which alter their conformation and/or stability. *In vivo*, posttranslational modifications frequently control activity and protein–protein interactions. This can be accomplished by oxidation, reduction, proteolysis of the polypeptide chain, or other chemical modifications that alter the surface properties and/or conformation of the protein. Spontaneous changes, such as deamidation, were proposed to function as internal timers signaling that the protein should be actively repaired or degraded.[168] Similarly, spontaneous chemical changes to proteins *in vitro* can alter the conformational dynamics and can affect protein stability and aggregation. The implications of chemical changes in proteins appear to be complex and depend not only on the amino acid affected but also on its sequential and structural neighbors as well. The major types of spontaneous chemical alterations that affect folding and aggregation are described in this section.

1.6.1 Length of the Polypeptide Chain

Cleavage of the protein backbone has been shown to alter the aggregation properties of both disease-related and therapeutic proteins. The Aβ(1-42) peptide is more prone to fibril formation than its slightly shorter counterpart Aβ(1-40).[169] In this case, the presence of residues 41 and 42 increases the overall hydrophobicity of the peptide, and this is thought to be responsible for enhanced aggregation. Proteolytic cleavage of blood factor proteins promotes aggregation in clot formation.[170] Here, it is the removal of a propeptide that initiates assembly. The clotting factor is protected from rapid aggregation because while covalently attached through the peptide bond, the propeptide region effectively interacts with the aggregation-prone region. Once the pro-moiety is cleaved, the peptide dissociates and no longer impedes intermolecular aggregation. Spontaneous hydrolysis can also occur during storage, often at Asp residues, and depending on the protein, this may promote aggregation.[171]

Addition of residues to the terminus of a normal protein also may facilitate aggregation. The most prevalent examples of this are found in the polyglutamine tract extensions associated with diseases that form nuclear plaques.[172] Modeling and *in vitro* studies suggest that an extension of a sufficiently long length increases the dynamics of the native protein such that partial unfolding results.[173] In contrast, polyglutamine sequences alone assemble into fibrils and amyloids.[174] A viable alternative proposal is that unfolding of the native protein ensues following aggregation of the polyglutamine segment.

1.6.2 Methionine Oxidation

Oxidation has been predicted to promote aggregation of proteins involved in amyloid plaques because oxidative stress is strongly correlated with protein misfolding diseases. Despite the fact that oxidation of numerous proteins embedded in amyloids is increased significantly, it appears that Met oxidation discourages amyloidosis. High-resolution NMR studies of Aβ peptides show that oxidation of Met35 to methionine sulfoxide substantially decreases aggregation rates.[21,175,176] Spectral changes suggest a conformational adjustment occurs in response to oxidation that disfavors formation of the turn involving Gly37 and Gly38, thereby inhibiting self-association.[177] Oxidation at methionine was also shown to disrupt fibril formation in the prion protein PrP^{SC}. Interestingly, modification did not prevent aggregation entirely, since prefilbrillar aggregates accumulated.[178] *In vitro* studies on recombinant factor VIIa (rFVIIa) also show that oxidation of its methionine residues fails to promote aggregation. In fact, conditions that stabilized rFVIIa against aggregation had increased methionine sulfoxide content.[179] Oxidation of methionine increases the polarity and hence solubility of the hydrophobic side chain. In keeping with this, studies of these and other amyloid proteins indicate that the site at which oxidation occurs does not encourage aggregation through direct participation in the intermolecular interface but may do so indirectly by promoting unique structural transitions that are responsible for generating low-molecular-weight oligomeric aggregates and large fibrillar forms.

1.6.3 Covalent Cross-Linking of Strands

Oxidative modification of proteins can lead to covalent cross-linking. This predominantly occurs through the formation of disulfide bonds between Cys residues, but covalent attachment through lysine and aromatic side chains has also been observed.

The involvement of disulfide bonds in protein stability and aggregation appears to depend in part on the biological environment in which the protein is normally found. Evolutionary selection occurs within a specific environment and protein sequences have been selected to avoid aggregation *in vivo*.[62] Extracellular proteins, which function in more oxidizing environments, often contain disulfide bonds that stabilize their native conformation. These proteins are often used as therapeutics and include insulin, growth hormones, interferons, and monoclonal antibodies. Cleavage of disulfide bonds by reducing agents leads to increased conformational mobility and uncondensed states that resemble MGs or random coils.[118] Mixing of disulfide bonds generates nonnative structures, including covalently linked intermolecular species that facilitate aggregation.[5,180] For example, in the absence of its metal ligand, an intramolecular disulfide bond forms in Cu, Zn superoxide dismutase (SOD1) that involves the Cys residues that normally participate in metal coordination. This leads to increased flexibility of a loop at the dimer interface and promotes aggregation.[181]

Cytosolic proteins are buffered from oxidation *in vivo* due to the presence of excess reducing capacity, largely in the form of glutathione (GSH).[182] As such, the vast majority of Cys residues in these proteins are in the reduced state. Oxidation rates depend on the pKa of the thiol, such that more easily ionized Cys are more susceptible to reaction.[183] Solvated Cys residues typically have pKa values about 8.7, but the pKa can be perturbed by several orders of magnitude as a result of the local chemical environment developed in a uniquely folded state.[57] Oxidation regulates the activity of some intracellular proteins. For example, the protein tyrosine phosphatase's (PTPase) catalytic action relies on the existence of a reactive thiolate anion in the active site.[184] Oxidation inactivates these enzymes and alters their conformation, facilitating their removal, possibly to prevent aggregation. Oxidative damage is strongly correlated with protein misfolding diseases, suggesting that the formation of disulfide bonds may play a role in intracellular aggregation. Studies of the oxidation of proteins that do not contain native disulfides often report that disulfide formation leads to aggregation or changes in solubility. This parallels the effect of disulfide mixing observed for proteins that contain native disulfide linkages.

Covalent dimerization of α-synuclein following oxidation promotes aggregation. This protein contains no Cys and so attachment is achieved not through disulfide bonding but instead as a result of tyrosine oxidation to form dityrosine. Cross-linking of these aromatics promotes fibril formation.[185] Selective Tyr cross-linking is achieved by employing the cytochrome c/H_2O_2 system, which has the added benefit of avoiding oxidation of other amino acids. Oxidation often is carried out using Fenton chemistry, which results in the modification of several residues, including Met, Cys, and His, but does not produce dityrosine.[186] Depending on the rate at which oxidation proceeded and the type of oxidant used, aggregate morphology for α-synuclein differed. Such results indicate the importance of spatial and temporal parameters on the aggregation mechanism.

Covalent linkages form between strands in collagen matrices that involve modified lysine residues. The normal structure of collagen consists of a triple helix, which constitutes the base unit in a higher-order network that composes basement membranes *in vivo*. This organized aggregate forms a semipermeable barrier with a small pore size, which limits infiltration of macromolecules into the tissue. The tight packing arrangement depends on covalent cross-linking and association of two hydrophobic regions. A staggered antiparallel orientation develops between triple helices, organizing the structure into tetrameric fibers. Tetramer formation is driven by interaction between two stretches of apolar residues and is stabilized by regular, covalent attachment through two disulfide bonds and two lysine-derived aldimine bonds per strand.[187] In the native matrix, aldimine linkages are found between specific lysine residues in the N-terminal region where hydroxylysine is found followed by a Gly-His/Glu-Arg sequence. The majority of lysine residues in collagen are modified to hydroxylysine or O-glycosylated hydroxylysine. These modifications sterically hinder association between helices on the outside face of the tetramer, which imparts both elasticity and strength to the polymer, supporting the function of the basement membrane. As the matrix ages, the number of glycosylated lysines increases and the extent of cross-linking increases. Glucose reacts with the ε-amino lysyl moiety and, through a series of steps, ultimately leads to the formation of a stable aldimine linkage. Exposure to increased concentrations of oxygen and glucose accelerates this reaction at other sites, causing structural rearrangements that increase the pore size and alter the permeability of the membrane. Although the details of the resulting structure are not fully described, available data suggest that not only more cross-linking occurs but that the positions of attachment also differ compared to the natural lysyl-aldehyde-derived linkages.[188]

1.6.4 Deamidation

Deamidation commonly abrogates function and may promote partial unfolding, which enables aggregate formation.[168] Degradation of asparagine is a spontaneous process, which shortens the shelf life of numerous protein therapeutics. The reaction typically leads to the formation of isoaspartic acid (IsoAsp) and Asp in a 3:1 ratio. IsoAsp is a beta amino acid, and the inclusion of an additional carbon in the backbone would prevent the direct hydrogen bonding required to support beta-sheet formation at the site of the modification. This effect would thus need to propagate aggregation through conformational changes or by negatively affecting surface properties, such as charge complementarity or hydrogen bonding. Studies of the amylin 20–29 peptide show that the native sequence does not aggregate, but amyloid formation results from the introduction of a small amount of deamidated species.[189] The mechanism resembles the template-driven aggregation postulated by Prusiner for prion proteins, which proceed in a pH-dependent manner like that observed with deamidation. Aggregates formed more rapidly at pH 5.7 than 2.25, suggesting that the ionization state of the new carboxyl moiety plays a role in assembly. In other cases, the introduction of a charged moiety, as would be the case at pH 5.7 for this peptide, has been shown to interfere with aggregation, In this case, however, the overall charge on the peptide becomes neutral following deamidation, lowering the barrier to intermolecular association. Although currently no structural data are available, it can also be speculated that the position of the carboxyl moiety near the positively charge N-terminus leads to a conformational change induced by electrostatic interaction.

Glutamine also deamidates spontaneously, albeit 100-fold more slowly than Asn. This modification has also been shown to alter the aggregation properties of proteins. In the case of the two-domain γD-crystallin proteins, deamidation of Gln at the interface between the domains diminished thermal stability due to the emergence of a charged side chain in a hydrophobic interface.[190] The charged moiety disrupts packing and increases exposure of the remaining apolar residues in the core. The larger accessible hydrophobic surface area then makes the protein more amenable to aggregation. In contrast, aggregation of α-zein is diminished by Gln deamidation. Native α-zein is highly insoluble due to a high hydrophobic amino acid content and subsequent extensive aggregation. Gln can be converted enzymatically by protein-glutaminase (PG) to Glu for *in vitro* analysis.[191] Deamidation by PG enhances solubility of the α-zein peptide by drastically reducing its propensity to self-associate.[192] The mechanism of action appears to be due to the development of negative charge, which is uncompensated, disfavoring intermolecular β interaction. This finding parallels a common approach employed by nature to protect edge strands in beta sheet-containing proteins from aggregation.[62]

1.6.5 Proline Isomerization

Protein folding studies have shown that a common rate-limiting step in arriving at the native structure is isomerization of proline.[118] As this is a spontaneous reaction, *in vivo* native-like states receive assistance in finding their correct conformation from peptyl-prolyl isomerases. β2m in its native monomeric state contains a *cis*-Pro at position 32. The native form is highly stable at physiological pH, resisting aggregation even at high concentrations, but oligomers form readily in the presence of Cu^{2+}. Cu binding induces a structural transition in which Pro32 converts to the *trans* isomer, initiating aggregation.[193] Mutations at this residue invoke inherently the *trans* conformation of the backbone and facilitate aggregation. β2m oligomers formed from the *trans*-Pro and mutant proteins eventually convert to amyloid fibrils, as occurs in the dialysis-induced disease state.

Despite the previous example, understanding the role of Pro in protein aggregation is not straightforward. A study of naturally occurring beta sheet-containing proteins found that one common mechanism by which native proteins prevent aggregation at the edge of their beta sheets is the incorporation of proline at a strategic position terminating the exposed strand. The sheet is terminated because the Pro creates the maximum allowable twist to support appropriate H-bond interactions on the concave side of the strand while preventing such interactions on the convex side.[62] Proline scanning mutagenesis has been used to identify regions within amyloidogenic proteins that promote aggregation. Substitution of Pro into Aβ revealed that residues 15–21 and 31–37 are critical determinants of aggregation as insertion within these regions impaired plaque formation. This approach has been used successfully in a therapeutic as well. Amylin, which is used to treat hyperglycemia in diabetics, forms amyloid fibrils in the pancreas. Variants of amylin in which Pro is substituted in positions 25, 28, or 29 lack the ability to polymerize and therefore do not cause amyloidosis.[194]

In many cases, chemical changes and mutations have little effect on protein stability. This is one reason that mutagenesis techniques are commonly employed to investigate protein function. In other cases, however, modification sufficiently alters the thermodynamic stability of the protein to encourage aggregation. Alterations to the surface are tolerated better by most proteins than those occurring in the core. This is probably because changes in the core synergistically affect packing efficiency by disrupting hydrophobic interactions that drive folding or remove hydrogen bonds within the core that stabilize polar or charged species. On the surface, changes that decrease stability may exert their effect by diminishing the net charge of the protein, disrupting specific electrostatic networks or increasing exposure of the apolar surface area to promote aggregation. It is difficult to predict the effect of a particular structural change on protein association because the site at which chemical modification occurs may not directly participate in the aggregation-inducing protein–protein interaction but instead may alter the energy landscape such that conformational dynamics allow the protein to more frequently sample less condensed states. This increases the probability that stable nonnative associations will result from intermolecular collisions.

1.7 APPROACHES TO PREDICTING AGGREGATION

Methods for predicting protein aggregation are based on fundamental physicochemical properties of the heteropolymer. The two most common approaches encompass evaluation of (1) the chemical composition or amino acid sequence and (2) structural features based on compatibility with the known fibrillar cross-β structure or propensity to form β-structure in the native state. Within each framework, several variations have been developed that have differing degrees of complexity, accuracy, and validity with respect to experimental data.

A sequence-based approach to predicting aggregation from the properties of individual amino acids was developed by Dobson and coworkers. It is based on the propensities of each residue to participate in aggregation and to adopt a beta structure.[125] The propensity to aggregate depends on the hydrophobicity of an individual side chain and also on the pattern of hydrophobic residues in which the residue is embedded. The propensity to form a beta structure is countered by the inclusion of a term for α-helical propensity. The net charge is considered and may be modulated to account for the effect of pH. An overall *Z*-score is assigned so that peptides of the same length with different sequences may be ranked relative to a randomly generated sequence. When the algorithm was applied to Aβ(1-42), α-synuclein, and tau, the highest *Z*-scores for aggregation corresponded to the regions of the sequence known to form amyloids, confirming that these fundamental properties of proteins correlate well with aggregation propensity.

The Chiti–Dobson equation was further developed to predict the rate of aggregation, specifically elongation or growth of the aggregate following nucleation, for mutated proteins. The assessment is based on the protein composition and compares the relative stability of individual mutants to the wild-type protein using first principles derived from protein folding and aggregation experiments. The equation includes terms for hydrophobicity, propensity to form α-helix and β-strand structure, and net charge. The equation used to calculate aggregation rate is

$$\ln(\nu_{\text{mut}}/\nu_{\text{wt}}) = 0.633(\Delta\text{Hydr}) + 0.198(\Delta\Delta G_{\text{coil-}\alpha} + \Delta\Delta G_{\beta\text{-coil}}) - 0.491(\Delta\text{charge})$$

where the change in hydrophobicity is ΔHydr, the difference in free energy between the random coil and structured state is $\Delta\Delta G_{\text{coil-}\alpha}$ for alpha and $\Delta\Delta G_{\beta\text{-coil}}$ for beta structure, and the net change in charge is Δcharge. Instability is defined as the inverse of the normalized ΔΔG or Tm. Several comparisons to epidemiological data sets have been performed, and the results validate this computational approach. On average, ΔΔG or Tm values correlate very well with the sum of the predicted instability and aggregation propensity, with instability being a better predictor of disease onset than aggregation propensity when each is considered independently.[195]

The Zyggregator method presented by Tartaglia and Vendruscolo also predicts protein aggregation propensity from the amino acid sequence using the same first principles as Dobson–Chiti, but this algorithm includes terms describing the pattern or spatial relationship between residues and takes gatekeeper residues into account.[196] The charge is also considered in a more distributed manner and is determined using a sliding window of 21 residues. Analyses of Aβ(1-42) and α-synuclein were performed, and the computational results obtained using only intrinsic factors correlate well with experimental data, identifying specific segments of the sequence responsible for aggregate formation. Moreover, when a competing term for the intrinsic propensity of the protein to form the native structure is included based on the CamP method, the approach yields higher accuracy results, enabling evaluation of different modes of aggregation within the same protein sequence. Zyggregator further includes linear correction terms to account for the extrinsic factors pH and ionic strength of the solution. The method was applied using a seven-residue window size to several additional peptides and proteins known to aggregate to more thoroughly investigate the relationship between structural protection and aggregation propensity.[197]

A model to predict aggregation-prone sequences from fundamental properties of the polypeptide chain and two environmental factors was also developed by Caflisch and coworkers.[198] The approach also predicts the orientation of the strands in the sheet as parallel or antiparallel. A genetic algorithm is used to optimize sequence space and conformational sampling is accomplished with MD. The computation yields the aggregation propensity along the sequence and the rate of aggregation. The absolute rate equation includes terms that account for aromaticity, polarity, solubility, and formal charges in addition to the terms used in the other computations described above. Orientation of the strands is primarily a reflection of the electric dipole moment of the peptide, such that polar residues favor an antiparallel arrangement due to the presence of unfavorable dipole–dipole interactions in an in-register parallel sheet organization. Concentration and temperature are included in the determination of the rate. This method achieves a high degree of accuracy in predicting both rate and structure.

TANGO is a statistical mechanical model designed to predict β-aggregation in peptides and proteins as a means of identifying nucleation sites for aggregation.[199] The approach is based on the observation that aggregates often contain an increased β-structure and the core regions of an aggregate are completely buried, such that nucleating sequences will have their hydrogen-bonding potentials largely satisfied. The algorithm takes into account competing conformations, including α-helix, beta sheet, the folded state, and β-aggregates, to calculate the partition function of the conformational phase space. The frequency at which a conformation is sampled depends on an energy function derived from statistical and empirical factors. Because packing interactions are considered, the approach differentiates between β-propensity and β-aggregation. TANGO accurately predicted *aggregation* or *no aggregation* for the majority of a set of 179 real peptides reported in the literature. Twenty-one false-positive and three false-negative results were found, and it was determined that peptides with low aggregation propensities were not identified using this method. The program can be interfaced after registering via the Web at http://tango.embl.de/ and is free for academic use.

Because many aggregates have been found to correspond to short peptide fragments derived from larger proteins, a library of 20-residue peptides was generated from which aggregation-prone elements were identified.[200] Based on the results, a computational approach known as peptide interaction matrix analyzer (PIMA) was developed to predict aggregation of proteins from sequences containing β-prone elements.[201] The protein sequence is divided into partially overlapping fragments of a specific length, and each peptide is threaded onto an in-register parallel or antiparallel beta-sheet structure. The energy associated with interaction between elements is calculated and used to predict aggregation tendencies. The study revealed that the prediction is best when 10-mer segments are used. Seven-mer fragments produce energies too small to effectively discriminate among and rank aggregation-prone sequences, whereas predictions were less accurate when based on 20 mers, presumably because not all 20 residues participated in the association. When peptides derived from amyloid-forming proteins were examined, the results matched experiments well.

A web-based server for prediction of amyloid structure aggregation (PASTA) was developed based on the in-register parallel alignment observed in most cross-β structures.[202] The assessment is made in two dimensions and involves determining first the probability of aggregation on a per-residue basis followed by the probability of cross-strand pairing with self-alignment. Cross-strand association is determined using a pairwise energy function to evaluate facing residues in the sheet. As a result, PASTA also provides an assessment of intermolecular hydrogen bond pairings between strands. The result is output as a contact map highlighting the probability of amyloid behavior. The freely accessible site is housed at http://protein.cribi.unipd.it/pasta/ and the source code is available for download.

AGGRESCAN is a sequence-based aggregation prediction tool available on the Web. The method relies on an aggregation propensity scale for each of the naturally occurring amino acids and is applied to short segments of the protein sequence to identify hot spots for aggregation.[203,204] The window size can be varied and has relatively little impact on the result using this approach, except at the extreme limits where very short sequences (less than five residues) generate numerous hot spots, many of which do not correspond to experimental findings, and with very long stretches relative to the length of the polypeptide being examined. The method was trained against 57 amyloidogenic proteins for which the aggregation-prone sequences were experimentally determined. It can be applied to proteins having up to 2000 residues, and multiple sequences can be input together for comparative analysis. Results are returned in minutes via the Web at http://bioinf.uab.es/aggrescan/.

The Dokholyan lab has developed the iFold server to enable large-scale simulations of folding and thermal unfolding using structure-based, coarse-grained models and DMD.[205] While the model is not specifically designed for analyzing aggregation, unfolding is a precursor to assembly, and assessing modes of unfolding for a protein may be useful in the determination of the accessibility of regions involved in aggregate formation. The primary goal of iFold is to investigate dynamics, and as some aggregate structures are known, their dynamics may be assessed. The server is found at http://iFold.dokhlab.org/. The same group developed H-predictor to identify hinge regions in proteins that are responsible for domain swapping.[91,206] A two-bead model estimates the enthalpy change upon hinge formation based on alterations in bonds and native contact distances. The method was applied to predict domain-swapped dimers of RNase A and focal adhesion targeting (FAT) domain proteins. H-predictor is available at http://dokhlab.unc.edu/tools/h-predictor.

Analysis tools have been developed to identify intrinsically disordered regions in proteins based on the amino acid sequence. Often, intrinsically disordered proteins resist aggregation and their sequences contain high percentages of charged residues. PONDR and DISEMBL are web-based approaches for predicting intrinsic disorder.[207,208] Disordered sequences may also be predicted to have a β-structure, but they are not prone to self-association because, similar to globular proteins, they too have endured evolutionary selection to avoid aggregation *in vivo*. This complementary analysis provides a means of identifying sequences within proteins that are unlikely to contribute to aggregation when unfolded.

1.8 CONCLUSIONS

A great increase in our understanding of protein aggregation has been gained as a result of comparisons to the more thoroughly studied and developed field of protein folding and unfolding. The models created to evaluate the importance of individual factors, both intrinsic and extrinsic, on kinetic and thermodynamic aspects of protein unfolding have been modified to investigate aggregating peptides and proteins and the mechanisms responsible for nucleation and elongation in aggregate formation. Examination of insoluble fibers using medium- to high-resolution experiments has provided molecular details about the patterns observed in regular aggregates as well as some specific interatomic distances between side chains. These details, along with the recognition that aggregates are often organized into beta sheets, have enabled *in silico* approaches that are beginning to elucidate the fundamental principles underlying assembly. So far, the majority of attention has been paid to fibrillar aggregates because they are frequently correlated with disease states and their repeating register is more amenable to analysis using existing experimental techniques. Computational approaches now need to be applied based on the insights gained from studying fibrillar systems, and new tools and analysis methods need to be developed to better characterize amorphous aggregates. Amorphous aggregates are more common, and their formation diminishes the efficacy and safety of protein therapeutics and also frustrates production and limits the shelf life of biotechnology products. An increased understanding of the molecular basis for protein aggregation is essential for our being able to stabilize proteins against aggregation *in vitro* and to develop effective treatments for protein misfolding diseases.

REFERENCES

1. Chiti F, Dobson CM. 2009. Amyloid formation by globular proteins under native conditions. Nature Chem Biol 5(1):15–22.
2. Dobson CM, Karplus M. 1999. The fundamentals of protein folding: Bringing together theory and experiment. Trends Biochem Sci 9:92–101.
3. Maas C, Hermeling S, Bouma B, Jiskoot W, Gebbink M. 2007. A role for protein misfolding in immunogenicity of biopharmaceuticals. J Biol Chem 282:2229–2236.
4. Kahn SE, Andrikopoulos S, Verchere CB. 1999. Islet amyloid: A long-recognized but underappreciated pathological feature of type 2 diabetes. Diabetes 48:241–253.
5. Cromwell MEM, Hilario E, Jacobson F. 2006. Protein aggregation and bioprocessing. AAPS J 8(3):E572–E579.
6. Rosenberg AS. 2006. Effects of protein aggregates: An immunologic perspective. AAPS J 8(3):E501–E507.
7. Alberts B, Johnson A, Lewis J, Raff M, Roberts K, Walter P. 2002. *Molecular Biology of the Cell*, 4th edition. New York: Garland Science.
8. Hartl FU, Hayer-Hartl M. 2002. Molecular chaperones in the cytosol: From nascent chain to folded protein. Science 295:1852–1858.
9. Meredith SC. 2005. Protein denaturation and aggregation: Cellular responses to denatured and aggregated proteins. Ann N Y Acad Sci 1066:181–221.
10. Bagola K, Sommer T. 2008. Protein quality control: On IPODs and other JUNQ. Curr Biol 18:R1019–R1021.
11. Kaganovich D, Kopito R, Frydman J. 2008. Misfolded proteins partition between two distinct quality control compartments. Nature 454:1088–1095.
12. Dinner AR, Karplus M. 1999. Is protein unfolding the reverse of protien folding? A lattice simulation analysis. J Mol Biol 292(2):403–419.
13. Patra AK, Mukhopadhyay R, Mukhija R, Krishnan A, Garg LC, Panda AK. 2000. Optimization of inclusion body solubilization and renaturation of recombinant human growth hormone from *Escherichia coli*. Protein Expr Purif 18:182–192.

14. Azakami H, Mukai A, Kato A. 2005. Role of amyloid type cross-β-structure in the formation of soluble aggregate and gel in heat-induced ovalbumin. J Agric Food Chem 53:1254–1257.
15. Sunde M, Blake C. 1997. The structure of amyloid fibrils by electron microscopy and X-ray diffraction. Adv Protein Chem 50:123–155.
16. Curtis-Fisk J, Spencer DS, Weliky DP. 2008. Native conformation at specific residues in recombinant inclusion body protein in whole cells determined with sold-state NMR spectroscopy. J Am Chem Soc 130(38):12568–12569.
17. Wille H, Michelitsch MD, Guenebaut V, Supattapone S, Serban A, Cohen FE, Agard DA, Prusiner SB. 2002. Structural studies of the scrapie prion protein by electron crystallography. Proc Natl Acad Sci U S A 99(6):3563–3568.
18. Antzutkin ON. 2000. Multiple quantum solid-state NMR indicates a parallel, not antiparallel, organization of β-sheets in Alzheimer's β-amyloid fibrils. Proc Natl Acad Sci U S A 97(24):13045–13050.
19. Petkova AT, Leapman RD, Guo Z, Yau W-M, Mattson MP, Tycko R. 2005. Self-propagating, molecular-level polymorphism in Alzheimer's β-amyloid fibrils. Science 307:262–265.
20. Paravastu NK, Leapman RD, Yau W-M, Tycko R. 2008. Molecular structural basis for polymorphism in Alzheimer's β-amyloid fibrils. Proc Natl Acad Sci U S A 105:18349–18354.
21. Hou L, Zagorski MG. 2004. Sorting out and driving forces for parallel and antiparallel alignment in the Aβ peptide fibril structure. Biophys J 86:1–2.
22. Dobson CM. 2006. The generic nature of protein folding and misfolding. In Uversky VN, Fink AL, eds., *Protein Misfolding, Aggregation, and Conformational Diseases, Part A: Protein Aggregation and Conformational Diseases*. New York: Springer, pp. 21–41.
23. Kayed R, Head E, Sarsoza S, Saing T, Cotman CW, Necula M, Margol L, Wu J, Breydo L, Thompson JL, Rasool S, Gurlo T, Butler P, Glabe CG. 2007. Fibril specific, conformation dependent antibodies recognize a generic epitope common to amyloid fibrils and fibrillar oligomers that is absent in prefibrillar oligomers. Mol Neuordegener 2:18.
24. Jayasinghe SA, Langen R. 2004. Identifying structural features of fibrillar islet amyloid polypeptide using site-directed spin labeling. J Biol Chem 279(46): 48420–48425.
25. Nelson R, Sawaya MR, Balbirnie M, Madsen AO, Riekel C, Grothe R, Eisenberg D. 2005. Structure of the cross-β spine of amyloid-like fibrils. Nature 435(9):773–778.
26. Wiltzius JJW, Sievers SA, Sawaya MR, Cascio D, Popov D, Riekel C, Eisenberg D. 2008. Atomic structure of the cross-β spine of islet amyloid poplypeptide amylin. Protein Sci 17:1467–1474.
27. Luca S, Yau W-M, Leapman RD, Tycko R. 2007. Peptide conformation and supramolecular organization in amylin fibrils: Constraints from SOLID-State NMR. Biochemistry 46(47):13505–13522.
28. Sumner Makin O, Serpell LC. 2004. Structural characterisation of islet amyloid polypeptide fibrils. J Mol Biol 335:1279–1288.

29. Kajava A, Aebi U, Steven A. 2005. The parallel superpleated beta-structure as a model for amyloid fibrils of human amylin. J Mol Biol 348(2):247–252.
30. Uversky VN, Fernandez A, Fink AL. 2006. Structural and conformational prerequisites of amyloidogenesis. In Uversky VN, Fink AL, eds., *Protein Misfolding, Aggregation, and Conformational Diseases, Part A: Protein Aggregation and Conformational Diseases*. New York: Springer, pp. 1–20.
31. Plakoutsi G, Bemporad F, Calamai M, Taddei N, Dobson CM, Chiti F. 2005. Evidence for a mechanism of amyloid formation involving molecular reorganisation within native-like precursor aggregates. J Mol Biol 351(4):910–922.
32. Glabe CG, Kayed R. 2006. Common structure and toxic function of amyloid oligomers implies a common mechanism of pathogenesis. Neurology 66(Suppl. 1):S74–S78.
33. Wei G, Mousseau N, Derreumaux P. 2007. Computational simulations of the early steps of protein aggregation. Prion 1(1):3–8.
34. Anfinsen CB. 1973. Principles that govern the folding of protein chains. Science 181(4096):223–230.
35. Dill KA, Ozkan SB, Shell MS, Weikl TR. 2008. The protein folding problem. Annu Rev Biophys 37:289–316.
36. Dobson CM. 2003. Protein folding and misfolding. Nature 426:884–890.
37. Jahn TR, Radford SE. 2008. Folding versus aggregation: Polypeptide conformations on competing pathways. Arch Biochem Biophys 469:100–117.
38. Dill KA, Chan HS. 1997. From Levinthal to pathways to funnels. Nature Struct Biol 4(1):10–19.
39. Dill KA, Fiebig KM, Chan HS. 1993. Cooperativity in protein-folding kinetics. Proc Natl Acad Sci U S A 90:1942–1946.
40. Chi EY, Krishnan S, Randolph TW, Carpenter JF. 2003. Physical stability of proteins in aqueous solution: mechanism and driving forces in nonnative protein aggregation. Pharm Res 20(9):1325–1336.
41. Dill KA. 1990. Dominant forces in protein folding. Biochemistry 29(31):7133–7155.
42. Kim PS, Baldwin RL. 1990. Intermediates in the folding reactions of small proteins. Annu Rev Biochem 59:631–660.
43. Rumbley J, Hoang L, Mayne L, Englander SW. 2002. An amino acid code for protein folding. Proc Natl Acad Sci U S A 98(1):105–112.
44. Fan H, Vitharana SN, Chen T, O'Keefe D, Middaugh CR. 2007. Effects of pH and polyanions on the thermal stability of fibroblast growth factor 20. Mol Pharm 4(2):232–240.
45. Thirumangaluthu R, Krishnan S, Brems DN, Randolph TW, Carpenter JF. 2006. Effects of pH, temperature, and sucrose on benzyl alcohol-induced aggregation of recombinant human granulocyte colony stimulating factor. J Pharm Sci 95:1480–1497.
46. Zhang Y, Roy S, Jones LS, Krishnan S, Kerwin BA, Chang BS, Manning MC, Randolph TW, Carpenter JF. 2004. Mechanism for benzyl alcohol-induced aggregation of recombinant human interleukin-1-receptor antagonist in aqueous solution. J Pharm Sci 93:3076–3089.

47. Dill KA, Alonso DOV, Hutchinson K. 1989. Thermal stabilities of globular proteins. Biochemistry 28:5439–5449.
48. Dwyer JJ, Gittis AG, Karp DA, Lattman EE, Spencer DS, Stites WE, Garcia-Moreno B. 2000. High apparent dielectic constants in the interior of a protein reflect water penetration. Biophys J 79:1610–1620.
49. Available at http://www.greenfluids.org/SubcriticalWater.htm (accessed January 2009).
50. Gallivan JP, Dougherty DA. 1999. Cation-π interactions in structural biology. Proc Natl Acad Sci U S A 96(17):9459–9464.
51. McGaughey GB, Gagnes M, Rappe AK. 1998. π-Stacking interactions: Alive and well in proteins. J Biol Chem 273(25):15458–15463.
52. Zhou Y, Linhananta A. 2002. Thermodynamics of an all-atom off-lattice model of the fragment B of staphylococcal protein A. J Phys Chem B 106(6):1481–1485.
53. Gazit E. 2002. A possible role for π-stacking in the self-assembly of amyloid fibrils. FASEB J 16:77–83.
54. Tartaglia GG, Cavalli A, Pellarin R, Caflisch A. 2004. The role of aromaticity, exposed surface, and dipole moment in determining protein aggregation rates. Protein Sci 13:1939–1941.
55. Flocco MM, Mowbray SL. 1994. Planar stacking interactions of arginine and aromatic side-chains in proteins. J Mol Biol 235:709–717.
56. Burley SK, Petsko GA. 1986. Amino-aromatic interactions in proteins. FEBS 203(2):139–143.
57. Wang S, Stauffacher CV, Van Etten RL. 2000. Structural and mechanistic basis for the activation of a low molecular weight protein tyrosine phosphatase by adenine. Biochemistry 39:1234–1242.
58. Tishmack PA, Bashford D, Harms E, Van Etten RL. 1997. Use of 1H NMR spectroscopy and computer simulations to analyze pKa changes in a protein tyrosine phosphatase: Experimental and theoretical determination of electrostatic properties in a small protein. Biochemistry 36:11984–11994.
59. Langer T, Lu C, Echols H, Flanagan J, Hayer-Hartl MK, Hartl FU. 1992. Successive action of DnaK, DnaJ and GroEL along the pathway of chaperone-mediated protein folding. Nature 356:683–689.
60. Marti DN, Bosshard HR. 2003. Electrostatic interactions in leucine zippers-Thermodynamic analysis of the contributions of Glu and His residues and the effect of mutating salt bridges. J Mol Biol 330:621–637.
61. Kumar S, Nussinov R. 2002. Relationship between ion pair geometries and electrostatic strengths in proteins. Biophys J 83:1595–1612.
62. Richardson JS, Richardson DC. 2002. Natural β-sheet proteins use negative design to avoid edge-to-edge aggregation. Proc Natl Acad Sci U S A 99(5):2754–2759.
63. Chiti F. 2006. Relative importance of hydrophobicity, net charge, and secondary structure propensities in protein aggregation. In Uversky VN, Fink AL, eds., *Protein Misfolding, Aggregation, and Conformational Diseases, Part A: Protein Aggregation and Conformational Diseases*. New York: Springer, pp. 43–59.
64. Janowski R, Kozak M, Jankowska E, Grzonka Z, Grubb A, Abrahamson M, Jaskolski M. 2001. Human cystatin C, an amyloidogenic protein, dimerizes through three-dimensional domain swapping. Nature Struct Biol 8(4):316–320.

65. Wang W, Hecht MH. 2002. Rationally designed mutations convert de novo amyloid-like fibrils into monomeric β-sheet proteins. Proc Natl Acad Sci U S A 99(5):2760–2765.

66. Eisenberg D. 2003. The discovery of the α-helix and β-sheet, the principal structural features of proteins. Proc Natl Acad Sci U S A 100(20):11207–11210.

67. Pauling LN. 1960. *The Nature of the Chemical Bond and the Structure of Molecules and Crystals: An Introduction to Modern Structural Chemistry*, 3rd edition. Ithaca, New York: Cornell University Press.

68. Ramachandran GN, Ramakrishnan C, Sasisekharan V. 1963. Stereochemistry of polypeptide chain configurations. J Mol Biol 7:95–99.

69. Shortle D. 2002. Composites of local structure propensities: Evidence for local encoding of lang-rang structure. Protein Sci 11:18–26.

70. Avbelj F, Baldwin RL. 2002. Role of backbone solvation in determining thermodynamic β propensities of the amino acids. Proc Natl Acad Sci U S A 99(3): 1309–1313.

71. Street AG, Mayo SL. 1999. Intrinsic β-sheet propensities result from van der Waals interactions between side chains and the local backbone. Proc Natl Acad Sci U S A 96:9074–9076.

72. Alexandrescu AT, Kammerer RA. 2003. Structure and disorder in the ribonuclease S-peptide probed by NMR residual dipolar couplings. Protein Sci 12(10): 2132–2140.

73. Feng L, Lee H-S, Prestegard JH. 2007. NMR resonance assignments for sparsely 15N labeled proteins. J Biomol NMR 38(3):213–219.

74. Buck M. 1998. Trifluoroethanol and colleagues: Cosolvents come of age. Recent studies with peptides and proteins. Q Rev Biophys 31(3):297–355.

75. Amani M, Yousefi R, Moosavi-Movahedi AA, Pintus F, Mura A, Floris G, Kurganov BI, Saboury AA. 2008. Structural changes and aggregation process of Cu/containing amine oxidase in the presence of 2,2,2′-trifluoroethanol. Protein Pept Lett 15(5):521–527.

76. Bieler S, Soto C. 2004. β-sheet breakers for Alzheimer's disease therapy. Curr Drug Targets 5:553–558.

77. Chou PY, Fasman GD. 1978. Empirical predictions of protein conformation. Annu Rev Biochem 47:251–276.

78. Penkett CJ, Redfield C, Dodd I, Hubbard J, McBay DL, Mossakowska DE, Smith RAG, Dobson CM, Smith LJ. 1997. NMR analysis of main-chain conformational preferences in an unfolded fibronectin-binding protein. J Mol Biol 274:152–159.

79. Shen Y, Bax A. 2007. Protein backbone chemical shifts predicted from searching a database for torsion angle and sequence homology J Biomol NMR 38:289–302.

80. Wishart D, Sykes B. 1994. The 13C chemical-shift index: A simple method for the identification of protein secondary structure using 13C chemical-shift data. J Biolmol NMR 4:171–180.

81. Creamer TP, Rose GD. 1992. Side chain entropy opposes alpha-helix formation but rationalizes experimentally-determined helix-forming propensities. Proc Natl Acad Sci U S A 89:5937–5941.

82. Creamer TP, Rose GD. 1994. Alpha-helix-forming propensities in peptides and proteins. Proteins 19:85–97.

83. Glyakina AV, Garbuzynskiy SO, Lobanov MY, Galzitskaya OV. 2007. Different packing of external residues can explain differences in the thermostability of proteins from thermophilic and mesophilic organisms. Bioinformatics 23:2231–2238.
84. Dominy BN, Minoux H, Brooks CLI. 2004. An electrostatic basis for the stability of thermophilic proteins. Proteins 57:128–141.
85. Takano K, Tsuchimori K, Yamagata Y, Yutani K. 2000. Contribution of salt bridges near the surface of a protein to the conformational stability. Biochemistry 39:12375–12381.
86. Lindman S, Xue W-F, Szczepankiewicz O, Bauer MC, Nilsson H, Linse S. 2006. Salting the charged surface: pH and salt dependence of protein G B1 stability. Biophys J 90:2911–2921.
87. Stigter D, Alonso DOV, Dill KA. 1991. Protein stability: Electrostatics and compact denatured states. Proc Natl Acad Sci U S A 88:4176–4180.
88. Lindman S, Linse S, Mulder FAA, Andre I. 2007. pKa values for side-chain caroxyl groups of a PGB1 variant explain salt and pH-dependent stability. Biophys J 92:257–266.
89. Spek EJ, Bui AH, Lu M, Kallenbach NR. 1998. Surface salt bridges stabilize the GCN4 leucine zipper. Protein Sci 7:2431–2437.
90. Strop P, Mayo SL. 2000. Contribution of surface salt bridges to protein stability. Biochemistry 39:1251–1255.
91. Sharma S, Ding F, Dokholyan NV. 2008. Probing protein aggregation using simplified models and discrete dynamics. Front Biosci 13:4795–4808.
92. Lodi PJ, Garrett DS, Kuszewski J, Tsang ML, Weatherbee JA, Leonard WJ, Gronenborn AM, Clore GM. 1994. High-resolution solution structure of the beta chemokine hMIP-1 beta by multidimensional NMR. Science 263:1762–1767.
93. Petkova AT, Ishii Y, Balbach JJ, Antzutkin ON, Leapman RD, Delaglio F, Tycko R. 2002. A structural model for Alzheimer's β-amyloid fibrils based on experimental constraints from solid state NMR. Science 99:6742–16747.
94. Mothana B, Roy S, Rauk A. 2009. Molecular dynamics study of the interaction of Aβ(13-23) with β-sheet inhibitors. ARKIVOC v:116–134.
95. Klimov DK, Newfield D, Thirumalai D. 2002. Simulations of β-hairpin folding confined to spherical pores using distributed computing. Proc Natl Acad Sci U S A 99(12):8019–8024.
96. Bellesia G, Shea J-E. 2007. Self-assembly of β-sheet forming peptides into chiral fibrillar aggregates. J Chem Phys 126:245104–245111,
97. Thurlkill RL, Grimsley GR, Scholtz JM, Pace CN. 2006. Hydrogen bonding markedly reduces the pK of buried carboxyl groups in proteins. J Mol Biol 362: 594–604.
98. Forsyth WR, Antosiewicz JM, Robertson AD. 2002. Empirical relationships between protein structure and carboxyl pKa values in proteins. Proteins 48:388–403.
99. Kawamura S, Tanaka I, Yamasaki N, Kimura M. 1997. Contribution of a salt bridge to the thermostability of DNA binding protein HU from *Bacillus stearothermophilus* determined by site-directed mutagenesis. J Biochem 121(3): 448–455.

100. Podust LM, Krezel AM, Kim Y. 2001. Crystal structure of the CCAAT box/enhancer-binding protein. J Biol Chem 276:505–513.

101. Carl N, Konc J, Janezic D. 2008. Protein surface conservation in binding sites. J Chem Inf Model 48:1279–1286.

102. Jones S, Thornton JM. 1996. Principles of protein-protein interactions. Proc Natl Acad Sci U S A 93:13–20.

103. Matsuo H, Shirakawa M, Kyogoku Y. 1995. Three-dimensional dimer structure of the λ-Cro repressor in solution as determined by heteronuclear multidimensional NMR. J Mol Biol 254(4):668–680.

104. Fermi G, Perutz MF, Shaanan B, Fourme R. 1984. The crystal structure of human deoxyhaemoglobin at 1.74 A resolution. J Mol Biol 175:159.

105. Jeong DG, Kim SJ, Kim JH, Son JH, Park MR, Lim SM, Yoon TS, Ryu SE. 2005. Trimeric structure of PRL-1 phosphatase reveals an active enzyme conformation and regulation mechanisms. J Mol Biol 345:401–413.

106. Pal SK, Peon J, Zewail AH. 2002. Biological water at the protein surface: Dynamical solvation probed directly with femtosecond resolution. Proc Natl Acad Sci U S A 99(4):1763–1768.

107. Zhang L, Wang L, Kao Y-T, Qiu W, Yang Y, Okobiah O, Zhong D. 2007. Mapping hydration dynamics around a protein surface. Proc Natl Acad Sci U S A 104(47):18461–18466.

108. Friedman R, Nachliel E, Gutman M. 2005. Molecular dynamics of a protein surface: Ion-residues interactions. Biophys J 89:768–781.

109. Baldwin RL. 1996. How Hofmeister ion interactions affect protein stability. Biophys J 71:2056–2063.

110. Timasheff SN. 2002. Protein hydration, thermodynamic binding, and preferential hydration. Biochemistry 41(46):13473–13482.

111. McLaurin J, Franklin T, Zhang X, Deng J, Fraser PE. 1999. Interactions of Alzheimer amyloid peptides with glycosaminoglycans. Effects on fibril nucleation and growth. Eur J Biochem 266:1101–1110.

112. Fan H, Zhang M, Middaugh CR. 2007. Effects of solutes on empirical phase diagrams of human fibroblast growth factor 1. J Pharm Sci 96(6):1490–1503.

113. Kamerzell TJ, Joshi SB, McClean D, Peplinkskie L, Toney K, Papac D, Li M, Middaugh CR. 2007. Parathyroid hormone is a heparin/polyanion binding protein; binding energetics and structure modification. Protein Sci 16(6):193–203.

114. Kamerzell TJ, Unruh JR, Johnson CK, Middaugh CR. 2006. Conformational flexibility, hydration and state parameter fluctuations of fibroblast growth factor -10: Effects of ligand binding. Biochemistry 45:15288–15300.

115. Salamat-Miller N, Feng J, Seidel CW, Assenov Y, Albrecht M, Middaugh CR. 2007. A network-based analysis of polyanion-binding proteins utilizing human protein arrays. J Biol Chem 282(14):10153–10163.

116. Vendruscolo M, Paci E, Karplus M, Dobson CM. 2003. Structures and relative free energies of partially folded states of proteins. Proc Natl Acad Sci U S A 100(25):14817–14821.

117. Margittai M, Langen R. 2006. Side chain-dependent stacking modulates tau filament structure. J Biol Chem 281(49):37820–37827.

118. Pain RH. 2000. *Mechanisms of Protein Folding*, 2nd edition. New York: Oxford University Press.
119. Fersht A. 1998. *Structure and Mechanism in Protein Science: A Guide to Enzyme Catalysis and Protein Folding*. London: W. H. Freeman Company.
120. Griffith JS. 1967. Self-replication and scrapie. Nature 215:1043–1044.
121. Prusiner SB. 1982. Identification of a protein that purifies with the scrapie prion. Science 216:136.
122. Come JH, Fraser PE, Lansbury PT Jr. 1993. A kinetic model for amyloid formation in the prion diseases: Importance of seeding. Proc Natl Acad Sci U S A 90:5959–5963.
123. Jarrett JT, Lansbury PT Jr. 1993. Seeding "one-dimensional crystallization" of amyloid: A pathogenic mechanism in Alzheimer's disease and scrapie? Cell 73:1055–1058.
124. Serio TR, Cashikar AG, Kowal AS, Sawicki GJ, Moslehl JJ, Serpell L, Arnsdorf MF, Lindquist SL. 2000. Nucleation conformational conversion and the replication of conformational information by a prion determinant. Science 289: 1317–1321.
125. Pawar AP, DuBay KF, Zurdo J, Chiti F, Vendruscolo M, Dobson CM. 2005. Prediction of aggregation-prone and aggregation-susceptible regions in proteins in neurodegenerative diseases. J Mol Biol 350:379–392.
126. Armen RS, Alonso DO, Daggett V. 2004. Anatomy of an amyloidogenic intermediate: Conversion of beta-sheet to alpha-sheet structure in transthyretin at acidic pH. Structure 12:1847–1863.
127. Lindberg MO, Oliveberg M. 2007. Malleability of protein folding pathways: A simple reason for complex behaviour. Curr Opin Struct Biol 17:21–29.
128. Karplus M, Weaver DL. 1994. Folding dynamics: The diffusion-collision model and experimental data. Protein Sci 3:650–668.
129. Beck C, Siemens X, Weaver DL. 2001. Diffusion-collision model study of misfolding in a four-helix bundle protein. Biophys J 81(6):3105–3115.
130. Liu D, Ren D, Huang H, Dankberg J, Rosenfield R, Cocco MJ, Li L, Brems DN, Remmele RL Jr. 2008. Structure and stability changes of human IgG1 Fc as a consequence of methionine oxidation. Biochemistry 47:5088–5100.
131. Schlunegger MP, Bennett MJ, Eisenberg D. 1997. Oligomer formation by 3D domain swapping: A model for protein assembly and misassembly. Adv Protein Chem 50:61–122.
132. Guo Z, Eisenberg D. 2006. Runaway domain swapping in amyloid-like fibrils of T7 endonuclease I. Proc Natl Acad Sci U S A 103:8042–8047.
133. Nilsson M, Wang X, Rodziewicz-Motowidlo S, Jankowski R, Lindstrom V, Onnerfjord P, Westermark G, Grzonka Z, Jaskolski M, Grubb A. 2004. Prevention of domain swapping inhibits dimerization and amyloid fibril formation of cystatin C. J Biol Chem 279:24236–24245.
134. Wahlbom M, Lindstrom V, Carlemalm E, Jaskolski M, Grubb A. 2007. Fibrillogenic oligomers of human cystatin C are formed by propagated domain swapping. J Biol Chem 282:18318–18326.
135. Udgaonkar JB. 2008. Multiple routes and structure heterogeneity in protein folding. Annu Rev Biophys 37:489–510.

136. Sali A, Shakhnovich E, Karplus M. 1994. How does a protein fold? Nature 369:248–251.
137. Bogatyreva NS, Finkelstein AV. 2001. Cunning simplicity of protein folding landscapes. Protein Eng 14(8):521–523.
138. Privalov PL. 1979. Stability of proteins: Small globular proteins. Adv Protein Chem 33:167–241.
139. Korzhnev DM, Lay LE. 2008. Probing invisible, low-population states of protein molecules by relaxation dispursion NMR spectroscopy: An application to protein folding. Acc. Chem Res 41(3):442–451.
140. Cellmer T, Bratko D, Prausnitz JM, Blanch H. 2007. Protein aggregation in silico. Trends Biotechnol 25(6):254–261.
141. Gsponer J, Vendruscolo M. 2006. Theoretical approaches to protein aggregation. Protein Pept Lett 13:287–293.
142. Chan HS, Dill KA. 1998. Protein folding in the landscape perspective- Chevron plots and non-Arrhenius kinetics. Proteins 30:2–33.
143. Dill KA, Bromberg S, Yue K, Fiebig KM, Yee DP, Thomas PD, Chan HS. 1995. Principles of protein folding—A perspective from simple exact models. Protein Sci 4(4):561–602.
144. Miyazawa S, Jernigan RL. 1985. Estimation of effective interresidue contact energies from protein crystal structures: Quasi-chemical approximation. Macromolecules 18(3):534–552.
145. Cellmer T, Bratko D, Prausnitz JM, Blanch H. 2005. Thermodynamics of folding and association of lattice-model proteins. J Chem Phys 122(174908):1–10.
146. Chan T, Jankovic B, Le V, Naverniouk I. 2004. *Comparative Study of Hydrophobic-Polar and Miyazawa-Jernigan Energy Functions in Protein Folding on a Cubic Lattice using Pruned-Enriched Rosenbluth Monte Carlo Algorithm.* Available at http://www.cs.ubc.ca/labs/beta/Div/CPSC545-WS-04/JanEtAl.pdf.
147. Nguyen HD, Hall CK. 2004. Molecular dynamics simulations of spontaneous fibril formation by random-coil peptides. Proc Natl Acad Sci U S A 101(46):16180–16185.
148. Ma B, Nussinov R. 2002. Stabilities and conformations of Alzheimer's β-amyloid peptide oligomers (Aβ16-22, Aβ16-35 and Aβ10-35): Sequence effects. Proc Natl Acad Sci U S A 99(22):14126–14131.
149. Pande VS, Rokhsar DS. 1999. Folding pathway of a lattice model for proteins. Proc Natl Acad Sci U S A 96:1273–1278.
150. Sutto L, Tiana G, Broglia RA. 2006. Heirarchy of events in protein folding: Beyond the Go model. Protein Sci 15:1638–1652.
151. Maity H, Maity M, Krishna MM, Mayne L, Englander SW. 2005. Protein folding: The stepwise assembly of foldon units. Proc Natl Acad Sci U S A 102:4741–4746.
152. Tozzini V. 2005. Coarse-grained models for proteins. Curr Opin Struct Biol 15:144–150.
153. Clementi C. 2008. Coarse-grained models of protein folding: Toy models or predictive tools? Curr Opin Struct Biol 18:10–15.
154. Zhou Y, Karplus M. 1999. Folding of a model three-helix bundle protein: A thermodynamic and kinetic analysis. J Mol Biol 293:917–951.
155. Zacharias M. 2003. Protein docking with a reduced protein model accounting for side-chain flexibility. Protein Sci 12:1271–1282.

156. Urbanc B, Cruz L, Ding F, Sammond D, Khare S, Buldyrev SV, Stanley HE, Dokholyan NV. 2004. Molecular dynamics simulation of amyloid β dimer formation. Biophys J 87:2310–2321.
157. Levy Y, Jortner J, Becker OM. 1998. Solvent effects on the energy landscapes and folding kinetics of polyalanine. Proc Natl Acad Sci U S A 98(5):2188–2193.
158. Nguyen HD, Hall CK. 2004. Phase diagrams describing fibrillization by polylanine peptides. Biophys J 87:4122–4134.
159. Pellarin R, Caflisch A. 2006. Interpreting the aggregation kinetics of amyloid peptides. J Mol Biol 360:882–892.
160. Hummer G, Grarde S, Garcia AE, Pohorille A, Pratt LR. 1996. An information theory model of hydrophobic interactions. Proc Natl Acad Sci U S A 93:8951–8955.
161. Kumar S, Yin X, Trapp BD, Hoh JH, Paulaitis ME. 2002. Relating interactions between neurofilametns to the structure of axonal neurofilament distributions through polymer brush models. Biophys J 82(5):2360–2372.
162. Steward RE, Thornton JM. 2002. Prediction of strand pairing in antiparallel and parallel beta-sheets using information theory. Proteins 48(2):178–191.
163. Naderi-Manesh H, Sadeghi M, Arab S, Moosavi Movahedi AA. 2001. Prediction of protein surface accessibilty with information theory. Proteins 42(4):452–459.
164. Ofran Y, Rost B. 2003. Analysing six types of protein-protein interfaces. J Mol Biol 325(2):377–387.
165. Bogan AA, Thorn KS. 1998. Anatomy of hot spots in protein interfaces. J Mol Biol 280:1–9.
166. Chakrabarti P, Janin J. 2002. Dissecting protein: Protein recognition sites. Proteins Struct Funct Genet 47:334–343.
167. Mintseris J, Weng Z. 2004. Optimizing protein representations with information theory. Genome Inform 15(1):160–169.
168. Aswad DW. 1995. *Deamidation and Isoaspartate Formation in Peptides and Proteins*. Boca Raton, FL: CRC Press.
169. Zbilut JP, Colosimo A, Conti F, Colafranceschi M, Manetti C, Valerio MC, Webber CLJ, Giuliani A. 2003. Protein aggregation/folding: The role of deterministic singularities of sequence hydrophobicity as determined by nonlinear signal analysis of acylphosphatase and Aβ(1-40). Biophys J 85:3544–3557.
170. Weisel JW, Veklich Y, Gorkun O. 1993. The sequence of cleavage of fibrinopeptides from fibrinogen is important for protofibril formation and enhancement of lateral aggregation in fibrin clots. J Mol Biol 232:285.
171. Shirley BA. 1995. *Protein Stability and Folding: Theory and Practice*. Totowa, NJ: Humana Press.
172. Perutz MF. 1996. Glutamine repeats and inherited neurodegenerative diseases: Molecular aspects. Curr Opin Struct Biol 6:848–858.
173. Bevivino AE, Loll PJ. 2001. An expanded glutamine repeat destabilizes native ataxin-3 structure and mediates formation of parallel β-fibrils. Proc Natl Acad Sci U S A 98(21):11955–11960.
174. Ross CA, Poirier MA, Wanker EE, Amzel M. 2003. Polyglutamine fibrillogenesis: The pathway unfolds. Proc Natl Acad Sci U S A 100(1):1–3.

175. Triguero L, Singh R, Prabhakar R. 2008. Comparative moelcular dynamics studies of wild-type and oxidized forms of full-length Alzheimer amyloid β-peptides Aβ(1-40) and Aβ(1-42). J Phys Chem B 112:7123–7131.
176. Watson AA, Fairlie DP, Craik DJ. 1998. Solution structure of methionine-oxidized amyloid β-peptide (1–40). Does oxidation affect conformational switching? Biochemistry 37:12700.
177. Yan Y, McCallum SA, Wang C. 2008. M35 oxidation induces Aβ-like structural and dynamical changes in Aβ42. J Am Chem Soc 130:5394–5395.
178. Breydo L, Bocharova OV, Makarava N, Salnikov VV, Anderson M, Baskakov IV. 2005. Methionine oxidation interferes with conversion of the prion protein into the fibrillar proteinase K-resistant conformation. Biochemistry 44:15534–15543.
179. Soenderkaer S, Carpenter JF, von de Weert M, Hansen LL, Flink J, Frokjaer S. 2004. Effects of sucrose on rFVIIa aggregation and methionine oxidation. Eur J Pharm Sci 21(5):597–606.
180. Ruiz L, Aroche K, Reyes N. 2006. Aggregation of recombinant human interferon alpha 2b in solution: Technical note. AAPS Pharm Sci Tech 7(4):E1–E5.
181. Ding F, Dokholyan NV. 2008. Dynamical roles of metal ions and the disulfide bond in Cu, Zn superoxide dismutase folding and aggregation. Proc Natl Acad Sci U S A 105(50):19696–19701.
182. Schafer FQ, Buettner GR. 2001. Redox environment of the cell as viewed through the redox state of the glutathione disulfide/glutathione couple. Free Rad Biol Med 30:1191–1212.
183. den Hertog J, Groen A, van der Wijk T. 2005. Redox regulation of protein-tyrosine phosphatases. Arch Biochem Biophys 434:11–15.
184. Chiarugi P, Buricchi F. 2007. Protein tyrosin phosphorylation and reversible oxidation: Two cross-talking posttranslation modifications. Antioxid Redox Signal 9:1–24.
185. Ruf RAS, Lutz EA, Zigoneanu IG, Pielak GJ. 2008. α-Synuclein conformation affects its tyrosine-dependent oxidative aggregation. Biochemistry 47:13604–13609.
186. Norris EH, Giasson BI, Ischiropoulos H, Lee VM. 2003. Effects of oxidative and nitrative challeneges on R-synuclein fibrillogenesis involve distinct mechanisms of protein modifications. J Biol Chem 278:27230–27240.
187. Siebold B, Qin R, Glanville RW, Hofmann H, Deutzmann R, Kuhn K. 1987. Construction of a model for the aggregation and cross-linking region (7S domain) of type IV collagen based upon an evaluation of the primary structure of the a1 and a2 chains in this region. Eur J Biochem 168:569–575.
188. Bailey AJ, Sims TJ, Avery NC, Miles CA. 1993. Chemistry of collagen cross-links: Glucose-mediated covalent cross-linking of type-IV collagen in lens capsules. Biochem J 296:489–496.
189. Nilsson MR, Driscoll M, Raleigh DP. 2002. Low levels of asparagine deamidation can have a dramatic effect on aggregation of amyloidogenic peptides: Implications for the study of amyloid formation. Protein Sci 11:342–349.
190. Flaugh S, Mills IA, King J. 2006. Glutamine deamidation destabiliizes human γD-crystallin and lowers the kinetic barrier to unfolding. J Biol Chem 281:30782–30793.

191. Yamaguchi S, Jeenes DJ, Archer DB. 2001. Protein-glutaminase from *Chryseobacterium proteolyticum*, an enzyme that deamidates glutaminyl residues in proteins: Purification, characterization and gene cloning. Eur J Biochem 268(5):1410–1421.

192. Yong YH, Yamaguchi S, Gu YS, Mori T, Matsumura Y. 2004. Effects of enzymatic deamidation by protein-glutaminase on structure and functional properties of α-zein. J Agric Food Chem 52(23):7094–7100.

193. Calabrese MF, Eakin CM, Wang JM, Miranker AD. 2008. A regulatable switch mediates self-association in an immunoglobulin fold. Nat Struct Mol Biol 15(9):965–971.

194. Goldsbury C, Goldie K, Pellaud J, Seelig J, Frey P, Muller SA, Kistler J, Cooper GJS, Aebi U. 2000. Amyloid fibril formation from full-length and fragments of amylin. J Struct Biol 130:352–362.

195. Wang Q, Johnson JL, Agar NYR, Agar JN. 2008. Protein aggregation and protein instability govern familial amyotrophic lateral sclerosis patient survival. PLoS Biol 6(7):1508–1526.

196. Tartaglia GG, Vendruscolo M. 2008. The Zyggregator method for predicting protein aggregation propensities. Chem Soc Rev 37:1395–1401.

197. Tartaglia GG, Pawar AP, Campioni S, Dobson CM, Chiti F, Vendruscolo M. 2008. Prediction of aggregation-prone regions in structured proteins. J Mol Biol 380:425–436.

198. Tartaglia GG, Cavalli A, Pellarin R, Caflisch A. 2005. Prediction of aggregation rate and aggregation-prone segments in polypeptide sequences. Protein Sci 14:2723–2734.

199. Fernandez-Escamilla AM, Rousseau F, Schymkowitz J, Serrano L. 2004. Prediction of sequence-dependent and mutational effects on the aggregation of peptides and proteins. Nat Biotechnol 22(10):1302–1306.

200. Tessier PM, Lindquist S. 2007. Prion recognition elements govern nucleation, strain specificity and species barrier. Nature 447:556–561.

201. Bui JM, Cavalli A, Gsponer J. 2008. Identification of aggregation-prone elements by using interaction-energy matrices. Angew Chem Int Ed 47:7267–7269.

202. Trovato A, Seno F, Tossato SCE. 2007. The PASTA server for protein aggregation prediction. Prot Eng Des Sel 20:521–523.

203. Conchillo-Sole O, de Groot NS, Aviles FX, Vendrell J, Daura X, Ventura S. 2007. AGGRESCAN: A server for the prediction and evaluation of hot spots of aggregation in polypeptides. BMC Bioinformatics 8:65.

204. Ventura S. 2005. Sequence determinants of protein aggregation: Tools to increase protein solubility. Microb Cell Fact 4:11–19.

205. Sharma S, Ding F, Nie H, Watson D, Unnithan A, Lopp J, Pozefsky D, Dokholyan NV. 2006. iFold: A platform for ineteractive folding simulations of proteins. Bioinformatics 22(21):2693–2694.

206. Ding F, Prutzman KC, Campbell SL, Dokholyan NV. 2006. Topological determinants of protein domain swapping. Structure 14:5–14.

207. Iakoucheva LM, Dunker AK. 2003. Order, disorder, and flexibility: Prediction from protein sequence. Structure 11(11):1316–1317.

208. Linding R, Jensen LJ, Diella F, Bork P, Gibson TJ, Russell RB. 2003. Protein disorder prediction: Implications for structural proteomics. Structure 11(11):1316–1317.

CHAPTER 2

Protein Aggregation Pathways, Kinetics, and Thermodynamics

YI LI and CHRISTOPHER J. ROBERTS

2.1 INTRODUCTION

Chapter 1 provided a review of the important chemistry and physics of protein structure and conformational stability that are relevant to protein aggregation; this includes the observation that nonnative or unfolded "monomer" species are typically implicated as being most reactive with respect to forming non-native, net irreversible protein aggregates.[1–8] It also reviewed the physics of protein–protein interactions and provided an overview of different types of aggregates that are typically observed—for example, fibrillar versus amorphous or afibrillar aggregates. This chapter builds upon these fundamentals to provide a perspective of protein aggregation as a global process in which multiple stages compete with each other, but only particular stages are of practical importance in controlling the kinetics or thermodynamics of aggregate formation.[8,9] In the remainder of this chapter, reversible and irreversible aggregation will be referred to, respectively, as self-association and aggregation in order to distinguish them.

Section 2.2 outlines a basic framework of (un)folding, self-association, and aggregation as a multistage process to help illustrate that the underlying physical attributes of a given protein, as well as the solvent and sample conditions of interest, all can potentially change which stage or stages are most important. This highlights potential pitfalls of assuming a given protein can follow only one pathway or is able to form only one type of aggregate. It also has implications for how or whether one can accurately predict aggregate characteristics and aggregation rates as part of product formulation and stabilization efforts.

Sections 2.3 and 2.4 provide quantitative and semiquantitative relationships between typical biophysical measurements and the resulting kinetics and/or

Aggregation of Therapeutic Proteins, Edited by Wei Wang and Christopher J. Roberts

thermodynamics of aggregate formation. These provide a mathematical and mechanistic framework for how different pathways are altered by changes in intrinsic (Chapter 3) and extrinsic (Chapter 4) factors of a protein product, as well as the environment(s) a protein is exposed to as part of manufacturing, storage, and shipping (Chapters 6–8). It also highlights the critical importance of selecting appropriate experimental assays to monitor and quantify self-association and aggregation (Chapter 5). Section 2.4 also summarizes key experimental signatures that aid in identifying which stage or stages are kinetically important for aggregation.

Section 2.5 briefly illustrates how the principles in Sections 2.2–2.4 can be applied within the context of aggregation mediated by physical adsorption or chemical modification;[10,11] the latter topic is of particular importance within the context of assuring that protein engineering efforts consider the effects of mutations on protein resistance to aggregation, in addition to the traditional focus on protein function and efficient expression.[12–14] Finally, Section 2.6 reviews the molecular and statistical thermodynamic basis for "weakly binding" cosolvent or cosolute effects on the thermodynamics of protein unfolding and self-association. Together with the principles reviewed in Sections 2.2–2.4, these provide theoretical underpinnings for interpreting the effects of excipients in mediating protein aggregation in, for example, Chapters 4 and 6–9. Table 2.1 provides a list of nomenclature used throughout this chapter.

TABLE 2.1. Nomenclature

Symbol	Definition	Section Introduced
a_{ij}	van der Waals *a* parameter for component *i* with respect to *j*	2.6
A_j	Irreversible aggregate (*A*) composed of *j* monomers	2.2
$A_{x,s}$	Surface-adsorbed nucleus	2.4
B_{22}	Second osmotic virial coefficient	2.3
$B_{22,HS}$	Theoretical value of B_{22} for a HS of diameter σ_{HS}	2.3
b_{ij}	van der Waals *b* parameter for component *i* with respect to *j*	2.6
c_0	Initial protein concentration (weight per volume)	2.4
c_1	Molar concentration of water	2.6
c_m	Concentration (mole per volume) of protein that assays as monomer under typical *ex situ* conditions (e.g., SEC)	2.3
F	Folded monomer	2.3
F'	Chemically modified folded monomer	2.5
F_j	Self-associated oligomer of *j* folded monomers	2.2
G_{ij}	Kirkwood–Buff integral for components *i* and *j*	2.6
$g_{ij}(r)$	Radial distribution function for components *i* and *j*	2.6
HMW	High molecular weight	2.2

TABLE 2.1. *Continued*

Symbol	Definition	Section Introduced
k'	First-order rate coefficient for chemical modification of monomer	2.5
$K_{1,i}$	Equilibrium constant for adding a monomer to a reversible oligomer composed of i monomers	2.3
$K_{1,x,surf}$	$K_{1,x}$ for surface-adsorbed protein	2.4
K_{ads}	Equilibrium constant for adsorption of R to a bulk interface	2.4
k_B	Boltzmann's constant	2.3
KB	Kirkwood–Buff	2.6
K_d	Equilibrium constant for dissociation ($F_2 \rightarrow 2F$)	2.2
K_{un}	Equilibrium constant for unfolding $\left(= \exp\left(-\Delta G_{un}^0/k_B T\right)\right)$	2.3
K_d^R	Equilibrium constant for dissociation ($R_2 \rightarrow 2R$)	2.3
k_{obs}	Observed or apparent rate coefficient for monomer loss	2.4
m_2	Molality of protein	2.6
m_3	Molality of cosolute or cosolvent	2.6
M	Monomer concentration (weight per volume) divided by initial protein concentration c_0	2.4
M'	Chemically modified monomer	2.5
M_1	Molecular weight of water [=] kilogram per mole	2.6
M_w^{agg}	Weight-average molecular weight of the sum over all soluble aggregates	2.4
N_{Av}	Avogadro's number	2.6
N_i	Number of molecules of component i	Appendix
Q	Canonical ensemble partition function	Appendix
Q_{HS}	Canonical ensemble partition function for equivalent HS system	Appendix
RDF	Radial distribution function	2.6
R_i	Reversible (self-associated) oligomer of i reactive (R) monomers	2.2
$R_{x,s}$	Surface-adsorbed prenucleus	2.4
T	Temperature	2.3
$u'_{ij}(r)$	Average attractive interaction energy or free energy between components i and j at a center-to-center distance r	2.6
U	Unfolded or partially unfolded monomer	2.3
U'	Chemically modified unfolded or partially unfolded monomer	2.5
$\bar{V}_1$	Partial molar volume of water	2.6
V	Volume	Appendix
x	Number of monomers in the smallest stable nonnative aggregate or "nucleus"	2.2
Y_a	Mass fraction of protein that exists *in situ* as a = monomer or oligomer but that is reversible upon simple dilution with buffer.	2.3
α	Exponent for K_{un} scaling of k_{obs}	2.4

(*Continued*)

TABLE 2.1. *Continued*

Symbol	Definition	Section Introduced
α_i^{bb}	Fractional surface area of the backbone of amino acid *i* that is exposed to solvent	2.6
α_i^{sc}	Fractional surface area of the side chain of amino acid *i* that is exposed to solvent	2.6
γ	Exponent for c_0 scaling of k_{obs}	2.4
ΔG_{un}^0	Standard free energy of monomer unfolding for a selected conformational step	2.3
$\Delta G_{R_2 \to 2R}^0$	Standard free energy for dissociation ($R_2 \to 2R$) at reference state concentration c_{ref}	
$\Delta g_{tr}^{sc,i}$	Transfer free energy of a side chain of amino acid *i* from water to a 1 M solution of a given osmolyte	2.6
Δg_{tr}^{bb}	Transfer free energy of the backbone of an amino acid from water to a 1 M solution of a given osmolyte	2.6
$\Delta\mu_a^{tr}$	Transfer free energy for protein from water to water–cosolute mixture	2.6
μ_2^{ex}	Excess chemical potential for protein monomer	2.6
ν	Apparent reaction order; determined from time course of monomer loss	2.4
ρ_i	Number of molecules of component *i* per unit volume	2.6, Appendix
σ_i	Effective HS radius of component *i*	2.6
$\sigma_{i,j}$	Effective HS radius of component *i* with respect to *j*	2.6
$\Gamma_{32}^{(\mu_3)}$	Preferential interaction parameter of cosolute with protein, relative to water with protein	2.6
$\Gamma_{32}^{sc,i}$	Preferential interaction parameter for the side chain of amino acid *i*	2.6
Γ_{32}^{bb}	Preferential interaction parameter for the backbone of an amino acid	2.6
Ξ	Grand canonical partition function	Appendix

2.2 NATIVE AND NONNATIVE AGGREGATION PATHWAYS

Figure 2.1 shows a schematic representation of aggregation pathways that illustrates the interplay between self-association, aggregation, folding, and unfolding. It also distinguishes between pathways that lead to soluble or insoluble aggregates. For aggregate formation during bulk manufacture (e.g., Chapter 6), the path(s) followed in Fig. 2.1 could reasonably "begin" with either folded or unfolded proteins, depending on what stages during manufacturing promote aggregation. In most cases, the large majority of proteins in the product immediately after purification and fill-finish will be folded. Therefore, when considering aggregate formation during storage, shipping, or

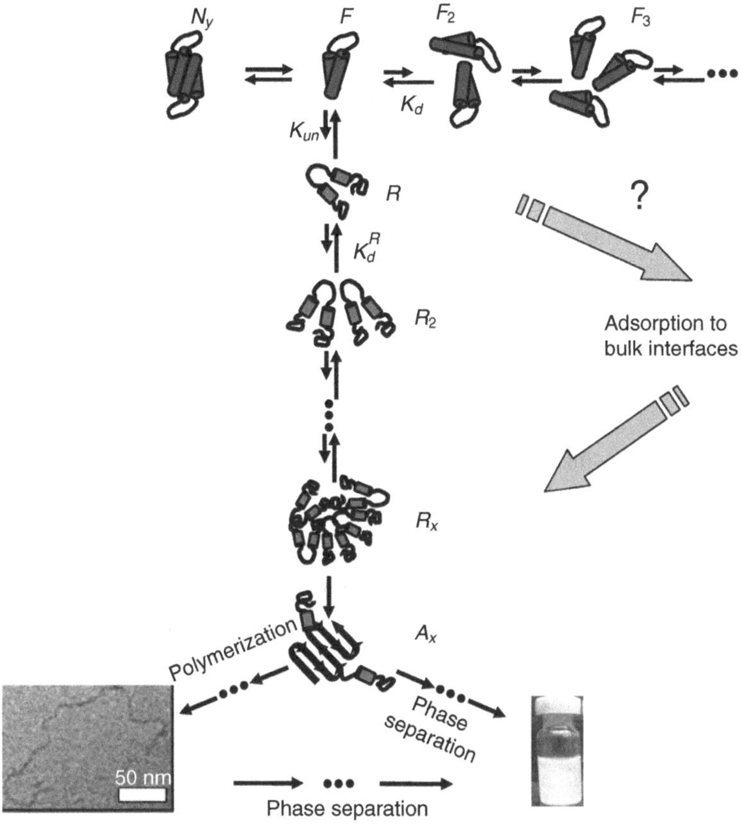

Figure 2.1. Schematic overview of multistage pathways for protein aggregation. Double arrows indicate steps that are reversible, with associated equilibrium constants defined for selected stages. Single arrows indicate steps that are effectively irreversible. The cryo-TEM image in the lower left corner is a high-molecular-weight soluble aggregate of bovine G-CSF (modified from Reference 54). See colour insert.

administration of a product, it is more typical for the path(s) in Fig. 2.1 to "begin" with folded proteins.

The top half of Fig. 2.1 distinguishes between two distinct types of self-association for folded proteins. The process of forming strong, highly specific structural contacts between two folded monomers is denoted as multimeric folding, with the multimer denoted as N_y (y = no. of monomers in the stable native multimer). Proteins such as trans-thyretin (TTR),[15] P22 tailspike,[16,17] and the Arc repressor[18] are all examples of this type of self-association. Numerous studies have shown that formation of such multimers thermodynamically stabilizes the folded structure of the constituent monomers,[16,17] provided that the total protein concentration is not so high as to also stabilize unfolded oligomers.[18] Values for the equilibrium dissociation constant(s) (K_d)

of such multimers are typically small compared to experimental conditions of interest (e.g., $K_d \sim 1\,\text{nM}$).

The other type of folded self-association in Fig. 2.1 is the formation of "weakly" self-associated oligomers (F_j) composed of j otherwise folded monomers (F). The monomers within these oligomers interact via colloidal forces, without highly specific structural arrangements or protein–protein surface contacts that are expected for multimeric native proteins. Thus, they are perhaps better considered as clusters in more of a colloidal sense.[8] Thermodynamically, it is still possible to formally define an equilibrium constant for the formation or dissociation of such oligomers.[8,19,20] Not surprisingly, K_d values for colloidal clusters are much higher than those for native multimers, with values that may be near or greatly above the experimental concentrations of interest.[8,19,21]

The bottom of Fig. 2.1 focuses on aggregation with a nonnative monomer state as a key intermediate. In Fig. 2.1, R denotes a "reactive" monomer with respect to forming nonnative aggregates that are effectively irreversible under the conditions that the aggregates are formed. R is depicted in Fig. 2.1 as a partially unfolded monomer. In general, R might be a stable unfolding intermediate,[22] a subset of all possible monomer conformations in the unfolded ensemble,[8,20] or possibly a compact unfolded state such as a molten globule[2,4] (see also Chapter 1). These pathways are distinct from spontaneous polymerization of folded proteins.[23–25] Modified versions of this scheme are included later in this chapter to highlight additional or alternative ways to populate R species besides unfolding in bulk solution.

In Fig. 2.1, the smallest effectively irreversible aggregate is termed a nucleus (A_x); x is an integer denoting the number of monomer chains in a nucleus. In this nomenclature, experimental nuclei range from dimers ($x = 2$)[26,27] to small oligomers ($x \sim 3$–10)[26,28,29] to large multimers or micelle-like species.[30,31] The process of creating a nucleus consists of at least two stages: reversible self-association of R monomers to form a prenucleus R_x, followed by a net irreversible structural rearrangement or another commitment step that creates the nucleus A_x. Rigorously, A_x does not need to be irreversible if its subsequent growth is so rapid that each A_x species does not exist long enough to reverse.[20] Because prenuclei are effectively reversible, the thermodynamics of prenucleation can be important in controlling prenuclei concentrations and, therefore, also in influencing the rates of nuclei formation. This will be discussed quantitatively in Section 2.3.

After nucleation, observed aggregation behaviors can often be grouped into one of three categories:[9,20,32] (1) aggregates remain soluble but do not grow substantially; (2) aggregates remain soluble but grow rapidly to high-molecular-weight (HMW) species; (3) aggregates that begin as dissolved molecular species rapidly self-assemble or coalesce with each other to undergo a macroscopic phase separation leading to insoluble, visible particulates or precipitates.

In the context of this chapter, and for any mathematical description of aggregation kinetics or thermodynamics, the term HMW aggregate refers to aggregates composed of many monomers (e.g., ~10 or more in References 20 and 32). It does not refer to a specific length scale or a particular type of experimental technique for characterization. However, many proteins of pharmaceutical interest have characteristic dimensions for a monomer that are ~1–10 nm. Therefore, within that context, the term HMW soluble aggregate is consistent with the definitions adopted elsewhere in this text and with the categories defined in Chapter 5. The discussion in Chapter 5 highlights the ambiguities of the term "soluble" when considering the choice of analytical assays and shows that defining aggregates based on their size range is more appropriate in that context.

For a mechanistic description of aggregation kinetics (Section 2.4), it is necessary to delineate between those aggregates that are molecularly dispersed (i.e., soluble) and those that are not (i.e., insoluble). Growth of soluble species occurs via molecular events such as one-at-a-time monomer addition (chain polymerization), whereas growth of insoluble species often occurs predominantly via coalescence of the phase-separated microdomains or particles. This is illustrated in mathematical terms in Section 2.4.

For concreteness, Fig. 2.1 is cast primarily in terms of nonnative aggregation occurring in bulk solution. However, most, if not all, of the principles can be applied to aggregation mediated by interfaces such as air–water, ice–water, and container–closure contacts with liquid products. Additional complexities arise, at least in part because of difficulties in experimentally monitoring the concentration and important structural attributes of such surface species (see also Chapter 7). Section 2.4 includes a subsection that illustrates how adsorption thermodynamics can be directly incorporated into current models of (net) irreversible aggregation; this also further highlights the need for appropriate experimental tools to monitor different stages of aggregation.

Section 2.5 illustrates how Fig. 2.1 can be adapted when chemical changes in the protein sequence alter the thermodynamics and/or kinetics of protein folding/unfolding and aggregation. Consideration of combined chemical changes and adsorption to bulk surfaces is straightforward by applying both illustrations simultaneously but is foregone here in the interest of space limitations. Section 2.6 reviews the effects of excipients (added solutes or cosolvents) on the thermodynamics of protein unfolding and self-association and includes simple molecular models for interpreting those effects in terms of different types of interactions.

2.3 THERMODYNAMICS OF REVERSIBLE SELF-ASSOCIATION

A key feature of Fig. 2.1, and of all net irreversible aggregation pathways, is that one or more commitment steps are required; that is, there is at least one key step that creates a species that is effectively irreversible. All steps that are "upstream" from such a commitment step are, in principle, reversible. There may be steps that are "downstream" that are also irreversible, but they are not kinetically relevant unless they affect the concentration of upstream species[9,33] (see also Section 2.4).

In the context of this section, one can consider separately the pool of monomers and reversible aggregates, that is, those that are depicted above the dashed line in Fig. 2.1. If the rates of interconversion between each of these species are fast compared to creation and growth of the irreversible species below the dashed line, then the concentration of each reversible species is determined solely by thermodynamics; that is, these steps pre-equilibrate relative to the rate-limiting step(s).[9,33] The concentration of reversible, folded aggregates may be important if they affect the physical properties of the solution such as visual clarity and pharmaceutical elegance (see Chapters 5, 7, and 8 in present text), as well as viscosity and parameters for product manufacture.[34] These examples are typically relevant at relatively high overall protein concentrations (c_m); what determines whether a c_m value is "high" is discussed below. That notwithstanding, reversible aggregates can be important even at lower c_m conditions, where concentrations of reversible aggregates may be negligible in terms of bulk solution properties. Even under such conditions, reversible aggregates may be precursors to irreversible species (e.g., as is R_x in Fig. 2.1). As a result, the thermodynamics of self-association can be important at low c_m values, in that they help to determine the concentration of key intermediates in the kinetic pathway(s) for irreversible aggregate formation (see also below and Section 2.4).

2.3.1 Self-Association of Folded/Native Proteins

For pre-equilibrated folded oligomers, the concentration (mole per volume) of each species (F_j, $j = 1, 2, 3, \ldots n_{max}$) is under thermodynamic control. At fixed temperature, pressure, and solvent conditions, the concentration of each pre-equilibrated species varies only with the total concentration of protein that is not already committed to irreversible aggregates (c_m, defined by Eq. 2.1). In what follows, all concentrations in square brackets have units of mole per volume unless otherwise stated.

Considering first the case where only folded, colloidally associated oligomers are appreciably populated, c_m can be expressed as

$$c_m = [F] + \sum_{i>1} i[F_i]. \tag{2.1}$$

Equilibrium constants for sequential oligomer formation can be defined as

$$K_{1,i} \equiv \exp\left(-\Delta G^0_{1,i}/k_B T\right) = \frac{[F_i]\gamma_{F(i)}}{([F]\gamma_F)^i}, \tag{2.2}$$

in which $\Delta G^0_{1,i}$ is the standard Gibbs free energy change for $iF \rightleftharpoons F_i$ at a selected reference state concentration (c_{ref}) for reactants and products; T is absolute temperature, k_B is Boltzmann's constant, and γ_α is the corresponding activity coefficient for species α. γ_α is defined such that $\gamma_\alpha \approx 1$ when the value of c_m is sufficiently small that protein "crowding" effects can be neglected. To a first approximation, this holds for total protein volume fractions less than ca. 0.1.[19,35,36] At higher protein concentrations γ_α values can be significantly greater than 1.[34,36] When $\gamma_\alpha \approx 1$, substituting Equation 2.2 in Equation 2.1 and rearranging gives

$$y_{Fj} = \frac{jK_{1,j}\left(y_F c_m\right)^{j-1}}{1+\sum_{i>1} iK_{1,i}\left(y_F c_m\right)^{i-1}}. \tag{2.3}$$

In Equation 2.3, $y_{Fj} \equiv j[F_j]/c_m$ is the mass fraction of folded protein chains that exist as F_j. Statistical mechanical calculations for simplified but exactly solvable systems[19,37] indicate that $K_{1,j} \approx K_{1,2}^{(j-1)} = K_d^{(1-j)}$ when $\gamma_\alpha \approx 1$; that is, the free energy of self-association is roughly linear with respect to the degree of association j. With this approximation, Equation 2.3 becomes

$$\left[F_j\right] = \frac{c_m\left(y_F c_m K_d^{-1}\right)^{j-1}}{1+\sum_{i>1} i\left(y_F c_m K_d^{-1}\right)^{i-1}}, \quad j = 1, 2, \ldots, \tag{2.4a}$$

with y_F calculated as the real positive root of

$$\left(y_F + \sum_{i>1} i\left(c_m K_d^{-1}\right)^{i-1} y_F^i\right) - 1 = 0. \tag{2.4b}$$

Equation 2.4 is a function of only c_m/K_d and holds simultaneously for all F_j ($j = 1, 2, 3, \ldots$). Figure 2.2A shows the solutions to Equation 2.4 as a function of c_m/K_d, illustrated there for the case of hexamers ($j = 6$) as the largest oligomer possible. For $c_m/K_d \sim 10^{-2}$ and lower, the concentrations of self-associated oligomers are too small in many cases for those species to be experimentally detectable relative to convoluting signal(s) from monomers. As argued below, K_d values will be large (~1–10 mM) for proteins that are not natively multimeric.

The results in Fig. 2.2A are cast in terms of the dissociation equilibrium constant K_d. However, unless there is strong binding ($c_m/K_d \gg 1$), it may be intractable to experimentally measure K_d. This is the case for weakly bound or colloidal oligomers, in which the constituent monomers interact through relatively nonspecific interactions such as discussed earlier in this section and in Chapter 1. In this case, alternative experimental measures of the average interactions between proteins are available, for example, the osmotic second virial coefficient (B_{22}).[8,38–40] Unfortunately, B_{22} and K_d are only indirectly related to one another and require more detailed statistical mechanical treatments to directly relate the two quantities.[8,19] Figure 2.2B displays illustrative

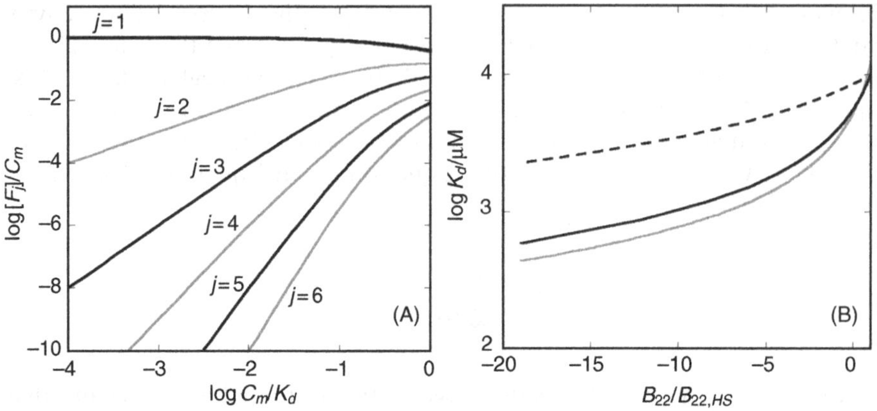

Figure 2.2. (A) Concentration profiles for self-association based on Equation 2.4 with hexamers as the largest oligomer included in the calculations. (B) Relationship between K_d and B_{22} for three different models of the colloidal interactions between proteins with monomer diameters ~4 nm; the different models are (1) hard sphere plus isotropic, short-ranged attraction (solid gray);[37] (2) hard dumbbell with screened electrostatics and dipolar distribution of charge (solid black);[37] and (3) isotropic nearest-neighbor attractions on cubic lattice (dashed black).[19]

results from such a treatment for three different choices of colloidal interaction models (additional details provided in figure caption). The curves show K_d as a function B_{22} scaled by its purely steric or hard-sphere (HS) value ($B_{22,HS}$). Using this dimensionless B_{22} value allows results from different proteins to be more easily compared on the same scale. The range of $B_{22}/B_{22,HS}$ used in Fig. 2.2B is similar to that reported experimentally for a number of proteins under conditions typical of liquid biopharmaceutial formulations.[21,38,41–43] The largest negative values of $B_{22}/B_{22,HS}$ in Fig. 2.2B are similar to the values estimated theoretically for interactions between highly hydrophobic proteins in molten-globule unfolded states.[44]

The results in Fig. 2.2B illustrate that although the value of K_d depends to some extent on the particular interaction model that is assumed, there are a number of common features. The first is that K_d values are of the order 10^3–10^4 μM for typical experimental virial coefficient values ($B_{22}/B_{22,HS}$ > ca. −20).[21,38,41–43] The magnitude of K_d is therefore relatively large compared to typical protein concentrations unless one is dealing with polypeptides or protein fragments.[30] Comparison with Fig. 2.2A shows that this result implies reversible oligomers are thermodynamically unfavorable even for conditions with strongly negative B_{22} values. This holds unless one works at very large protein concentrations (e.g., total protein volume fractions ~0.1 or greater)[19] where excluded volume or crowding effects are expected to significantly promote self-association.[34]

Furthermore, changing from positive to negative B_{22} values does not imply a qualitative change in oligomer formation. That is, a change in the sign of B_{22} does not require a change from favorable oligomer formation ($c_m/K_d \sim 1$ or higher) to unfavorable oligomer formation ($c_m/K_d \ll 1$).[8,19] Thermodynamically, this seemingly nonintuitive behavior occurs because B_{22} values neglect contributions to K_d from communal entropy, and those contributions are large and thermodynamically unfavorable for self-association at typical protein concentrations.[8,45]

In this context, it should be made clear that the above arguments hold so long as one does not directly observe stable, reversible oligomers, for example, via nonlinear Debye behavior in laser light scattering.[21] In those contrasting cases, it is not appropriate to treat monomer–monomer interactions in the context of a second-order virial expansion (i.e., B_{22}); instead, K_d values should be accounted for directly, or more advanced treatments of experimental data must be considered.[21,35,46]

2.3.2 Nonnative Self-Association

To illustrate how the discussion in Section 2.3.1 translates to nonnative self-association, we consider the case where self-association of folded monomers is negligible (i.e., $c_m/K_d \ll 1$ in Eq. 2.4); for simplicity, monomer unfolding is also treated as two-state ($F \leftrightarrow U$, and $U = R$ in Fig. 2.1). These assumptions simplify the resulting mathematical expressions and discussions below. It is straightforward to relax one or both of them[33,47] but unnecessary for much of the discussion later in this chapter. In what follows in this section, all self-association and unfolding steps are treated as pre-equilibrated. In Section 2.4, this is equivalent to the case of nucleated kinetics with a prenucleus stoichiometry of $i = x$.

With the above assumptions, the relevant process to consider from Fig. 2.1 is

$$F \underset{}{\overset{K_{un}}{\rightleftarrows}} R \underset{}{\overset{1/K_d^R}{\rightleftarrows}} R_2 \rightleftarrows \cdots \rightleftarrows R_x.$$

In this case, the analogous expressions to Equations 2.1, 2.2, and 2.4 are given below with equivalent expressions elsewhere:[20]

$$c_m = [F] + [R] + \sum_{i=2}^{x} i[R_i]. \tag{2.5}$$

The equilibrium constant for monomer–dimer formation is defined as

$$K_d^R \equiv \exp\left(-\Delta G^0_{R_2 \to 2R}/k_B T\right) = \frac{[R]^2}{[R_2]}, \tag{2.6}$$

and the resulting $[R_j]$ concentrations are given by

$$[R_j] = \frac{f_R c_m \left(y_m f_R c_m / K_d^R\right)^{j-1}}{1 + f_R \sum_{i=2}^{x} i\left(y_m f_R c_m / K_d^R\right)^{i-1}} \quad (\text{for } j = 1, 2, \ldots, x), \tag{2.7a}$$

with y_m calculated as the real positive root of

$$\left(y_m + f_R \sum_{i=2}^{x} i\left(f_R c_m / K_d^R\right)^{i-1} y_m^i\right) - 1 = 0. \tag{2.7b}$$

In Equation 2.7, y_m is the mass fraction of c_m that is truly monomer (i.e., F or R), and f_R is defined as

$$f_R = \frac{[R]}{[F]+[R]} = \frac{K_{un}}{1+K_{un}} \tag{2.8}$$

with

$$K_{un} \equiv \exp\left(\frac{-\Delta G_{un}^0}{k_B T}\right). \tag{2.9}$$

In Equation 2.9, ΔG_{un}^0 is the standard free energy change for the $F \rightarrow R$ transition at a given temperature (T), pressure, and solvent composition.

Under many practical conditions for product formulations or manufacturing conditions, $\Delta G_{un}^0 >> k_B T$. That is, sample conditions strongly favor F, and therefore f_R in Equation 2.7 can be replaced with K_{un}. Comparison with Equation 2.4 and Fig. 2.2A then indicates that under folding-favoring conditions, an effective dissociation constant is given by K_d^R / K_{un}. From Fig. 2.2B, liberal estimates of K_d values at strongly attractive (i.e., hydrophobic) B_{22} values are no smaller than the order 10–100 μM. Typical values of K_{un} are ~10^{-4} or smaller under folding-favoring conditions for globular proteins or for similarly sized subdomains of larger proteins.[11,14,48–50] Therefore, provided that c_m is not extremely high (~0.1 M), the solution to Equation 2.7b is $y_m \approx 1$, and the solution to Equation 2.7a becomes

$$\frac{[R_j]}{c_m} \approx K_{un}\left(K_{un} c_m / K_d^R\right)^{j-1}. \tag{2.10}$$

Inserting representative experimental values for K_{un} and c_m in Equation 2.10, along with estimates of K_d^R outlined above, clearly shows that R_j ($i = 2, \ldots, x$) are expected to be well below detection limits with available experimental techniques.

As described in Section 2.4, the rate of forming irreversible nonnative aggregates is proportional to the concentration of R_x. In this context, Equation 2.10 and the discussion above highlights that although low-level intermediates such as R_x are likely undetectable experimentally, they can remain important species from the perspective of controlling aggregation rates. This is independent of whether c_m is near or above K_d, despite commonly applied arguments to the contrary.[30,51] Those arguments typically are valid only when one considers very large x values (~10^2 or more). In that special case, it can be shown that all $[R_j]$ are effectively zero below a threshold or critical c_m value.[20,30] Experimentally, that situation is apparently more typical when monomers are hydrophobic fragments of full-length proteins[30,52] or synthetic proteins with large contiguous patches of hydrophobic surface.[31]

2.4 AGGREGATION KINETICS AND DISTINGUISHING KINETIC PATHWAYS

Much of the discussion in this section is a summary of results from previous analysis of experimental kinetics and mechanistic models of nonnative aggregation for globular proteins.[20,32,33,53] As such, only the salient results for the purposes of this chapter will be presented here. The reader is referred to the original references for details regarding derivations and/or experimental aspects.

2.4.1 Aggregation Kinetics

Kinetics of nonnative aggregate formation are often monitored experimentally via techniques such as size-exclusion chromatography (SEC) or field-flow fractionation (FFF), and in some cases by laser light scattering, turbidity, or spectroscopic probes of secondary or tertiary structure of the constituent monomers[8] (see also Chapter 5). In what follows, kinetics are cast in terms of the time dependence of m, the monomer concentration scaled by its initial value. For cases where aggregates remain in solution throughout the time course of an experiment, the weight-average molecular weight (M_w^{agg}) of the total aggregate size distribution is also a useful quantity in terms of distinguishing different kinetic pathways of aggregation.[9,20,32,54] As highlighted elsewhere,[8] most other commonly used techniques for monitoring aggregation can either be cast in terms equivalent to m and/or M_w^{agg}, or they are qualitative or semi-quantitative surrogates for these two quantities. In the context of Table 2.2 and the discussion below, the term soluble aggregate denotes any aggregate species that is available for a monomer to be directly incorporated. The term is not used to imply a given threshold size or dimension of an aggregate, except in so far as empirical observations suggest that the solubility of aggregates may decrease with increasing aggregate molecular weight.[29,55]

TABLE 2.2. Summary of Experimental Signatures to Distinguish Candidate Pathways Based on Aggregation Kinetics Monitored by Aggregate Molecular Weight (M_w) and/or Mass Fraction of the Remaining Monomer (m)

Kinetic Pathway	Rate-Limiting Step	Soluble Aggregates				Insoluble Aggregates		
		ν (t basis)[a]	γ (c_0 basis)[b]	α, $k_{obs} \sim (K_{un})^{\alpha}$	β,[c] $M_w^{agg} \sim (1-m)^{\beta}$	ν (t basis)[a]	γ (c_0 basis)[b]	α, $k_{obs} \sim (\Delta G_{un})^{\alpha}$
Unfolding-limited	$F \rightarrow R$	1	0	<1	[d]	1	0	<1
Association-limited	$2R \rightarrow R_2$							
	Slow growth	2	1	2	~0	2	1	2
	Fast chain polymerization	1	1	3/2	1	2	1	2
	Fast condensation polymerization	2	1	2	>>1	2	1	2
Nucleation (and growth)	$R_{x-1} + R \leftrightarrows R_x \rightarrow A_x$							
	Slow growth	$x \geq 2$	$x - 1$	x	~0	$x \geq 2$	$x - 1$	x
	Fast chain polymerization	1	≥1	≥3/2	1	$x \geq 2$	$x - 1$	x
	Fast condensation polymerization	>1	≥1	≥2	>>1	$x \geq 2$	$x - 1$	x
Chemically controlled	$M \rightarrow M'$							
	Non-covalent aggregation	1	0	[f]	[d]	1	0	[f]
	Disulfide polymerization	3/2	0	0	[e]	–	–	–

[a]Reaction order (ν) based on $dm/dt \approx -k_{obs}m^{\nu}$.
[b]Reaction order (γ) based on $k_{obs} \sim c_0^{\gamma}$.
[c]Scaling exponent for M_w of total soluble aggregates as a function of the mass fraction of monomer already coverted to aggregate (=1 – m).
[d]To the best of the authors' knowledge, this analysis has not been shown in published work.
[e]Roefs and de Kruif[65] proposed expressions for the pseudo-steady-state M_w at a given (steady) monomer concentration, but only solved for the kinetic behavior in the limit of $t \rightarrow 0$.
[f]See text in Section 2.5.

With the exception of kinetics that show a true lag phase for aggregate formation,[9] the kinetics of monomer loss at fixed T, pressure, and solvent composition can often be well described by the seemingly simple rate expression

$$\frac{dm}{dt} = -k_{obs}m^{\nu}, \tag{2.11}$$

in which ν is the apparent reaction order. k_{obs} is an observed or effective rate coefficient that may include contributions from the equilibrium constant (K_{un}) for unfolding to form the R monomer state, as well as $K_{1,x}$ for prenuclei formation ($xR \rightleftharpoons R_x$). Only in the case of a true unfolding-limited aggregation is k_{obs} equated with a single rate coefficient (see also Table 2.2).

The precise relationship between k_{obs} and different stages in Fig. 2.1 depends on which stage or stages in Fig. 2.1 are rate limiting. Table 2.2 summarizes many of the possible scenarios, some of which are discussed in more detail below. In the interest of brevity, they will not be described in detail here except to illustrate the points below. So long as folding/unfolding is fast compared to the downstream stages of self-association and any subsequent conformational rearrangements to form stable, nonnative contacts within aggregates, k_{obs} will scale according to Equation 2.12 under folding-favoring conditions:

$$k_{obs} \sim (K_{un})^{\alpha} K_{1,x} c_0^{\gamma} \approx \frac{(K_{un})^{\alpha} c_0^{\gamma}}{(K_d)^{x-1}}. \tag{2.12}$$

The precise values of α and γ depend on the rate-limiting step(s)—cf. Table 2.2. The rightmost expression in Equation 2.12 follows from the isodesmic approximation introduced in Section 2.3.

Physically, the relationships in Equation 2.12 arise in the following way. Increasing the free energy of unfolding decreases K_{un} and thus decreases the concentration of R monomers at a given total monomer concentration. As Equation 2.10 illustrated, under native-favoring conditions, this concomitantly reduced concentrations of all nonnative oligomers, including those that can convert to irreversible aggregates. Decreasing the concentration of R monomers also decreases the rate of addition of those monomers to existing aggregates. As a result, α is typically significantly larger than 1, and therefore changes in unfolding free energy have large effects on k_{obs} when R monomers are thermodynamically disfavored (see also Chapters 1, 4, and 6–9). The observation[26,29,54] that k_{obs} depends on the initial protein concentration c_0 follows because K_{un} and K_d are constant at fixed T, pressure, and solvent composition; however, the concentration of reversible oligomers depends on c_m (Fig. 2.2A), and c_m equals the product of c_0 and m.[9,20]

Finally, k_{obs} depends on K_d because increasing K_d disfavors formation of the reversible precursors to irreversible aggregate formation (cf. Section 2.3).

Taken together, Equation 2.12 provides quantitative or semiquantitative guidance for interpreting the effects of changes in protein concentration and/or sample conditions on aggregation rates. A change in solvent conditions will likely affect both K_{un} and K_d. Empirically, cases exist where only one is dominant,[56] as well as where there is a balance between contributions from each.[1,57]

In order to quantitatively interpret experimental aggregation kinetics in terms of Equation 2.12, one must be able to identify which step(s) in the overall aggregation process are rate limiting. This is the subject of Section 2.4.2 and Table 2.2. A brief discussion is also included in Section 2.4.2 regarding the question of aggregation mediated by bulk interfaces such as those between water and air or solid surfaces. Table 2.2 also includes entries for limiting cases where the rate-limiting step(s) for aggregation are due to breaking or forming covalent bonds within or between protein monomers. These are briefly discussed as part of Section 2.5.

2.4.2 Distinguishing Aggregation Pathways and Rate-Limiting Steps

Table 2.2 summarizes a series of experimental signatures of different rate-limiting steps in the overall aggregation process. A key conclusion from inspection of Table 2.2 is that it is often not possible to distinguish different cases unless one analyzes multiple signatures of experimental kinetics. This is particularly the case when aggregates do not remain soluble, as the only kinetic information available is often monomer concentration versus time ($m(t)$ and dm/dt).[8,9]

In qualitative terms, the results in Table 2.2 can be summarized as follows. Independent of whether aggregates are soluble, one can determine the apparent reaction order (ν) from the time course of $m(t)$, provided it is measured over more than one half-life.[9,20,26] This is repeated at multiple c_0 values to obtain γ from k_{obs} versus c_0. If aggregates remain soluble, M_w^{agg} can be measured along with m as aggregation proceeds, and β can be determined.[32,58] Furthermore, one can obtain quantitative information regarding both nucleation and growth rates if m and M_w^{agg} are fit using mechanistic mathematical models.[28,32,54,58–60] Although not shown explicitly in Table 2.2, the values of α, ν, and γ are not arbitrary; values of one or more of them can be used to predict the value of the other(s) if one has identified which type of aggregation kinetics are relevant to a protein and sample conditions of interest. Alternatively, one can use those relationships as self-consistency checks regarding whether a given model accurately describes a given experimental system.[9]

It is not currently possible to reliably predict a priori what physical factors predominantly determine which aggregation pathway is followed for a given protein. That is, it remains an empirical exercise to measure aggregation kinetics as a function of protein concentration, temperature, and solvent conditions and then to fit the data or analyze it within the context of different models to determine, for example, if the rate-limiting step(s) include unfolding, monomer self-association or nucleation, and different forms of nonnative aggregate growth. In this context, it is notable that recent results from our laboratory have shown that the type of growth mechanism and the resulting form of aggregates (soluble vs. insoluble and globular vs. extended or polymeric) for a-chymotrypsinogen A (aCgn) correlates well with the value of B_{22} for native monomers over a relatively broad range of solution pH and ionic strength.[61]

This is consistent with a physical scenario where electrostatic repulsions and attractions between native monomers reflect some of the important interactions between aggregates and between monomers and aggregates. As such, when $B_{22}/B_{22,HS}$ approaches zero or negative values, aggregates in solution readily associate with each other and rapidly form large, insoluble particles or precipitates. Conversely, under solvent conditions with $B_{22}/B_{22,HS} \gg 1$ due to electrostatic repulsions, nonnative aggregates occur as soluble HMW polymers or as smaller soluble oligomers. Figure 2.3 illustrates this by showing a state diagram for aCgn adapted from Reference 61. Points that are labeled as type Ib are those for which essentially all aggregates are insoluble and lie almost exclusively at $B_{22}/B_{22,HS}$ values that are less than 1. The opposite extreme are points labeled as type II, which correspond to the HMW soluble aggregates in which aggregate–aggregate association is negligible, as illustrated by the inset TEM image adapted from Reference 54.

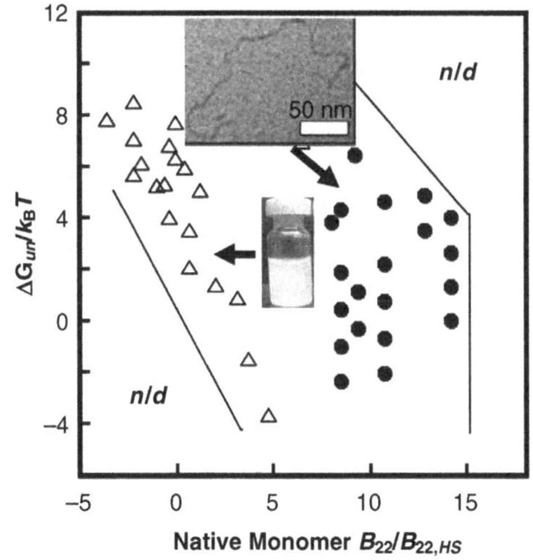

Figure 2.3. Partial state diagram for aggregation of α-chymotrypsinogen A, adapted from Li et al.[61] Symbols denote different conditions of pH, temperature, and NaCl concentration that lead to either soluble polymeric aggregates (circles, type II) or readily visible haze and particle/precipitate formation (triangles, type Ib). The images are included as insets to help clarify the different aggregate types. The TEM image is excerpted from Weiss et al.[54] for conditions that bG-CSF forms analogous polymeric soluble aggregates (same image as in Fig. 2.1).

2.4.3 Influence of Adsorption to Macroscopic Surfaces

All of the relationships in Table 2.2 were derived from mechanistic models that implicitly treat creation and growth of stable aggregates as occurring in bulk solution. As Fig. 2.1 suggests, this assumption can be relaxed to allow for adsorption of monomer and/or oligomers at bulk interfaces between water and solid or vapor surfaces. As an example, consider an adsorption-mediated aggregation path such as

$$F \underset{}{\overset{K_{un}}{\rightleftarrows}} R \underset{}{\overset{K_{ads}}{\rightleftarrows}} R_s \underset{}{\overset{K_{1,x,surf}}{\rightleftarrows}} R_{x,s} \rightarrow A_{x,s} \rightarrow A_x \rightarrow \cdots,$$

where K_{ads} is the adsorption equilibrium constant for the linear regime of the adsorption isotherm of R monomers; R_s is the surface-adsorbed R monomer species; and $K_{1,x,surf}$, $R_{x,s}$, and $A_{x,s}$ are surface analogues to $K_{1,x}$, R_x, and A_x. In the limiting case that the irreversible step(s) for aggregate formation are slow compared to adsorption and surface diffusion, the mathematical derivations leading to Equation 2.12 instead give

$$k_{obs} \sim (K_{un})^{\alpha} K_{ads} K_{1,x,surf} c_0^{\gamma}. \tag{2.13}$$

An analogous expression can be derived if one instead considers adsorption of F to occur first, followed by surface-induced unfolding (see Chapter 7), with K_{ads} describing adsorption of F and K_{un} describing surface unfolding.

A comparison of Equations 2.12 and 2.13 shows that if one is considering only a single set of experimental conditions, it may not be possible to distinguish between adsorption-mediated aggregation and bulk aggregation. In order to show that adsorption is important, one can instead vary experimental parameters that affect mass transfer, such as the degree of mixing of the bulk solvent and the surface area-to-volume ratio between bulk phases.[62,63] Alternatively, if one finds that those variables do not affect aggregation rates, then it may be sufficient to neglect the effects of interfaces.

2.5 CHEMICAL MODIFICATIONS

This section illustrates two different types of effects that chemical modifications are anticipated to have on nonnative aggregation kinetics. The first situation is when a protein is originally expressed and folded in its wild-type form or as some other desired variant. It can then undergo one or more chemical reactions or posttranslational modifications that create a chemically different monomer species (M'). One example that has been shown to affect aggregation rates is deamidation of Asn or Gln side chains to form Asp or Glu residues, also resulting in a change in protonation at pH > ca. 4.[11] An alternative example is deliberate polyethylene glycol (PEG)-ylation to slow clearance in the blood stream and also potentially to impede protein self-association.[64]

The second limiting situation is when one considers variants that have been deliberately engineered using molecular biology to have altered enzymatic or other biological function(s). In both situations, the new protein may differ from the parent monomer (M) in terms of one or more of the following aspects that Sections 2.2–2.4 highlighted as important for aggregation: altered K_{un} and/or K_d, a change in nucleus stoichiometry (x), a change in the location of the rate-limiting step(s), a change in the intrinsic nucleation rate for $R_x \rightarrow A_x$, and altered chemical reactivity of the monomer toward other monomers. The lattermost case is illustrated by examples in which interprotein disulfide formation propagates similarly to radical chain polymerization.[65] This is also the case summarized by the last row in Table 2.2 and is that studied in detail for β-lactoglobulin at neutral pH. Unlike the cases considered below, the disulfide polymerization mechanism does not necessarily require unfolding from a kinetic perspective, although one may anticipate that significant disruption of secondary structure would occur as a result of the successive reactions forming and disrupting inter- and intraprotein disulfides, respectively.

If the unfolded form of M' (i.e., R') is much more reactive to aggregation than the parent monomer, such that formation of R' is rate limiting for aggregation, the following two limiting scenarios exist: both are considered under the assumption that F is the favored conformational state of the monomer under the sample conditions of interest.

If the chemical reaction occurs readily with F monomers and $F \xrightarrow{k'} F'$ is rate limiting, then

$$\frac{\mathrm{d}m}{\mathrm{d}t} = -k'm. \tag{2.14}$$

Simple first-order kinetics are recovered, with $k_{obs} = k'$, and there is no dependence of k_{obs} on c_0, K_d, or K_{un}. Alternatively, if $F \leftrightarrow R$ equilibrium establishes quickly compared to conversion of R' to aggregates, or if $R \xrightarrow{k'} R'$ is rate limiting, then

$$\frac{\mathrm{d}m}{\mathrm{d}t} = -k'K_{un}m. \tag{2.15}$$

In this case, $k_{obs} = k'K_{un}$, and there is a strong dependence on unfolding free energy.

Finally, if the rate-limiting steps for aggregation involve association and or nucleation of R' monomers, then Equations 2.11 and 2.12 are recovered except that K_{un}, K_d, ν, α, and γ will be those for R' rather than R.[11] In some cases, this may be desirable, if one can rationally identify amino acid replacements that will alter K_{un} and/or K_d to therefore deliberately decrease k_{obs}. In addition, the intrinsic stability of the aggregates may be altered by changing only a few amino acids in the sequence. It is not yet mechanistically well understood how changes of specific amino acid contacts involved in steps such as nucleation

($R_x \rightarrow A_x$) translate mechanistically to the kinetics of aggregate creation and growth, and thus a detailed discussion of this topic is foregone here. However, this is an active area of research[12,60,66–70] and is inherently a factor within the context of Chapter 3.

2.6 EFFECTS OF COSOLVENTS OR COSOLUTES

As illustrated elsewhere in this book (cf. Chapters 1, 4, 6–8), adding cosolvent or cosolutes (e.g., excipients) to protein formulations can significantly affect protein aggregation. Although the large majority of published reports in the pharmaceutical literature dealing with cosolute–protein interactions focus on the direct effects on protein folding, self-association and aggregation are also impacted directly by such interactions.[71–74]

In this section, we briefly review the classical and statistical thermodynamics of solvation for protein folding and self-association in binary mixed aqueous solvents and provide a context for qualitative and semiquantitative analysis of pharmaceutical protein systems. Based on the principles reviewed in Sections 2.2–2.4, the effects of cosolutes on solvation thermodynamics translate to rates of nonnative aggregation because changes in the free energy of any "upstream" step in Fig. 2.1 potentially cause net changes in the concentration of the intermediates involved in the rate-limiting step(s) for aggregation. For example, if dimerization is rate limiting in Fig. 2.1, then increasing the unfolding free energy (decreasing K_{un}) leads to a lower $[R]$ value at fixed initial protein concentration c_0. This in turn produces a lower rate of aggregate (dimer) formation. By analogy, if nucleation is rate limiting, then increasing the unfolding free energy and/or the self-association free energy leads to lower $[R_x]$ and therefore slower overall rates of aggregate formation. Quantitatively, Equation 2.12 and Table 2.2 illustrate mathematically how changes in K_{un} and/or K_d^R produce changes in aggregation rates (i.e., k_{obs}).

To be consistent with common practice when treating the thermodynamics of protein solutions, we adopt the Scatchard notation,[75] in which water is denoted as component 1, protein is component 2, and (neutral) cosolute is component 3. For the sake of illustration, only neutral cosolutes are treated here. The basic treatments below hold if the cosolute is dissociable and charged (e.g., a salt), but under some conditions, additional considerations, such as Donnan equilibrium, may also need to be included.[43]

2.6.1 Transfer Free Energies and Preferential Interaction Parameters

Changes in solvent composition lead to changes in K_{un} and/or K_d^R because of changes in the excess chemical potential of protein (μ_2^{ex}) in the folded (F), unfolded (R), and/or associated state ($R_{i=2,3,\ldots}$); that is, for a given temperature (T), pressure (p), and molality of cosolute (m_3), K_{un} and K_d^R can be expressed as

$$-k_BT\ln K_{un}=\mu_{2,R}^{ex}(T,p,m_3)-\mu_{2,F}^{ex}(T,p,m_3) \tag{2.16a}$$

and

$$-k_BT\ln\frac{K_d^R}{K_d^{HS}}=2\mu_{2,R}^{ex}(T,p,m_3)-\mu_{2,R_2}^{ex}(T,p,m_3). \tag{2.16b}$$

Equation 2.16a implicitly assumes the protein concentration is sufficiently dilute to neglect the effects of protein–protein interactions on $\mu_{2,F}^{ex}$ and $\mu_{2,R}^{ex}$. Protein–protein interactions enter implicitly in Equation 2.16b via μ_{2,R_2}^{ex}. The equilibrium dissociation constant for a dimer of equivalent hard spheres is denoted as K_d^{HS} in Equation 2.16b. K_d^{HS} accounts for purely steric or excluded volume effects of converting from a monomer to a dimer and depends solely on the volume fraction or on the protein concentration of the standard state used to evaluate K_d^R.[8,19,37] For concreteness, the following derivation is given explicitly for $\mu_{2,F}^{ex}$. Extension to $\mu_{2,R}^{ex}$ and μ_{2,R_2}^{ex} is carried out by analogy.

In general, one can express $\mu_{2,F}^{ex}$ as

$$\mu_{2,F}^{ex}(T,p,m_3)=\mu_{2,F}^{ex}(T,p,m_3=0)+\Delta\mu_{2,F}^{tr}(T,p,m_3), \tag{2.17}$$

with the transfer free energy $\Delta\mu_{2,F}^{tr}(T,p,m_3)$ defined as

$$\Delta\mu_{2,F}^{tr}(T,p,m_3)=\int_0^{m_3}\left(\frac{\partial\mu_{2,F}^{ex}}{\partial m_3}\right)_{T,p,m_2\to 0}dm_3. \tag{2.18}$$

This can be rewritten equivalently as

$$\Delta\mu_{2,F}^{tr}(T,p,m_3)=-\int_0^{m_3}\Gamma_{32,F}^{(\mu_3)}\left(\frac{\partial\mu_3^{ex}}{\partial m_3}\right)_{T,p,m_2\to 0}dm_3. \tag{2.19}$$

Equation 2.19 follows from Equation 2.18 based on thermodynamic identities and the definition of the preferential interaction parameter ($\Gamma_{32,F}^{(\mu_3)}$) for cosolvent (3) with protein (2) at fixed T, p, and number of molecules of folded monomers and water:[71]

$$\left(\partial\mu_{2,F}^{ex}/\partial\mu_3^{ex}\right)_{T,p,m_2}=\frac{\left(\partial\mu_{2,F}^{ex}/m_3\right)_{T,p,m_{i\neq 3}}}{\left(\partial\mu_3^{ex}/m_3\right)_{T,p,m_{i\neq 3}}}\equiv-\Gamma_{32,F}^{(\mu_2)}. \tag{2.20}$$

Commonly used nomenclature neglects the superscript (μ_3) on Γ_{32}; it is retained here to make it clear that different values for Γ_{32} will be obtained, depending on what experimental variables are held fixed when one measures preferential interaction parameters.[71]

To help see this more clearly, $\Gamma_{32,i}^{(\mu_3)}$ ($i=F,R,R_x$, etc.) and $\partial\mu_3^{ex}/\partial m_3$ can be expressed using results from statistical mechanics, particularly the Kirkwood–Buff fluctuation theory.[76,77] Using results from Ben-Naim[76] (after some algebra) gives

$$\Gamma_{32}^{(\mu_3)} = \rho_3 (G_{23} - G_{12}) = M_1 N_{Av} c_1 m_3 (G_{23} - G_{12}) \tag{2.21}$$

and

$$\left(\frac{\partial \mu_2^{ex}}{\partial m_3}\right)_{T,p,m_2 \to 0} = \frac{k_B T M_1 N_{Av} c_1^2 \bar{V}_1 (G_{12} - G_{23})}{1 - M_1 N_{Av} c_1 m_3 (G_{13} - G_{33})}. \tag{2.22}$$

In the above equations, M_1 is the molecular weight of water (in kilogram per mole), N_{Av} is Avogadro's number, c_1 is the molar concentration of water, and $\bar{V}_1$ is the partial molar volume of water. If one makes the approximation that $\bar{V}_1 \approx 1/c_1$, Equation 2.22 reduces to the form used by Rösgen et al.[78]

The above expressions also make use of the so-called Kirkwood–Buff integrals (G_{ij}) between components i and j (i.e., water, protein, or cosolute). G_{ij} for any given i–j pair is defined by[76,77]

$$G_{ij} = \int_{r=0}^{\infty} [g_{ij}(r) - 1] 4\pi r^2 dr. \tag{2.23}$$

In the above equation, $g_{ij}(r)$ is the radial distribution function (RDF) for component i with respect to component j for a given value of the center-to-center distance (r) between i and j. Physically, $g_{ij}(r)$ is the normalized probability of finding a randomly selected i–j pair of molecules with a given value of r, after ensemble averaging over all relative orientations of the i–j pair, as well the translational and rotational degrees of freedom of all other molecules in the system. Each RDF is normalized by the probability of finding an i–j pair at a distance, r, in a random ideal mixture at the bulk composition of interest. Thus, at large r, all $g_{ij} = 1$, and the integral in Equation 2.23 is convergent. KB integrals have units of volume (or volume per mole, depending on convention).

If $G_{ij} \gg 0$, component i tends to accumulate in the vicinity of component j to a greater degree than in a spatially random mixture and vice versa for $G_{ij} \ll 0$. Care must be taken not to overinterpret the latter result, as there are always steric contributions (excluded volume) in any G_{ij} value for real molecules, and this must be factored in when trying to interpret G_{ij} in terms of numbers of "bound" or "excluded" molecules of j around a molecule of i.[79]

Physically, Equations 2.21–2.23 show that if a cosolute is differentially accumulated near a protein to a greater degree than is water, both relative to their bulk concentrations, then $\Gamma_{32,i}^{(\mu_3)} > 0$ and $\partial\mu_2^{ex}/\partial m_3 < 0$. This is the standard interpretation of protein interactions with cosolutes such as urea and guanidine. However, the same result can occur if both cosolute and water are excluded from the vicinity of the protein, but water is excluded to a greater degree (see also discussion below).

Similarly, if a cosolute is differentially excluded from a protein to a greater degree than is water, then $\Gamma_{32,i}^{(\mu_3)} < 0$ and $\partial\mu_2^{ex}/\partial m_3 > 0$. This is the standard interpretation of protein interactions with large hydrophilic cosolutes such as sucrose and PEG. However, the same result can occur if both cosolute and water are accumulated near the vicinity of the protein, but water is accumulated to a greater degree. Although contrary to the historical view for sucrose–protein interactions, recent results indicate that this latter scenario occurs for at least some proteins in sucrose–water solutions (see also Section 2.6.3).

Thus, at a molecular level, the sign and magnitude of $\Gamma_{32}^{(\mu_3)}$ (and equivalently, the sign but not necessarily the magnitude of $\partial\mu_2^{ex}/\partial m_3$) is related to the difference in accumulation and/or exclusion of component 3 (cosolute) in the vicinity of component 2 (protein), relative to that for component 1 (water). Thus, it reflects a balance of the protein–cosolute and protein–water interactions. This is part of the reason that $\Gamma_{32}^{(\mu_3)}$ is termed a preferential interaction parameter rather than simply an interaction parameter. These interactions are averaged over all protein–water, protein–cosolvent, water–water, water–cosolvent, and cosolvent–cosolvent interactions at a given T and p. Although typical usage in the literature often implies that G_{ij} are functions of only i–j interactions, their values for protein solutions are of course intimately tied to all solvent–solvent interactions—this is perhaps more easily realized by noting that G_{23} includes contributions from favorable association of solvophobic regions on a protein with those on a cosolvent molecule. Such solvophobic attractions arise because of the interactions of water with the protein and with the cosolvent, but also because of water–water interactions.

Returning to the process of unfolding a protein, it follows from Equations 2.16–2.19 that the dependence of K_{un} on excipient type and concentration is determined by the difference in $\Gamma_{32}^{(\mu_3)}$ between folded and unfolded states:

$$\left(\frac{\partial \ln K_{un}}{\partial m_3}\right)_{T,p,m_{j\neq3}} = -\frac{\left[\Gamma_{32,F}^{(\mu_3)} - \Gamma_{32,R}^{(\mu_3)}\right]}{k_B T}\left(\frac{\partial \mu_3^{ex}}{\partial m_3}\right)_{T,p,m_2\to0} = \frac{\Delta\Gamma_{32}^{(\mu_3)}}{k_B T}\left(\frac{\partial \mu_3^{ex}}{\partial m_3}\right)_{T,p,m_2\to0}. \tag{2.24}$$

The rightmost equality defines $\Delta\Gamma_{32}^{(\mu_3)} = \left[\Gamma_{32,R}^{(\mu_3)} - \Gamma_{32,F}^{(\mu_3)}\right]$. Based upon thermodynamic stability criteria,[80] $(\partial \ln \mu_3^{ex}/\partial m_3)_{T,p,m_{j\neq3}}$ must be positive at equilibrium. Therefore, the sign of $\Delta\Gamma_{32}^{(\mu_3)}$ determines the sign of $\left(\frac{\partial \ln K_{un}}{\partial m_3}\right)_{T,p,m_{j\neq3}}$. As is apparent in Equation 2.24, the magnitude of $\left(\frac{\partial \ln K_{un}}{\partial m_3}\right)_{T,p,m_{j\neq3}}$ depends on both $\Delta\Gamma_{32}^{(\mu_3)}$ and $(\partial \ln \mu_3^{ex}/\partial m_3)_{T,p,m_{j\neq3}}$. In what follows below, a simple but qualitatively realistic molecular model is provided, which links the sign and magnitude of $\Delta\Gamma_{32}^{(\mu_3)}$ to molecular properties such as size and intermolecular attractions and repulsions.

2.6.2 Relating $\Gamma_{32}^{(\mu_3)}$ to Molecular Interactions

From the discussion in the previous section, it is apparent that the value and sign of $\Gamma_{32}^{(\mu_3)}$ is determined not simply by whether protein–cosolvent interactions are attractive or favorable; rather, $\Gamma_{32}^{(\mu_3)}$ follows from the balance between protein–cosolvent and protein–water interactions, averaged over all water–

water, water–cosolvent, and cosolvent–water interactions. In this section, a simple but qualitatively realistic molecular model is considered, which may provide useful insights into qualitatively interpreting $\Gamma_{32}^{(\mu_3)}$ in terms of molecular interactions. Below, we consider a simple but analytically solvable three-component molecular model to qualitatively illustrate some aspects of the underlying physics of preferential exclusion and the relationship between $\Gamma_{32}^{(\mu_3)}$, molecular size, and intermolecular interactions.

2.6.2.1 Simple Perturbation Theory for $\Gamma_{32}^{(\mu_3)}$

A van der Waals (vdW) fluid is one in which all molecules have an effective HS repulsion between each other and may have net attractive or repulsive non-steric interactions between each other. The steric interactions are captured by so-called vdW b parameters:[81]

$$b_{ij} = b_{ji} = \frac{\pi}{12}(\sigma_i + \sigma_j)^3 = \frac{2\pi}{3}\sigma_{i,j}^3, \tag{2.25}$$

with σ_i denoting the effective HS diameter of component i, and $\sigma_{i,j} \equiv (\sigma_i + \sigma_j)/2$ defining the average HS diameter for an i–j pair. The non-steric interactions are captured by so-called vdW a parameters:

$$a_{ij} = a_{ji} = -2\pi \int_{(\sigma_i+\sigma_j)/2}^{\infty} u'_{ij}(r) g_{ij}^{HS}(r) r^2 dr, \tag{2.26}$$

in which $u'_{ij}(r)$ is the non-steric contribution to the potential energy between components i and j at a center-to-center distance r; the superscript HS denotes that the RDF between components i and j is evaluated in the HS reference system. The Appendix provides a derivation of the key results for this section. Based on the discussion above, the key quantities of interest from the Appendix are

$$\Gamma_{32}^{(\mu_3)} \approx \frac{k_B T \rho_1^2 b_{23} b_{33} M_1 m_3 - 2\rho_1 a_{23} [1 - \rho_1 (b_{13} + b_{33} M_1 m_3)]^2}{k T \rho_1^2 b_{33}^2 - 2\rho_1 a_{33} [1 - \rho_1 (b_{13} + b_{33} M_1 m_3)]^2} \tag{2.27}$$

and

$$\left(\frac{\partial \mu_2^{ex}}{\partial m_3}\right)_{T,V,m_2 \to 0} = -2\rho_1 M_1 a_{23} + k_B T M_1 \left\{ \frac{\rho_1^2 b_{12} b_{13}}{[1 - \rho_1 (b_{11} + b_{13} M_1 m_3)]^2} + \frac{\rho_1^2 b_{23} b_{33} M_1 m_3}{[1 - \rho_1 (b_{13} + b_{33} M_1 m_3)]^2} \right\}. \tag{2.28}$$

ρ_i denotes the concentration (number or mole per unit volume) of species i.

In what follows, it is explained how Equations 2.27 and 2.28 can provide physical insight into the balance between molecular contributions to $\Gamma_{32}^{(\mu_3)}$ and the change in protein chemical potential upon addition of cosolutes. The terms in brackets in Equation 2.28 arise solely because of steric effects; that is, they are due only to the loss of configurational entropy that a protein experiences when cosolutes are added that occupy space in the system (i.e., the "excluded volume" effect). Note that these terms are necessarily positive, and so act to increase the protein chemical potential upon addition of cosolutes, no matter what cosolute is added.

The terms proportional to a_{23} and a_{33} are due to contributions, respectively, from solvent-averaged protein–cosolute (23) and cosolute–cosolute (33) interactions other than steric repulsions. For example, if solvent-averaged protein–cosolute interactions are attractive, then $a_{23} > 0$, and this term contributes a favorable effect (decreasing μ_2^{ex}) upon the addition of a cosolute. Conversely, if these interactions are repulsive, then $a_{23} < 0$, and this will make the addition of a cosolute even more unfavorable than the case with excluded volume effects alone.

Equations 2.27 and 2.28 also illustrate that excluded volume effects are always present for any physically realistic values of b_{ij}. Essentially all realistic cosolutes are larger than water, and therefore they would be preferentially excluded if there were no attractions between protein and cosolute. That is, in the limit that $a_{23} = 0$, Equations 2.27 and 2.28 simplify to

$$\Gamma_{32}^{(\mu_3)} \approx -\frac{b_{23}}{b_{33}} M_1 m_3 \tag{2.29}$$

and

$$\left(\frac{\partial \mu_2^{ex}}{\partial m_3}\right)_{T,V,m_2 \to 0} = k_B T M_1 \left\{ \frac{\rho_1^2 b_{12} b_{13}}{[1-\rho_1(b_{11}+b_{13}M_1 m_3)]^2} + \frac{\rho_1^2 b_{23} b_{33} M_1 m_3}{[1-\rho_1(b_{13}+b_{33}M_1 m_3)]^2} \right\}. \tag{2.30}$$

Furthermore, considering that $b_{23} \gg b_{12}$ and $b_{33} \gg b_{13}$ for most realistic solutes, Equation 2.30 simplifies to

$$\left(\frac{\partial \mu_2^{ex}}{\partial m_3}\right)_{T,V,m_2 \to 0} = k_B T M_1 \left\{ \frac{\rho_1^2 b_{23} b_{33} M_1 m_3}{[1-\rho_1(b_{13}+b_{33}M_1 m_3)]^2} \right\}, \tag{2.31}$$

showing that the addition of cosolutes without protein–cosolute attractions results in an increase in protein chemical potential.

Preferentially accumulated cosolutes $\left(\Gamma_{32}^{(\mu_3)} > 0 \text{ and } \left(\partial \mu_2^{ex}/\partial m_3\right)_{T,p,m_2} < 0\right)$ are therefore those for which protein–cosolute interactions are sufficiently favorable to overcome the excluded volume effects that would otherwise make all realistic cosolutes preferentially excluded, relative to water, at the protein surface. In the alternative limit of cases where protein–cosolute and cosolute–cosolute interactions greatly dominate over excluded volume effects (i.e., large a_{23} and a_{33}), Equations 2.27 and 2.28 simplify to

$$\Gamma_{32}^{(\mu_3)} \approx \frac{a_{23}}{a_{33}} \tag{2.32}$$

and

$$\left(\frac{\partial \mu_2^{ex}}{\partial m_3}\right)_{T,V,m_2 \to 0} = -2\rho_1 M_1 a_{23}. \tag{2.33}$$

Thus, in this "strongly interacting" limit the resulting expressions are rather simple and intuitive. Noting that a_{33} must be positive for realistic (neutral) cosolutes, the sign of $\Gamma_{32}^{(\mu_3)}$ is determined by a_{23}. Perhaps more importantly from the perspective of changes in μ_2^{ex} with added cosolute is that the sign and magnitude of $(\partial \mu_2^{ex}/\partial m_3)_{T,V,m_2 \to 0}$ are determined by a_{23}. Finally, a subtle but important point is that a_{23} and a_{33} depend implicitly on interactions with water (component 1); that is, they are not simply reflective of direct 2–3 and 3–3 interactions. This follows because a_{ij} parameters depend on ensemble averages over the degrees of freedom of all solvent components in the reference ensemble (i.e., via the RDF in Eq. 2.26).

2.6.2.2 Size Dependence of b_{23} and a_{23} The discussion in Section 2.6.1 highlighted that it is not the value of $\Gamma_{32}^{(\mu_3)}$ or $(\partial \mu_2^{ex}/\partial m_3)_{T,V,m_2 \to 0}$ for one particular protein state or species that determines experimentally observable behaviors such as cosolute effects on protein folding or aggregation. Rather, it is the change in $\Gamma_{32}^{(\mu_3)}$ or $(\partial \mu_2^{ex}/\partial m_3)_{T,V,m_2 \to 0}$ upon converting from one protein state (e.g., unfolded) to another (e.g., folded or associated). In terms of the simple perturbation theory illustrated here, this translates to changes in b_{23} and a_{23} with all solvent conditions held fixed. For example, consider the case of a change from F to U, with Equations 2.27 and 2.28 applied to both F and U. The change in $\Gamma_{32}^{(\mu_3)}$ upon unfolding $\left(\Delta\Gamma_{32}^{(\mu_3)} = \Gamma_{32,U}^{(\mu_3)} - \Gamma_{32,F}^{(\mu_3)}\right)$ is given by

$$\Delta\Gamma_{32}^{(\mu_3)} \approx -\frac{(k_B T \rho_1^2 b_{33} M_1 m_3)\Delta b_{23} - 2\rho_1 [1-\rho_1 (b_{13} + b_{33} M_1 m_3)]^2 \Delta a_{23}}{kT\rho_1^2 b_{33}^2 - 2\rho_1 a_{23} [1-\rho_1 (b_{13} + b_{33} M_1 m_3)]^2}, \tag{2.34}$$

and the derivative of the change in protein chemical potential with respect to m_3 is given by

$$\left(\frac{\partial \Delta\mu_2^{ex}}{\partial m_3}\right)_{T,V,m_2 \to 0} = -2\rho_1 M_1 \Delta a_{23} + k_B T M_1 \left\{\frac{\rho_1^2 b_{13} \Delta b_{12}}{[1-\rho_1 (b_{11} + b_{13} M_1 m_3)]^2} + \frac{(\rho_1^2 b_{33} M_1 m_3)\Delta b_{23}}{[1-\rho_1 (b_{13} + b_{33} M_1 m_3)]^2}\right\}. \tag{2.35}$$

Based on the definition of b_{ij} (Eq. 2.25), the value of Δb_{23} is given by $\Delta b_{23} = \int_{\sigma_{2,U}}^{\sigma_{2,F}} \frac{\pi}{4}(\sigma_2 + \sigma_3)^2 \, d\sigma_2$.

In the limit that the protein diameter is much larger than that of the cosolute, and the change in protein diameter ($\Delta\sigma_2 = \sigma_{2,U} - \sigma_{2,F}$) is small compared to the value for the folded state ($\sigma_{2,F}$), this expression reduces to

$$\Delta b_{23} \approx \frac{\pi}{4}(\sigma_{2,F})^2 \, \Delta\sigma_2 \tag{2.36}$$

and shows that (to a first approximation) the change in b_{23} is proportional to the change in molecular surface area.

Similarly, based on the definition of a_{23} (Eq. 2.26), the change in a_{23} upon unfolding is given by

$$\begin{aligned}\Delta a_{23} &= a_{23,U} - a_{23,F} \\ &= -2\pi\left[\int_{(\sigma_{2,U}+\sigma_3)/2}^{\infty} u'_{23}(r) g_{23}^{HS}(r) r^2 dr - \int_{(\sigma_{2,F}+\sigma_3)/2}^{\infty} u'_{23}(r) g_{23}^{HS}(r) r^2 dr\right].\end{aligned} \tag{2.37}$$

Again, we take the limit that $\sigma_{2,F} \gg \sigma_3$ and now also approximate that $\mu'_{23} \approx -\varepsilon_{23}$ if the cosolute is next to the surface of the protein (i.e., within some short distance δ); otherwise, $\mu'_{23} \approx 0$. This gives

$$\Delta a_{23} = a_{23,U} - a_{23,F} \approx \varepsilon_{23} 2\pi\delta\left[\sigma_{2,U}^2 - \sigma_{2,F}^2\right] \tag{2.38}$$

Equation 2.38 shows that, to a first approximation, Δa_{23} is also proportional to the change in molecular surface area of the protein upon unfolding. This result, along with Equation 2.36, illustrates the molecular basis for the empirical observation that changes in the unfolding "*m* value" is often proportional to the change in exposed surface area of unfolding.[82,83] Independent of the vdW model results, the observation that the magnitude, but not the sign, of $\Gamma_{32}^{(\mu_3)}$ and $\left(\partial\mu_2^{ex}/\partial m_3\right)_{T,V,m_2\to 0}$ is roughly proportional to surface area highlights that cosolutes that favor *F* over *U* for monomers also presumably favor aggregates or self-associated protein states over *U* monomers. For example, preferentially excluded cosolutes such as PEG have long been known to drive aggregation in the form of "salting out" of folded proteins.[84,85]

2.6.3 An Additive Approach to $\Delta\mu_2^{tr}$, $\partial\mu_2^{ex}/\partial m_3$, and $\Gamma_{32}^{(\mu_3)}$

Bolen and coworkers have developed a molecular thermodynamic approach to interpreting, and even potentially predicting, $\Delta\mu_{2,F}^{tr}$ as a function of cosolute identity and composition.[86–88] Using the nomenclature of Auton and Bolen,[86,87] the approach is based on empirical additivity of the transfer free energy values for the amide backbone (Δg_{tr}^{bb}) and for side chains ($\Delta g_{tr}^{sc,i}$) of individual amino acids (i is an index running across each type of amino acid). The values of Δg_{tr}^{bb} and $\Delta g_{tr}^{sc,i}$ are determined experimentally for the transfer from pure water to a 1M solution of a given cosolute, at fixed T and p. The contribution from each amino acid to the solvation free energy for a given protein conformational state (e.g., F and U) is based on the fraction of the amino acid's surface area that is solvent exposed in that conformational state. This fractional surface area is denoted as α_i^{sc} and α_i^{bb} (for the side-chain and backbone contributions) and is referenced versus the value for each amino acid in a synthetic polypeptide. Given these quantities, combining with Equations 2.17–2.19, and adding the simplifying approximation here that the transfer free energy value for each amino acid is linear in osmolyte concentration gives

$$\left(\frac{\partial \mu_{2,F}^{ex}}{\partial m_3}\right)_{T,p,m_2 \to 0} \approx \sum_{i \in \{a.a.\}} n_i \Delta g_{tr}^{sc,i} \alpha_i^{sc} + \Delta g_{tr}^{bb} \sum_{i \in \{a.a.\}} n_i \alpha_i^{bb}, \qquad (2.39)$$

with the indices on each summation running over each type of amino acid, and n_i denoting the number of that type of amino acid in the protein sequence. In writing Equation 2.3, we have used Δg_{tr}^{bb} and $\Delta g_{tr}^{sc,i}$ evaluated from solubility data on a molal scale.[86]

The values of α_i^{sc} and α_i^{bb} must be known or estimated from structural information on the protein. While this can be carried out robustly for the folded state of proteins with known three-dimensional structures, doing so for proteins without such detailed structural knowledge remains a phenomenological exercise. However, with judicious choices of algorithms to estimate the changes in solvent-accessible surface areas, reasonable quantitative and semi-quantitative predictions may be possible.[87]

Independent of exact knowledge of α_i^{sc} and α_i^{bb}, the results and approach of Auton and Bolen,[86,87] combined with additional results from Auton et al.,[89] can also provide useful insights regarding $\Gamma_{32}^{(\mu_3)}$ for each amino acid, the relative contributions from side chains versus backbone, and the relative magnitude of $\Gamma_{32}^{(\mu_3)}$ and $\partial \mu_2^{ex}/\partial m_3$. Applying Equation 2.17–2.19 and the above approximations to experimental values for Δg_{tr}^{bb} and $\Delta g_{tr}^{sc,i}$,[87,89] along with values of $(\partial \mu_3^{ex}/\partial c_3)_{T,p}$ for binary 1–3 mixtures,[89] one can obtain values for the preferential interaction parameter of individual amino acid side chains ($\Gamma_{32}^{sc,i}$) and for the backbone (Γ_{32}^{bb}) using

$$\Gamma_{32}^{sc,i} \cong -\frac{-\Delta g_{tr}^{sc,i}}{(\partial \mu_3^{ex}/\partial c_3)_{T,p}} \qquad (2.40)$$

and

$$\Gamma_{32}^{bb} \cong -\frac{\Delta g_{tr}^{bb}}{(\partial \mu_3^{ex}/\partial c_3)_{T,p}}. \qquad (2.41)$$

In writing Equations 2.40 and 2.41 we have implicitly switched the transfer free energy values to the molar scale to be consistent with the treatment and data of Auton et al.[89]

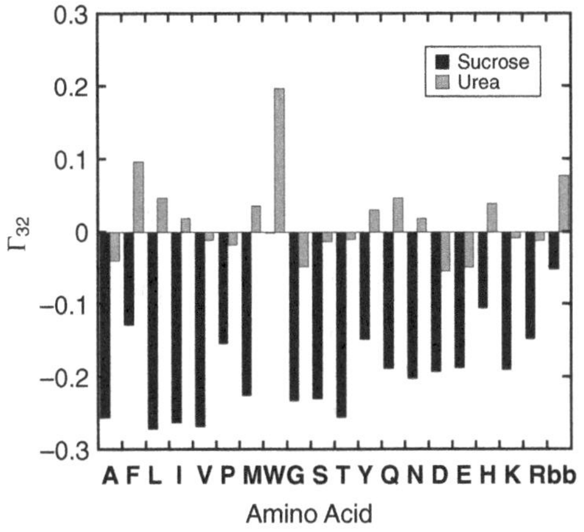

Figure 2.4. Preferential interaction parameter values between different cosolutes (sucrose = black, urea = gray) and each amino acid side chain or the peptide backbone (labeled as *bb* on the far right). Values were calculated using Equations 2.40 and 2.41, with data taken from Auton et al.[89]

Figure 2.4 shows the values of the preferential interaction parameter for each amino acid side chain (labeled with its single-letter amino acid code) and for the amide backbone (labeled *bb*), based on the data in Auton et al.[89] for sucrose solutions and for urea solutions at 25°C and physiological pH. The results show the expected trend for sucrose, in that $\Gamma_{32}^{sc,i}$ is negative (preferential exclusion of sucrose) for all amino acid side chains and for the backbone. The backbone has the smallest $\Gamma_{32}^{sc,i}$ value, with the exception of Trp. It is not possible to determine conclusively why Trp has a $\Gamma_{32}^{sc,i}$ value so close to zero for sucrose, but considering the relatively large size of sucrose and Trp (i.e., large b_{ij} in the vdW model in Section 2.6.2), this indicates significantly favorable sucrose–Trp interactions (i.e., $a_{23} \gg 0$ in Section 2.6.2). Assuming that the unfolded state has more solvent-accessible side chains than the folded state or an oligomer, this leads to the expected result that sucrose should act as a nonspecific stabilizer for folding and should also promote self-association.

For urea, the backbone instead shows the largest magnitude for $\Gamma_{32}^{sc,i}$, when compared to all side chains except for Trp and Phe. The backbone, along with a number of side chains, shows significantly positive values for $\Gamma_{32}^{sc,i}$, hence its canonical behavior as a preferentially accumulated cosolute. Given that the unfolded state has more side chains and backbone exposed to solvent than either the folded or aggregated states, this provides a consistent molecular viewpoint as to why adding urea leads to protein unfolding and, in some cases, aggregate dissociation. The fact that the backbone contribution is relatively large is also consistent with its relatively ubiquitous (i.e., nonspecific) ability to serve as a chemical denaturant for proteins.[88,89] It is notable that none of the $\Gamma_{32}^{sc,i}$ values are very large (all lie between ±0.3), which shows that these cosolutes are not strongly or specifically "bound," and thus for them to have large effects on the chemical potential of a given solute (component 2) they must be present at relatively high concentrations and/or be acting on relatively large molecules (e.g., proteins).

To put the results in terms of free energy per amino acid transferred from water to a 1 M cosolute–water mixture, each of the Γ_{32} values in Fig. 2.4 must be multiplied by the value of $(\partial\mu_3^{ex}/\partial c_3)_{T,p}$ for the water–cosolvent mixture (cf. Eq. 2.20). The values for $(\partial\mu_3^{ex}/\partial c_3)_{T,p}$ under typical conditions are less than 1 kcal/mol for transfer to a 1 M cosolute solution.[89] Therefore, the transfer free energies per amino acid are on the order of 10–10^2 cal/mol. Thus, it is the additive effect across multiple amino acids and the balance between favorable and unfavorable transfer of different amino acids in a given protein sequence that ultimately give rise to measurable effects (on the order of kilocalories per mole).

Finally, using the methods explained in Reference 89, it is possible to further separate the preferential interaction parameters into their individual KB integrals (i.e., G_{12} and G_{23} in Eq. 2.21). A particularly instructive result from such an inverse-KB approach is to consider G_{12} and G_{23} for the peptide backbone for a range of different cosolutes. Figure 2.5 (adapted from Reference 89) shows such a comparison for a number of cosolutes, including canonical formulation excipients such as sucrose and sorbitol, as well as for urea and other natural osmolytes that help to mediate osmotic pressure and solvation within biological systems.

To orient the reader, Fig. 2.5 is a plot of G_{12} versus G_{23} and places a single point where this pair of values lies for the peptide backbone, for a given choice

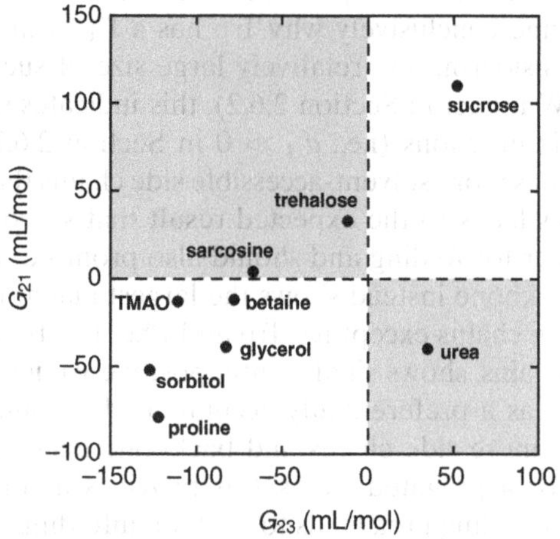

Figure 2.5. Comparison of KB integrals for water–peptide (G_{12}) and cosolute–peptide (G_{23}) for the peptide backbone and a range of naturally occurring, neutral cosolutes (osmolytes). See main text for additional details and discussion. Data values are from Reference 89.

of cosolute. Points that lie in the upper left quadrant fit the classic picture of preferential exclusion of a cosolute, in that $G_{23} < 0$ (solute is excluded from the protein surface) and $G_{12} > 0$ (water is accumulated at the protein surface). All such cases necessarily have negative $\Gamma_{32}^{sc,i}$ values. Points that lie in the lower right corner show the expected behavior of a classic, preferentially accumulated cosolute. G_{23} is positive (cosolute is accumulated), while G_{12} is negative (water is excluded or depleted) from the protein surface. All points in that quadrant necessarily have positive $\Gamma_{32}^{sc,i}$ values. Points that lie in the other two quadrants indicate cases where both water and cosolute are either excluded or accumulated to some degree near the protein surface. In these cases, the net result for $\Gamma_{32}^{sc,i}$ being positive or negative depends on which solvent species is excluded or accumulated to a greater degree.

Of the osmolytes investigated by Auton et al.,[89] only trehalose clearly shows classic behavior that is typically assumed for preferentially excluded cosolutes. Only urea shows classic behavior for a preferentially accumulated solute (i.e., a denaturant). Most of the other cosolutes are excluded from the protein surface to some degree, with water also excluded. This is perhaps not surprising because the steric contributions to G_{ij} are always negative for any physically plausible solvent or solute species. Sucrose is quite interesting, as it clearly lies in the upper right quadrant. Physically, this means that sucrose is accumulated near the peptide backbone, in contrast to what is historically assumed or argued regarding the effect of sucrose on protein folding thermodynamics.[90–92] However, sucrose ultimately has a positive Γ_{32}^{bb} because water is accumulated to an even greater extent, and thus it is correct to state that sucrose is preferentially excluded; however, it appears to be incorrect to assume it has a lower surface concentration than in the bulk solution. KB theory and KB integrals cannot indicate conclusively why such a result occurs, but it was speculated[89] that this seemingly counterintuitive result was because favorable sucrose–backbone interactions, combined with sucrose having strong favorable interactions with water, that help water to accumulate near the protein surface to an even greater degree. Such a situation again highlights that molecular interpretation of cosolvent effects on protein thermodynamics inherently involves a balance of interactions between all components in the solution.

2.6.4 Predicting Cosolvent/Cosolute Effects on Unfolding, Association, and Aggregation

From the above discussion, it is clear that available theory and most current approaches for understanding the effects of added cosolutes (excipients) on protein unfolding and self-association are most useful for rationalizing, in hindsight, why a given excipient increases or decreases the free energy for a given process. Only the additive transfer free energy (ATFG) results of Bolen and coworkers appear to have been reasonably successful in quantitatively predicting the effects of added cosolutes for protein unfolding. Even for that ATFG approach, there are a number of current limitations.[88,89] These include the following: one must know the three-dimensional structure of the folded state; estimates for the three-dimensional structure of the unfolded state are currently empirical or at best phenomenological; extending to excipients other than those tested requires a relatively large amount of experimental effort for each new excipient; results thus far have been confined to neutral osmolytes; and it remains unclear how well the additivity and weak binding assumptions will hold for excipients such as surfactants, different Hofmeister salts and polyelectrolytes, and other polymeric excipients that may be of interest in drug delivery vehicles.

In addition, all of the results to date have been tested extensively only for conditions that are best considered "dilute" solutions when compared, for example, to the conditions encountered during freeze- or spray-drying of biopharmaceutical products. It is not even clear at this point how well one should expect the high-concentration solvent conditions encountered during those unit operations to follow equilibrium behavior (see also Chapter 9). Finally, the ideas of preferential accumulation of water and/or cosolutes have, until recently, received much less attention in direct applications to understanding and predicting their effects on weak self-association and/or on aggregation of proteins. This is due in part to difficulties with reliably measuring equilibrium constants for weakly self-associating systems, as well as the inherent difficulties in interpreting competing effects of excluded volume and non-steric interactions between protein, water, and cosolute on measurable quantities such as virial coefficients.[41,93,94]

APPENDIX—DERIVATION OF Γ_{32} FOR VAN DER WAALS (vdW) MIXTURE

The vdW treatment of fluids can be considered as a simple but qualitatively and semiquantitatively effective perturbation theory that captures a balance of steric repulsions between molecules and short-ranged attractions.[81] We begin with the canonical partition function (Q) for a vdW fluid composed of three components ($Q(N_1, N_2, N_3, V, T)$); N_i = number of molecules of species i, V = volume, and T = absolute temperature. The derivation is an adaptation of that for the single-component vdW fluid,[81] with details included here to allow one to better see the source of excluded volume effects versus direct attractions or non-steric repulsions between different species. Using the nomenclature adopted earlier, 1 = water, 2 = protein, and 3 = cosolvent or cosolute, the resulting partition function is given by

$$Q(N_1, N_2, N_3, V, T) = Q_{HS}(N_1, N_2, N_3, V, T) \times \exp\left[\frac{-\langle U' \rangle_0}{k_B T}\right], \quad (2.A1)$$

in which k_B is Boltzmann's constant and $Q_{HS}(N_1, N_2, N_3, V, T)$ is the corresponding canonical partition function for the three-component mixture of hard spheres with each component having an effective HS diameter σ_i ($i = 1, 2, 3$). The HS term accounts for all steric interactions between molecules and thus accounts for all excluded volume effects in the results below.

The second term on the right-hand side of Equation 2.A1 accounts for direct attractions and (non-steric) repulsions that contribute to the potential energy (U) for the system. The prime on U denotes that steric repulsions have been subtracted out. The brackets $\langle \cdots \rangle_0$ in Equation 2.A1 indicate an ensemble average within the HS mixture reference state at the temperature and composition dictated by the values of (N_1, N_2, N_3, V, T). Using the standard approximation[81] for $Q_{HS}(N_1, N_2, N_3, V, T)$ gives

$$Q(N_1, N_2, N_3, V, T) = \frac{1}{\prod_{i=1}^{3} N_i!\Lambda_i^{3N_i}} \prod_{i=1}^{3}\left(V - \sum_{j=1}^{3} N_j b_{ij}\right)^{N_i} \exp\left[\frac{-\langle U' \rangle_0}{k_B T}\right], \quad (2.A2)$$

in which Λ_i is a generalized deBroglie wavelength for species i, and b_{ij} accounts for the excluded volume of species j that species i "sees" when one considers an isolated i–j pair of molecules (and vice versa; $b_{ij} = b_{ji}$). b_{ij} was defined in Equation 2.25.

Equation 2.A2 is accurate at low to moderate number densities and is sufficient to illustrate the physical results in Section 2.6 without adding unnecessary mathematical complexity. More accurate treatments of the excluded volume contributions at high number densities are available, provided one considers only relatively simple molecule or particle geometries.[95,96]

The contribution from non-steric interactions is given by $\langle U' \rangle_0$. With the typical assumption of pairwise additive interactions between molecules, this can be expressed as[81]

$$\langle U' \rangle_0 = \frac{1}{2V} \sum_{i=1}^{3} \sum_{j=1}^{3} N_i N_j \int_{\sigma_{i,j}}^{\infty} 4\pi r^2 g_{ij}^{HS}(r) u'_{ij}(r)\, dr. \quad (2.A5)$$

In Equation 2.A5, $u'_{ij}(r)$ is the value of the orientationally averaged interaction energy (with steric repulsions subtracted out) between a molecule of i and a molecule of j at a center-to-center distance r.

The standard definition of the vdW a parameters follows from rearranging Equation 2.A5 to read

$$\langle U' \rangle_0 = -\frac{1}{V} \sum_{i=1}^{3} \sum_{j=1}^{3} N_i N_j a_{ij}, \quad (2.A6)$$

with a_{ij} defined by Equation 2.26.

The grand canonical ensemble (Ξ) for the system at fixed temperature (T), volume (V), and chemical potentials of each component (μ_i, $i = 1, 2, 3$) is defined as

$$\Xi(\mu_1, \mu_2, \mu_3, V, T) = \sum_{N_1} \sum_{N_2} \sum_{N_3} e^{\left(\sum_{i=1}^{3} N_i \mu_i\right) k_B T} Q(N_1, N_2, N_3, V, T). \quad (2.A7)$$

Using the identity $p = V^{-1}k_{B}T\ln\Xi$ and substituting Equations 2.A2–2.A6 in Equation 2.A7, and simplifying the algebra gives

$$\frac{p}{kT}=\frac{\sum_{i=1}^{3}\mu_i^{ex}\rho_i}{k_BT}-\sum_{i=1}^{3}\rho_i\ln\rho_i+\sum_{i=1}^{3}\rho_i\ln\left(1-\sum_{j=1}^{3}b_{ij}\rho_j\right)+\sum_{i=1}^{3}\rho_i+\frac{\sum_{i=1}^{3}\sum_{j=1}^{3}\rho_i\rho_j a_{ij}}{k_BT} \tag{2.A8}$$

In the above expression, $\rho_i = N_i/V$, and $\mu_i^{ex}=\mu_i - k_BT\ln\Lambda_i^3$ is the excess chemical potential of component i. The equilibrium values for chemical potentials and pressure are obtained by solving $(\partial p/\partial\rho_i)_{T,V,\mu_j,\rho_{k\neq i}}=0$ simultaneously for each i = 1, 2, 3. Doing this for i = 2 and for i = 3 gives (after some rearrangement)

$$\frac{\mu_2^{ex}}{k_BT}=\ln\rho_2+\frac{b_{12}\rho_1}{1-\sum_{j=1}^{3}b_{1j}\rho_j}+\frac{b_{22}\rho_2}{1-\sum_{j=1}^{3}b_{2j}\rho_j}+\frac{b_{23}\rho_3}{1-\sum_{j=1}^{3}b_{3j}\rho_j}-\ln\left(1-\sum_{j=1}^{3}b_{2j}\rho_j\right)-2\sum_{j=1}^{3}\frac{\rho_j a_{2j}}{k_BT} \tag{2.A9}$$

and

$$\frac{\mu_3^{ex}}{k_BT}=\ln\rho_3+\frac{b_{13}\rho_1}{1-\sum_{j=1}^{3}b_{1j}\rho_j}+\frac{b_{23}\rho_2}{1-\sum_{j=1}^{3}b_{2j}\rho_j}+\frac{b_{33}\rho_3}{1-\sum_{j=1}^{3}b_{3j}\rho_j}-\ln\left(1-\sum_{j=1}^{3}b_{3j}\rho_j\right)-2\sum_{j=1}^{3}\frac{\rho_j a_{3j}}{k_BT}. \tag{2.A10}$$

The effects of changing cosolvent composition or molality (m_3) in the limit of low concentration of protein ($m_2 \to 0$) then follows (after some algebra) as

$$k_BTM_1\left\{\frac{b_{23}\rho_1}{1-\rho_1(b_{12}+b_{23}M_1m_3)}+\frac{b_{23}\rho_1}{1-\rho_1(b_{13}+b_{33}M_1m_3)}\right\}-2M_1\rho_1a_{23} \tag{2.A11}$$

and

$$\left(\frac{\partial\mu_3^{ex}}{\partial m_3}\right)_{T,V,m_2\to0}=k_BTM_1\left\{\frac{1}{M_1m_3}+\frac{b_{13}^2\rho_1^2}{[1-\rho_1(b_{11}+b_{13}M_1m_3)]^2}+\frac{b_{33}^2\rho_1^2M_1m_3}{[1-\rho_1(b_{13}+b_{33}M_1m_3)]^2}\right\}+k_BTM_1\left\{\frac{2b_{33}\rho_1}{1-\rho_1(b_{13}+b_{33}M_1m_3)}\right\}-2M_1\rho_1a_{33}, \tag{2.A12}$$

with M_1 denoting the molecular weight of water (units of kilogram per mole). $\Gamma_{32}^{(\mu_3)}$ can be defined formally as[71,76]

$$\Gamma_{32}^{(\mu_3)}=\frac{(\partial\mu_2^{ex}/\partial m_3)_{T,p,m_2}}{(\partial\mu_3^{ex}/\partial m_3)_{T,p,m_2}}=-\frac{(\partial\mu_2^{ex}/\partial m_3)_{T,V,m_2}-\frac{x_1M_1\bar{V}_2\bar{V}_3}{\underline{V}_{mix}\kappa_T}}{(\partial\mu_3^{ex}/\partial m_3)_{T,V,m_2}-\frac{x_1M_1\bar{V}_3^2}{\underline{V}_{mix}\kappa_T}}, \tag{2.A13}$$

with x_i denoting the mole fraction of component i. The second equality follows from transforming from derivatives at constant pressure to derivatives at constant volume, with $\bar{V}_i$ = partial molar volume of component i, and with κ_T and $\underline{V}_{mix}$ denoting, respectively, the isothermal compressibility and molar volume of the mixture. For illustration purposes, we approximate that $\left(\partial\mu_i^{ex}/\partial m_3\right)_{T,p,m_2} \approx \left(\partial\mu_i^{ex}/\partial m_3\right)_{T,V,m_2}$ for $i = 2, 3$. Doing so, and keeping only the largest terms in Equations 2.A11 and 2.A12 when substituting into Equation 2.A13 gives Equations 2.27 and 2.28.

ACKNOWLEDGMENTS

T. M. Young is thanked for assistance in constructing Fig. 2.2. Financial support from Merck & Co. (YL) is gratefully acknowledged.

REFERENCES

1. Chi EY, Krishnan S, Kendrick BS, Chang BS, Carpenter JF, Randolph TW. 2003. Roles of conformational stability and colloidal stability in the aggregation of recombinant human granulocyte colony-stimulating factor. Protein Sci 12(5): 903–913.
2. Fink AL. 1998. Protein aggregation: Folding aggregates, inclusion bodies and amyloid. Fold Des 3(1):R9–R23.
3. Goers J, Permyakov SE, Permyakov EA, Uversky VN, Fink AL. 2002. Conformational prerequisites for a-lactalbumin fibrillation. Biochemistry 41(41):12546–12551.
4. Grillo AO, Edwards KL, Kashi RS, Shipley KM, Hu L, Besman MJ, Middaugh CR. 2001. Conformational origin of the aggregation of recombinant human factor VIII. Biochemistry 40(2):586–595.
5. Khurana R, Gillespie JR, Talapatra A, Minert LJ, Ionescu-Zanetti C, Millett I, Fink AL. 2001. Partially folded intermediates as critical precursors of light chain amyloid fibrils and amorphous aggregates. Biochemistry 40(12):3525–3535.
6. Raso SW, Abel J, Barnes JM, Maloney KM, Pipes G, Treuheit MJ, King J, Brems DN. 2005. Aggregation of granulocyte-colony stimulating factor in vitro involves a conformationally altered monomeric state. Protein Sci 14(9):2246–2257.
7. Wang W. 2005. Protein aggregation and its inhibition in biopharmaceutics. Int J Pharm 289(1–2):1–30.
8. Weiss WFI, Young TM, Roberts CJ. 2009. Principles, approaches, and challenges for predicting protein aggregation rates and shelf life. J Pharm Sci 98:1246–1277.
9. Roberts CJ. 2007. Non-native protein aggregation kinetics. Biotechnol Bioeng 98(5):927–938.
10. Binger KJ, Griffin MD, Howlett GJ. 2008. Methionine oxidation inhibits assembly and promotes disassembly of apolipoprotein C-II amyloid fibrils. Biochemistry 47(38):10208–10217.

11. Flaugh SL, Mills IA, King J. 2006. Glutamine deamidation destabilizes human gammaD-crystallin and lowers the kinetic barrier to unfolding. J Biol Chem 281(41):30782–30793.
12. Famm K, Hansen L, Christ D, Winter G. 2008. Thermodynamically stable aggregation-resistant antibody domains through directed evolution. J Mol Biol 376(4): 926–931.
13. Nieba L, Honegger A, Krebber C, Pluckthun A. 1997. Disrupting the hydrophobic patches at the antibody variable/constant domain interface: Improved in vivo folding and physical characterization of an engineered scFv fragment. Protein Eng 10(4):435–444.
14. Worn A, Pluckthun A. 2001. Stability engineering of antibody single-chain Fv fragments. J Mol Biol 305(5):989–1010.
15. Wiseman RL, Powers ET, Kelly JW. 2005. Partitioning conformational intermediates between competing refolding and aggregation pathways: Insights into transthyretin amyloid disease. Biochemistry 44(50):16612–16623.
16. Gage MJ, Lefebvre BG, Robinson AS. 2006. Determinants of protein folding and aggregation in P22 tailspike. In Murphy RM, Amos T, eds., *Misbehaving Proteins*. New York: Springer.
17. Gage MJ, Robinson AS. 2003. C-terminal hydrophobic interactions play a critical role in oligomeric assembly of the P22 tailspike trimer. Protein Sci 12(12):2732–2747.
18. Robinson CR, Rentzeperis D, Silva JL, Sauer RT. 1997. Formation of a denatured dimer limits the thermal stability of Arc repressor. J Mol Biol 273(3):692–700.
19. Young TM, Roberts CJ. 2007. A quasichemical approach for protein-cluster free energies in dilute solution. J Chem Phys 127:165101/165101–165101/165110.
20. Andrews JM, Roberts CJ. 2007. A Lumry-Eyring nucleated polymerization model of protein aggregation kinetics: 1. Aggregation with pre-equilibrated unfolding. J Phys Chem B 111(27):7897–7913.
21. Bajaj H, Sharma VK, Kalonia DS. 2007. A high-throughput method for detection of protein self-association and second virial coefficient using size-exclusion chromatography through simultaneous measurement of concentration and scattered light intensity. Pharm Res 24(11):2071–2083.
22. Finke JM, Roy M, Zimm BH, Jennings PA. 2000. Aggregation events occur prior to stable intermediate formation during refolding of interleukin 1b. Biochemistry 39(3):575–583.
23. Ferrone FA, Hofrichter J, Eaton WA. 1985. Kinetics of sickle hemoglobin polymerization. II. A double nucleation mechanism. J Mol Biol 183(4):611–631.
24. Ferrone FA, Hofrichter J, Eaton WA. 1985. Kinetics of sickle hemoglobin polymerization. I. Studies using temperature-jump and laser photolysis techniques. J Mol Biol 183(4):591–610.
25. Ho MR, Lou YC, Lin WC, Lyu PC, Huang WN, Chen C. 2006. Human pancreatitis-associated protein forms fibrillar aggregates with a native-like conformation. J Biol Chem 281(44):33566–33576.
26. Andrews JM, Roberts CJ. 2007. Non-native aggregation of alpha-chymotrypsinogen occurs through nucleation and growth with competing nucleus sizes and negative activation energies. Biochemistry 46(25):7558–7571.

27. Speed MA, King J, Wang DIC. 1997. Polymerization mechanism of polypeptide chain aggregation. Biotechnol Bioeng 54(4):333–343.

28. Pallitto MM, Murphy RM. 2001. A mathematical model of the kinetics of beta-amyloid fibril growth from the denatured state. Biophys J 81(3):1805–1822.

29. Roberts CJ, Darrington RT, Whitley MB. 2003. Irreversible aggregation of recombinant bovine granulocyte-colony stimulating factor (bG-CSF) and implications for predicting protein shelf life. J Pharm Sci 92(5):1095–1111.

30. Lomakin A, Teplow DB, Kirschner DA, Benedek GB. 1997. Kinetic theory of fibrillogenesis of amyloid beta-protein. Proc Natl Acad Sci U S A 94(15):7942–7947.

31. Top A, Kiick KL, Roberts CJ. 2008. Modulation of self-association and subsequent fibril formation in an alanine-rich helical polypeptide. Biomacromol 9(6):1595–1603.

32. Li Y, Roberts CJ. 2009. A Lumry-Eyring nucleated-polymerization model of protein aggregation kinetics: 2. Competition between chain- and condensation-polymerization. J Phys Chem B 113:7020–7032.

33. Roberts CJ. 2006. Non-native protein aggregation: Pathways, kinetics, and shelf-life prediction. In Murphy RM, Tsai A, eds., *Misbehaving Proteins: Protein (Mis) Folding, Aggregation, and Stability*. New York: Springer, pp. 17–46.

34. Saluja A, Kalonia DS. 2008. Nature and consequences of protein-protein interactions in high protein concentration solutions. Int J Pharm 358(1–2):1–15.

35. Jimenez M, Rivas G, Minton AP. 2007. Quantitative characterization of weak self-association in concentrated solutions of immunoglobulin G via the measurement of sedimentation equilibrium and osmotic pressure. Biochemistry 46(28):8373–8378.

36. Minton AP. 2008. Effective hard particle model for the osmotic pressure of highly concentrated binary protein solutions. Biophys J 94(7):L57–L59.

37. Young TM, Roberts CJ. 2009. Structure and thermodynamics of colloidal protein cluster formation: Comparison of square-well and simple dipolar models. J Chem Phys 131:125104/1–9.

38. Bajaj H, Sharma VK, Kalonia DS. 2004. Determination of second virial coefficient of proteins using a dual-detector cell for simultaneous measurement of scattered light intensity and concentration in SEC-HPLC. Biophys J 87(6):4048–4055.

39. Chi EY, Krishnan S, Randolph TW, Carpenter JF. 2003. Physical stability of proteins in aqueous solution: Mechanism and driving forces in nonnative protein aggregation. Pharm Res 20(9):1325–1336.

40. Tessier PM, Lenhoff AM, Sandler SI. 2002. Rapid measurement of protein osmotic second virial coefficients by self-interaction chromatography. Biophys J 82(3):1620–1631.

41. Dumetz AC, Snellinger-O'brien A M, Kaler EW, Lenhoff AM. 2007. Patterns of protein protein interactions in salt solutions and implications for protein crystallization. Protein Sci 16(9):1867–1877.

42. Velev OD, Kaler EW, Lenhoff AM. 1998. Protein interactions in solution characterized by light and neutron scattering: Comparison of lysozyme and chymotrypsinogen. Biophys J 75(6):2682–2697.

43. Asthagiri D, Paliwal A, Abras D, Lenhoff AM, Paulaitis ME. 2005. A consistent experimental and modeling approach to light-scattering studies of protein-protein interactions in solution. Biophys J 88(5):3300–3309.
44. Cheung JK, Truskett TM. 2005. Coarse-grained strategy for modeling protein stability in concentrated solutions. Biophys J 89(4):2372–2384.
45. Amzel LM. 1997. Loss of translational entropy in binding, folding, and catalysis. Proteins 28(2):144–149.
46. Alford JR, Kendrick BS, Carpenter JF, Randolph TW. 2008. Measurement of the second osmotic virial coefficient for protein solutions exhibiting monomer-dimer equilibrium. Anal Biochem 377(2):128–133.
47. Marky LA, Breslauer KJ. 1987. Calculating thermodynamic data for transitions of any molecularity from equilibrium melting curves. Biopolymers 26(9):1601–1620.
48. Becktel WJ, Schellman JA. 1987. Protein stability curves. Biopolymers 26(11): 1859–1877.
49. Privalov PL. 1979. Stability of proteins: Small globular proteins. Adv Protein Chem 33:167–241.
50. Vermeer AW, Norde W. 2000. The thermal stability of immunoglobulin: Unfolding and aggregation of a multi-domain protein. Biophys J 78(1):394–404.
51. Harper JD, Lansbury PT Jr. 1997. Models of amyloid seeding in Alzheimer's disease and scrapie: Mechanistic truths and physiological consequences of the time-dependent solubility of amyloid proteins. Annu Rev Biochem 66:385–407.
52. Lee CC, Walters RH, Murphy RM. 2007. Reconsidering the mechanism of polyglutamine peptide aggregation. Biochemistry 46(44):12810–12820.
53. Roberts CJ. 2003. Kinetics of irreversible protein aggregation: Analysis of extended Lumry-Eyring models and implications for predicting protein shelf life. J Phys Chem B 107(5):1194–1207.
54. Weiss WF IV, Hodgdon TK, Kaler EW, Lenhoff AM, Roberts CJ. 2007. Nonnative protein polymers: Structure, morphology, and relation to nucleation and growth. Biophys J 93(12):4392–4403.
55. Bartkowski R, Kitchel R, Peckham N, Margulis L. 2002. Aggregation of recombinant bovine granulocyte colony stimulating factor in solution. J Protein Chem 21(3):137–143.
56. Bajaj H, Sharma VK, Badkar A, Zeng D, Nema S, Kalonia DS. 2006. Protein structural conformation and not second virial coefficient relates to long-term irreversible aggregation of a monoclonal antibody and ovalbumin in solution. Pharm Res 23(6):1382–1394.
57. Mulkerrin MG, Wetzel R. 1989. pH dependence of the reversible and irreversible thermal denaturation of gamma interferons. Biochemistry 28(16):6556–6561.
58. Li Y, Weiss WF IV, Roberts CJ. 2009. Characterization of high molecular-weight nonnative aggregates and aggregation kinetics by size exclusion chromatography with inline multi-angle laser light scattering. J Pharm Sci 98:3997–4016.
59. Andrews JM, Weiss WF IV, Roberts CJ. 2008. Nucleation, growth, and activation energies for seeded and unseeded aggregation of alpha-chymotrypsinogen A. Biochemistry 47(8):2397–2403.
60. Trovato A, Seno F, Tosatto SCE. 2007. The PASTA server for protein aggregation prediction. Protein Eng Des Sel 20(10):521–523.

61. Li Y, Ogunnaike BA, Roberts CJ. 2010. Multi-variate approach to global protein aggregation behavior and kinetics: Effects of pH, NaCl, and temperature for α-chymotrypsinogen A. J Pharm Sci 99:645–662.
62. Jones LS, Kaufmann A, Middaugh CR. 2005. Silicone oil induced aggregation of proteins. J Pharm Sci 94(4):918–927.
63. Sluzky V, Tamada JA, Klibanov AM, Langer R. 1991. Kinetics of insulin aggregation in aqueous solutions upon agitation in the presence of hydrophobic surfaces. Proc Natl Acad Sci U S A 88(21):9377–9381.
64. Frokjaer S, Otzen DE. 2005. Protein drug stability: A formulation challenge. Nat Rev Drug Discov 4(4):298–306.
65. Roefs SP, de Kruif KG. 1994. A model for the denaturation and aggregation of beta-lactoglobulin. Eur J Biochem 226(3):883–889.
66. Conchillo-Sole O, de Groot NS, Aviles FX, Vendrell J, Daura X, Ventura S. 2007. AGGRESCAN: A server for the prediction and evaluation of "hot spots" of aggregation in polypeptides. BMC Bioinformatics 8:65.
67. Gsponer J, Vendruscolo M. 2006. Theoretical approaches to protein aggregation. Protein Pept Lett 13(3):287–293.
68. Galzitskaya OV, Garbuzynskiy SO, Lobanov MY. 2006. Prediction of amyloidogenic and disordered regions in protein chains. PLoS Comput Biol 2(12):1639–1648.
69. Caflisch A. 2006. Computational models for the prediction of polypeptide aggregation propensity. Curr Opin Chem Biol 10(5):437–444.
70. Fernandez-Escamilla AM, Rousseau F, Schymkowitz J, Serrano L. 2004. Prediction of sequence-dependent and mutational effects on the aggregation of peptides and proteins. Nat Biotechnol 22(10):1302–1306.
71. Anderson CF, Courtenay ES, Record MTJ. 2002. Thermodynamic expressions relating different types of preferential interaction coefficients in solutions containing two solute components. J Phys Chem B 106:418–433.
72. Baynes BM, Trout BL. 2004. Rational design of solution additives for the prevention of protein aggregation. Biophys J 87(3):1631–1639.
73. Timasheff SN. 1993. The control of protein stability and association by weak interactions with water: How do solvents affect these processes? Annu Rev Biophys Biomol Struct 22:67–97.
74. Zhou HX, Rivas G, Minton AP. 2008. Macromolecular crowding and confinement: Biochemical, biophysical, and potential physiological consequences. Annu Rev Biophys 37:375–397.
75. Scatchard G. 1946. Physical chemistry of protein solutions. I. Derivation of the equations for the osmotic pressure. J Am Chem Soc 68:2315–2319.
76. Ben-Naim A. 1992. *Statistical Thermodynamics for Chemists and Biochemists*. New York: Plenum Press.
77. Kirkwood JG, Buff FP. 1951. The statistical mechanical theory of solutions. I. J Chem Phys 19:774–777.
78. Rösgen J, Pettitt BM, Bolen DW. 2005. Protein folding, stability, and solvation structure in osmolyte solutions. Biophys J 89(5):2988–2997.
79. Shulgin IL, Ruckenstein E. 2008. Local composition in solvent + polymer or biopolymer systems. J Phys Chem B 112(10):3005–3012.

80. Debendetti PG. 1996. *Metastable Liquids: Concepts and Principles*. Princeton, NJ: Princeton University Press.
81. McQuarrie DA. 2000. *Statistical Mechanics*. Sausalito, CA: University Science Books.
82. Itzhaki LS, Neira JL, Fersht AR. 1997. Hydrogen exchange in chymotrypsin inhibitor 2 probed by denaturants and temperature. J Mol Biol 270(1):89–98.
83. Myers JK, Pace CN, Scholtz JM. 1995. Denaturant m values and heat capacity changes: Relation to changes in accessible surface area of protein unfolding. Protein Sci 4(10):2138–2148.
84. Arakawa T, Timasheff SN. 1985. Mechanism of poly(ethylene glycol) interaction with proteins. Biochemistry 24(24):6756–6762.
85. Middaugh CR, Tisel WA, Haire RN, Rosenberg A. 1979. Determination of the apparent thermodynamic activities of saturated protein solutions. J Biol Chem 254(2):367–370.
86. Auton M, Bolen DW. 2004. Additive transfer free energies of the peptide backbone unit that are independent of the model compound and the choice of concentration scale. Biochemistry 43(5):1329–1342.
87. Auton M, Bolen DW. 2005. Predicting the energetics of osmolyte-induced protein folding/unfolding. Proc Natl Acad Sci U S A 102(42):15065–15068.
88. Auton M, Holthauzen LM, Bolen DW. 2007. Anatomy of energetic changes accompanying urea-induced protein denaturation. Proc Natl Acad Sci USA 104(39):15317–15322.
89. Auton M, Bolen DW, Rosgen J. 2008. Structural thermodynamics of protein preferential solvation: Osmolyte solvation of proteins, aminoacids, and peptides. Proteins 73(4):802–813.
90. Kendrick BS, Chang BS, Arakawa T, Peterson B, Randolph TW, Manning MC, Carpenter JF. 1997. Preferential exclusion of sucrose from recombinant interleukin-1 receptor antagonist: Role in restricted conformational mobility and compaction of native state. Proc Natl Acad Sci USA 94(22):11917–11922.
91. Lee JC, Timasheff SN. 1981. The stabilization of proteins by sucrose. J Biol Chem 256(14):7193–7201.
92. Liu Y, Bolen DW. 1995. The peptide backbone plays a dominant role in protein stabilization by naturally occurring osmolytes. Biochemistry 34(39):12884–12891.
93. Liu W, Cellmer T, Keerl D, Prausnitz JM, Blanch HW. 2005. Interactions of lysozyme in guanidinium chloride solutions from static and dynamic light-scattering measurements. Biotechnol Bioeng 90(4):482–490.
94. Weatherly GT, Pielak GJ. 2001. Second virial coefficients as a measure of protein–osmolyte interactions. Protein Sci 10(1):12–16.
95. Berg OG. 1990. The influence of macromolecular crowding on thermodynamic activity: Solubility and dimerization constants for spherical and dumbbell-shaped molecules in a hard-sphere mixture. Biopolymers 30(11–12):1027–1037.
96. Boublik T. 1974. Statistical thermodynamics of convex molecule fluids. Mol Phys 27(5):1415–1427.

CHAPTER 3

Identification and Impact of Aggregation-Prone Regions in Proteins and Therapeutic Monoclonal Antibodies

SANDEEP KUMAR, XIAOLING WANG, and SATISH K. SINGH

The formation of protein aggregates can be considered a natural phenomenon that is influenced by the environment in which the proteins are maintained and by their composition. A number of computational tools have been developed to look for sequence and structural factors that contribute toward aggregate formation in a protein. Many of these tools predict aggregation on the basis of the presence of short sequences that promote amyloid formation as well as the presence of exposed hydrophobic regions. Our work on therapeutic monoclonal antibodies (mAbs) suggests that, in many instances, aggregation-prone regions (APRs) are present in complementarity-determining region (CDR) loops and in adjoining β-strands. Since these parts of the molecule also contribute significantly toward the binding between antibody and target, in these cases, there can be an unintended and potentially undesirable link between a propensity toward aggregation and molecular function. The findings communicated here provide insights that can potentially be applied to the rational design of more "druggable" therapeutic candidates exhibiting potency and specificity but also stability and solubility.

3.1 INTRODUCTION

Biotherapeutics are large macromolecules derived mostly from biological sources using recombinant DNA technologies. Such therapeutics include a diverse range of products including growth factors, cytokines, hormones,

Aggregation of Therapeutic Proteins, Edited by Wei Wang and Christopher J. Roberts

receptors, enzymes, clotting factors, mAbs, and protein-based vaccines. The broader category of biotherapeutics also includes cell-based vaccines and therapies, tissues, intact viruses, bacteria, and other biological materials. More than 200 biotherapeutic drug products and vaccines are available today. Greater than 400 new products are currently in various stages of clinical testing for more than 200 diseases (http://www.bio.org/). This chapter, however, is limited to consideration of protein biotherapeutics including mAbs.

Biotherapeutic drugs typically undergo a complex series of processing steps during production, harvest, purification including viral clearance, and packaging to obtain the final product seen by the patient or the caregiver. These processes stress the protein molecule in many ways.[1] Some of the stresses involved are high concentrations, variable temperatures, pH extremes, varying ionic strengths, shear stresses, air–water as well as a variety of solid–liquid interfaces. These stresses can impact the potency and quality of the product through multiple physicochemical degradation pathways.[2] Among the various degradation pathways, aggregation is the most common and least understood at the molecular level. Aggregates are considered a potential risk factor for immunogenicity.[3] Aggregation is therefore a fundamental attribute for assessing the quality of biotherapeutic drugs. A basic understanding of the aggregation mechanism(s) and of why some molecules are more prone to aggregation than others would go a long way toward reducing or even eliminating this risk factor. One step toward this goal would be *in silico* identification of the regions in the biotherapeutic molecules that may make them vulnerable to aggregation. This knowledge could suggest strategies to mitigate aggregation through appropriate design and selection of the molecule itself.

In this chapter, we review the various computational tools available to identify potential APRs in biotherapeutics. We also present our recent work in understanding the relationship between aggregation and protein function. These observations are important for successful biotherapeutic drug candidate design and selection because they suggest strategies to improve solubility without compromising potency. An important point to note is that these computational tools are designed to predict sequence and structural regions prone to non-covalent aggregation. Pathways of aggregation through covalent modifications are not considered here.

We cast this article under the concept of energy landscapes and argue that aggregation is a natural consequence of the response of a protein molecule to changes in itself (e.g, sequence mutations or chemical degradation such as deamidation, oxidation, and clipping) and its environment (e.g., temperature, pH, ionic strength, freeze/thaw, and agitation/shear). We also hypothesize that non-covalent aggregation in protein therapeutics has many parallels with that seen in amyloid formation. This enables us to "port over" the tools used to detect, study, and predict aggregation in amyloids to the realm of biotherapeutics. Experimentally, it has been shown that aggregates formed by proteins in the bacterial inclusion bodies and by biopharmaceuticals toward the end of their expiry dates also bind thioflavin T (ThT) and Congo red, the marker dyes for detecting amyloids.[4,5]

Not all parts of the whole protein contribute equally toward aggregate formation. Instead, proteins have been shown to contain short regions in their sequences that are particularly prone to aggregation. These APRs contribute significantly toward the tendency of the protein to aggregate[6–8] and may not be evolutionarily conserved among homologous proteins. These observations have been based on the examination of small proteins and peptides. Recently, we have shown that the same might also apply to therapeutic mAbs.[9]

Several computational methods have been designed to predict APRs in proteins. Most of the methods are based on phenomenology and use only the protein sequences to identify short APRs of five to nine residues.[8] Other methods based on pattern recognition, three-dimensional (3-D) profiles, and molecular simulations are also emerging.[10–15] Most of the work reviewed here is from sequence-based predictions of the APRs. Such regions can then be mapped on to the 3-D structures of the protein molecules to obtain a structural perspective. Molecular dynamics simulations have been recently published to identify potential APRs in human IgG antibodies.[13,14] At the conclusion of this chapter, we summarize and present the utility of prediction of APRs in guiding the design or selection of biotherapeutic candidates that are both potent and developable. "Developable" in the context of this chapter implies molecules that do not pose significant solubility/aggregation problems during production (e.g., expression), purification, formulation, or storage.

3.2 ENERGY LANDSCAPES, PROTEIN FOLDING, AND AGGREGATION

It is a central dogma in biochemistry that the amino acid sequence of a protein contains all the necessary information to code for its "native" 3-D structure, which enables the biological function. This cardinal belief has been held as "truth" by biochemists while much progress has been made in our understanding of conformational ensemble heterogeneities, multiple protein folding pathways, role of chaperones in *in vivo* protein folding, intrinsically disordered proteins, misfolding, and aggregation over the past couple of decades. Proteins can populate multiple ordered structures besides the native state, and there are multiple routes that protein molecules can take to transit from one state to another.[16] Several years ago, we postulated that the energy landscapes of the proteins are not static. Instead they are *dynamic*. The energy landscapes change as the proteins respond to the perturbations in their environment such as temperature, pH, salt concentrations, pressure, presence or absence of ligands, or interacting partners via conformer population shifts.[17,18] Extending this concept, we propose that changes in a protein itself (e.g., mutations and chemical modifications like deamidation, oxidation, and clipping) can also move its energy landscape. Aggregation-competent partially

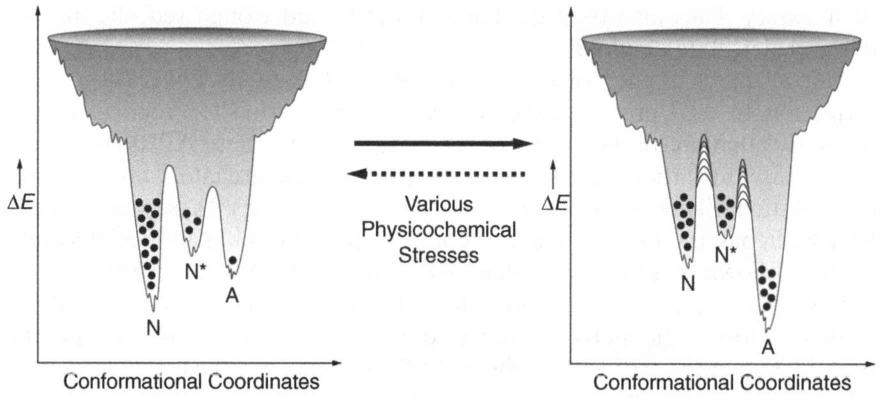

Figure 3.1. The energy landscape of a biotherapeutic molecule is dynamic. It moves in response to the changes in its environment and in itself as a consequence of various physicochemical stresses and degradations. N denotes native state; N* denotes native-like aggregation-competent state, and A denotes the aggregated state of the molecule. Different stresses may change the barriers between these states differently and may lead to conformer population shifts differently among the three states.

folded intermediates and aggregated conformations represent different basins in the energy landscape that a protein visits under appropriate conditions. Moreover, we propose that this energy landscape dynamics must be encoded in specific sequence and structural motif(s) that determine the proteins' response toward the change. Along these lines, a biotherapeutic molecule also responds to stresses faced during production and storage by moving its energy landscape via conformational population shifts (Fig. 3.1). The information on how a biotherapeutic candidate responds to the various stresses is contained in its sequence and molecular structure. An understanding of this connection through bioinformatics and biophysical computations may help address some of the challenges faced during the development of biotherapeutics.

3.3 PREDICTION OF APRs IN PROTEINS AND BIOTHERAPEUTICS

3.3.1 Computational Tools

Some regions of protein sequences have been found to be more aggregation prone than others.[8] Monsellier et al. have studied this by constructing sequence-scrambled variants of the first 29 residues of horse heart apomyoglobin.[19] The amino acid composition for the variants was kept the same as the wild type. The clustering of the most amyloidogenic residues in one sequence region significantly reduced the lag phase and increased the elongation rate of fibril growth. Early studies on the small proteins and peptides led to characterization of the physicochemical properties of aggregation associated with neurodegenerative diseases.[20] Pawar et al. defined intrinsic aggregation propensities of individual amino acids and developed algorithms for detecting regions "sensitive" to aggregation in protein sequences.[6]

Numerous proteins aggregate via formation of a cross-β steric zipper motif as seen in amyloid fibrils.[11,21–23] In such proteins, one can often detect short sequence stretches that drive aggregation, regardless of the rest of the protein sequence or structure. Typically, these sequence stretches have unique features with respect to charge, hydrophobicity, aromaticity, and secondary structural preference. A number of computer programs have been developed to predict these APRs in proteins due to their involvement in neurodegenerative diseases. A few of these programs are reviewed below:

3.3.1.1 PAGE PAGE makes predictions for absolute aggregation rates and aggregation propensities of proteins by sliding a small window of five to nine residues over the length of the sequence, one residue at a time.[24] The regions are then ranked according to their aggregation propensity. The aggregation propensity is calculated based on aromaticity, β-strand propensity, charge, solubility, and average polar and nonpolar accessible surface area of each residue in the given window. The absolute aggregation rate is related to aggregation propensity via a factor that is a function of concentration and temperature. More information for this program can be found at the website http://www.biochem-caflisch.uzh.ch/. PAGE predicts APRs for both parallel and antiparallel β-sheet fibrillar organization in a large set of natural protein sequences.[24] The phenomenological equations used in PAGE have factors similar to the one derived earlier by DuBay et al.[25] for the prediction of absolute rates of protein aggregation.

3.3.1.2 TANGO TANGO is a statistical mechanics-based algorithm that identifies β-aggregating regions in a given protein sequence.[26] TANGO is based on physicochemical principles of β-sheet formation and assumes that the core regions of the aggregates are fully buried. It also takes into account the competition between other conformational states such as α-helix, β-strand, turn, random coil, and β-aggregates for the given sequence region. TANGO is a web-based application and more information on this algorithm can be found at the website http://tango.crg.es/. Fernandez-Escamilla et al. collected 179 peptides from 21 different proteins where experimental evidence for aggregation is available in literature.[26] TANGO correctly predicted 155 out of 179 peptides (87%) to be aggregation prone, with 21 false positives and 3 false negatives. They also experimentally tested their algorithm on an independent set of 71 peptides. TANGO correctly predicted 65 out of the 71 peptides (92%) to be aggregating or non-aggregating. Fernandez-Escamilla et al. also distinguished between β-sheet aggregation and β-sheet propensity and made the point that peptides with high β-sheet propensity may not necessarily aggregate.

3.3.1.3 Zyggregator Zyggregator predicts relative propensities for both folding and aggregation for a given protein sequence and identifies the regions where aggregation propensities are significantly higher.[8,25] This is a web-based application and more information can be found at the following website: http://www-vendruscolo.ch.cam.ac.uk/software.html. Zyggregator is based on the concept that amino acid sequences determine both folding and misfolding behaviors of the proteins. It also shows the influence of structural context on the ability of a sequence region to promote or to suppress aggregation.[8]

3.3.1.4 Aggrescan Aggrescan is another web-based application for the prediction of APRs in protein sequences.[7,27] Here, the aggregation propensities for the amino acids are derived from *in vivo* experiments as opposed to the *in vitro* experiments. This program also computes the aggregation propensity by sliding over the sequence with a small window size. More information can be found at the website http://bioinf.uab.es/aggrescan/. This method was validated by comparing experimental and predicted APRs in 24 fibrillar deposition disease-linked polypeptides (table 1 in Reference 27).

3.3.1.5 PASTA PASTA seeks to identify regions in a protein sequence that may contribute toward the formation of a cross-β structural motif associated with amyloid fibrils.[28,29] Using a data set of 500 high-resolution globular protein crystal structures, the authors have derived pairwise interaction energy potentials for the interacting residues to be found within β-sheets facing one another on neighboring strands. The algorithm predicts parallel/antiparallel β-sheet pairing as well as a registry of hydrogen bonds in the amyloidogenic sequence stretches. PASTA is also a web-based application and more information can be found at the website http://protein.cribi.unipd.it/pasta/. As in the case of TANGO,[26] the authors used the data on 179 peptides from literature to benchmark predictions from their methods. Their algorithm yields approximately 80% correct predictions at a PASTA interaction energy threshold of –4.0 (fig. 1 in Reference 29).

3.3.1.6 SALSA SALSA calculates β-strand contiguity from Chou–Fasman secondary structural propensity values.[30] Latent β-strand propensity as measured by β-strand contiguity peaks located by this algorithm corresponded very well with the observed location of β-sheet structure in fibrillar forms of α-synuclein, amyloid β (Aβ), and tau, the model proteins in amyloid research. SALSA uses only a single parameter, namely, β-strand propensity, and focuses on the intrinsic ability of a given sequence region to form an amyloid fibrillar type of aggregation. Unlike the other programs, the authors of SALSA[30] state that their method does not take into account the influence of external environmental factors such as pH, temperature, ionic strength, and protein concentration.

3.3.1.7 AMYLPRED AMYLPRED is a consensus prediction tool.[31] It utilizes five different methods, namely, TANGO, AGGRESCAN, Zyggregator,

SALSA, and PASTA to arrive at the consensus amyloidogenic segment predictions. The consensus algorithm is freely available to academic users at http://biophysics.biol.uoa.gr/AMYLPRED. Interestingly, the authors found that the majority of the consensus amyloidogenic segments in proteins are solvent exposed and have the ability to act as conformational switches.

3.3.1.8 3-D Profile Methods Besides the above primarily amino acid sequence-based methods, structural 3-D profile methods have also been developed. The input for these programs is the protein sequence.

The 3-D profile method from the Eisenberg lab identifies fibril-forming segments in proteins based on the crystal structure of the cross-β spine formed by the hexapeptide NNQQNY.[11] In this method, each six-residue segment of a protein sequence is mapped onto an ensemble of templates obtained from the NNQQNY crystal structure by small atomic displacements. The energy of each mapping sequence is evaluated by RosettaDesign, and the low-energy sequence matches are identified. This method showed an accuracy of approximately 80% in distinguishing between fibril and non-fibril-forming peptides.[11] Recently, Zhang et al. have also developed a similar method for identifying the fibril formers in protein sequences.[10] However, in this approach, the fibril-forming propensity of the peptide is evaluated by the interaction energy score of the central strand with other strands. The authors found that the method performs at an accuracy level similar to the one by Thompson et al.[11] Since their algorithm uses residue-based statistical potentials, it is computationally faster.[10]

3.3.2 Experimentally Studied Aggregation-Prone Sequences and Sequence Patterns

Lopez de la Paz and Serrano have studied the link between amino acid sequence and amyloid fibril formation.[23] They have used a de novo designed amyloid hexapeptide STVIIE and mutated each of the six positions with all the possible 19 natural amino acids and studied amyloid fibril formation. They discovered that some positions are very tolerant to mutations, while the others are restrictive. Mutations also affect the rate of aggregation. Mutations that accelerate the β-sheet polymerization do not always end up as amyloid fibrils. Instead, the mutations that slow down the polymerization rate yield "abundant" amyloid fibrils. The authors summarize their results in the form of two sequence patterns stated below:

1. {P}1-{PKRHW}2-[VLS(C)WFNQE]3-[ILTYWFNE]4-[FIY]5-{PKRH}6 for acidic pH and
2. {P}1-{PKRHW}2-[VLS(C)WFNQ]3-[ILTYWFN]4-[FIY]5-{PKRH}6 for neutral pH.

These sequence patterns are written in PROSITE format. The numbers 1, 2, 3 … 6 represent the positions in the hexapeptide. The curly ({ }) and the straight ([ ]) brackets indicate disallowed and allowed residues at a given position.

We have mined the literature and collected as exhaustively as possible all known sequences where there is experimental evidence for amyloid-like aggregation. We relied on the studies that were positive in at least two of the following essential components that are considered markers for the presence of amyloids: ThT binding curves, ThT stain pictures, atomic force microscopy (AFM)/transmission electron microscopy (TEM) fibril pictures, X-ray diffraction patterns, Congo red staining pictures, Congo red absorbance curves, and turbidity curves. Additional pieces of evidence such as enrichment of β-strand conformation by circular dichroism (CD)/Fourier transform infrared spectroscopy (FTIR) and resistance to proteolysis were also considered. We found more than 400 nonidentical short sequences and three sequence patterns based on the experimental studies involving approximately 70 different proteins and numerous synthetic peptides. We also found several polar zipper sequences (Q/N). Two of the three sequence patterns are already listed above. The third sequence pattern is

3. [KE]1-[FV]2-[FV]3-[EK]4, where the residue at position 1 is not the same as the one on position 4.[32]

The sequences based on Aβ peptide, prion protein (PrP), α-synuclein, β2-microglobulin, insulin, amylin (islet amyloid polypeptide), and de novo or designed peptides constitute the bulk of the library. Most of the short sequences contain β-branched hydrophobic and aromatic residues. Charged residues are infrequent in these sequences and Arg is especially rare. This observation is consistent with the early parameterization performed for detecting APRs.[24–26] Many sequences are repetitive and are of lower complexity, indicating their inherent ambivalence toward different conformational states. The smallest amyloidogenic sequence in our database is just three residues long: GYE from Aβ 9–11.[33] The longest one contains 51 residues: SYGGEGIGNVAVAGELPV AGKTAVAGRVPIIGAVGFGGPAGAAGAVSIAGR from Chlorion A.[11]

3.3.3 Prediction of APRs in Therapeutic mAbs

As discussed above and elsewhere in this book, therapeutic mAbs can aggregate due to several reasons. At the molecular level, these mAbs can self-associate in several different ways. Schematically, one can think of Fab:Fab, Fab:Fc, or Fc:Fc associations involving two or more mAbs (Fig. 3.2). Knowledge of potential APRs in such mAbs can yield clues to possible modes of mAb self-association. This knowledge can be useful in the structure-based design of mAbs with reduced propensity to aggregate.

Recently, we have used TANGO and PAGE to identify the potential APRs in the commercially available therapeutic mAbs.[9] We found that the commercially available therapeutic mAbs contain two to eight sequence regions

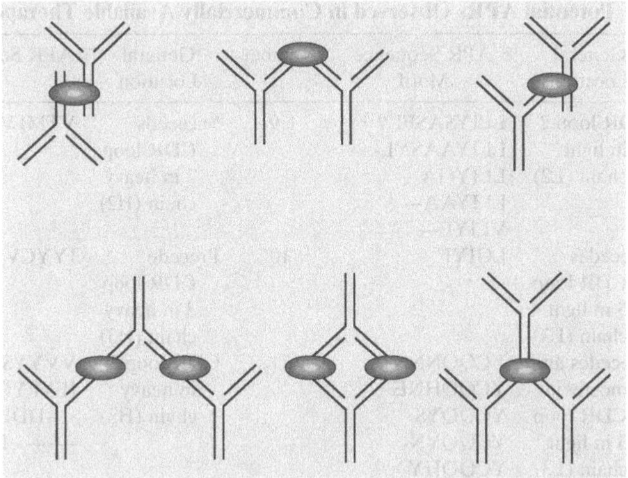

Figure 3.2. A schematic diagram illustrating some of the modes by which therapeutic mAbs may self-associate at the molecular level to produce aggregates under different physicochemical stresses and degradations. Two or more mAb molecules can come together via Fab:Fab, Fc:Fc, or Fab:Fc associations. More complex mAb self-association modes can also be conceived. Knowledge of potential APRs in individual mAb sequences and structures may yield useful clues to potential association modes that seed aggregation in mAbs.

that are strongly predicted (high probability) to be aggregation prone in a heavy- and light-chain pair (TANGO aggregation score ≥50% and/or Z-score for PAGE aggregation propensity >1.96; see Wang et al.[9] for details). Table 3.1 summarizes the APRs found in commercial therapeutic mAbs. The APRs in mAbs have similar characteristics as those seen in small proteins and peptides. These APRs are 5–10 residues long. Most of them are rich in β-branched aliphatic and aromatic residues. Hydroxyl-containing Ser/Thr residues are also found in several mAb APRs, while charged residues are rare. The APRs found in light-chain CDR3 are glutamine (Q)/asparagine (N) rich. These findings imply that aggregation of mAbs may also occur via formation of cross-β structure motif. In line with this idea, Maas et al. have shown that biopharmaceuticals bind ThT at the end of their shelf life.[4] The biopharmaceuticals studied by Maas et al.[4] also contain strong APRs.[9]

Another interesting finding of our study is that some of these APRs are located in variable domains, primarily in CDR loops and in adjoining framework β-strands. This work couples aggregation with molecular recognition in the Fab part of mAbs. Curiously, APRs do not coincide with the various protein binding sites found in the Fc portion of the mAbs. Such information and the tools described here can be used for optimization of structures of therapeutic mAb candidates to reduce their aggregation propensity while maintaining/enhancing potency.

TABLE 3.1. Potential APRs Observed in Commercially Available Therapeutic mAbs [9]

Number	General Location	APR Sequence Motif	Number	General Location	APR Sequence Motif
1	CDR loop 2 in light chain (L2)	LLIYSASFLY LLIYAASYL- LLIYGA— LLIYAA— VLIYF—	9	Succeeds CDR loop 2 in heavy chain (H2)	VTMLV
2	Precedes CDR loop 3 in light chain (L3)	LGIYF	10	Precede CDR loop 3 in heavy chain (H3)	IYYCV
3	Precedes and mostly in CDR loop 3 in light chain (L3)	YCQQNNN YCQQHNE YCQQYS- YCQQYN- YCQQHY- YCLQYD- YCQQS–	11	CDR loop 3 in heavy chain (H3)	VVYYSNSYWYF— IFYFYGTTY-F— —DDHYC— ——FAVWG
4	C_L domain in light chain	SVFIFP SVFIF- TVFIFP	12	C_H2 domain	SVFLFPP SVFLFP- SVFIFP-
5	C_L domain in light chain	VVCLL VVCFL	13	C_H2 domain	VLMISL
6	CDR loop 1 in heavy chain (H1)	YIFSNYWIQWV -IFTDF— —YYWTWI	14	C_H2 domain	IVTCVVV
7	CDR loop 2 in heavy chain (H2)	IGAIY LGVIW IGYIS IGYIY	15	C_H2 domain	VVSVLTVL VVSVLTVV
8	CDR loop 2 in heavy chain (H2)	TTEYN- -TEYNQ -TNYNQ	16	C_H3 domain	GSFFLYS GSFFLY- -SFFLYS -SFFLY-

The APRs were detected using TANGO and PAGE programs.[24,26]

Using molecular dynamics simulations on the computer model of a full IgG1 antibody and a combination of accessible surface area and hydrophobicity, specific aggregation-prone structural motifs were identified in the constant regions by Chennamsetty et al.[13,14] The authors have devised a measure called spatial aggregation propensity (SAP). SAP localizes the surface-exposed hydrophobic patches in proteins. The concept behind SAP is similar to that behind surface aggregation propensity used by Pechmann et al.[34] Mutation of specific hydrophobic residues located in the hinge and Fc regions to Lys improved both thermodynamic stability of the mAbs and also their resistance to aggregation. These surface-exposed hydrophobic patches can act as structural hot spots for aggregation. However, these patches are not contiguous in sequence and are thus unrelated to the APRs identified by the sequence-based approaches.

3.3.4 Other Useful Studies

Besides the various computational tools described above, a few other studies are worthy of mention.

3.3.4.1 Packing Density May Identify Potential APRs Galzitskaya et al. studied factors that influence protein conformational changes.[35] Such factors are important for identification of conformationally unstable regions in proteins. Using the residue-wise mean packing density as a parameter, the authors were able to correctly predict potential amyloidogenic regions in 8 out of 12 amyloid-forming peptides and proteins with experimentally determined amyloidogenic regions. The authors found that the regions with strong expected packing density may be amyloidogenic and those with weak expected packing density might be inherently disordered.

3.3.4.2 APRs and Structural Context The structural context around the APRs may impact the ability of these motifs to form amyloid-like fibrils. For example, a series of peptides corresponding to individual β-strands in β-lactoglobulin can form amyloid aggregates and can also seed amyloidosis in the full length β-lactoglobulin with reduced disulfide bonds. However, when the disulfide bonds are intact, only the strand βA promotes amyloid formation by β-lactoglobulin.[36] Natural proteins also often interrupt contiguous aggregation-prone stretches of hydrophobic and aromatic amino acids by using "gatekeeper" charged residues in their primary sequences.[37] This natural design feature is thought to serve the purpose of minimizing the deleterious effect of the presence of APRs.[36,38,39] A systematic analysis of the crystal structures of the protein complexes shows that the interface regions of proteins are more prone to aggregation than the rest of surface.[34] It follows that the physicochemical forces behind the functional (native) association of proteins and nonfunctional aggregation are similar. The key lies in the close regulation of functional protein interfaces by specific interactions that stabilize the native states.

3.3.4.3 Molecular Simulations Can Identify Potential APRs and Seeds Molecular simulations have been used to understand the seeding and polymerization of aggregation in proteins and peptides. These studies have been recently reviewed.[40,41] Here, we highlight two interesting studies that are relevant to the topic of this review.

Different overlapping sequence stretches from three amyloidogenic polypeptides, namely, $A\beta_{1-42}$, human amylin (human islet amyloid polypeptide), and N-terminal domain of yeast prion Ure2p ($Ure2p_{1-94}$) were studied.[12] For all the sequence stretches, parallel in-register β-aggregate systems were constructed and subjected to equilibrium implicit solvent molecular dynamics (MD) simulations. Subsets of these were also examined by explicit solvent MD simulations. 12-VHHQKLVFFAE-22 in $A\beta_{1-42}$, 10-RLANFLVHSSNN-22 in human amylin and three different regions, 33-SNINFEF-39, 45-NNNNN-49, and 55-NNNVQNN-61 in Ure2p were found to have maximal aggregation propensities. Using these observations, Cecchini et al. designed a double mutant Ure2p-N4748S_{1-94}.[12] MD simulations showed the elimination of the aggregation hot spot at residues 45–49 in the double mutant. Consistently, the ThT binding kinetic assays showed slower and reduced aggregation in the double mutant.

Polyglutamine (PolyQ) sequence stretches can be highly aggregation prone. These stretches have been linked to several neurodegenerative diseases. Using Monte Carlo simulations, Vitalis et al. have shown that PolyQ peptides of length 15 or more can spontaneously dimerize at concentrations typically used in the *in vitro* experiments.[15]

3.3.4.4 Mechanistic Features of Seeding and Propagation of Amyloid-Like Fibrils Nelson and Eisenberg reviewed various models for mechanism of seeding and propagation of amyloid-like fibrils.[22] They classified different models into refolding, gain of interaction, and natively disordered classes. The refolding models assume structurally different native and fibrillar conformations for the proteins/peptides. These models suggest that backbone interactions drive fibril formation. In contrast, the gain-of-interaction models propose largely native-like conformation for the proteins and focus on the role of short sequence motifs (APRs) in fibril formation. The natively disordered models appear to fill the gaps between the above two alternate model classes. The authors concluded that no single class of models is able to explain all of amyloid fibril formation data. However, gain of interaction models with cross-β spine appear to account for a wider range of fibril properties.[22]

3.4 CONCLUSIONS AND FUTURE DIRECTIONS

The molecular origins of aggregation are similar between small peptides/proteins and large biotherapeutic molecules such as mAbs. Since APRs in therapeutic mAbs also lie in short sequence stretches, it follows that even one or a few changes in the sequences of the mAbs may significantly alter their solubility and therefore the aggregation propensity. From a product formulation and developability standpoint, improvement in mAb solubility is highly desirable. It may also lead to greater expression levels in the cell lines and may help achieve high-concentration dosage forms. Improvements in protein native state stability via elimination/mitigation of APRs may also help increase the shelf life of the product.

However, any developability-related sequence mutations should not adversely affect the potency of the biotherapeutic. In this regard, the observed incidence of putative APRs in the CDR loops and in adjacent framework β-strands is significant. A rational approach that considers mAb stability and solubility along with potency and specificity at the early discovery stage may potentially lead to more druggable therapeutic candidates. This approach is also in sync with the initiative on quality by design being emphasized by the regulatory agencies.[42]

Using computational approaches, one can rapidly screen a large number of candidates for appropriate properties. A smaller set of candidates can then be experimentally studied for the identification of a candidate for further development.[43–46] Using the available sequence data, we have shown that the tools developed to identify APRs in small peptides and proteins can also be applied to therapeutic mAbs.[9] Experimentally, it remains to be seen if disruption of APRs via mutagenesis shall lead to decreased aggregation rate and, hence, improved solubility in mAbs. Due to their speed and relatively little computational resource demands, the sequence-based aggregation prediction tools can prove very useful. Moreover, one can obtain valuable insights for rational design when these predictions are combined with molecular models of the candidates. The molecular models can help guide experiments by pinpointing the positions to mutate.

The solubility of a biotherapeutic protein may also be impacted due to reasons other than the presence/absence of APRs. A complementary strategy is to use the knowledge generated from the studies on extremophilic proteins. Proteins from organisms adapted to extreme environmental conditions such as heat, cold, and high salinity have also evolved strategies to modulate their stability and solubility via changes in their surface electrostatic properties.[47] Similar "nature-inspired" strategies can also be devised to eliminate or mitigate aggregation in therapeutic mAbs in a rational fashion.

ACKNOWLEDGMENTS

This work is supported by the predictive stability initiative at Pfizer Inc. The authors thank Drs. Tapan Das, Sandeep Nema, Graeme Bainbridge, and Sa V. Ho for their interest in this work. Bradley Lohrum of Global Creative Solutions is thanked for help with graphics.

REFERENCES

1. Cromwell MEM, Hilario E, Jacobson F. 2006. Protein aggregation and bioprocessing. AAPS J 8(3):E572–E579.
2. Manning MC, Patel K, Borchardt RT. 1989. Stability of protein pharmaceuticals. Pharm Res 6(11):903–918.
3. Rosenberg AS. 2006. Effects of protein aggregates: An immunologic perspective. AAPS J 8(3):E501–E507.

4. Maas C, Hermeling S, Bouma B, Jiskoot W, Gebbink MFBG. 2007. A role for protein misfolding in immunogenicity of biopharmaceuticals. J Biol Chem 282(4): 2229–2236.
5. Wang L, Maji SK, Sawaya MR, Eisenberg D, Riek R. 2008. Bacterial inclusion bodies contain amyloid-like structure. PLoS Biol 6(8):1791–1801.
6. Pawar AP, Dubay KF, Zurdo J, Chiti F, Vendruscolo M, Dobson CM. 2005. Prediction of "aggregation-prone" and "aggregation-susceptible" regions in proteins associated with neurodegenerative diseases. J Mol Biol 350(2):379–392.
7. de Groot N, Pallares I, Aviles F, Vendrell J, Ventura S. 2005. Prediction of "hot spots" of aggregation in disease-linked polypeptides. BMC Struct Biol 5(1):18.
8. Tartaglia GG, Pawar AP, Campioni S, Dobson CM, Chiti F, Vendruscolo M. 2008. Prediction of aggregation-prone regions in structured proteins. J Mol Biol 380(2): 425–436.
9. Wang X, Das TK, Singh SK, Kumar S. 2009. Potential aggregation prone regions in biotherapeutics: A survey of commercial monoclonal antibodies. mAbs 1(3):1–14.
10. Zhang Z, Chen H, Lai L. 2007. Identification of amyloid fibril-forming segments based on structure and residue-based statistical potential. Bioinformatics 23(17): 2218–2225.
11. Thompson MJ, Sievers SA, Karanicolas J, Ivanova MI, Baker D, Eisenberg D. 2006. The 3D profile method for identifying fibril-forming segments of proteins. Proc Natl Acad Sci U S A 103(11):4074–4078.
12. Cecchini M, Curcio R, Pappalardo M, Melki R, Caflisch A. 2006. A molecular dynamics approach to the structural characterization of amyloid aggregation. J Mol Biol 357(4):1306–1321.
13. Chennamsetty N, Helk B, Voynov V, Kayser V, Trout BL. 2009. Aggregation-prone motifs in human immunoglobulin G. J Mol Biol 391(2):404–413.
14. Chennamsetty N, Voynov V, Kayser V, Helk B, Trout BL. 2009. Design of therapeutic proteins with enhanced stability. Proc Natl Acad Sci U S A 106(29):11937–11942.
15. Vitalis A, Wang X, Pappu RV. 2008. Atomistic simulations of the effects of polyglutamine chain length and solvent quality on conformational equilibria and spontaneous homodimerization. J Mol Biol 384(1):279–297.
16. Dill KA, Chan HS. 1997. From Levinthal to pathways to funnels. Nat Struct Biol 4(1):10–19.
17. Kumar S, Ma B, Tsai C-J, Sinha N, Nussinov R. 2000. Folding and binding cascades: Dynamic landscapes and population shifts. Protein Sci 9(1):10–19.
18. Tsai C-J, Kumar S, Ma B, Nussinov R. 1999. Folding funnels, binding funnels, and protein function. Protein Sci 8(6):1181–1190.
19. Monsellier E, Ramazzotti M, de Laureto PP, Tartaglia G-G, Taddei N, Fontana A, Vendruscolo M, Chiti F. 2007. The distribution of residues in a polypeptide sequence is a determinant of aggregation optimized by evolution. Biophys J 93(12):4382–4391.
20. Murphy RM. 2002. Peptide aggregation in neurodegerative disease. Annu Rev Biomed Eng 4(1):155–174.

21. Nelson R, Sawaya MR, Balbirnie M, Madsen AO, Riekel C, Grothe R, Eisenberg D. 2005. Structure of the cross-β spine of amyloid-like fibrils. Nature 435(7043): 773–778.
22. Nelson R, Eisenberg D. 2006. Recent atomic models of amyloid fibril structure. Curr Opin Struct Biol 16(2):260–265.
23. Lopez de la Paz M, Serrano L. 2004. Sequence determinants of amyloid fibril formation. Proc Natl Acad Sci U S A 101(1):87–92.
24. Tartaglia GG, Cavalli A, Pellarin R, Caflisch A. 2005. Prediction of aggregation rate and aggregation-prone segments in polypeptide sequences. Protein Sci 14(10):2723–2734.
25. DuBay KF, Pawar AP, Chiti F, Zurdo J, Dobson CM, Vendruscolo M. 2004. Prediction of the absolute aggregation rates of amyloidogenic polypeptide chains. J Mol Biol 341(5):1317–1326.
26. Fernandez-Escamilla A-M, Rousseau F, Schymkowitz J, Serrano L. 2004. Prediction of sequence-dependent and mutational effects on the aggregation of peptides and proteins. Nat Biotechnol 22(10):1302–1306.
27. Conchillo-Sole O, de Groot N, Aviles F, Vendrell J, Daura X, Ventura S. 2007. AGGRESCAN: A server for the prediction and evaluation of "hot spots" of aggregation in polypeptides. BMC Bioinformatics 8(1):65.
28. Trovato A, Chiti F, Maritan A, Seno F. 2006. Insight into the structure of amyloid fibrils from the analysis of globular proteins. PLoS Comput Biol 2(12):e170.
29. Trovato A, Seno F, Tosatto SCE. 2007. The PASTA server for protein aggregation prediction. Protein Eng Des Sel 20(10):521-523.
30. Zibaee S, Makin OS, Goedert M, Serpell LC. 2007. A simple algorithm locates beta-strands in the amyloid fibril core of alpha-synuclein, Abeta, and tau using the amino acid sequence alone. Protein Sci 16(5):906–918.
31. Frousios KK, Iconomidou VA, Karletidi C-MK, Hamodrakas SJ. 2009. Amyloidogenic determinants are usually not buried. BMC Struct Biol 9(1):44.
32. Tjernberg L, Hosia W, Bark N, Thyberg J, Johansson J. 2002. Charge attraction and beta propensity are necessary for amyloid fibril formation from tetrapeptides. J Biol Chem 277(45):43243–43246.
33. Naskar J, Drew MGB, Deb I, Das S, Banerjee A. 2008. Water-soluble tripeptide Aβ (9-11) forms amyloid-like fibrils and exhibits neurotoxicity. Org Lett 10(13): 2625–2628.
34. Pechmann S, Levy ED, Tartaglia GG, Vendruscolo M. 2009. Physicochemical principles that regulate the competition between functional and dysfunctional association of proteins. Proc Natl Acad Sci U S A 106(25):10159–10164.
35. Galzitskaya OV, Garbuzynskiy SO, Lobanov MY. 2006. Prediction of amyloidogenic and disordered regions in protein chains. PLoS Comput Biol 2(12):e177.
36. Hamada D, Tanaka T, Tartaglia GG, Pawar A, Vendruscolo M, Kawamura M, Tamura A, Tanaka N, Dobson CM. 2009. Competition between folding, native-state dimerisation and amyloid aggregation in β-lactoglobulin. J Mol Biol 386(3): 878–890.
37. Reumers J, Maurer-Stroh S, Schymkowitz J, Rousseau F. 2009. Protein sequences encode safeguards against aggregation. Hum Mutat 30(3):431–437.

38. Otzen DE, Kristensen O, Oliveberg M. 2000. Designed protein tetramer zipped together with a hydrophobic Alzheimer homology: A structural clue to amyloid assembly. Proc Natl Acad Sci U S A 97(18):9907–9912.
39. Richardson JS, Richardson DC. 2002. Natural β-sheet proteins use negative design to avoid edge-to-edge aggregation. Proc Natl Acad Sci U S A 99(5):2754–2759.
40. Bratko D, Cellmer T, Prausnitz JM, Blanch HW. 2007. Molecular simulation of protein aggregation. Biotechnol Bioeng 96(1):1–8.
41. Cellmer T, Bratko D, Prausnitz JM, Blanch HW. 2007. Protein aggregation in silico. Trends Biotechnol 25(6):254–261.
42. Paulson B. 2008. Forum assesses biotech QbD progress. Int Pharm Qual 2(5):1–5.
43. Marshall SA, Lazar GA, Chirino AJ, Desjarlais JR. 2003. Rational design and engineering of therapeutic proteins. Drug Discov Today 8(5):212–221.
44. Carter PJ. 2006. Potent antibody therapeutics by design. Nat Rev Immunol 6(5):343–357.
45. Hwang I, Park S. 2008. Computational design of protein therapeutics. Drug Discov Today Tech. doi: 10.1016/j.ddtec.2008.11.004 (in press).
46. Trevino SR, Scholtz JM, Pace CN. 2007. Amino acid contribution to protein solubility: Asp, Glu, and Ser contribute more favorably than the other hydrophilic amino acids in RNase Sa. J Mol Biol 366(2):449–460.
47. Kumar S, Singh SK, Gromiha MM. 2010. Temperature dependent molecular adaptations, microbial proteins. In Flickinger MC, Drew SW, Spier RE, eds, *Encyclopedia of Industrial Biotechnology*. Hoboken, NJ: John Wiley & Sons.

CHAPTER 4

External Factors Affecting Protein Aggregation

WEI WANG, NING LI, and STAN SPEAKER

4.1 INTRODUCTION

The behavior of biotherapeutic proteins is significantly different from small molecule drugs both in liquid and solid states, especially in terms of physical stability. One such property is the high tendency of protein molecules to aggregate under a variety of conditions. The aggregation tendency is arguably the most common and troubling manifestation of protein instability during the development of protein biotherapeutics.[1] Protein aggregates usually exhibit either reduced or, in many cases, no biological activity.[2–5] A clear correlation was established between loss of recombinant factor VIII (rFVIII) and its degree of aggregation as shown in Fig. 4.1.[6] To achieve effective prevention or inhibition of protein aggregation, it is of paramount importance to understand all the possible factors that affect or control the protein aggregation process.

Many factors have been identified and reported in the literature affecting the process and rate of protein aggregation. These factors can be roughly divided into two major categories: internal (protein structure related) and external (protein environment related). Chapter 3 in this book has focused extensively on factors that belong to the first category. This chapter focuses on the external factors that affect the process and rate of protein aggregation. To facilitate discussion, these external factors are further divided into the following sections: (1) temperature; (2) solution conditions and composition (pH, buffer type and concentration, ionic strength, excipients and level, protein concentration, metal ions, denaturing and reducing agents, impurities, organic solvents, containers/closures, sources of proteins, sample treatment, and analytical methodologies); (3) processing steps (fermentation/expression,

Aggregation of Therapeutic Proteins, Edited by Wei Wang and Christopher J. Roberts

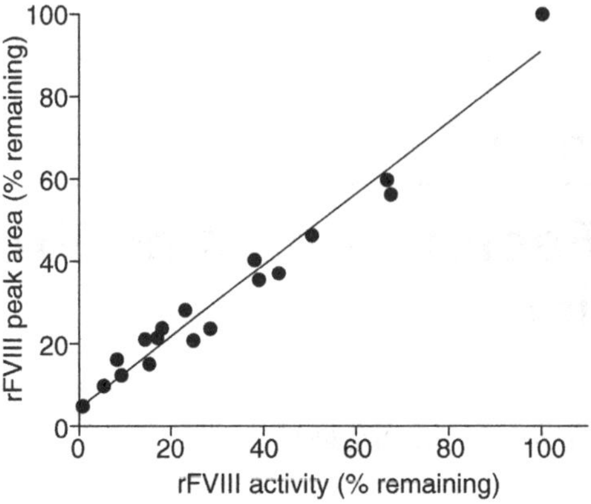

Figure 4.1. Correlation of loss of rFVIII activity (by the one-stage clotting assay) with aggregation of rFVIII (by SEC-HPLC assay) for stability samples at 40°C.

unfolding/refolding, purification, freeze-thaw, shaking and shearing, pressurization, formulation/filling, drying, preparation of modified protein or delivery systems, etc.); and (4) solid-state condition and composition (solid-state pH, excipients and level, physical state of the solid, and moisture content). These factors are summarized in Table 4.1.

4.2 PROTEIN AGGREGATION PATHWAYS

There are many external factors that influence protein aggregation. To facilitate examination of these factors, protein aggregation pathways are briefly discussed here. Figure 4.2 shows the major aggregation pathways: (1) indirect physical aggregation through formation of unfolding intermediates (pathway 1), (2) direct aggregation through protein self-association (pathway 2a) or chemical linkages (pathway 2b), and (3) indirect aggregation through chemical degradations (pathway 3).

4.2.1 Physical Aggregation through Formation of Unfolding Intermediates

Under normal conditions, native proteins in solution are in an equilibrium with their unfolding intermediates (I state), which are also called molten globules.[7] These folding intermediates are also in equilibrium with the completely unfolded/denatured states (D state). In terms of physical aggregation processes, significant evidence suggests that the poorly populated protein folding/unfolding intermediates (I state in Fig. 4.2) are precursors of the aggregation process.[8–12] This is because these intermediates expose more hydrophobic patches and have a high flexibility relative to the folded state.[13] In contrast,

TABLE 4.1. External Factors Affecting Protein Aggregation

Factors	Proteins/Conditions	% Aggregates or Aggregate Rate	References
General factors			
Temperature	Incubation of IL-1β at 0.1 mg/mL in acetate, pH 5.0, 39–60°C, or bovine/porcine insulin at 30 mg/mL, pH 2.5, 35–75°C	Increased aggregation (IL-1β)/fibrillation (insulin) with temperature	23,60
Light	UV (297–320 nm) irradiation of γ-crystallin at 1–2 mg/mL	Increased protein aggregation/turbidity	284
Solution factors			
pH	Incubation of a botulinum neurotoxin at 0.55 mg/mL at pH 5.0 and 8.0 at 50°C for a day or an IgG2 at 30 mg/mL at pH 4.0, 5.0, and 6.0 at 37°C for 15 weeks	<5% and >90% (neutotoxin); 52.9%, 0.9%, and 0.2% (IgG2)	61,159
Buffer species	Incubation of IFN-tau in a solution of 100 mM NaCl and 20 mM phosphate, Tris, or histidine buffer at pH 7 at 50°C	0.0133/min, 0.0086/min, or 0.0076/min (IFN-tau)	31
Buffer concentration	Incubation of aFGF at 0.1 mg/mL in 4.6 or 9.2 mg/mL phosphate buffer, pH 7.4, 50°C	Aggregation rate dA/min = 100 or ~10	76
Ionic strength	Incubation of RNase A at 5 mg/mL in 100 mM citrate containing 0, 250, and 500 mM NaCl, pH 3.0, 75°C	Aggregation increases with increasing ionic strength.	209

(*Continued*)

TABLE 4.1. *Continued*

Factors	Proteins/Conditions	% Aggregates or Aggregate Rate	References
Additives—salt (negative charge)	Incubation of 4 μM chymotrypsin with 10 nM polyvinyl sulfonate at 25°C in 50 mM citrate buffer at pH 2.5 in the absence or presence of 100 mM NaF, NaCl, or NaBr	*K* = 1.27E-4/s, 1.29 E-4/s, 1.66 E-4/s, and 2.2E-4/s, respectively	75
Additives—preservative	Incubation of recombinant human interleukin-1 receptor antagonist (rhIL-1ra) at 100 mg/mL at 37°C in the presence of 0.9 wt/vol % benzyl alcohol	Nucleation rate was increased by approximately four orders of magnitude.	113
Additives—surfactant	Agitation of bovine insulin at 0.6 mg/mL in PBS, pH 7.4 at 37°C at 100 rpm without or with 10 mM n-Octyl-β-D-glucopyranoside/n-Dodecyl-β-D-maltoside	Aggregated in 2 days or no significant change for 40 days	85
Protein concentration	Incubation of bovine insulin at 7.5, 15.0, 30.0 and 50.0 mg/mL, pH 2.5, 21°C	Aggregation increases with increasing conc.	23
Surface	Agitation of bovine insulin at 0.6 mg/mL in PBS, pH 7.4, 37°C with beads	Aggregation: Teflon > siliconized glass > polypropylene > glass	85
Organic solvent	Addition of equal volume of ethanol to β-lactoglobulin solution at 50 mg/mL, pH 7–9, 25°C	Aggregated/gelated within a few hours	423
Metal ions	Incubation of deoxyribonuclease I at ~18 mg/mL in 150 mM NaCl, 40°C for 30 days without or with 1 mM $CaCl_2$	Complete aggregation or no visible aggregation	410

TABLE 4.1. *Continued*

Factors	Proteins/Conditions	% Aggregates or Aggregate Rate	References
Protein source	Incubation of bovine insulin at 15 mg/mL, pH 2.1, 21°C, lots A and B	Aggregation $A > B$	23
Processing factors			
Refolding	Refolding rhGH at 0.667 mg/mL in 4.5 M GdnHCl in the absence or presence of Tween 20 (10:1 = molar Tween:rhGH)	Aggregation $A_{350} \approx 0.6$, or 0.2	22
Freeze-thawing	Freeze-thawing hemoglobin for zero, three, and five times at 50 mg/mL in PBS, pH 7.4	Aggregation: 5 > 3 > 0 times	155
Shaking	Vortexing hGH for 1 min at 0.5 mg/mL in phosphate, pH 7.4	67% precipitated	261
Shearing	Shearing (10^5) hGH at 2 mg/mL in 5 mM phosphate, pH 7.4	30%	424
Stirring	Stirring a monoclonal antibody (IgG1) at 10 mg/mL in 125 mM NaCl, and 25 mM sodium acetate at pH 6.0 at 200 rpm at 5°C in the presence of 0, 0.005, or 0.01% Tween 20	Turbidity ≈ 2000, 350, or 1000 FTU	348
Pressure	Increasing pressure from 0 to 2.3 kbar to a solution of H^+-ATPase at 0.05 mg/mL in 3 mM $MgCl_2$, 5% glycerol, and 50 mM Tris–HCl, pH 7.0	Increased aggregation with increasing pressure	425
Vacuum-drying	Vacuum-drying of a human IgG1:mannitol powder at 32°C at 24 h at 0.1 mbar at weight ratios of 2:8, 3:7, and 4:6	0%; 2.8%, and 3.3%	184

(*Continued*)

TABLE 4.1. *Continued*

Factors	Proteins/Conditions	% Aggregates or Aggregate Rate	References
Freeze-drying	Freeze-drying of a botulinum neurotoxin in 20 mM sodium succinate buffer at pH 5 with no other excipient, 0.01% Tween 20, 4% mannitol, or 5% hydroxyethyl starch	~33% ~14% ~17% ~3%	159
Spray-drying	Spray-drying of a human IgG1 : mannitol at weight ratios of 4:0, 4:1, and 1:4	2.7%, 0.8%, and 8.3%	184
Spray-freeze drying	Spray-freeze drying Darbepoetin α at 3.8 mg/mL, 2.1 mg/mL trehalose, and 20 mM phosphate at pH 6.0 using ultrasonic atomization at 120 and 25 kHz	9% insoluble	382,383
Reconstitution	Reconstitution of KGF at 0.4–0.65 mg/mL in 10 mM phosphate, pH 7.0, with and without heparin	Aggregation: water > heparin	139
Reconstitution medium	Reconstitution of lyophilized recombinant human interlukin-1 receptor antagonist (rhIL-1ra) at 15 mg/mL in 10 mM potassium phosphate buffer at pH 7.0	11% and 19%, respectively, with water or 0.9% benzyl alcohol	237
Other processes	Emulsification (o/w) of γ-chymotrypsin at 40 mg/mL in methylene chloride containing PLGA	34% aggregated	386

TABLE 4.1. *Continued*

Factors	Proteins/Conditions	% Aggregates or Aggregate Rate	References
Solid-state factors			
Drying method	Incubation of freeze-dried or film-dried Met-hGH at 50°C with 3X (wt) trehalose and 0.5X (wt)Tween 20	1.1%/month or <0.35%/month	364
Solid-state "pH"	Incubation of a lyophilized IgG1 antibody at 50°C at pH 5 or 7	7.0 $month^{-1/2}$ or 10 $month^{-1/2}$	395
Excipient types—small	Incubation of lyophilized factor XIII at 2 mg/mL, pH 8, 40°C for 1 month in the absence or presence of 100 mM mannitol, trehalose, or sucrose	Aggregated (soluble) ~60%, ~40%, <10%, or <5%	375
Excipient types—large	Incubation of freeze-dried Met-hGH at 50°C with (1) 3X (wt) trehalose, (2) 3X (wt) dextran 40, and (3) 3X (wt) trehalose and 0.5X (wt) Tween 20	(1) <0.35%/month, (2) 7.3%/month, (3) 1.1%/month	364
Excipient levels	Storage of spray-dried IgG1:mannitol at 40°C for 52 weeks at weight ratios of 4:0, 4:1, and 1:4	15%, 10%, 18%	184
Similar sugars	Incubation of lyophilized hGH with sucrose or trehalose (1:2 weight ratio), pH 7.4, 50°C	2.4% or 2.0% insoluble aggregates per √week	365
Buffer species	Incubation of lyophilized factor IX with phosphate or histidine buffer at 30°C for 3 months	>2.5% or <0.2%	345

(*Continued*)

TABLE 4.1. *Continued*

Factors	Proteins/Conditions	% Aggregates or Aggregate Rate	References
Moisture	Incubation of the lyophilized hGH with a trehalose/GH ratio of 6 at 40°C for 6 months at a moisture levels of 0.3%, 3.3%, and 7.0%;	0.5%, 0.28%, 0.54%	397
	Incubation of the lyophilized hGH with a sucrose/GH ratio of 6 at 40°C for 6 months at a moisture levels of 0.5%, 3.1%, 4.9%	0.26%, 0.35%, 1.11%	397

IFN, interferon; PBS, phosphate buffered saline; FTU, formazin turbidity units; PLGA, poly(lactic-co-glycolic acid).

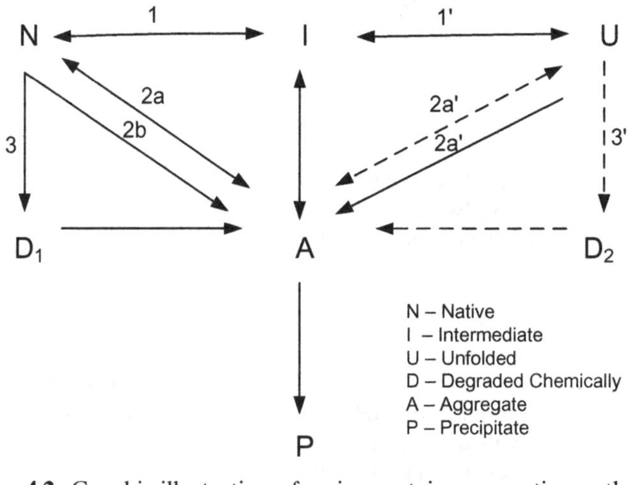

Figure 4.2. Graphic illustration of major protein aggregation pathways.

completely folded or unfolded proteins do not aggregate easily as the hydrophobic side chains are either mostly buried out of contact with water or randomly scattered.[14] Espargaro et al. evaluated the aggregation behavior of a domain of α-spectrin and 23 mutants as model globular proteins. It was demonstrated that the *in vivo*/*in vitro* aggregation propensities of these proteins do not correlate with the predicted aggregation propensities of the unfolded state but correlate with the conformational stability of these proteins, implying that partially folded intermediates are the aggregation-prone species.[15] Recent simulation studies indicated that aggregation of native proteins with accessible hydrophobic areas on its surface was intensified when the solution conditions favor the partially unfolded conformation as opposed to either the native or fully unfolded conformations.[16]

If the unfolding is the rate-limiting step, the aggregation kinetics are considered as unfolding-limited aggregation ($k_{obs} = k_{unf}$), one of the four major types of aggregation kinetics.[17] Even if unfolding is not rate limiting, partially unfolded conformers may still be key intermediates in the overall aggregation process.[17] Many proteins have been shown to aggregate through the initiation of unfolding intermediates (pathway 1 in Fig. 4.2). These proteins include the granulocyte colony-stimulating factor (GCSF),[18] scrapie amyloid (prion) protein (PrP27-30),[19] carbonic anhydrase B (CAB),[20,21] recombinant human growth hormone (rhGH),[22] insulin,[23] human lysozyme variants,[24] P22 tailspike polypeptide,[25,26] and phage P22 wild-type coat protein and its mutants (A108V, G232D, and F353L),[27] recombinant human granulocyte stimulating factor (rhGCSF),[28] bovine serum albumin (BSA),[29] β-lactoglobulin,[30] and IFN-tau.[31]

The initial protein aggregates (A state in Fig. 4.2) are soluble oligomers but gradually become insoluble as they exceed certain size and solubility limits (P state in Fig. 4.2). The P state can take on different forms such as amorphous precipitates or ordered structures such as fibrils, depending on the protein structures and experimental conditions. The amorphous precipitation and fibril formation have been described, respectively, as disordered and ordered aggregation pathways.[32] It should be noted that oligomerization may not be an obligate intermediate in the process of fibril formation, as there are compounds that can inhibit oligomerization but not fibrillization and compounds that inhibit fibrillization but not oligomerization of $A\beta_{42}$ at room temperature.[33]

4.2.2 Direct Aggregation through Self-Association or Chemical Linkages

Many proteins can directly associate into protein aggregates physically from the native (folded) state (pathway 2a in Fig. 4.2) without going through the intermediate state (I). Such associations can be simply electrostatic or both electrostatic and hydrophobic depending on the experimental conditions.[34] Others believe that such associations are mainly induced by two major types of interactions: electrostatic and van der Waals forces.[28] Protein self-association may or may not be accompanied by subtle conformational changes.[35] Such association often leads to formation of reversible oligomers/aggregates,[36] and these can be considered the precursors of irreversible aggregates/precipitates.[37] This process has been described as association-limited aggregation.[17] In the association process, multiple sites can participate in a single protein.[36]

A key parameter to measure the tendency of protein–protein self-association is the second virial coefficient—B_{22}. A positive value indicates protein–protein net repulsion while a negative value indicates protein–protein net attraction. In the latter case, protein–protein interactions are favored over protein–solvent interactions, potentially dominating protein aggregation.[38] The B_{22} values can be significantly different depending on the solution pH and ionic strength, as both affect the charge density/distribution of proteins. Even nonionic species could have a strong influence on the B_{22} value. Chi et al.[28] demonstrated that decreasing the salt concentration from 160 to 10 mM in PBS did not significantly affect the B_{22} value of rhGCSF, while addition of sucrose, a nonionic sugar, above 0.25 M at pH 7 increased the B_{22} value to a positive territory. On the other hand, a protein with a large and positive B_{22} value can still aggregate if the B_{22} value does not reflect the true value of the monomer form, as in the case of GCSF.[28] Another similar example is the self-association of insulin at room temperature.[39]

It is known that protein self-association is mainly related to colloidal stability, while formation of partially unfolded intermediates is mainly related to conformational stability. Either conformational or colloidal stability could be potentially rate limiting, depending on the solution conditions, such as the aggregation of rhGCSF.[28] In reality, it is sometimes difficult to differentiate between the two pathways (aggregation pathways 1 and 2a). Often, the relative B_{22} values do not indicate the relative aggregation tendency. For example, Bajaj et al.[40] demonstrated that ovalbumins readily aggregate during incubation at 37°C at pH 4.0, but no aggregation was observed at pH 7.4. However, similar B_{22} values were obtained in all solution conditions irrespective of the different aggregation tendencies. Therefore, it was concluded that the structural changes were sufficient to lead to long-term aggregation and it is unlikely that B_{22} will correlate with long-term aggregation behavior. This observation on ovalbumin aggregation was later confirmed by another study by Broersen et al.[41] It was demonstrated that changing the number of ovalbumin net charges from –1 to –26 did not change the aggregation behavior significantly and the dominant factor determining ovalbumin aggregation propensity was the rate of denaturation. In general, it suffices to say that minimization of protein surface charge will likely lead to increased aggregation. Recent experimental results suggest that the formation of particulates is a common property of all proteins when surface charge on the molecules is minimal, regardless of the specific amino acid sequence.[42]

Many chemical degradations directly cross-link protein chains, leading to aggregation (pathway 2b in Fig. 4.2). The most common cross-linking reaction is the intermolecular disulfide bond formation/exchange.[43–46] This is clearly the case for the heat-induced aggregation of a 1:1 protein mixture of β-lactoglobulin B (β-Lg) and κ-casein A (κ-CN).[47] It was found that Cys121 or Cys119 in β-Lg formed disulfide bonds with cysteines of other β-Lg and κ-CN molecules. Similar results were reported for heat-induced aggregation of pure β-Lg B or its 1:1 molar mixture with α-lactalbumin (α-La) in solutions.[48] Formation of disulfide linkages can further promote physical aggregation of proteins such as Z19 α-zein aggregation under nonaqueous conditions.[49] It is obvious that surface-located cysteines are more involved than internally located cysteins in participation in disulfide bond formation/exchanges.[48]

Disulfide-bonded proteins without any free cysteines can still undergo aggregation through disulfide exchanges via β-elimination. This has been reported for lyophilized insulin during storage.[50] Other non-disulfide cross-linking pathways leading to protein aggregation include formaldehyde-mediated cross-linking,[51,52] dityrosine formation,[53] oxidation,[54] and the Maillard reaction.[55,56]

4.2.3 Indirect Aggregation through Chemical Degradation

The last major aggregation pathway (pathway 3 in Fig. 4.2) is chemical degradation-induced protein aggregation. Chemical degradations often change the physical properties of a protein. Any degradation-associated change in protein hydrophobicity or association tendency could potentially lead to increased protein aggregation. Many chemical degradation pathways have been shown to increase the aggregation tendency of proteins. The most prevalent one seems to be oxidation. Oxidation of the well-protected Met4 residue in the variable domain of an IgG light chain induced noticeable secondary and tertiary structural changes and made this protein more sensitive to stirring-induced aggregation.[57] The metal-catalyzed oxidation of His and Met residues in human relaxin led to an increased aggregation tendency.[58] Exposure of soybean trypsin inhibitor (STI) to low levels of oxidant hypobromous acid (≤5-fold molar excess) results in selective oxidation of His, Trp, and Tyr residues, followed by protein aggregation.[59] Autoxidation of the two cysteine residues in the denatured interleukin-1β (IL-1β) led to hydrophobic/covalent aggregation and precipitation in aqueous solution at or above 39°C.[60] Other chemical degradations can also increase the aggregation tendency. Cleavage between Asp and Pro residues in the CH2 domain of an IgG2 molecule at 37°C induced protein aggregation.[61] Deamidation of mutant βA3 crystallins altered the tertiary structure as monitored by circular dichroism (CD) and led to the formation of 15–20% soluble aggregates as monitored by sedimentation velocity experiments in all deamidated proteins.[62] Similarly, deamidation of Gln-54 and/or Gln-143 in human γ D-crystallin lowered both the thermodynamic and kinetic barrier to the unfolding of the N-terminal domain and made the protein more susceptible to GdnHCl-induced aggregation at pH 7.0.[63]

4.2.4 Protein Aggregation from the Unfolded State

Theoretically, the completely unfolded state (U) of a protein can form aggregates directly (pathway 2a′ in Fig. 4.2). This is true for many proteins that have been shown to be largely in an unfolded state naturally or to conform only to two apparent states: native and unfolded. For example, von Bergen et al.[64] demonstrated that the microtubule-associated protein tau, a natively unfolded protein in solution, is able to polymerize into ordered paired helical filaments (PHFs). α-Synuclein is a natively unfolded molecule that can self-aggregate to form oligomers and fibrillar intermediates.[65] Similarly, a mutant transthyretin (ΔN50) adopts a largely unfolded conformation and shows a strong tendency to aggregate.[66] In the study of aggregation of a glycoprotein *Peniophora lycii* phytase (Phy) and a deglycosylated form (dgPhy), Hoiberg-Nielsen et al.[67] demonstrated that the aggregation rate of Phy at the denaturation temperature is clearly proportional to the concentration of thermally denatured (unfolded) protein, supporting a proposed model of protein aggregation $N \leftrightarrow D \rightarrow A$. On the other hand, a similar but less pronounced correlation was found for dgPhy, suggesting that the aggregation process for the deglycosylated form is dominated by denatured proteins, but the process also involves contributions from unfolding intermediates and/or native proteins.

The unfolded state can also undergo chemical degradation and form aggregates directly (pathway 2b′ in Fig. 4.2) or indirectly (pathway 3′ in Fig. 4.2). For example, Hu et al.[57] demonstrated that the Met4 residues in the variable domain of an IgG light chain is well protected in the native state but can be oxidized in the denatured state in the presence of 4M GdnHCl, and the oxidized protein is more sensitive to stirring-induced aggregation.

4.2.5 Aggregation Nucleation

The process from N to A via pathway 1 or 2a can be considered as the nucleation step, which is often rate limiting, that is, nucleation-dependent aggregation. The aggregation of bovine carbonic anhydrase has been suggested to follow a nucleation-dependent polymerization mechanism.[68] This kind of aggregation kinetics has been described as aggregation by nucleation and growth,[17] and alternatively, nucleation–polymerization.[69] Nucleation has been demonstrated or suggested to be the initial step for the fibrillation of a sequence of *Escherichia coli* OsmB protein,[70] aggregation of K97I IL-1β during refolding,[71] aggregation of β-amyloid peptide,[72,73] and thermally induced aggregation of α-chymotrypsinogen A and bovine GCSF in acidic solutions.[74] Protein aggregation in these cases could follow apparent first-order kinetics due to rate-limiting nucleation, followed by a monomer addition process. For example, first-order kinetics of protein aggregation have been observed for bovine pancreatic chymotrypsin[75] and IFN-tau.[31] The rate-limiting step in IFN-tau aggregation is a conformational change from the compact native state to a slightly expanded aggregation-competent state.

A lag time before aggregation often signifies the nucleation step. Such a lag time has been observed in the aggregation of many proteins, including acidic fibroblast growth factor (aFGF) at pH 7.4,[76] recombinant human interleukin-1 receptor antagonist (rhIL-1ra) at pH 6.5,[77] α-chymotrypsin (CT) under thermal stress,[78] apomyoglobin in aqueous urea solutions,[79] and firefly luciferase at pH 7.5.[80] The lag phase can take as long as several days depending on the solution conditions, such as aggregation of α-synuclein at 1 mg/mL.[81]

The definition of an aggregation nucleus has not been described consistently. In the study on the aggregation of protein L in 25% (vol/vol) 2,2,2-trifluoroethanol (TFE) and 10 mM HCl, Cellmer et al.[82] demonstrated that the aggregation kinetics is a linear elongation of polymers (leading to fibril formation) with a critical nucleus size of either two or three monomers. Baynes et al.[68] suggested that a small multimer is the nucleus for the aggregation of bovine carbonic anhydrase.

4.2.6 Multiplicity in Protein Aggregation

It is worth noting that multiple aggregation pathways often occur for a single protein. For example, Sun and Hayakawa[83] demonstrated that complete aggregation of ovalbumins was observed upon incubation at 80°C for 15 min, and aggregates were both disulfide bonded and physically linked. Insulin can undergo both a physical aggregation process, leading to the formation of either soluble hexamers or insoluble fibrils, and a chemical aggregation process, leading to formation of either soluble dimers via cyclic anhydride intermediate or insoluble disulfide-bonded aggregates.[84–88] The aggregation of an IgG2 molecule (at pH 5.0 at 37°C) consists of two major aggregation pathways. The first of these involves formation of both dimers and high-molecular-weight (HMW) aggregates with altered secondary and tertiary structures, containing cleaved antibody molecules in the CH2 domain. The second major aggregation pathway for IgG2 molecules entails the formation of a dimer mainly through both disulfide and nondisulfide (approximately 26%) linkages, having identical secondary and tertiary structures as the intact antibody molecule.[61] Other similar examples include basic fibroblast growth factor (bFGF)[89] and β-galactosidase.[45]

Physical aggregation processes alone can also take different forms in a single protein—amorphous aggregation and fibril formation, depending on the experimental conditions. It has been suggested that under conditions of high net charge, amyloid fibrils form, whereas under conditions of low net charge, particulates form.[42] This has been clearly demonstrated in the study of the aggregation behavior of seven different proteins, including bovine insulin, bovine β-lactoglobulin, human transthyretin, horse heart myoglobin, hen egg white lysozyme (HEWL), BSA, and human α-synuclein.[42] All of these proteins are also known to form amyloid fibrils, but a substantial number of spherical particulates formed with relatively monodisperse diameters upon incubation of the protein solutions at 90°C in the presence and absence of 0.1 M NaCl.[42] Therefore, prevention of protein aggregation needs to target all aggregation pathways.[90]

4.3 EFFECTS OF TEMPERATURE

Temperature is a measure of the average kinetic energy of a certain kind of vibrational motion of the constituent particles—molecules, atoms, and subatomic particles. The mean kinetic energy (E_{mean}) of the translational motion of an individual atom or molecule (particle) in a gas state is a proportional function of temperature as described below:[91]

$$E_{\mathrm{mean}} = \frac{3}{2} k_{\mathrm{B}} T, \tag{4.1}$$

where k_B is the Boltzmann constant and T is the absolute temperature. In a more relevant liquid state, the Stokes–Einstein equation relates the diffusion coefficient of a species (D) directly to the absolute temperature (T):[92]

$$D = k_{\mathrm{B}} T / (6\pi\eta R_h), \tag{4.2}$$

where R_h is the hydrodynamic radius of the species, and η is the fluid viscosity. The diffusion coefficient is not only proportional to the absolute temperature but is also inversely proportional to the hydrodynamic radius. Therefore, changing the temperature of a protein solution or solid would alter the diffusion of the molecules—a necessary step for physical protein aggregation. In addition, any change in molecular motion would change the chemical reactivity of the molecules, which could affect protein aggregation through chemical degradations. Therefore, temperature is arguably the most common and critical factor for consideration when proteins are handled during the entire development and commercialization processes.

4.3.1 Relationship between Temperature and Protein Stability

The thermodynamic stability of a protein can be measured by its unfolding free energy change (ΔG_{unf}) for the transition from the native (folded) to the unfolded state (Fig. 4.2). Generally, the higher the ΔG_{unf}, the more likely the protein is to remain in its native state and the less likely the protein will aggregate. However, many proteins have very limited stability, with ΔG_{unf} in the range of only 5–20 kcal/mol.[93–96] The relatively complex relationship between ΔG_{unf} and temperature (T) has been derived in different forms and can be described in the following equation under certain pressure and solution conditions:[97,98]

$$\Delta G_{\mathrm{unf}} = \Delta H_{\mathrm{vH}} (1 - T/T_{\mathrm{m}}) - \Delta C_{\mathrm{p}} (T_{\mathrm{m}} - T + T \ln(T/T_{\mathrm{m}})), \tag{4.3}$$

where ΔH_{vH} is the van't Hoff enthalpy of thermal unfolding, T_m is the melting temperature, and ΔC_p is the heat capacity change of thermal unfolding (assuming temperature independency of ΔC_p). When $T = T_m$, $\Delta G_{unf} = 0$. The relation-

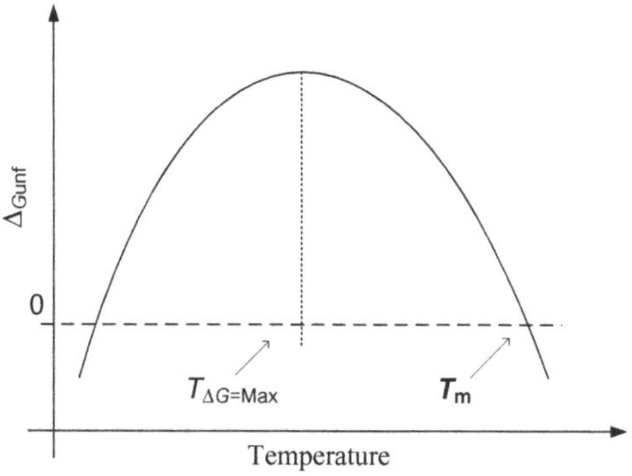

Figure 4.3. Relationship between temperature and protein thermodynamic stability.

ship between ΔG_{unf} and T is also graphically illustrated in Fig. 4.3.[98] It is clear that the maximum free energy change of protein unfolding resides in a narrow temperature range and proteins are usually relatively stable in this narrow temperature range. Decreasing or increasing temperature will destabilize the protein and can potentially promote protein aggregation. Several globular proteins have maximum free energy change ΔG_{unf} (8–17 kcal/mol) in a temperature range between 10 and 40°C, including basic pancreatic trypsin inhibitor, CT, cytochrome c, lysozyme, metmyoglobin, RNase A, and trypsin.[99] Therefore, some proteins could aggregate easily at room temperature, such as insulin[100] and a therapeutic immunoglobulin.[101]

4.3.2 Effect of High Temperature on Protein Aggregation

High temperature can have several effects on proteins, leading to accelerated formation of aggregates. These potential effects include thermodynamic instability, hydrophobic interaction, diffusion of proteins, and chemical reactions. According to Fig. 4.3, increasing the temperature would eventually pass the temperature of maximum thermodynamic stability and would gradually reduce protein stability. Destabilization promotes generation of protein unfolding intermediates, leading to aggregation, such as in the case of ovalbumin,[102] and a mutant of Cro repressor protein.[103] Thus, high temperature is a common parameter to be selected for accelerating protein aggregation, albeit this can be convoluted with non-Arrehenius behavior, which makes quantitative predictions more difficult.[17] In general, thermally induced protein aggregation is often irreversible, such as in the case of thrombin,[104] recombinant porcine growth hormone (pGH),[105] and recombinant human megakaryocyte growth and development factor (rhMGDF).[106] The reversibility does depend on the stage of the aggregation process and on the size of the aggregates.[107]

As discussed above, proteins are stable only in a narrow temperature range. Increasing the temperature slightly above room temperature may cause some proteins to aggregate on much shorter timescales. At 37°C, many proteins aggregate easily, including GCSF 1.0–1.5 mg/mL at pH 7,[18,108] recombinant human keratinocyte growth factor (rhKGF) at 0.5 mg/mL at pH 7,[109] bovine insulin at pH 2.5,[110] tubulin at pH 6.4,[111] recombinant human platelet-activating factor acetylhydrolase (rhPAF-AH, pI = 6.7) after reconstitution (also at room temperature),[112] rhIL-1ra,[113,114] IgG molecules at different pH values,[61,115,116] and mutant transthyretin (ΔN50) at pH 7.5.[66] Other examples include aggregation of human serum albumin (HSA) at 40°C,[117] aggregation of firefly luciferase in 25 mM tricine at pH 7.5 at 42°C,[80] and aggregation of recombinant factor VIIa (rFVIIa) at different pH values at 45°C.[118]

In addition to the influence on protein stability, higher temperature increases protein diffusion, frequencies of molecular collisions, and hydrophobic interactions, promoting protein aggregation.[13,25] Roy et al.[114] demonstrated that increasing the temperature from 25 to 37°C changes the nature of the interaction between 8-anilinonaphthalene-1-sulfonate and rhIL-1ra from electrostatic to both electrostatic and hydrophobic, and increasing temperature also increased the degree of hydrophobic interaction between benzyl alcohol and rhIL-1ra. Vetri and Militello[30] found that the aggregation rate of β-lactoglobulin gradually increased with incubation temperature from 10 to 80°C mainly through hydrophobic interactions. The aggregation lag time observed at low temperature can be reduced at high temperatures such as in the example of the aggregation of human IFN-γ aggregation at 1 μM,[119] and insulin at 20 μM,[120] and holo-α-lactalbumin.[121] The decreased lag time with increasing temperature is likely a result of a combination of decreased stability, increased hydrophobic interaction, and increased protein diffusion/collision.

Proteins become partially unfolded when the temperature is close to the T_m. Therefore, proteins often aggregate rapidly near the T_m.[41] Further increasing the temperature would eventually cause the protein to melt/unfold at T_m (Fig. 4.3). Thermally induced protein unfolding is often followed/accompanied by immediate aggregation due to exposure of the hydrophobic residues. Examples of such proteins include hemoglobin,[122] *Artocarpus hirsuta* lectin,[123] horse liver alcohol dehydrogenase (ADH),[124] asparaginase-2 or some of its mutants,[125] aminotransferase from pig heart mitochondria,[126] lysozyme,[127] and glycoprotein *P. lycii* phytase (Phy) and a deglycosylated form (dgPhy).[67] A further increase in temperature above the T_m often leads to a more rapid aggregation. Examples include KGF-2 (fibroblast growth factor [FGF]-10),[128] lysozymes,[129,130] BSA,[131] holo-α-lactalbumin,[121] recombinant factor VIII,[132] and β-Lg and κ-CN.[47]

When the aggregation process is composed of two distinct steps, nucleation and growth, temperature may affect each step to a different degree. Andrews et al.[133] found that the aggregation of α-chymotrypsinogen A is a function of temperature under acidic conditions and the nucleation has a large, negative activation energy (ca. −76 kcal/mol), while growth has a weak dependence on temperature. In comparison, formation of large nonsoluble aggregates of bovine pancreatic chymotrypsin at 25°C at pH 2.5 fits first-order kinetics with a low activation energy value of 4.2 ± 0.2 kcal/mol.[75] Increasing temperature may also change the pathways of protein aggregation depending on the solution conditions. Fedurkina et al.[134] found that the first-order rate constant for creatine kinase aggregation did not depend on protein concentration at 51°C, but did at 60°C. Therefore, a change in temperature from 51 to 60°C apparently changed the kinetic regime of thermal creatine kinase aggregation. While the aggregation rate of creatine kinase at 51°C was limited by the stage of protein denaturation, it is limited at 60°C by the stage of aggregate growth.[134]

High temperature can affect protein aggregation indirectly by changing the relative composition of secondary structures, resulting in different aggregation behavior.[135] High temperature can certainly accelerate chemical degradations in proteins to promote protein aggregation indirectly. For example, formation of covalent insulin oligomers and polymers was increased at temperatures >25°C.[88]

4.3.3 Melting Temperature (T_m) and Protein Aggregation

The melting temperature of proteins generally represents their relative thermal stability.[136] The higher the T_m, the less likely the protein will unfold. Indeed, Rees and Robertson[98] analyzed a large number of globular and water-soluble proteins in terms of thermodynamic protein stability and T_m and found a general positive relationship between protein thermodynamic stability and T_m. An increase in stability by 0.008 kJ/mole/residue is, on the average, associated with a 1°C increase in T_m, while it is 1.01 kJ/mole for a series (a total of 15) of staphylococcal nuclease variants. Apparently, a protein of a higher T_m would, in general, have a less tendency to aggregate.[28]

In many cases, however, proteins do not aggregate easily at or even above their T_m. This is the case for HEWLs, which have a T_m of 55°C at pH 2, but aggregation does not occur immediately at this temperature.[129] Thermal unfolding of an IL-1ra mutein (K93A, T_m = 63°C) did not lead to immediate aggregation and was almost completely reversible as monitored by CD.[77] Therefore, prediction of protein aggregation based on T_m values will not be reliable. Because of this, a higher T_m does not necessarily mean a lower tendency in protein aggregation.[137] For example, Derrick et al.[128] found that a variety of polyanions (e.g., heparin, sucrose octasulfate [SOS], and inositol hexaphosphate) increased the T_m of KGF-2 (T_m = 37°C) by approximately 9–15°C, but above 40°C, aggregation occurred to a greater extent in the presence of these polyanions. Changing the pH of a model monoclonal antibody (mAb) (IgG_2) solution from 4.0 to 9.0 increased the thermal stability with more compact conformations, but the rate of aggregation increased at 37°C.[138] In reality, changing the experimental conditions can dramatically change the aggregation tendency of a thermally unfolded protein.[137] It is the actual solution conditions that often control protein aggregation.

4.3.4 Effects of Low Temperature

As shown in Fig. 4.3, decreasing temperatures also destabilize or denature proteins eventually. RNase undergoes cold denaturation at −22°C under 3000 bar (to prevent freezing)[139] and serum albumin undergoes cold denaturation at a temperature lower than 0°C in solution.[140] Destabilization at low temperature may facilitate generation of unfolding intermediates, promoting protein aggregation. On the other hand, due to weaker hydrophobic interactions at low temperatures, aggregation may not occur easily at these conditions. Probably due to such a reason, low-temperature-induced protein aggregation is often reversible, unlike high-temperature-induced protein aggregation. The well-known low-temperature-induced aggregation (below 37°C) of serum cryoglobulins is reversible.[141] Human IgM cryoglobulin preparations easily precipitate or gel at temperatures below 10–12°C, and the process is reversible at higher temperatures[142] The low-temperature-induced aggregation of cryoglobulins is poorly understood and was thought to involve sites within both Fab and Fc.[142] In a recent study, aggregation of IgG1 at low temperature was found to be reversible as measured by light scattering.[143]

Low temperatures below the freezing point can also promote protein aggregation by changing the physical properties of the frozen solutions. A frequent observation is the crystallization of a crystalline excipient, leading to phase separation, loss of protein protection, and protein aggregation during low-temperature storage above the glass transition temperature (Tg′) of the maximally freeze-concentrated solution. Piedmonte et al.[144] demonstrated that an Fc fusion protein formulated in sorbitol formed aggregates upon frozen storage at −30°C due to crystallization of sorbitol, as the Tg′ of the solution is as low as −45°C. Similarly, sorbitol crystallization is also the cause of a 2.4% increase in dimer and aggregate content of a Fc fusion protein at −30°C after 68 weeks.[144] In the presence of a polymeric excipient, freezing can easily cause phase separation, promoting protein aggregation.[145,146]

4.4 EFFECTS OF SOLUTION CONDITIONS AND COMPOSITION ON PROTEIN AGGREGATION

Proteins are expressed/purified generally in an aqueous environment and many commercial protein products are in a liquid dosage form. Therefore, the solution condition and its composition need to be carefully controlled, as the tendency of protein aggregation can be significantly different under various conditions. These solution conditions/factors could potentially influence protein aggregation directly or could indirectly contribute to the overall rate of protein aggregation in solution. These include solution pH, buffer species and concentration, ionic strength, excipients, protein concentration, metal ions, nonaqueous solvents, impurity, and container/closures.

4.4.1 Solution pH

Proteins are made of amino acids and are therefore amphoteric in nature. The solution pH dictates the type and distribution of surface charges on proteins.[147,148] These charges not only affect intramolecular folding interactions but also intermolecular protein–protein interactions. Due to the delicate nature of the folded states balancing both intra- and intermolecular interactions, proteins are usually stable only in a narrow pH range.[149,150] Therefore, the type and distribution of protein surface charge is a key parameter in determining the rate of protein aggregation, in addition to proteins/polypeptide sequence hydrophobicity and propensity of forming secondary structures, and so on.[151,152] Changing the solution pH affects the aggregation of many proteins, including an interleukin-2 mutein,[153] mAbs,[116,154] basic FGF,[43] deoxy hemoglobin,[155] IL-1β,[60] rhIL-2,[156] human relaxin,[58] bovine pancreatic ribonuclease A (RNase A),[2,157] and bovine insulin.[23]

4.4.1.1 Protein Aggregation under Neutral and Moderate Acidic/Basic Conditions Many proteins are prone to aggregation under moderate acidic/basic conditions or even at a completely neutral pH. Proteins aggregate at a neutral pH mainly because protein–protein interactions or surface hydrophobicities are favored. Sun and Hayakawa[83] found that the surface hydrophobicity of ovalbumins markedly increased when the pH was decreased from 8.0 to pH 7.0. Additional examples include aggregation of rhGCSF at 37°C at pH 7[158] and W7FW14F apomyoglobin at pH 7.0 at room temperature.[120] Other proteins are prone to aggregation at a pH slightly below or above the neutral condition, including firefly luciferase at pH 5.5 at 30°C,[80] a botulinum neurotoxin (50 kD) at pH 8 at 50°C,[159] an Fc fusion protein at pH 3.6 at room temperature,[160] a mAb and ovalbumin at pH 4.0 at 37°C,[40] and a mouse mAb (mIgG1) at pH 3.9 at or above room temperature.[161]

Protein aggregation caused by pH changes can be simply an effect of altered charge–charge interactions and/or pH-induced protein destabilization/partial unfolding. For example, the aggregation tendency of GCSF is significantly higher at pH 6.1 or 7.0 than at pH 3.5.[28] Although GCSF remains conformationally native and compact from pH 2 to 7, B_{22} values of the protein changed from positive to negative as the pH was increased from 3.5 to 6.1 or 7.0, indicating that pH-induced protein–protein attractions are likely responsible for the rapid aggregation at neutral pH.[28] Several other studies have shown that pH-induced protein destabilization is a frequent factor. Bajaj et al.[40] found that aggregation of a mAb was accelerated at pH 4.0 compared to pH 7.4, apparently due to partial unfolding of the protein at pH 4.0. In the case of ovalbumin, a low pH (4.0) caused not only partial unfolding of the protein but also complete unfolding of a certain fraction of the protein, while similar B_{22} values were obtained in all solution conditions.[40] Similarly, the slower aggregation of rhPAF-AH (pI = 6.7) in 50 mM sodium bicarbonate at pH 8 than in water at pH 6.5 was probably due to pH-induced protein destabilization because the B_{22} value was lower in the bicarbonate solution.[112] These studies suggest that B_{22} value alone would not be able to correlate well with long-term protein aggregation.

4.4.1.2 Protein Aggregation at Extreme pH values At extreme pH values, proteins are heavily charged. The dense charges on the protein surface significantly increase repulsive intramolecular/intermolecular interactions. The increased intramolecular interactions often lead to at least partial protein unfolding, exposing hydrophobic patches and potentially increasing protein aggregation. For example, protein L can aggregate and form fibrils rapidly after a lag phase in a 10 mM HCl solution.[82] Glycodelin A (GdA, pI = 4.7) has a stable dimeric structure at the alkaline pH, but a low pH of 2 increases its tendency for aggregation.[162]

In many cases, pH-induced protein unfolding does not lead to immediate protein aggregation due to the strong intermolecular charge–charge repulsions. Fibrillation of bovine core histones (unfolded protein) in the presence of 100 mg/mL PEG-3500 was enhanced at pH 7.0 but was inhibited at pH 2.0.[163] Gaikwad and Khan[123] studied pH-dependent aggregation of oligomeric *A. hirsuta* lectin and found that thermal denaturation (45°C for 15 min) of the protein led to aggregation, but aggregation of the denatured protein did not occur in the pH ranges of 1–3 at 10–12°C. Aggregation of bovine pancreatic chymotrypsin alone leads to formation of large soluble aggregates in 9 h at 25°C at pH 3.5, but no detectable change was observed at pH 2.5.[75]

Another reason for failed protein aggregation at extreme pH values is the failure to alter the protein conformation or formation of even "acid-stable" conformations. For example, recombinant factor XIII (rFXIII) maintains a compact structure and activity in a wide pH range (between pH 3.6 and 10.4).[164] Ejima et al.[161] demonstrated that incubation of a humanized mAb, hIgG4-A, at pH 2.7 and 3.5 at 4°C for 24 h caused little change in the near-UV CD spectra, indicating that the acid conformation is stable. β-Lactoglobulin A (pI = 5.1) has a similar β-sheet conformation both at pH 7 and 2.[162]

4.4.1.3 Protein Aggregation versus pI Theoretically, protein solubility is minimal at its pI due to minimal protein charge–charge repulsions. Recent results suggest that significant protein–protein interactions can still exist from the marked charge anisotropy of protein around its pI, such as insulin.[100] Nevertheless, the rate and extent of protein aggregation should be maximal around the pI and should be negatively correlated with the absolute value of charge.[151,152] Indeed, Militello et al.[29] found that the aggregation of BSA increased as pD is adjusted closer to its pI (about 5) at 58°C. Giger et al.[100] studied aggregation of insulin in the pH range of 3–9 and found that aggregation (turbidity) of insulin at room temperature was fastest at pH 5.6 (pI = 5.5) in a 10 mM NaCl solution.

However, this is certainly not the case for all proteins. Some proteins actually aggregate at a lower rate around their pI values. Majhi et al.[165] found that the initial aggregation rate of β-lactoglobulin at 1 mg/mL in 4.5 mM NaCl is bell shaped at ambient temperature with a maximum rate at pH 4.6–4.7 in the pH range of 3.8–5.2, and the aggregation rate at its pI (pH 5.2) is 30 times lower than that at pH 4.7. Similarly, nonnative aggregation of rhGCSF in PBS solution can take place easily at pH 6.9 at 1.5 mg/mL (with loss of 96% monomer), but only 30% is lost at pH 6.1 (the protein's pI) upon incubation at 37°C for 5 days.[28] Therefore, the aggregation behavior of a protein should also be examined at its pI, and it should not be assumed that this will necessarily be the worst case.

4.4.1.4 Indirect Effect of pH on Protein Aggregation The pH of a solution affects not only the aggregation behavior of a protein directly but also the protein interaction with excipients, indirectly affecting protein aggregation. Chung et al.[166] found that dextran sulfate interacts strongly and electrostatically with BSA and that the interaction between dextran sulfate and BSA varied significantly with the pH of the solution. This led to a different extent of aggregation prevention by dextran sulfate. Dextran sulfate suppressed BSA aggregation at pH 5.1 and 6.2, while it had no effect at pH 7.5. Another example is the aluminum-induced aggregation of BSA, where increasing pH from 7.4 to 9.0 gradually increased the lag time and rate of aggregation.[167]

4.4.1.5 pH and Aggregate Morphology Protein aggregates come in a variety of forms, and the solution pH plays a significant, if not dominant, role in controlling the morphology of those aggregates. In a study on the aggregation of HEWLs, Arnaudov and de Vries[129] found that at pH 2.0 and 57°C, both fibrillar and spherical aggregates were formed, but at pH 3 and 4, only spherical aggregates were observed. Rana et al.[168] demonstrated that the incubation of bovine carbonic anhydrase II (BCA II) at 37°C under mildly denaturing conditions at pH 3.5 formed ordered amyloid aggregates, but amorphous aggregates were formed upon thermal incubation of the protein at 75°C for 5 min at pH 6.5.

Recent investigations have shown that the morphology of aggregates depends on the closeness of the solution pH to its pI. Krebs et al.[42] studied seven widely different, nonrelated proteins that are known to form amyloid fibrils and demonstrated that around their pI values, a different generic form of aggregates could also occur. Under these conditions, gels consisting of relatively monodisperse spherical particulates are formed. Thus, under conditions of high net charge, amyloid fibrils form, whereas under conditions of low net charge, particulates form. Therefore, these results suggest that the formation of particulates in the regime where charge on the molecules is minimal is a common property of all proteins.[42] It was proposed that protein misfolding into clearly defined aggregates is a generic process, and the outcome depends solely on the general properties of the protein state during aggregation rather than on the specific amino acid sequence. Similar results were obtained in a different study where Militello et al.[29] found that the aggregation of BSA proceeds through the formation of ordered aggregates (oligomers) at pD far from pI (about 5), but disordered structures increase as the pD is closer to its pI upon incubation at 58°C for 2h.

4.4.1.6 pH versus Aggregation Mechanisms Changing the solution pH could potentially change the aggregation mechanisms, as pH can affect protein aggregation through several possible mechanisms—including altered charge–charge interactions, partial or complete unfolding of proteins, and altered chemical degradation rates and pathways. Van Buren et al.[61] found two major aggregation pathways in an IgG2 molecule at pH 4.0 and 6.0 at 37°C. The major aggregation pathway at pH 4.0 is the formation of both dimers and HMW containing cleaved antibody molecules (in the CH2 domain), while at higher pH 6.0, the major pathway is the formation of a covalent dimer while formation of higher-molecular-weight aggregates was largely suppressed. The covalent dimer consisted of both disulfide-linked antibody molecules and another species (approximately 26%). Since the thiolate ions are often the initiating species for disulfide formation/exchange, increasing formulation pH usually leads to increased formation of disulfide-linked aggregates.[169]

4.4.2 Type and Concentration of Buffering Agents

As discussed above, pH strongly affects the rate of protein aggregation and many proteins are stable only in a narrow pH range. Therefore, a buffering agent is often used in solutions to maintain an optimum pH. Although many buffering agents are available for pH adjustment, the aggregation behavior of proteins can be significantly different in different buffer systems.[43,106,170,171]

Phosphate buffers have been shown to have a negative effect on protein aggregation in several studies. Katayama et al.[31] studied the effects of buffering species on the aggregation of IFN-tau at pH 7 at 50°C and found that the protein aggregation rate was fastest in phosphate buffer, slower in Tris buffer, and slowest in histidine buffer. The rate constants for the formation of soluble aggregates were 0.0133/min, 0.0086/min and 0.0076/min, respectively, in the above buffers. Therefore, it appears that histidine enhances the thermal stability of the protein relative to other buffering agents, and the stability enhancement/aggregation inhibition was apparently due to its weak binding to the native state of IFN-tau. Similarly, Raibekas et al.[77] demonstrated that aggregation of rhIL-1ra at 107 mg/mL at 40°C with a lag time (about 50 min) was significantly faster in a 10 mM phosphate buffer than in a 10 mM citrate buffer.

Buffer concentration can also influence the aggregation behavior of proteins to a great extent.[43,76,170–172] Raibekas et al.[77] demonstrated that the aggregation rate of rhIL-1ra at 140 mg/mL at pH 6.5 in the presence of 80 mM NaCl was reduced with increasing concentrations of both phosphate or citrate buffers from 0 to 80 mM. At 40 mM, the aggregation rate was reduced by 40% and 98% for phosphate and citrate, respectively. In the same study, it was found that phosphate has approximately fourfold weaker aggregation–inhibition than either citrate or pyrophosphate, and this is apparently in agreement with the strength of protein–anion interactions at the IL-1ra single anion-binding site as judged by the estimated dissociation constant values of 2.9, 3.8, and 13.7 mM for pyrophosphate, citrate, and phosphate, respectively.

4.4.3 Ionic Strength

Ionic strength is one of the key solution parameters that potentially affect protein aggregation. Both positive and negative ions in a solution can interact eletrostatically with proteins. Such interactions can lead to different conformational states, such as β-Lg, leading to different aggregation behavior. Some aggregation processes contain two clear stages—nucleation and growth—and ionic strength may have different effects in different aggregation stages.[35] The salt type can also make a significant difference in protein aggregation.[173]

4.4.3.1 Ionic Strength and Protein Aggregation Increasing ionic strength often neutralizes surface charges of proteins, and reduces intra- and intermolecular charge–charge interactions due to electrostatic screening.[100] If the interactions favor protein folding/stability, reduction of such interactions would destabilize the protein, partially expose the hydrophobic patches, and lead to increased protein aggregation. For example, the rate of thermally induced holo-α-lactalbumin aggregation at pH 7.0 increased with increasing NaCl concentrations from 0 to 60 mM.[121] Baussay et al.[174] found that increasing the ionic strength of β-lactoglobulin solution increased protein aggregation by reducing the critical association concentration (CAC) and by reducing the critical gel concentration, Cg. Other examples include enhancement of rhGCSF aggregation in the presence of 150 mM NaCl[28] and increased rate of β-lactoglobulin aggregation with increasing ionic strength.[175,176] Therefore, protein solubility is often reduced at a higher ionic strength such as chicken egg white lysozyme.[177] It should be noted that such aggregates cover a range of different types (reversible, irreversible, native, nonnative, etc.).

In contrast, if intra- and intermolecular charge–charge interactions do not favor protein folding/stability, reduction of such interactions would stabilize the protein and inhibit protein aggregation. Indeed, some proteins are stabilized and are less likely to aggregate at higher ionic strengths. For example, Alford et al.[35] found that incubation of rhIL-1ra at 37°C causes protein aggregation and that the rate of aggregation decreases with increasing ionic strength from 0.025 to 0.184 M (4.5 mM citrate). The initial aggregation rate of β-lactoglobulin at pH 5.0 at ambient temperature decreased with increasing ionic strength from 0.0045 to 0.1 M.[165] The aggregation rate of both glycosylated and deglycosylated *P. lycii* phytase (Phy) at the denaturation temperature was reduced when NaCl was increased from 0 to 0.5 M.[67]

More complex relationships have also been reported between protein aggregation and ionic strength. Giger et al.[100] studied the aggregation of insulin at pH 5.6 (pI = 5.5) and found that the aggregation (turbidity) of insulin at

0.1 mg/mL at room temperature was fastest in a 10 mM NaCl solution and was reduced at other ionic strengths such as 0.5, 3.0, 30.0, or 100.0 mM. The aggregation of apoferritin at pH 5.0 is bell shaped with increasing Na^+ concentrations, which could be explained by a well-shaped relationship between B_{22} and Na^+ concentration.[37] Therefore, the effect of ionic strength is protein dependent and can vary depending on the solution conditions.

Ionic strength is commonly adjusted with NaCl. Other inorganic salts can also be used and may exert a significant effect on protein aggregation, possibly through binding to proteins. Both Na_2SO_4 (0.22 or 0.44 M) and NaSCN (0.36 or 0.73 M) have been found to reduce both the size and number (concentration) of aggregate particles of BSA under thermal stress with a greater effect at a higher concentration.[178] While the chaotropic anion SCN^{-1} considerably diminished the aggregation of CT in the presence of 1 mM phenyl methane sulfonyl fluoride (PMSF) at 65°C at pH 7.0, two other negative ions, Cl^{-1} and especially SO_4^{-2}, at identical ionic strengths, enhanced the heat-induced aggregation of CT.[78]

4.4.3.2 Ionic Strength and Chemical Aggregation Ionic strength can affect chemical reactions and can indirectly affect protein aggregation. For example, Livney et al.[48] found that addition of 100 mM NaCl increased the amount of precipitates, in the form of flakes, in pure β-Lg solution and its 1:1 molar mixture with α-La at pH 6 upon incubation at 85°C. The precipitates were formed mainly through disulfide bond formation.

4.4.3.3 Ionic Strength and Other Stability-Influencing Factors Ionic strength may have a different degree of effect on protein aggregation depending on the presence of other stability-influencing factors. One such factor is the glycosylation state of a protein. Hoiberg-Nielsen et al.[67] studied the kinetics of thermally induced aggregation of a glycoprotein *P. lycii* phytase (Phy) and its deglycosylated form (dgPhy). It was found that increasing the ionic strength from 0 to 0.5 M NaCl at the denaturation temperature slowed the aggregation rate of Phy by a factor of 2 and slowed the aggregation rate of dgThy by a factor of 10.

The ionic strength also affects the aggregation of proteins induced by other agents. Rezaei-Ghaleh et al.[179] found that the log initial rate of TFE-induced aggregation of CT at 25°C decreased rather linearly with the square root of ionic strength (10–40 range), using several salts including NaCl, KCl, NH_4SCN, and so on. Certain lipids do not influence the aggregation of insulin at 0.05% (w/w) at pH 1.9 at 60°C, but in the absence of 0.1 M NaCl at pH 1.9, they enhanced the rate of aggregation.[180]

Ionic strength can influence the physical state of the aggregates. Baussay et al.[174] found that increasing the ionic strength of β-lactoglobulin solution increased the local density of the aggregates. Addition of 0.1 M NaCl increased the size of particulates of several proteins (bovine insulin, bovine β-lactoglobulin, human transthyretin, horse heart myoglobin, HEWL, BSA, or human α-synuclein) induced by incubation of the protein at 90°C.[42]

4.4.4 Excipients/Additives

Proteins can encounter a variety of excipients/additives in solution, which could affect the aggregation behavior of proteins. Some additives stabilize proteins, while others destabilize proteins and promote aggregation (Table 4.2). The degree of protein stabilization or destabilization can be significant. In the study on the thermal stability of HEWL at 100 μM, Petersen et al.[127] found that the melting temperature was significantly increased in a wide pH range of 3.5–9.5 in the presence of 1 M sorbitol, and an increase of as high as 9.5°C was observed at pH 9.5. These excipients/additives include small neutral additives like sugars and polyols, amino acids, amines, polymers, surfactants, and preservatives, which are discussed in sequence in the following section.

4.4.4.1 Small Neutral Additives Many small neutral compounds have a profound effect on protein stability or aggregation.[181] A major category of these neutral compounds is sugars/polyols. These sugars/polyols normally have a stabilization effect and inhibit protein aggregation under different experimental conditions. They can help to maintain the correct conformation of proteins, which enables them to be called chemical chaperones or to be more widely recognized as osmolytes.[181,182] Many proteins are stabilized by these compounds, which differ from traditional chaperones, in that they possess limited or no specific molecular recognition for facilitation of protein folding. For example, the rate of aggregation of IFN-tau at 50°C was reduced gradually with increasing sucrose concentrations at 0.25, 0.5, 0.75, and 1.0 M.[31] Sucrose at 10% reduced the amount of HMW aggregates to a significant degree during elution of a Fc fusion protein at pH 3.6.[160] Sorbitol at 1 M inhibited the self-aggregation of HEWL upon thermal unfolding at pH values at and above the lysozyme's pI (9.3).[127] Their protective effect on protein aggregation generally increases with increasing concentration, such as sucrose or sorbitol for BSA[178] and trehalose for citrate synthase (CS).[182]

These sugars/polyols also protect proteins under different processing or even culturing conditions. Xia et al.[181] demonstrated that addition of sucrose at 0.5 M almost completely suppressed the aggregation of arginine kinase during refolding at 25°C. Beranger et al.[183] demonstrated that trehalose treatment of prion-infected cells decreased the size of de novo produced PrP^{Sc} aggregates, while treatment with glucose did not have this effect. It also significantly protected prion-infected cells from induced oxidative damage.[183]

These sugars/polyols do not seem to work on all stages of the aggregation process. Vilasi et al.[120] found that trehalose (50–600 mM) did not suppress the early phase of the aggregation process (formation of prefibrillar oligomeric structures) of a apomyoglobin analogue but did inhibit the formation of fibrils from early aggregates in a dose-dependent manner. Because of this effect, tre-

TABLE 4.2. Effect of Solution Excipients/Additives on Protein Aggregation

Categories/ Additives	Proteins	Experimental Conditions	Aggregation	References
Small neutral molecules—sucrose	Recombinant factor VIIa (rFVIIa)	1 mg/mL in 10 mM bis-Tris, 10 mM $CaCl_2$, 150 mM NaCl, 0.05% sodium azide, pH 5, 45°C, 0–10 M sucrose	Decreased aggregation (but increased oxidation) with increasing sucrose concentrations	118
Small charged molecules—spermidine	Lysozyme	Normal refolding-induced aggregation; thermal-assisted refolding-induced aggregation; heat-induced aggregation	Enhanced Enhanced Inhibited	190
Surfactants	PEG-GCSF	1 mg/mL at 29°C, 59 weeks in the presence of 0, 0.004, and 0.001 mg/mL Tween 20	1.2% 2.7% 5.9%	232
Polymers—heparin	aFGF	0.1 mg/mL in PBS, pH 7.2, 55 °C; above + 0.033 mg/mL heparin	A_{350} reaches >1 instantly; A_{350} reaches 0.1 in ~20 min	426
Proteins—BSA	β-lactoglobulin	Total amount of protein = 5% (w/w), 78°C, 30 min; above + BSA at 3:1 weight ratio	2.57E-3 mg/mL/s; 5.07E-3 mg/mL/s	212
Preservatives—benzyl alcohol	rhGCSF	1.5 mg/mL in 150 mM NaCl, and 10 mM sodium phosphate, pH 7.0, 37°C, 5 days; above + 0.9% benzyl alcohol	About 60%, 100%	158

halose seems to maintain the aggregates in a nonfibrillar, highly cytotoxic state. In contrast, trehalose (300 or 600 mM) delayed the formation of seed aggregates of insulin at 20 μM at 50°C but did not inhibit amyloid fibril formation.

In a few cases, these sugars/polyols destabilize proteins and promote protein degradation, or even aggregation. Schule et al.[184] found that preparation of an IgG1 solution at 10% led to a detection of 0.8% aggregates and that addition of mannitol at an IgG1/mannitol ratio of 20/80 led to the formation of 2.6% aggregates. These sugars/polyols could also potentially react with proteins, depending on the solution pH and temperature, and the modified protein may have a different aggregation behavior.[185]

4.4.4.2 Small Charged Additives Several types of small charged additives have been shown to affect the aggregation of proteins.[186] They influence the aggregation behavior of proteins either as an effect of ionic strength and/or as a result of specific interactions between the positive or negative ions with proteins. Such bindings have been observed and used to explain the different effects of buffering species on IL-1ra stability[77] and the negative effect of negatively charged chloride ions on the aggregation of β-lactoglobulin-enriched whey protein isolates.[187] These small charged additives, as ionic species, can have a beneficial or adverse effect on aggregation behavior, depending on their concentration and the protein.

A major category of small charged additives includes natural amino acids and their derivatives. Recent data show that amino acids may have multiple effects—destabilization of the partially folded states and early aggregates and solubilization of the native state.[188] Some of these compounds behave like chemical chaperones and can stabilize proteins, inhibiting protein aggregation. Particularly, some positively charged amino acids are very effective in inhibiting protein aggregation. For example, histidine has been shown to inhibit the aggregation of IFN-tau at 50°C relative to phosphate or Tris through weak binding to the native state of the protein.[31] Histidine has also been shown to inhibit precipitation of a fully human anti-IL8 mAb (IgG2) relative to citrate upon thermal incubation.[115] Lysine is able to enhance the chaperone function of α-B-crystallin against the aggregation of α-synucleinA53T.[32]

Arginine has been used for many years to assist protein refolding and to inhibit protein aggregation.[189] For example, arginine has been shown to effectively inhibit heat- and refolding-induced aggregation of HEWL.[190] In different processes, arginine has been shown to protect ADH and insulin from heat- and dithiothreitol (DTT)-induced aggregation in a concentration-dependent manner.[191] Arginine is also able to enhance the chaperone function of α-B-crystallin against aggregation of α-synucleinA53T.[32] On the other hand, the apparent positive effect of amino acids can be overshadowed by the effect of negative ions. For example, the negative Cl^- ions apparently interact with the positively charged β-lactoglobulin-enriched whey protein isolate for increased aggregation. The minimum temperature for thermally induced aggregation of the protein at pH 4.0 was reduced when arginine HCl concentration was increased from 10 to 200 mM but increased when arginine HCl concentration was higher than 200 mM.[187]

The mechanism for the protective effect of arginine has been studied in several labs. The protection effect was thought to be due to its protein solubilization properties by weakening protein–protein interactions rather than a protein stabilization effect.[189] Additional evidence suggests that arginine deters aggregation by slowing protein–protein association reactions through preferential exclusion from the protein–protein encounter complexes (not from dissociated protein molecules), with only a small concomitant effect on protein folding.[68] Recent evidence suggests that it effectively suppresses intermolecular interactions among aggregation-prone molecules.[191] In a more detailed study, Das et al.[192] demonstrated that arginine exists as clusters in solution and that the molecular clusters display a hydrophobic surface by the alignment of its three methylene groups, which interact with the hydrophobic surfaces on the proteins, masking the hydrophobic surface and inhibiting protein–protein aggregation.

Proline is another amino acid that has received more attention recently. It was shown to stabilize proteins in several reports. For example, addition of proline at 3 M almost completely suppressed the arginine kinase aggregation during refolding at 25°C.[181] Ignatova and Gierasch[188] found that proline abrogated aggregation of a model aggregation-prone protein, P39A cellular retinoic acid-binding protein, both *in vitro* and *in vivo* (in an *E. coli* expression system). The protective effect of proline on this protein was confirmed in a later study, where proline was shown to be the most effective in the solubilization of native proteins, while trehalose was ineffective.[193]

The observed protective effect of proline is believed to be nonspecific, as protection is presumed through a combination of unfavorable solvophobic backbone interactions and favorable side-chain interactions that are not specific to a particular sequence or structure.[188,193] Indeed, Kar and Kishore[194] studied the effect of hydroxyproline on the thermal stability and aggregation of lysozyme and found that the exceptional capability of hydroxyproline was due to the induction of tertiary structures of lysozyme through preferential hydration in increasing the thermal stability of lysozyme (up to 26.4°C) and in preventing lysozyme fibril formation *in vitro*. In addition, they compared the effect of hydroxyproline to that of proline and observed a similar increase in the activity and suppression of protein aggregation.

Several amine compounds have been shown to inhibit protein aggregation. Rezaei-Ghaleh et al.[195] found that both putrescine and spermidine reduced the aggregation (based on turbidity) of CT at 25°C induced by TFE (32%) and that a clear stabilizing effect against its thermal unfolding and toward formation of alcohol-induced alpha helices was considered due to their kosmotropic properties. Schlieben et al.[196] found that the purified lipase from *Bacillus thermocatenulatus* started to precipitate upon concentration above 0.2 mg/mL in a Tris buffer, but the solubility could be elevated to 1.6 mg/mL in the presence of 50–200 mM imidazole.

On the other hand, some amino acids and amine compounds have been shown to exert a negative effect on protein aggregation. Histidine was found not only to increase the rate of aggregation of recombinant humanized monoclonal antibody (rhuMAb) anti-CD20 during storage but also to induce coloration of the solution compared to an acetate-buffered solution at pH 5.[197] Both glycine and trimethylamine-N-oxide have been shown to increase the aggregation rate of rabbit skeletal muscle glycogen phosphorylase b induced by GdnHCl.[198] Dopamine causes aggregation of both wild type and A53T mutant α-synuclein in a temperature-dependent manner.[199] Addition of spermidine dramatically induced aggregation of lysozyme under both normal and thermal-assisted refolding processes at all pH values (4–8), although it inhibited heat-induced aggregation among tested compounds.[190]

Some organic acids have also been shown to affect protein aggregation. The effect seems to result from weak binding of the anions with the proteins such as growth hormones[200] and FGFs.[201] Phytic acid and SOS have been shown to inhibit aggregation of several types of growth hormone (GH).[200] Considine et al.[202] demonstrated that inclusion of myristic acid or conjugated linoleic acid in a β-Lg solution at pH 7.2 stabilized the native monomer and inhibited dimerization of β-Lg during heat treatment above 70°C and stabilized the protein during pressure treatment above 500 MPa. On the other hand, several polyanions (SOS and inositol hexaphosphate) increased the extent of aggregation of KGF-2 above 40°C, although the thermal unfolding temperature was increased by approximately 9–15°C.[128]

4.4.4.3 Polymers Polymers have been extensively used with proteins. They may inhibit protein aggregation through one or more of these properties of polymers—surface activity, preferential exclusion, weak binding, steric hindrance of protein–protein interactions, and increased viscosity limiting protein structural movement, particularly when included as part of solid-state formulations. These polymers include neutral polymers, amphoteric polymers, polycations/polyanions, and cyclic polymers.

4.4.4.3.1 Neutral Polymers Many neutral polymers are able to protect proteins from rapid aggregation. Ficoll-70000 at 150 mg/mL was shown to reduce fibrillation of human insulin upon stirring at 37°C.[163] Hydroxyethyl (heta) starch or PEG 4000 were found to be concentration-dependent inhibitors of heat-induced aggregation of low-molecular-weight urokinase (LMW-UK), presumably by preferential exclusion of the polymers from the protein surface.[149] This stabilization mechanism conforms to that proposed by Baynes and Trout[203] for a class of large solution additives—"neutral crowders." Their experimental results suggest that for a protein of 20-Å radius, an additive of 8 Å (in comparison, guanidinium and urea have apparent radii of 2–3 Å) can increase the free energy barrier for association and dissociation by as much as 3–6 kcal/mol, and increasing the size of neutral crowders (excluded effect) increases the gap effect and increases the barrier. In contrast, Ficoll-70000 at 150 mg/mL was shown to accelerate fibrillation of S-carboxymethyl-α-lactalbumin upon stirring at 37°C and to accelerate the fast fibrillation of human insulin at pH 2 under the same stirring conditions.[163]

Recent simulation studies suggest that native proteins with accessible hydrophobic areas on its surface are prone to aggregation, which can be inhibited by introducing polymers of appropriate hydrophobicity and chain length into the solution, such that these polymer molecules wrap around the hydrophobic regions of both unfolded and folded proteins, segregating the protein molecules and preventing protein aggregation.[16] Polyethylene glycols (PEGs) are such polymers. PEG 4000 was found to be a concentration-dependent inhibitor of heat-induced aggregation of LMW-UK.[149] PEGs of different molecular weights were found to inhibit aggregation of rhKGF at 45°C,[204] thermally induced aggregation of LMW-UK,[149] and aggregation of α_1-antitrypsin aggregation at 4°C.[205] Not all proteins are sensitive to PEG protection. PEG 300 was found to be ineffective in the inhibition of heat-induced aggregation of LMW-UK.[149]

On the other hand, a strong hydrophobic interaction between the polymer and protein may cause denaturation, that is, through loss of the hydrophobic core.[16] The degree of interactions is certainly dependent on proteins and experimental conditions. While PEG-3500 at 100 mg/mL significantly inhibited fibrillation of human insulin at 37°C at pH 7.5 (mostly hexameric insulin), it actually accelerated the fast fibrillation of human insulin (rigid α-helical protein) at pH 2 (monomeric insulin) (reduced nucleation time—shortened lag time by two and four times) and also accelerated fibrillation of S-carboxymethyl-α-lactalbumin at pH 7.5.[163] It seems that the overall effect is a balance between protein binding and macromolecular crowding effects (more discussion later). Among these interaction forces, hydrophobic interactions appear to be more effective in controlling protein aggregation than charge effects.[206]

4.4.4.3.2 Polycations/Polyanions A variety of polycations/polyanions were found to bind to proteins and to influence the aggregation behavior. Weak binding was demonstrated between polyanions (including both heparin and dextran sulfate) and bovine, porcine, and human growth hormones (hGHs) and was considered to be responsible for the inhibition of aggregation of the proteins.[200] Almost 100% inhibition of aggregation by heparin and dextran sulfate was achieved for porcine somatotropin (pST) and bovine somatotropin (bST) at certain concentrations. It was also demonstrated that FGFs strongly bind to low-molecular-weight heparin (LMWH, M(r) = approximately 5000) and are thereby stabilized against deactivation and proteolytic cleavage.[207] The results suggest that the binding complexes contain an average of 2.5 ± 0.3 bFGF molecules per heparin.[207]

Due to the binding effect, several polyanions were found to inhibit aggregation of rhKGF at 37°C, including dextran sulfate (8 and 500 kD), pentosan polysulfate, polyphosphoric acid, poly-L-glutamic acid, poly(acrylic acid), poly(methacrylic acid), and heparin.[109,204] Some polyanions are effective to inhibit the lyophilization/reconstitution-induced aggregation of rhKGF in 5% mannitol and 10 mM phosphate buffer (pH 7.0), including heparin, dextran sulfate, polyphosphoric acid, and poly-L-glutamic acid.[204] Heparin suppressed the aggregation of insulin in the pH range from 3 to 9 by binding to the positive domain of one protein and inhibiting its interaction with the negative domain of another.[100] Heparin was also found to inhibit protein aggregation by direct dissociation of protein aggregate particles through formation of soluble complexes.[100] Other proteins that are stabilized by multiple electrostatic interactions include epidermal growth factor (EGF)[76,208] and RNase.[209]

Inhibition of protein aggregation by polycations/polyanions can be strongly pH dependent, as pH affects their binding affinities and mechanisms. Dextran sulfate was found to suppress BSA aggregation at pH 5.1 and 6.2, while it had no effect at pH 7.5, as the interaction between dextran sulfate and BSA varied significantly with the solution pH, which led to a different extent of aggregation prevention by dextran sulfate.[166] It was also found that the complex formation of the intermediate state of denatured BSA with dextran sulfate is a prerequisite to suppression of aggregation by preventing further oligomerization/aggregation of denatured protein.

However, when binding is strong, these polycations/polyanions may adversely affect the aggregation process. It was found that strong electrostatic interaction between dextran sulfate and BSA rather facilitated the denaturation of BSA.[166] Boeris et al.[75] demonstrated that bovine pancreatic chymotrypsin alone at pH 3.5 forms soluble aggregates at 25°C in 9 h but addition of a strong polyanion, polyvinyl sulfonate, facilitated formation of soluble or even nonsoluble complexes between a positively charged protein and the polymer depending on the relative ratio. Stoichiometric ratios of 500–780 (protein:polyanion) were found in the precipitate in the pH range of 1.0–3.5. A variety of polyanions (e.g., heparin, SOS, and inositol hexaphosphate) increased the extent of aggregation of KGF-2 above 40°C, although the thermal unfolding temperature was increased by approximately 9–15°C.[128] The observed differences in aggregation behavior in the presence of polycations/polyanions could also be a result of a concentration-dependent effect.[200]

4.4.4.3.3 Amphoteric Polymers/Proteins Amphoteric polymers, including proteins, can have a significant effect on protein aggregation, likely through the same mechanism as for the neutral and polycationic/polyanionic polymers. Song et al.[210] demonstrated that the valosin-containing protein (VCP) could prevent the aggregation of aggregation-prone luciferase *in vitro* through mediation of the D1 domain of VCP, which does not require ATP binding or hydrolysis. Giese et al.[81] demonstrated that formation of large α-synuclein aggregates could be inhibited almost completely upon addition of β-synuclein at a molar ratio of 1:1. Gelatin (up to 1%) was effective in a concentration-dependent manner in inhibiting heat-induced aggregation of LMW-UK.[149,211]

In contrast, replacing β-lactoglobulin with increasing amounts of BSA while keeping the total protein concentration constant at 5% (w/w) significantly increased the denaturation (and aggregation) rate of β-lactoglobulin at 78°C.[212] Results clearly indicate that BSA contributed to the formation of intermolecular disulfide bonds in heat-induced denaturation/aggregation. Similarly, BSA at 30 mg/mL was shown to accelerate fibrillation of S-carboxymethyl-α-lactalbumin during stirring at 37°C.[163]

Some proteins inhibit/prevent aggregation of other proteins, likely through chaperone-like activities.[213] Heat shock proteins (HSPs) exhibit such activities and their protective effects have been widely reported. For example, Pivovarova et al.[214,215] have shown that the small HSPs (Hsp24 or Hsp27) effectively prevented aggregation of thermally denatured F-actin, although it did not affect the thermal unfolding of F-actin. Similarly, heat shock protein 105 (Hsp105) was shown to suppress the formation of mutant superoxide dismutase-containing aggregates in cultured cells through mutual interactions.[216] Overexpression of Hsp70 *in vitro* reduced detergent-insoluble HMW α-synuclein species and protected cells from α-synuclein-induced cellular toxicity.[65] Hsp20, Hsp27, or carrot Hsp17.7 were all able to prevent A_β aggregation.[217] Other examples include suppression of thermal aggregation of creatine kinase at pH 8.0 by small HSP Hsp 16.3 (mycobacterium tuberculosis)[134] and significant inhibition of heat-induced aggregation of CS by small HSP P26.[182] This effect of aggregation inhibition by chaperone proteins has been shown to act through the binding of the protein to denatured monomers or short oligomers upon heating, forming relatively small and highly soluble complexes,[214] or specifically through hydrophobic interactions.[78,213]

Another commonly examined chaperone protein is the family of crystallins. α-Crystallin was shown to inhibit thermal (65°C) aggregation of CT in a concentration-dependent manner.[78] It was also shown to significantly inhibit aggregation of purified β_H-crystallin and precipitation of purified β_L- and γ-crystallin at pH 7.0 at 37°C.[218] Similar chaperone activity was also reported for α-B-crystallin, which can inhibit the aggregation of amyloid fibril forming protein α-synucleinA53T.[32]

Protein aggregation can be inhibited by rationally designed peptides.[219] These designs include insertion in the sequence of a β-sheet breaker, formation of N-methyl derivative, insertion of a bulky group, or addition of a charge to reduce hydrophobicity. Indeed, such strategies are effective. Etienne et al.[220] demonstrated that amyloid beta-protein aggregation can be inhibited with peptides containing alternating α, α-disubstituted amino acids.

4.4.4.3.4 Cyclic Polymers Cyclodextrin and its derivatives are cyclic polymers commonly used in controlling protein aggregations. α-Cyclodextrin has been shown to effectively inhibit the aggregation of rhGH during refolding.[221] It was also shown to diminish effectively temperature-induced aggregation of horse liver ADH with a delay in unfolding of secondary structures and an increase in T_m by more than 5°C.[124] HP-β-CD, a derivative of cyclodextrin, was found to inhibit both heat-induced and lyophilization/reconstitution-induced aggregation of rhKGF.[109,204] Maltosyl-β-cyclodextrin was found to suppress insulin aggregation.[222]

It seems that protein stabilization results from coverage of hydrophobic surfaces of proteins through binding to the internal hydrophobic cavity of α-cyclodextrin. This conclusion is based on the observation that addition of phenylalanine would cause reaggregation of the α-cyclodextrin-stabilized protein, presumably through competition of phenylalanine for binding to the internal hydrophobic cavity of α-cyclodextrin.[124]

4.4.4.4 Surfactants Surfactants can be generally divided into two classes: nonionic and ionic. Both classes have been shown to affect protein aggregations induced by different stresses. Their direct or indirect effects are multifaceted.[38,223]

4.4.4.4.1 Nonionic Surfactants Many nonionic surfactants have been tried or used in protein formulations. These nonionic surfactants have hydrophobic tails, which can bind to hydrophobic patches on protein surfaces.[224] It has been shown that Tween 20 can bind to rhGH, and the binding stoichiometry between Tween 20 and rhGH has been determined to be 2.5–3.5 (Tween 20:rhGH) by electron paramagnetic resonance.[225] Tween 20 and Tween 80 have shown saturable binding to Albutropin™ (hGH genetically fused to human albumin) with a molar binding stoichiometry of 10:1 and 9:1 (surfactant:protein), respectively.[226] Such an interaction would block or partially block the aggregation-prone hydrophobic sites on the protein surface, preventing protein–protein interactions. Such interactions could also increase the free energy of unfolding, leading to protein stabilization.[226]

Therefore, these nonionic surfactants are especially effective against shaking/shearing-induced protein aggregation. A commonly used type of nonionic surfactant for this purpose is Tween.[227] Tween 20 was shown to completely block shaking-induced rhGH aggregation at a Tween 20:rhGH molar ratio of 4:1.[228] This molar ratio is close to the binding stoichiometry, suggesting that the inhibition of rhGH aggregation is probably due to Tween 20 binding. Both Tween 20 and Tween 80 prevented agitation-induced aggregation of a recombinant fusion protein, Albutropin, at pH 7 at the molar binding stoichiometry (two binding sites on Albutropin), even though the concentrations of surfactants were well below their critical micelle concentrations (59 and 12 μM, respectively, for Tweens 20 and 80).[226]

Nonionic surfactants have also been tried or used against thermally induced protein aggregation. Tween 80 was shown to inhibit thermally induced aggregation of BSA.[229] Poloxamers 407 and 188 (Pluronic F68) were shown to inhibit thermally induced aggregation of rhGH and methionyl pGH.[230] On the other hand, the effects of surfactants on protein aggregation have not been consistent among different proteins. Tween 80 at 0.1% showed no effect on the aggregation of IL-1β at 100 μg/mL at 60°C[60] and POE(10)L [$CH_3(CH_2)_{11}(OCH_2CH_2)_{10}OH$)] did not inhibit aggregation of CAB upon heating.[231]

In fact, the negative effect of surfactants was observed in several accelerated or real-time stability studies. For example, increasing concentrations of nonionic surfactants (Tween 20, Tween 80, and Triton X-100) accelerated the aggregation of aFGF at 50°C.[76] In a study on the thermal gelation of β-lactoglobulin at 14% (w/v), Kerstens et al.[176] found that Tween 20 at surfactant/protein molar ratios between $R = 0.5$ and 2.0 induced formation of spherical aggregates during heating at pH 6.8 and formation of more coarse/opaque gels at pH 5.5 in the presence of 0.05 M NaCl. Tween 20 was found to promote aggregation of PEG-GCSF in a concentration-dependent manner during storage at 29°C.[232] The increased aggregation could be due to destabilization of the protein through binding. Such a binding could induce partial denaturation of proteins. It was shown that Triton X-100 can induce a partial denaturation of the tobacco mosaic virus coat protein, and surfactant–protein complexes aggregate though hydrophobic interactions.[233]

4.4.4.4.2 Ionic Surfactants Ionic surfactants or surfactant-like compounds were shown to affect protein aggregation more intensely than nonionic surfactants. This is because these compounds generally bind to proteins more strongly than nonionic surfactants due to binding to both polar and nonpolar groups in proteins. Cationic surfactant cetyltrimethylammonium bromide (CTAB) was shown to bind the subunit of the tobacco mosaic virus coat protein in a 4:1 molar ratio, and binding does seem to alter the native structure of the protein.[234] Like nonionic surfactants, the binding between the nonpolar chain/structure of ionic surfactants and protein may shield the nonspecific interactions between protein hydrophobic patches/groups. In addition, the high charge density of ionic surfactants on the protein surface could possibly induce a charge-screening effect between the protein polar groups.[235] The overall effect on protein stability/aggregation would be dependent on the binding strength and the concentrations of these surfactants.

Several ionic surfactants have been shown to inhibit protein aggregation. CTAB, a cationic surfactant, was shown to effectively inhibit refolding-induced aggregation of rhGH.[221] Sulfobetaine, 3-(1-pyridinio)-1-propanesulfonate can delay or prevent thermal aggregation of xylanase by differential scanning calorimetry (DSC), depending its concentration.[235] Octanoic acid at 5 mM completely abolished BSA aggregation at 70°C with an 18°C increase in T_m.[131] Sodium dodecyl sulfate (SDS) was found to reduce aggregation of aFGF in a concentration-dependent manner at 50°C[76] and to prevent aggregation of heat-denatured RNase[209] or CAB.[231] Sodium cholate at 1% (w/v) inhibited solubility-limited aggregation of purified lipase from *B. thermocatenulatus*.[196]

In several cases, ionic surfactants were found to induce protein aggregation. CTAB at low micromolar concentrations (start at about 15 μM) induced aggregation of tobacco mosaic virus coat protein at 25°C, but no aggregation was observed at higher surfactant concentrations (above 300 μM).[234]

4.4.4.5 Preservatives Many protein drug products need to be made with preservatives to accommodate multiple administrations. Yet these preservatives may not be compatible with some proteins. One phenomenon of incompatibility is accelerated protein aggregation in the presence of preservatives. Preservatives often contain a hydrophobic component, which can bind to proteins as discussed above. Such a binding was detected, albeit at weak levels, by isothermal calorimetry between benzyl alcohol and rhIL-1ra, which did not detectably change the secondary structure but caused minor perturbation of the tertiary structure of the protein toward partially unfolded states.[236] Presumably due to the hydrophobic nature of the interaction, the degree of interactions increased with temperature.[114]

Benzyl alcohol is perhaps the most studied preservative in protein formulation literature. Its effect was evaluated in several studies on the aggregation of rhIL-1ra. Benzyl alcohol at 0.9% was found to accelerate aggregation and precipitation of rhIL-1ra at pH 6.5 at 37°C.[236] This observation was later confirmed in a different study, where benzyl alcohol at 0.9% (w/v) increased the nucleation rate of rhIL-1ra at 37°C by approximately four orders of magnitude without changing the secondary structures of the protein.[113] The effect of benzyl alcohol was concentration dependent and the apparent T_m of rhIL-1ra was reduced from 56 to 51, 48, and 44°C, respectively, in the presence of 0.9%, 1.5%, and 2.0% benzyl alcohol. The adverse effect of benzyl alcohol was also observed during reconstitution of lyophilized rhIL-1ra, where reconstitution of the dried solid with 0.9% (w/v) benzyl alcohol caused a greater degree of protein aggregation than reconstitution with water.[237] Later, Roy et al.[114] found the effect of benzyl alcohol is temperature dependent. While benzyl alcohol at 0.9% accelerated aggregation of rhIL-1ra at 37°C (almost no monomers after 7 days), it showed no effect at 25°C, corresponding to temperature-dependent perturbation in the tertiary structure of rhIL-1ra in the presence of benzyl alcohol. It was found that the reactivity of buried free cysteines in rhIL-1ra was enhanced by benzyl alcohol at 37°C but not at 25°C.[114]

Benzyl alcohol was also able to influence aggregation of other proteins. rhGCSF at 1.5 mg/mL lost about 40% monomer at pH 7.0 in 5 days at 37°C and addition of benzyl alcohol (0.9% w/v) resulted in complete aggregation of rhGCSF under the same conditions.[158] Similarly, spectroscopic studies showed that 0.9% benzyl alcohol altered the tertiary structure of rhGCSF without detectably altering the secondary structure, and the adverse effect of benzyl alcohol was also found to be greater at 37°C than at 25°C, suggesting hydrophobic interactions between benzyl alcohol and the protein.[158]

4.4.4.6 Antioxidants Antioxidants are often used to inhibit oxidation reactions. Ascorbic acid is a commonly used antioxidant for small molecule drugs. However, for proteins, it was shown to have a negative effect on protein stability/aggregation in several studies. For example, Linetsky et al.[218] showed that incubation of calf lens proteins (crystallins) at pH 7.0 with 20 mM ascorbic acid caused aggregation of the proteins to sizes ranging from 2.2×10^6 up to 3.0×10^8 Da due to the chemical reaction of ascorbic acid with the protein (ascorbylation) upon incubation at 37°C for 4 weeks. Detailed examination showed that it was the purified β_H-, β_L- or γ-crystallin (not α-crystallin) that aggregated upon the reactions of ascorbate with these proteins.

4.4.5 Protein Concentration

Many studies have been conducted to evaluate the effect of protein concentration on aggregation. Because of the existence of different aggregation pathways (e.g., partial unfolding-induced versus simple native association-initiated aggregation), increasing protein concentration could potentially result in these consequences: (1) increased aggregation due to increased chance of association, (2) decreased aggregation due to crowding effect (see discussions below), and (3) precipitation due to solubility limit. Table 4.3 lists some examples.

4.4.5.1 Protein Behavior at High Concentrations High protein concentration solutions can be crowded (>100 mg/mL). Under such conditions, the chances of protein–protein associations/interactions are higher.[34,36,106] Even so, protein structures may not change much, even under associated states.[35] The structure of each antibody was suggested to be physically similar in a wide concentration range of 0.1–190.0 mg/mL in 40 mM citrate–phosphate buffer at pH 6.[238]

The concept of macromolecular crowding has been proposed elsewhere and has been used to describe the crowded environment in intracellular compartments.[239] According to this theory, crowding should favor formation of compact conformations over extended conformations of flexible macromole-

TABLE 4.3. Effect of Protein Concentration on Protein Aggregation

Proteins	Experimental Conditions	Concentration Range	Aggregation Trend	References
Fc fusion protein	Elution with a 50 mM citrate buffer at pH 3.6	0.5–4.0 mg/mL	Increased rate and extent	160
GCSF	Incubation, pH 7.0, 37°C	1–20 mg/mL	Increased rate and extent	18
Glycogen phosphorylase b	Incubation in 0.1 M NaCl and 0.2 mM EDTA, pH 6.8, 37°C	1–3 mg/mL	Increased rate with shortened lag time	249
Holo-α-lactalbumin	Heating a mixture (1:1 molar ratio of Ca^{2+} and protein) in 10 mM sodium phosphate buffer, pH 7.0, 95°C for 30 min	5–60 mg/mL	Increased extent	121
Glyceraldehyde 3-phosphate dehydrogenase	Incubation in 10 mM phosphate, pH 7.5, 45°C	1–3 mg/mL	Reduced rate with prolonged lag time	249
Hemoglobulin	Thermally induced aggregation	Increasing concentrations	Reduced aggregation	241

cules and, at the same time, favor both specific macromolecular associations (formation of well-defined oligomers) and nonspecific macromolecular associations (formation of large aggregates).[240] For example, the T_m of two model proteins, human hemoglobin A_0 and bovine fibrinogen, are approximately 2–10°C higher in more concentrated than in dilute solutions, although the opposite trend was observed for lysozyme and BSA.[241] Self-association of an IgG in saline buffer was found to favor formation of predominantly trimers at high concentrations, and one-third of the total protein was trimer at 100 mg/mL.[242]

While formation of more compact protein structures could increase protein stability, increased associations could favor aggregation over time at higher protein concentrations. Therefore, the increased protein stability at higher concentrations may not translate into a reduced rate of aggregation. It appears that a variety of factors would determine the overall effect of high protein concentrations on the colloidal and conformational stability of a particular protein.[241] The ultimate effect would be the result of a balance between the two factors. For example, increasing the concentration of two IgG1 mAbs to 100 mg/mL reduced the aggregation temperature by about 10°C for both antibodies but increased the melting temperature by 2–3°C at pH 6.[238] Thus, different proteins may have different behaviors at high concentrations.

4.4.5.2 Enhanced Protein Aggregation at High Concentrations

In most cases, protein aggregation increases with increasing protein concentration because of an increase in the chances of protein–protein association.[34,36,106] It is also possible that proteins need a critical concentration to form the initial nucleus for initiation of the aggregation process.[72] The mean field lattice model predicts that proteins will aggregate/precipitate at sufficiently high concentrations.[10] In a more recent study, the *in vitro* aggregation rates of a wide range of unstructured polypeptides were found to be positively correlated with protein concentration over five orders of magnitude with a correlation coefficient of 0.92.[152]

Many proteins have been shown to aggregate more easily at higher concentrations. This is often reflected in the rate and/or extent of protein aggregation. For example, increasing the concentration of an Fc fusion protein from 0.5 to 4.0 mg/mL gradually increased the percentage of HMW aggregates during elution at pH 3.6[160] Increasing the concentration of GCSF from 1 to 20 mg/mL at pH 7.0 increased both the rate and extent of protein aggregation at 37°C.[18] The aggregate levels of α-La and β-lactoglobutin mixture after 90 min at 75°C at pH 3.9 were about 10% and over 40%, respectively, at a total protein concentration of 3 and 16 mg/mL.[243] Other examples of enhanced protein aggregation at higher concentrations under different experimental conditions include an IgG1,[244] an IgG2 mAb,[138] a rabbit IgG,[245] a botulinum neurotoxin (50 kD),[159] CT,[78] recombinant factor VIII,[246] lysozyme,[130] aFGF,[76] insulin,[23,88] β-lactoglobulin,[247] IL-1β,[60] LMW-UK,[149] and apomyoglobin.[79]

Increased tendency of protein aggregation at high protein concentrations can also be reflected in the aggregation temperature (at which protein aggregation is first observed during a thermal scan), the aggregation lag time, or the direction of protein–protein association equilibrium. For example, increasing the concentration of two IgG1 mAbs to 100 mg/mL reduced the aggregation temperature by about 10°C for both antibodies at pH 6.[238] Similarly, concentrated BSA gels at a much lower temperature (~15°C) than in a dilute solution at pH 5.6.[241] Increasing the concentration of cold shock protein A reduced the lag time (to the power of 2.1 of the protein concentration).[248] Increasing the concentration of glycogen phosphorylase b from 1 to 3 mg/mL not only shortened the lag time but also increased the aggregation rate at pH 6.8 at 37°C.[249] Increasing the concentration of rhIL-1ra shifts the rhIL-1ra monomer–dimer equilibrium significantly to favor dimerization at low ionic strength at 37°C.[35]

The aggregation dependency on protein concentration may vary depending on the experimental conditions. The first-order rate constant for the aggregation of creatine kinase at pH 8.0 was found to increase with increasing protein concentration in the range 0.1–0.4 mg/mL at 60°C but was found to be protein concentration independent in the range 0.05–0.2 mg/mL at 50.6°C.[134]

The aggregation kinetics may or may not follow second order, even though high protein concentrations accelerate protein aggregation. The aggregation rate of β-lactoglobulin at 78°C at pH 7.0 in water increased 31 times with increasing protein concentrations from 1% to 6% with second-order kinetics.[250] On the other hand, aggregation of GCSF above 2.5 mg/mL at pH 7.0 at 37°C follows apparent first-order kinetics, as the rate-limiting step involves a unimolecular process.[18] Similarly, the rate-controlling step in the aggregation of rFVIII is also a unimolecular reaction involving protein conformation changes, leading to first-order aggregation kinetics under thermal stress.[246] Apparent first-order kinetics was also observed for the aggregation of creatine kinase at 50.6 and 60.0°C[134] and for the aggregation of firefly luciferase both under thermal (pH 7.5, 42°C) and acidic (pH 5.5, 30°C) conditions.[80]

4.4.5.3 Unaltered or Reduced Protein Aggregation at High Concentrations Some proteins have a similar aggregation tendency in a limited concentration range. Fibrillation of BSA at elevated temperatures is independent of seeding aggregates and shows only moderate concentration dependence (0.5–10.0 mg/mL).[131] Broersen et al.[41] studied the aggregation of ovalbumin and its charged variants in solution at 7°C below the thermal transition temperature and found that all four model proteins at concentrations ranging from 5 to 40 mg/mL at pH 7.0 aggregated within 7 days without apparent concentration dependency.

In a few cases, increasing the protein concentration led to reduced aggregation. As discussed before, proteins become more compact and slightly more stable in a crowded environment. If such an effect dominates protein–protein association, aggregation may be reduced at higher concentrations. Increasing the concentration of glyceraldehyde 3-phosphate dehydrogenase from 1 to 3 mg/mL not only prolonged lag time but also reduced the aggregation rate at pH 7.5 at 45°C.[249] Increasing the concentration of PEG-GCSF from 1 to 10 mg/mL gradually inhibited the agitation-induced aggregation at 23°C.[232] If the aggregation tendency of monomers is higher than that of polymers, reduced aggregation could also be seen at higher concentrations due to polymer formation. The aggregation tendency of human IFN-γ dimer was found to be less competent in aggregation than the monomer, and increasing human IFN-γ concentration from 1 to 4 μM delayed the time of reaching maximum aggregation.[119] For some proteins, nucleated polymerization (fibril formation) can be accompanied by off-pathway aggregation, and the off-pathway aggregation could be slowed down at high protein concentrations.[251]

4.4.5.4 Solubility-Controlled Protein Aggregation in Folded States

Each protein has a solubility limit, which is strongly dependent on the solution conditions. Proteins can easily aggregate, precipitate, or crystallize when protein concentration exceeds this limit. For example, purified lipase from *B. thermocatenulatus* showed a strong tendency to aggregate and began to precipitate in a Tris buffer above a concentration of 0.2 mg/mL, and the solubility could be increased to 1.6 mg/mL in 50–200 mM imidazole.[196] A sharp increase in the aggregation of K97I IL-1β was observed during refolding when its concentration exceeded a critical value.[71]

4.4.6 Metal Ions

Metal ions play a critical role in maintaining the correct conformations for many proteins. An excessive amount of zinc is needed to maintain hexameric insulin in a stable state.[252] Calcium chloride is needed to maintain the normal function and to minimize protein aggregation of FVIII.[253] Copper ion plays an essential role in stabilizing the blue copper protein (BCP) rusticyanin (Rc), and the oxidized form (Cu(II)) is more efficient than the reduced form in this respect.[254] Copper (II) ions at nanomolar concentrations can also block strong proaggregatory effect of Mn^{2+} at micromolar concentrations on prion protein (PrP) aggregation.[255]

In many cases, metal ions can facilitate aggregation of proteins (Table 4.4). The proaggregation effects are generally mediated through binding to proteins.[256,257] Binding can result in small changes in tertiary structure in proteins, leading to formation of aggregation-competent conformers. For example, aluminum was found to bind to BSA, inducing protein aggregation through tertiary conformational changes,[167] and binding of aluminum to rhFVIII leads to both secondary and tertiary structural changes.[256] Another mechanism of metal-induced aggregation is the influence of metal on the colloidal stability of proteins. For example, binding of calcium to β-lactoglobulin reduced the zeta potential of the protein to a minimum of −5 mV, accelerating aggregation

TABLE 4.4. Effect of Multivalent Metal Ions on Protein Aggregation

Metal Ions	Proteins	Experimental Conditions	Aggregation Effect	References
Zinc	β-lactoglobulin	Heating β-lactoglobulin at 3.3 mM at a neutral pH, 50°C with 1 mM zinc in solution	Promoted aggregation	427
Aluminum	BSA	Incubation of BSA at 2.5 μM at pH 7.4 in the presence of >0.1 mM aluminum(III), room temperature	Induced aggregation	167
Calcium	β-lactoglobulin	Incubation of β-lactoglobulin with 5 mM $CaCl_2$, pH 7.0, 78°C	Promoted aggregation	250
Copper (II)	Human muscle acylphosphatase	Incubation of human muscle acylphosphatase at pH 5.5, 25°C with 0.1 mM copper (II)	2.5-fold increase in aggregation	258
Manganese(II)	Prion protein (PrP)	Incubation of PrP with Mn^{2+} at low μM	Induced aggregation	255

at pH 7.0.[250] Metal ions can also act at the level of intermolecular interactions critical for the formation of multimeric aggregates.[255]

Different metal ions may have different degrees of effect on the same proteins. The slow aggregation of rhFVIII under physiological conditions was significantly accelerated in the presence of several metal ions, including Al^{+3}, Tb^{+3}, Co^{+2}, and Fe^{+3} with Al^{+3} being the most efficient.[256] Several metal ions, including Co^{2+}, Ca^{2+}, Cu^{2+}, Fe^{3+},Cd^{2+}, Mg^{2+}, Mn^{2+}, Ni^{2+}, and Zn^{2+}, increased the aggregation rate of human muscle acylphosphatase (AcP) at pH 5.5 at 25°C with Cu^{2+}, producing the most remarkable acceleration of aggregation (2.5-fold higher at 0.1 mM).[258] The effect of metal ions is often concentration dependent.[255]

Metal ions also influence the kinetics of protein aggregation and properties of aggregates. The heat-induced aggregation of β-lactoglobulin at pH 7.0 at 78°C follows second-order kinetics, but stoichiometric addition of calcium leads to first-order aggregation kinetics.[250] Simmons et al.[259] studied the influences of temperature (70–90°C) and shear rate (111–625/s) upon the rate of growth and size of aggregates in whey protein concentrate solutions (WPC35), and the addition of minerals resulted in the formation of much smaller aggregates and increased deposition onto the surface of the Couette apparatus. This was attributed to interactions between Ca and β-lactoglobulin both in the bulk and at the heated surface.

4.4.7 Denaturant and Reducing Agents

Denaturants and/or reducing agents are widely used in the protein production process. Most proteins completely denature at a high enough denaturant concentration, such as 8 M urea.[260] Completely unfolded proteins are not easy to aggregate, and therefore most proteins remain soluble in these denaturants at high concentrations. In fact, these protein denaturants could potentially dissolve protein aggregates (mainly physical). Urea at 8 M was shown to dissociate apomyoglobin precipitates formed in 0.5–4.0 M urea solutions[79] and to break the IgG aggregates present initially in buffer.[260] SDS can easily dissociate insoluble aggregates of hGH[261] and fibrillar aggregates of HEWL.[129] Different denaturants may have different degrees of disaggregation/dissociation effect. SDS was found to be more effective in dissolving α-synuclein aggregates than Triton X-100,[65] and the solubility of whey protein aggregates is significantly different in solutions containing different denaturants.[262]

Some aggregates, however, are not easily solubilized in denaturants. Noncovalent insulin fibrils could not be dissolved in several solvent systems including 7 M urea from pH 2 to 8, 5 M GdnHCl, 50% acetonitrile or other detergent solutions due to strong hydrophobic interactions in insulin fibrils.[23] The heat-induced (boiling) aggregates of a membrane protein could not be dissociated by 6 M urea or by 2% SDS.[263] The ease of aggregate dissolution also depends on the density of protein aggregates.[264]

At an intermediate concentration, denaturants can either inhibit or induce protein aggregation because of the formation of partially unfolded states.[198] Urea in the concentration range of 0.5–2.0 M reduced the amount of HMW Fc fusion protein aggregates formed during elution at pH 3.6 relative to the buffer control.[160] Urea at 0.37 or 0.74 M reduced the initial rate of the thermally induced aggregation of BSA (compared to pure buffer).[178] Guanidine (Gdn) at 0.5 M suppressed the aggregation of HEWL above pH 5.5 during thermal-assisted refolding (98°C for 1 min) with a reduction in T_m from 77.5 to 71.9°C.[190] In contrast, GdnHCl at concentrations ranging from 1.0 to 2.5 M increased aggregation of GCSF at 1 mg/mL at 37°C because of the enrichment of a highly aggregation-prone structure (no substantial loss of secondary structure but increased hydrophobic surface area).[18] An intermediate GdnHCl concentration of 1.5 M was found to be optimal for the generation of ordered amyloid aggregates of BCA II at pH 3.5 at 37°C.[168] Similarly, GdnHCl at 1–3 M can induce aggregation of apo, Cu(I), or Cu(II) rusticyanin (Rc).[254]

Often, denaturants have opposite effects on protein aggregation in a narrow concentration range. For example, PrP at 5 nM was in a non-aggregated state in a 0.2% SDS solution, but significant aggregation occurred when SDS concentration was diluted below 0.05%.[255] GdnHCl at ≥2 M can prevent thermally induced (45°C for 15 min) aggregation of oligomeric *A. hirsuta* lectin, but aggregation was enhanced at 1 M GdnHCl.[123]

The use of reducing agents can have a significant effect on protein aggregation, as free cysteines are generated for potential cross-linking reactions after reduction. For example, increasing the concentration of DTT, a commonly used reducing agent, promotes formation of labile and probably unstructured RNase A aggregates, leading to possible precipitation.[265] DTT was also shown to induce aggregation of ADH and insulin[191] and of porcine and bST at 37°C.[200] Protein disulfide isomerase (PDI) catalyzes reduction of disulfide bonds in proteins and PDI-catalyzed insulin reduction leads to aggregate formation.[266]

4.4.8 Impurities

Many impurities/contaminants that are carried with proteins during various production processes (from protein expression to manufacturing of final products) can potentially affect protein stability/aggregation. Such potential impurities include host cell proteins, enzymes, cell membrane components, and residual metal ions.

Protein contaminants can promote aggregation by participating directly in the aggregation process. The wild-type α-synuclein can form coaggregates with a A30P mutant or PrP, and the aggregation rate is dependent on the relative ratio of the two proteins.[81] While no aggregation was observed upon heating β-lactoglobulin A or β-Lg alone at 75°C for 30 min, heating the mixture of α-La and β-lactoglobulin A or β-Lg under the same condition led to the formation of aggregates with α-La incorporated into both the disulfide-bonded and the hydrophobically associated aggregates, as well as the formation α-La dimers and other oligomers.[267]

Lipids are hydrophobic molecules and can easily bind to proteins, causing protein aggregation. Lysozymes were shown to bind to liposomes through both electrostatic and hydrophobic protein–lipid interactions and to promote aggregation.[268] Some lipids in a 10-fold molar excess can enhance the rate of aggregation of insulin at pH 1.9 in the absence of 0.1 M NaCl (no effect in the presence 0.1 M NaCl).[180]

Protein aggregates in a solution can be treated as an impurity. The presence of such aggregates can potentially increase the protein aggregation rate, as protein aggregation is often initiated through the formation of "seeds." Andrews et al.[133] demonstrated that addition of seed aggregates increased the aggregation rate of α-chymotrypsinogen A at 1.2 mg/mL (47 μM) at 65°C. Addition of preformed seeds eliminated the time dependence of the lag phase in the aggregation of protein L in 10 mM HCl.[82]

The presence of contaminant enzymes can lead to increased protein aggregation. Glutamyl endopeptidase can cause hydrolysis of whey proteins and aggregation of generated peptides increases with the degree of hydrolysis.[269]

4.4.9 Organic Solvents

Organic solvents can be encountered during the production of proteins. Such solvents can have positive or negative effects depending on the protein and relative quantity of the solvents. An analysis of protein stabilization mechanisms led to the conclusion that proteins will be unstable in most polar organic solvents such as ethanol and will be extremely stable in nonpolar solvents such as cyclohexane.[270] Solubility studies suggest that protein solubility will be markedly lower in polar organic solvents such as ethanol and that proteins will be essentially insoluble in nonpolar solvents such as cyclohexane.[270]

Therefore, pure organic solvents should promote protein aggregation/precipitation. In reality, organic solvents are often used as cosolvents and their effects vary depending on the relative amount in the solvent mixture. The most commonly used organic solvent in protein production is glycerol. This solvent can stabilize proteins and can inhibit aggregation under different experimental conditions/processes. Glycerol has been shown to suppress aggregation of a mutant creatine kinase[271] and arginine kinase[181] during refolding. Glycerol at 2–30% gradually increased the lag time, slowed the aggregation rate, and reduced the final extent of aggregation of a mutant creatine kinase at 50°C by stabilizing the native state.[271] While glycerol at 10% reduced the amount of aggregates of Fc fusion protein during elution at pH 3.6, 10% propylene glycol significantly increased the rate of aggregation.[160]

Trifluoroethanol as an alpha-helix enhancer induces aggregation of several proteins, such as CT,[179] protein L,[82] and human muscle AcP.[258] Its effects on protein aggregation vary significantly depending on its concentration. TFE above 35% promoted a nonnative structure of lentil seedling amine oxidase with a high alpha-helix content and between 10% and 35% facilitated aggregation of the protein.[272] TFE at 5–70% (v/v) promoted aggregation of CT at pH 7.0 at 25°C, and maximal aggregation was observed around 32% TFE.[179] TFE at 25% (v/v) induced partial unfolding (within a few seconds) of human muscle AcP at 25°C with subsequent formation of globular aggregates (60–200 nm in diameter), and 5X dilution (to 5% TFE) led to rapid refolding of partially unfolded monomeric proteins and disaggregation of 60–200 nm globular aggregates.[107] TFE at 10% did not promote aggregation of BSA at 70°C even with a significant reduction in T_m from 65 to 53°C but at 20% inhibited the aggregation process dramatically.[131]

Other polar organic solvents have also been used and their effects vary. Zeins, the major storage proteins, are soluble in alcoholic solution with a high tendency to aggregate, and incubation of Z19 α-zein in 90% ethanol at 80°C for an hour generated about 30% aggregates with diameters between 4 and 5 μm.[49] It has been shown that increasing ethanol concentration from 0% to 60% progressively increases solvent exposure of the hydrophobic core of hGH, potentially facilitating protein aggregation.[273] DMSO effectively inhibited arginine kinase aggregation in a concentration-dependent manner during refolding at pH 8.4,[181] and heating fibrillated human calcitonin in 50% AcOH/water could reconvert the fibrils to soluble monomers.[274]

4.4.10 Containers/Closures

Proteins have a tendency to be adsorbed onto a variety of surfaces of containers. Such adsorption may cause loss of proteins and even protein aggregation with or without structural changes. Adsorption of rhFVIII to hydrophobic silica surface (silanized) caused changes in tertiary structure, but a smaller degree of change in structure was seen on negatively charged hydrophilic surfaces.[275] The reversible adsorption of rhPAF-AH to nanosized hydrophilic silica particles induced no detectable change in protein secondary and tertiary structures but resulted in rapid and extensive aggregation of rhPAF-AH at pH 6.5.[112] Based on these observations, it was assumed that protein adsorption causes a reduction in the energy barrier to the formation of aggregate seeds (also see Chapter 7).[112]

Proteins may have different affinities to different types of surfaces. rhFVIII can be adsorbed to hydrophobic silica surface (silanized) at 0.13 μg/cm² in the presence of 8 ppm Tween 80 mainly through hydrophobic interactions and to negatively charged hydrophilic surface at 0.3 μg/cm² mainly through electrostatic attractions.[275] rhPAF-AH was also found to be adsorbed to nanosized hydrophilic silica particles with a maximum protein adsorption of 1.89 mg/m² at pH 6.5, primarily due to electrostatic interactions between anionic SiO groups and amine groups on proteins.[112] An IgG in a PBS solution was found to be adsorbed to a polyethylene terephthalate surface in a concentration-dependent manner at 37°C and maximum adsorption was 8 μg/cm², which is theoretically higher than calculated for monolayer adsorption.[276] Santos et al.[277] studied whey protein adsorption on stainless steel surfaces (316 L) by a single pass of a protein solution through a cell at 3 L/min at 72°C (T_m of β-lactoglobulin = 75°C) and after 7 min, protein surface adsorption amounted to 3.6 mg/m² with surface-induced aggregation as the main mechanism for the formation of multilayers. Protein surface adsorption can be strongly affected by solution pH, salt concentration, and temperature.[112]

Surfactants may or may not affect protein adsorption depending on the type of surfaces and the protein. Adsorption of rhFVIII at 50 μg/mL to a hydrophobic silica surface (silanized) was reduced from 0.13 μg/cm² in the presence of 8 ppm Tween 80 to about 0.10 μg/cm² in the presence of 28 or 88 ppm Tween 80 (below and above critical micelle concentration [CMC]), but adsorption to a negatively charged hydrophilic surface was unaffected in the presence of 88 ppm Tween 80.[275] Similarly, lysozyme adsorption to hydrophobic silica surface (silanized) was reduced, but adsorption to hydrophilic surface was not affected in the presence of Tween 80.[278] The reduced adsorption in the presence of a higher concentration of Tween 80 was due to several possibilities: (1) faster movement of Tween 80 to the interface, (2) higher affinity of Tween 80 at the interface than FVIII, or formation of Tween 80–FVIII complex.[275]

In comparison, adsorption of rhPAF-AH to nanosized hydrophilic silica particles was reduced from a maximum of 1.89–0.11 mg/m^2 in the presence of 0.1% Pluronic-F68 at pH 6.5 (CMC was found to be 0.001%), suggesting competitive adsorption.[112]

The effect of container material on protein aggregation could be amplified under stressed processing conditions. It was found that freeze-thawing of a mAb induced formation of less soluble aggregates at pH 3 and 4 in Flexboy bags or Teflon containers but more loss of proteins (up to ~55% at 0.5 mg/mL) than in polypropylene containers.[279] Even the fine structure and/or degree of cleanness of containers of the same material may impact the surface adsorption and aggregation of proteins. The degree of shaking-induced aggregation of insulin (Novolin R) was found to be significantly different among different glass vials, and the aggregate level after 72 h spread between 4% and 41% in seven vials.[280]

Rubber stoppers are often siliconized for ease of processing during production. A trace amount of silicone may cause protein aggregation due to its hydrophobic nature. The aggregation propensity of four model proteins of various molecular weights and isoelectric points (concanavalin A, BSA, ribonuclease A, and lysozyme) was found to increase significantly in the presence of 0.5% silicone oil in solution.[281] The silicone-induced aggregation was examined at three pH values (4.5, 6.5, and 7.2), and the strongest effect was at pH 6.5 for ConA, 4.5 for RNase A, 7.2 for BSA, and 4.5 for lysozyme.[281]

4.4.11 Sources of Proteins

The stability/aggregation behavior of proteins may be significantly different depending on the source and quality of proteins. The aggregation behavior of three commercially available holo-α-lactalbumin—Sigmaα-La, IEXα-La, and Cα-L—was found to be significantly different in at pH 7.0 at 95°C.[121] The relative extent of aggregation after a 30-min treatment is Sigmaalpha-La > IEXalpha-La > Calpha-La with different lag times, number of aggregate intermediates, and degree of reversibility. Similarly, the aggregation behavior of two other commercial preparations (A and B) of α-La was found to be significantly different at 95°C at neutral pH in a complex mineral salt environment.[282] Preparation A had a higher solubility at pH 6.75 and 6.8, yielded more reactive thiol groups, had a (25%) faster first-order rate constant, and formed spherical aggregates with much higher-molecular-weight species than preparation B.[282] While preparation A showed much higher aggregation tendency at pH 4.6, preparation B aggregated significantly more than preparation A at pH 6.75 and 6.8.

4.4.12 Light

Proteins can be sensitive to light exposure, leading to many types of light-induced degradations.[283] It has been shown that UV (297–320 nm) irradiation of γ-crystallin at 1–2 mg/mL causes protein aggregation and solution turbidity, possibly due to a photocross-linking reaction(s) and/or protein insolubilization of the photolytic products.[284] Recent investigations indicate that UV irradiation can induce cross-linking reactions between Cys and Trp residues and photolysis of disulfide bonds, leading to new disulfide bond formation/exchange.[285,286]

4.5 EFFECTS OF PROCESSING STEPS ON PROTEIN AGGREGATION

Protein aggregation can be induced in various processing steps. Such processing steps include fermentation/expression, unfolding/refolding, purification, freeze-thaw, shaking and shearing, pressurization, drying, preparation of protein drug delivery systems, and administration.

4.5.1 Fermentation/Expression

Fermentation or cell culturing is the first step in large-scale protein production. In this process, proteins can be expressed in a variety of cell systems under different conditions, and protein aggregation can occur to a different degree at the same time.[71,287] In fact, many proteins aggregate into so-called inclusion bodies during expression in bacterial systems. For example, during the expression of the SH3 domain of α-spectrin and its 23 mutants as model globular proteins in the cytoplasm of *E. coli* cells, both soluble and insoluble proteins (aggregated form) were obtained.[15] Since protein activity of inclusion bodies correlate negatively with the predicted aggregation rates of proteins,[288] controlling protein aggregation is important during expression of inclusion body-oriented proteins.

Protein aggregation during fermentation/expression can be inhibited by adjusting the fermentation/expression conditions. For example, lowering the fermentation temperature from 30 to 20°C inhibited the aggregation of recombinant human consensus interferon-alpha mutant (cIFN) during *Pichia pastoris* fermentation.[289] The use of surfactants is a simple and often effective way to control protein aggregation during expression. The addition of Tween 20 at 0.2 mg/mL or Tween 80 at 0.1 mg/mL successfully inhibited aggregation of cIFN during expression in yeast.[289,290] Both Tween 20 and Tween 80 were effective in inhibiting aggregation of rhGCSF during expression in *P. pastoris*.[291] Other additives could also be used to inhibit protein aggregation. Use of proline in an *E. coli* expression system abrogated aggregation of a model aggregation-prone protein, P39A cellular retinoic acid-binding protein, during expression.[188] The addition of copper sulfate (50 or 100 μM) as an oxidizing agent to the Chinese hamster ovary (CHO) cell culture production medium facilitated disulfide bond formation and reduced the level of free thiol by more than 10-fold without any effect on cell growth or antibody production.[292]

A key factor in controlling protein aggregation during fermentation/expression is the effect of the glycosylation state of the protein. Glycosylation generally increases the solubility of proteins and leads to less tendency for aggregation.[293] The reduced aggregation can be attributed to an increase in the overall molecular solvent-accessible surface area (SASA)[293] and/or to steric hindrance of the glycans in the aggregation process.[67,294] The glycosylation pattern of expressed proteins can be significantly different, depending on the construct, cell lines, and expression conditions.[295–297] The O-linked N-acetylglucosamine (O-GlcNAc) has been shown to play a critical role in controlling protein aggregation and 6-diazo-5-oxonorleucine (DON, known to decrease O-GlcNAc levels) treatment or O-GlcNAcase overexpression in cell culture reduced the O-GlcNAc levels on Sp1 transcription factor, leading to accelerated aggregation *in vitro* (44°C for 30 min).[294] Preincubation of cells with tunicamycin caused production of nonglycosylated subunits of thyroid-stimulating hormone (TSH) in cell culture, which showed a high degree of aggregation, especially after heating at 37°C under nonreducing conditions.[298]

4.5.2 Unfolding/Refolding

Unfolding/refolding is a key step after protein expression in a microbial expression system during large-scale protein production. Unfolded proteins can easily aggregate during the refolding process, as the protein goes through the unfolding intermediate state such as rhGH[221] and CS.[299] Protein aggregation can significantly reduce the yield of correctly folded proteins during refolding. Therefore, a vast body of literature addressing protein aggregation during refolding is available.[300,301] Many factors have been found to affect protein aggregation during refolding, including temperature, protein concentration, type and concentration of denaturant, pH, ionic strength, refolding catalysts, thiol/disulfide agents, and miscellaneous additives.

4.5.2.1 Temperature Since increasing temperature increases both hydrophobic interactions and chances of intermolecular intermediate interactions, an increase in protein aggregation is usually associated with a higher rate of folding at a higher temperature. A folding temperature of higher than 35°C often increases protein aggregation significantly relative to protein folding, such as in the example of protein CheY,[302] β-lactamase,[303] and phage P22 coat protein mutants.[27] In reality, the optimal refolding temperature is protein dependent. For example, at pH 8.5, the relative renaturation recovery for carbonic anhydrase was 37°C (32%) > 25°C (30%) > 4°C (19%) > 50°C (0%).[304] The refolding yield (about 60%) for bovine CAB at 150 min at 20°C was greater than at either 4 or 36°C.[305] Among the four temperatures (10, 15, 20, and 25°C) studied in the refolding of P22 tailspike polypeptide, refolding at 20°C produced the highest yield of native tailspike trimer.[25] Decreasing the temperature from 20 to 10 or 0°C resulted in an increase in the renaturation yield of a fusion protein at high salt concentrations.[306] For an unknown protein, a refolding temperature of 15°C would be a safe bet.[300]

4.5.2.2 Protein Concentration Increasing protein concentration during refolding usually increases protein aggregation due to increased chances of intermolecular interactions.[71,72] Simulation of *in vitro* protein refolding and aggregation indicates that aggregation dominates over refolding at high protein concentrations, and the size of protein aggregates increases with increasing protein concentration.[307] The refolding yield of several proteins dropped gradually with increasing protein concentrations, including β-lactamase,[303] recombinant pGH,[308] α_1-antitrypsin,[309] lysozyme,[310,311] immunoglobulin G,[312] rhDNase,[313] and CAB.[314]

Therefore, proteins are refolded usually at relatively low concentrations to avoid or reduce protein aggregation. For example, refolding of carbonic anhydrase II was carried out at low protein concentrations (0.11–1.0 μM) such that significant aggregation did not take place on the timescale of the experiment.[315] Optimal refolding concentrations have been found to be <1 mg/mL for bovine CAB,[3] at 5 μM for reduced lysozyme[316] and <18 μM for K97I IL-1β.[71] On the other hand, refolding of proteins at very low concentrations is not efficient, and loss or denaturation of proteins from surface adsorption may become relatively significant.[316] Thus, refolding of unknown proteins is recommended in a narrow concentration range of 10–50 μg/mL.[300,301,317]

4.5.2.3 Denaturants In close relation to protein concentration, the denaturant concentration strongly affects both the rate and extent of on-pathway folding and off-pathway aggregation during refolding.[317] Low concentrations of denaturants may decrease the solubility of a protein as the denaturants convert native to denatured and aggregation-prone species, while high denaturant concentrations should solubilize the protein as it weakens the protein–protein attractions in water.[10,318] Indeed, protein aggregation is usually favored at low denaturant concentrations as in the case for the refolding of CAB[3] and K97I IL-1β.[71] High concentrations often favor protein refolding. Addition of 0.5 M GdnHCl increased the refolding yield of lysozyme at 1 mg/mL from 8% to 40% at pH 8.0.[319]

Many studies showed V-shaped relationships between the rate of protein folding (ln k) and denaturant concentration, including refolding of the N-terminal domain of the protein L9[320] and protein L (Y43W)[321] and modeling the kinetic folding profile of two hypothetical proteins.[322] Therefore, a desirable concentration of a denaturant during refolding depends on the nature of the denaturant, the protein and its concentration, and other refolding conditions. Optimal refolding concentrations of denaturants are found to be 3.5 M urea for rpGH,[308] 1.5 M GdnHCl for lysozyme (0.02 mg/mL),[311] and higher or lower than 4.5 M GdnHCl for rhGH.[22]

4.5.2.4 Refolding Catalysts The well-known function of molecular chaperones in bacteria led to the use of such molecular systems to inhibit protein aggregation during protein refolding.[323,324] They suppress protein aggregation by alternate binding and releasing of the folding intermediates.[325] However, these chaperones may not be as effective as small protein stabilizers. Both glycerol and the chaperone system, GroEL–GroES, enhanced the refolding yield (less aggregation) of CS, but glycerol was more effective than the chaperone system.[299]

Some nonprotein additives are called "artificial chaperones" because they have a chaperone-like effect. These include detergents and cyclodextrins, which are used to suppress formation of protein aggregates during protein refolding.[22,231,304,314,326,327]

4.5.2.5 Thiol/Disulfide Agents In the oxidative refolding of reduced proteins, a proper redox pair is often used to promote disulfide bond formation or to catalyze the –SH and –SS– interchange. A common pair is the reduced and oxidized glutathione (GSH/GSSG).[300] Such a redox pair (2-mercaptoethanol and oxidized glutathione) was shown to facilitate the refolding of reduced immunoglobulin G by catalyzing the SH–SS interchange reaction at pH 8.0 and 4°C.[312] An optimal ratio may exist, depending on the redox pair and on the protein concentration.[326] In the oxidative renaturation of HEWL at 1 mg/mL, the optimum ratio of GSH/GSSG was between 0.8 and 3.0, and for DTT/GSSG, the ratio was between 0.3 and 0.6.[317]

4.5.2.6 Miscellaneous Additives Many additives were found to be effective in inhibiting protein aggregation during refolding. Suppression of protein aggregation often correlates with improved refolding yields.[299] Arginine is the commonly used additive in inhibiting protein aggregation and in increasing the yield of protein refolding. Arginine was shown to inhibit aggregation and to increase the refolding yield of hen egg lysozyme in several studies.[310,317,319,328] Recent studies have shown that other structurally similar compounds can be more effective than arginine. Hamada and Shiraki[319] found that L-argininamide (ArgAd) at 500 mM is 1.7-fold more effective than arginine in increasing the refolding yield of hen egg lysozyme. Thermal unfolding experiments indicate that ArgAd has a greater denaturing effect than Arg. Similar results were also obtained in the refolding of bovine carbonic anhydrase.[319] In a different study, a large variety of N′-substituted N-methylimidazolium chlorides were found to be effective in inhibiting aggregation during oxidative refolding of lysozyme at 25°C, and N′-ethyl or N′-butyl derivatives were found to be more effective.[328]

Other additives were also shown to be effective in inhibiting protein aggregation during refolding. These additives include sucrose, Ficoll 70,[303,315] glycerol,[310,317,329] PEG,[313] organic solvents (acetone, acetamide, or dimethyl sulfoxide),[311] Na_2SO_4,[231] sarcosine, ammonium sulfate, N-acetyl glucosamine (NAG), and glucose.[329] Possible mechanisms for aggregation inhibition during refolding include stabilization of the native state, destabilization of incorrectly folded molecules or intermediates, and solubilization of folded molecules or folding intermediates.[301,330,331]

A few additives are shown to promote protein aggregation during refolding. Addition of 500 mM NaCl dramatically induced aggregation of reduced lysozyme at 1 mg/mL under both normal (20°C) and thermal-assisted refolding processes (98°C for 1 min) at all pH values (4–8).[190] Addition of 500 mM ornitine or glycine reduced the refolding yield of lysozyme at 1 mg/mL from 8% to 1% and 4%, respectively, at pH 8.0.[319]

4.5.3 Purification

Purification is another key step in protein production, where residual host cell proteins and other contaminants are removed. This process could potentially induce protein aggregation if the purification conditions are not friendly. Protein A affinity chromatography is a highly efficient method for protein purification and generally requires low pH for protein elution. The low pH could induce significant protein aggregation. IgM was found to aggregate to an unacceptable level during purification in this step.[332] In a different study, protein A resin was found to increase the aggregation rate of an Fc fusion protein during elution at pH 3.6 when compared to solution alone, suggesting that protein A–Fc interactions can destabilize the protein structure and can increase the tendency to aggregate.[160]

Protein aggregation can also occur under mild purification conditions. Processing of Eprex bulk drug substance through an ion exchange HiTrap Q HP column using 10 mM Tris buffer at pH 6.9 with a step gradient of NaCl increased the aggregate level from 1.4% to 3.6% as monitored by analytical ultracentrifugation (AUC).[333]

Protein aggregation during purification could be minimized by adjusting the detailed purification conditions. For example, the rate constant for the aggregation of an Fc fusion protein during elution was about 0.11/h at room temperature but could be dropped to about 0.01/h at 4°C. Use of proper additives could significantly reduce the aggregation of mAb during the ultrafiltration/diafiltration (UF/DF) process.[334]

4.5.4 Freeze-Thaw

Freezing a protein solution is often required for preservation of protein activity/stability during the production processes. Such a process could potentially cause protein aggregation, depending on the type and concentration of proteins, the presence of other formulation additives, and the parameters of the freezing process (temperature, rate, etc.). Freeze-thaw-induced aggregation was reported for lactate dehydrogenase (LDH) and glutamate dehydrogenase (GDH) in 10 mM citrate at pH 6.5,[335] an antibody fragment (27 kD) in PBS at pH 7.3,[336] and hemoglobin at 50 mg/mL in PBS at pH 7.4.[155] Freezing of a protein solution may cause protein aggregation due to one or more of the following freezing-induced stresses: low temperature, solute concentration, formation of ice–water interfaces, and potential pH changes or phase separation.[337]

A frequent cause of protein aggregation during freezing is the formation of ice–water interfaces. Proteins interact with such interfaces and expose hydrophobic regions, leading to increased aggregation. Therefore, the size of ice crystal will influence the degree of aggregation due to differences in surface area. In this case, a slower rate of freezing should inhibit aggregation of these proteins as a smaller ice–water interface will be generated.[335] It should be noted that the size and homogeneity of ice crystals will differ significantly, depending on the type and concentration of other additives present in the protein solution.[338] On the other hand, a lower freezing rate may facilitate crystallization of other solution components, promoting potential crystallization-induced protein aggregation.[339]

Freezing-induced pH change is another major cause for protein aggregation due to selective crystallization of certain buffering species.[340] This could happen even in the presence of stabilizers such as sucrose and trehalose.[341] The solution pH has been shown to affect freeze-thaw-induced aggregation of many proteins to a different degree, such as a mAb IgG2[279] and a chimeric antibody.[172] Therefore, proper selection of a suitable buffer would prevent/inhibit protein aggregation.[337] Other factors also influence the aggregation behavior of proteins, such as freezing/thawing time, which varies significantly depending on the configuration of containers. It was found that the distance from the edge of a container to its center has a major impact on freezing and thawing times.[342] A threefold increase in the freezing distance could lengthen the freezing time 10-fold and the thawing time 20-fold.[214]

Protein aggregation during the freezing process can be minimized by the addition of a variety of additives in protein solutions. These additives may inhibit protein aggregation by minimization of protein–ice interactions, suppression of pH changes, preferential interaction and/or other mechanisms such as increased solution viscosity, and steric hindrance of protein–protein interactions.[115,172,197,343,344] Among these additives, surfactants seem to be the most commonly used as they are effective in minimizing protein–ice interactions.[335,345,346]

4.5.5 Agitation

Agitation (often referred to as shaking or shearing) is often encountered during protein manufacturing and shipping. Agitation can potentially induce protein aggregation via a number of proposed mechanisms. Agitation in the form of shaking creates air–water interfaces. The hydrophobic property of air relative to water promotes protein adsorption at the interface, maximizing exposure of the hydrophobic residues to the air and potentially initiating aggregation.[257,347] Thus, agitiation may increase the amount of air–water interface that is available to facilitate surface-mediated unfolding and/or aggregation (see also Chapters 2 and 7). Stirring, without creation of new interfaces, can also increase the rate of mass transfer between bulk solution and existing interfaces, thereby increasing the turnover rate of protein adsorbing from bulk solution to interfaces between water and air or container–closure contacts. However, different forums of agitation could potentially generate different species of aggregates both qualitatively and quantitatively.[348] Many proteins easily aggregate during agitation; examples include an antibody,[349] a mAb—IgG1,[348,350] an IgG4,[351] the variable domain of an IgG light chain (114 aa),[57] IL-2 mutein,[223] Albutropin,[226] rFXIII,[346] hGH,[228,230,251] and insulin,[23,352] deoxy hemoglobin,[353] and α-synuclein (natively unfolded protein).[163] Surfactants are most commonly used and are generally effective in reducing shaking/shearing-induced protein aggregation, such as Tweens for IL-2 mutein,[223] factor XIII[346] or rhGH,[261] Pluronic F68, or Brij 35 for rhGH.[261]

Agitation could also influence protein aggregation indirectly by facilitating chemical degradations. For example, agitating solutions of a mAb IgG4 (equivalent to $2.89 \times 10^4\,S^{-1}$ at 20°C for 7h) resulted in the formation of a small portion of fragments. The fragmentation is likely due to agitation-induced exposure of sensitive amino acids for faster degradation, which, under normal conditions, do not occur readily. The variable domain of an IgG light chain (114 aa) contains one Met4, well protected from oxidation in its native state; shaking exposes this amino acid for faster oxidation, and the oxidized protein is more sensitive to stirring-induced aggregation.[57]

4.5.6 High Hydrostatic Pressure

High pressure has been shown to be detrimental to certain proteins. This is purportedly due to increased hydrophobic interactions at high pressure.[354] It was found that ultra-high-pressure homogenization induced aggregation of dispersions of whey protein isolate containing 6% or 10% (w/w) protein at pH 6.5 at 250–300MPa (not below 225MPa), and the average sizes of aggregates increased from 7 to 50nm when the pressure was increased from 250 to 300MPa.[354] Similarly, high pressure treatment of whey protein concentrate solutions (12% w/v, pH 6.65) at 800MPa for 20–120min led to formation of aggregates of different sizes.[355] Application of hydrostatic pressures gradually caused aggregation of rhIL-1ra at pH 7.0 above 180MPa (midpoint)[356] and aggregation of β-Lg at pH 7.2 above 500MPa.[202]

Pressure treatment can also generate chemically linked aggregates, as high pressure can expose reactive groups, which are normally well covered under normal pressure.[357] For example, when IL-1ra begins to denature at pressures near 140MPa, cysteines that are normally buried in the native state become exposed.[356] Similarly, pressure-induced aggregates of whey proteins contain both non-covalent and covalent, disulfide-linked aggregates.[355]

Different proteins have different levels of sensitivity to pressure-induced aggregation. The sensitivity of the major whey proteins to pressure-induced aggregation is in this order: β-Lg B > β-Lg A > BSA > α-LA, and the large internal hydrophobic cavity of β-Lg may have been partially responsible for its high sensitivity.[355] The distribution of aggregate species can be different depending on the pressure and its duration.[202]

Pressure-induced aggregation could be inhibited by lowering the temperature in some cases. Kolakowski et al.[358] demonstrated that pressurizing a solution of β-Lg isolate for 15 min at 300 MPa at 25°C led to detectable formation of aggregates, but aggregates were not detected when pressurization was conducted at 4°C or at −2°C. On the other hand, in the presence (or even absence) of moderate concentration of chaotropic agents, high pressure has been shown to foster disaggregation of proteins such as GH and lysozyme.[359–361]

4.5.7 Drying

Generally, the amount of water covering the surface of a protein in a fully hydrated state is around 0.3 g/g protein.[362,363] In comparison, the water content of a dried protein product is usually less than 10%. Therefore, the drying process removes part of the hydration layer, which may disrupt the native state of a protein and may cause protein aggregation. Several drying processes have been used/tried, including vacuum-drying, freeze-drying, spray-drying and spray-freeze drying (SFD). All of these processes could potentially cause protein aggregation.

4.5.7.1 Vacuum-Drying The simplest method for rapid dehydration of proteins is vacuum-drying. Due to the removal of the hydration layer, the protein may aggregate during this process. Vacuum-drying of CS or LDH caused significant aggregation with loss of activity.[182] Vacuum-drying of spray-dried IgG1/mannitol mixtures at 32°C for 24 h at 0.1 mbar promoted aggregate formation by 2.8% and 3.3% for formulations at 30/70 and 40/60 ratios, respectively, due to additional crystallization of mannitol during the drying process.[184] Evaporative drying has a similar effect on proteins and may cause protein aggregation, such as methionyl hGH.[364] Trehalose alone or in combination with late embryogenesis abundant (LEA) proteins is able to inhibit vacuum-drying-induced aggregation of CS.

4.5.7.2 Freeze-Drying The freeze-drying process is the technique most often used to prepare dried proteins, because drying under the frozen state could potentially protect the protein from temperature-induced aggregation during the long drying process and could produce a desired cake structure in the presence of a suitable excipient(s). As with the vacuum-drying process, freeze-drying could also promote protein aggregation. Many proteins have been shown to aggregate to a different degree during the freeze-drying process, including hGH,[365] Met-hGH,[364] a botulinum neurotoxin,[159] β-galactosidase,[366] lysozyme,[367] human or mouse IgGs,[368–371] rhIL-1ra,[237] and bovine ribonuclease A.[265] The freeze-drying-induced aggregates generally consist of non-covalently linked monomers and, in few cases, covalently linked oligomers.[169,372,373]

Many formulation excipients have been used and have been found to be very effective in inhibiting protein aggregation during freeze-drying, especially sugars. These excipients include sucrose for rhIL-1ra,[237] both trehalose and sucrose for IL-6[374] and human factor XIII,[375] mannitol for a botulinum neurotoxin,[159] Tween 80 for bovine IgG[368] and rhIL-1ra,[237] Tween 20 and hydroxyethyl starch for a botulinum neurotoxin,[159] hydroxyethyl cellulose for aFGF,[376] and PEG for human FXIII.[375] Stabilization of protein by polymers is likely due to their capability of forming a glass with a higher transition temperature.

Some excipients are found to increase protein aggregation during freeze-drying. In the lyophilization of a human mAb at 1 mg/mL, it was demonstrated that the addition of 5% mannitol has no effect on the lyophilization-induced aggregation in PBS or in physiological saline, and that the addition of 10% mannitol actually accelerated aggregation during the lyophilization process.[370] Lyophilization of rhIL-1ra at 15 and 35 mg/mL at pH 7.0 resulted in the formation of insoluble aggregates of 11% and 5%, respectively, and the addition of 140 mM NaCl in the prelyophilization solution increased the aggregate level to 15% and 7%.[237]

Annealing during freeze-drying was also found to play a critical role in controlling protein aggregation.[377]

4.5.7.3 Spray-Drying Spray-drying is another widely examined method for dehydration of proteins. During this process, proteins are exposed to elevated temperature and the air–water interface for a short period of time during atomization. Both stresses may cause protein destabilization and aggregation. Many proteins have been spray-dried and aggregated to a different degree during this process. These include IgG1,[184] IgG,[378] mAbs,[379] hGH,[380] and Met-hGH.[364] Due to high temperature and different kinds of stress, spray-drying can be more damaging to proteins than the freeze-drying process, such as Met-hGH.[364]

Since the main stress during spray-drying is the exposure of proteins to the air–water interface, the use of surfactants is often effective in inhibiting protein aggregation, such as Tween 20 for hGH[380] and Met-hGH.[364] Other excipients were also found effective, such as trehalose or sorbitol for an IgG,[378] trehalose, lactose, or mannitol for a human mAb.[379] These excipients are able to stabilize proteins by replacing water during drying.[378]

On the other hand, some excipients may cause protein aggregation during this process. Spray-drying of IgG1 alone generated 1.9%, and addition of mannitol in a 20/80 ratio of IgG1/mannitol increased the aggregate level to 5.7% after spray-drying.[184] Mannitol was found to crystallize during spray-drying and caused protein aggregation.

4.5.7.4 SFD SFD eliminates the use of elevated temperature during spray-drying, but air–water interfaces are still created and remain a potential cause of protein aggregation. A few proteins have been shown to aggregate to a different degree during SFD, including interferon-γ,[381] lysozyme,[367] darbepoetin α,[382,383] and BSA.[384]

This process could be more damaging to a protein than the simple spray-drying process. For example, SFD of darbepoetin α by ultrasonic atomization at 120 and 25 kHz generated 9% insoluble aggregates, while spray-drying at benchtop and pilot scales did not lead to detectable formation of insoluble or HMW soluble aggregates using a two-fluid nozzle.[382,383]

4.5.7.5 Supercritical Fluid (SCF) Drying SCF drying is another drying method that has been tested. Jovanovic studied SCF drying. Lysozyme solutions with and without added sucrose or trehalose were dried by spraying them in an SCF composed of CO_2 and ethanol or CO_2 only.[385] Protein aggregation was detected by infrared after drying pure lysozyme and, after the addition of sugars, in protein-to-sugar ratios between 1:0 and 1:10 (w/w), reduced aggregation significantly.[385]

4.5.8 Preparation of Protein Delivery Systems

Proteins can be incorporated into a drug delivery system to achieve a prolonged or targeted therapeutic effect. Such a process could potentially destabilize a protein and could cause protein aggregation. The use of organic solvents in such processes is a frequent factor for inducing protein aggregation.[352,386–388] The proper use of a stabilizing agent may inhibit or prevent protein aggregation under these processing conditions.[387,389]

4.5.9 Analytical Methodologies

The level of protein aggregates in samples can be significantly different depending on the sample treatment and on the analytical procedures. The aggregates (with an average size of 3.2 μm) of a therapeutic IgG formed can be easily disrupted during field flow fractionation (FFF) analysis.[101] Thermal protein denaturation/aggregation can be significantly different depending on the heating rate.[126]

4.5.10 Miscellaneous Processes

In addition to the above processes, other processes were also reported to destabilize proteins and to cause aggregation. Nebulization processes caused rapid aggregation of a variety of interferons in several studies.[381,390,391] Irradiation caused aggregation of ceruloplasmin and hemoglobin.[392] Filling of a mAb solution through a radial piston pump (shearing effect) facilitated the formation of aggregates and particulates.[334] Iodine (I^{125}) labeling of albumin with chloramine-T as an oxidizing agent caused enhanced aggregation and fragmentation of the proteins, and the labeled protein also formed a greater amount of aggregates relative to unlabeled proteins during freeze-thaw (–20°C for 12 h).[276]

4.6 EFFECTS OF SOLID-STATE CONDITION AND COMPOSITION ON PROTEIN AGGREGATION

There are many proteins that aggregate to a different degree in solid states under different experimental conditions. Both physically and chemically linked protein aggregates can be formed in solid state.[262] Major factors that influence protein aggregation in the solid state include solid-state pH, excipients and their levels, the physical state of the solid, and moisture content.

4.6.1 Solid-State "pH"

The solid-state pH is the pH of a solution after a dehydration process. As for a liquid formulation, this solid-state "acidity/basicity" can also affect the aggregation of the dried protein product during storage. This has been demonstrated in several lyophilized proteins, including botulinum neurotoxin,[159] BSA,[393] recombinant human albumin (rHA),[394] IgG1 antibody,[395] IL-1ra,[335] Rnase,[2] antibody–vinca conjugate,[396] and insulin.[50] For example, the aggregation rate constant of a lyophilized IgG1 antibody at 50°C is 7.0 $month^{-1/2}$ at pH 5 but 10 $month^{-1/2}$ at pH 7.[395]

It should be noted that the most stable or unstable pH in a solution state can be different from that in the solid state. For example, a botulinum neurotoxin (50 kD) at 0.1 and 0.55 mg/mL was stable at pH 5 during incubation at 50°C for 4 h, but the protein rapidly aggregated at both concentrations at pH 8.0; in comparison, the protein was stable during both lyophilization and storage when freeze-dried from a prelyophilized solution of either pH 5 or 8.[159] Therefore, the optimal pH of a solid protein product needs to be evaluated in addition to evaluation of pH stability in a liquid state.

4.6.2 Excipients and Excipient Levels

Proteins are often dehydrated with a variety of excipients/additives. Excipients and their levels have a profound effect on solid-state stability. Generally, amorphous excipients replacing water for the required formation of hydrogen bondings with proteins may provide protein stabilization and aggregation inhibition, while crystallizing excipients are not able to form hydrogen bondings and may accelerate protein aggregation during storage.[184]

Among these amorphous excipients, sugars are the most commonly used for stabilizing proteins against aggregation in the solid state. Sucrose and trehalose are two disaccharides that have been widely used against the aggregation of many proteins, including hGH,[365,397] BSA,[398] a botulinum neurotoxin,[159] β-galactosidase,[366] IgG1,[395] IL-1ra,[399] and tumor necrosis factor.[400] It should be noted that the relative stabilization effect of sucrose and trehalose has not been consistent in the literature even for the same protein. While sucrose was found to be more effective in protecting lyophilized hGH against aggregation at several sugar/protein ratios at 50°C,[397] trehalose was more effective in protecting lyophilized hGH against formation of insoluble aggregates at the same temperature in a different study.[365]

Sugar-protected proteins in the solid state also include IL-6,[374] IL-11,[401] insulin,[402] *Humicola lanuginosa* lipase,[403] tetanus toxoid,[404] and antibodies.[372,378,379,405,406] Other small molecule excipients that have been shown to inhibit protein aggregation include histidine in a lyophilized mAb,[115] salts in lyophilized BSA[407] or recombinant human serum albumin (rHSA)[394] or carboxylic acids in lyophilized rHSA,[408] amino acids in lyophilized IgG,[184] IL-2,[409] hGH,[170] or IL-1ra,[399] and metal ions in lyophilized rhDNase.[410]

Some polymers are also effective in the inhibition of protein aggregation during storage.[159,371,394,400,401,407,409,411] In general, low-molecular-weight stabilizers provide better stability than large ones. For example, the aggregation rate of Met-hGH in a freeze-dried trehalose-based formulation was approximately four times smaller than that in the corresponding Ficoll 70-based formulation at 50°C.[364]

The protective effect of excipients is often concentration dependent. For example, both sucrose and trehalose significantly protected lyophilized IgG1 against aggregation, and the aggregation rate of the protein gradually decreased when the weight ratio of sugar to protein increased.[372,395] Similarly, both sugars stabilized lyophilized hGH against aggregation at 50°C with increasing sugar/hGH ratios from 1 to 6.[397] Increasing the amount of NaCl from 0 to 0.45 M in the buffer gradually reduced the formation of insoluble aggregates of dried whey protein at 35°C in the model system.[262] It was shown that sugars protect lyophilized mAbs against aggregation effectively when the sugar-to-protein molar ratio was ≥500:1.[372,412–414]

Some excipients that protect proteins from aggregation in the liquid state or during lyophilization may not be effective in the solid state. Tween 20 was shown to stabilize a botulinum neurotoxin during lyophilization against aggregation in different formulations but did not change the aggregation rate during storage at 40°C for 8 weeks.[159] In fact, Tween 20 was shown to accelerate the aggregation of freeze-dried Met-hGH formulations at 50°C.[364] Mannitol could not stabilize effectively or even destabilizes antibodies such as aggregation of a lyophilized model rhuMAb (HER2) in the presence of mannitol during storage.[369,413]

The physical and chemical properties of excipients may change with time in solid state, which can indirectly influence the aggregation behavior of proteins. Crystallization of excipients can occur during storage, leading to loss of protection of proteins, affecting protein aggregation in solid state. Excipient crystallization depends not only on the type but also on the level of excipients in the protein formulation. For example, SCF drying of lysozyme with protein-to-sugar (sucrose or trehalose) ratios between 1:0 and 1:10 (w/w) led to formation of amorphous products, but crystallization during storage only occurred with a protein-to-sugar ratio of 1:10.[385] Crystallization at 1:10 ratio occurred in 30 days at 4°C with sucrose but in 90 days with trehalose. Incubation of lyophilized β-lactoglobulin : glucose mixture (about 1:100 molar ratio at pH 8) at 60°C for 5h led to glucosylation at 14 of 16 potential glucosylation sites, and the glucosylated form was found to inhibit urea-induced aggregation of the protein at 10mg/mL.[415]

4.6.3 Physical State of the Solid

The physical state of the solid protein product is another factor that can potentially influence protein aggregation. Several parameters can be used to describe the physical state of the solid product—amorphous or crystalline state, tendency for excipient recrystalization, glass transition temperature, molecular mobility, and so on.

As mentioned above, excipients need to be in an amorphous state so that they may form hydrogen bonds with proteins for stabilization.[184] Generally, proteins have a higher tendency to aggregate with crystalline excipients than with amorphous excipients. For example, it has been shown that the spray-dried IgG1/mannitol mixture with a weight ratio of 20/80 up to 40/60 generated crystalline IgG1/mannitol formulations, which failed to prevent antibody aggregation, while amorphous formulations with higher ratios 60/40 up to 80/20 IgG1/mannitol provided stabilization.[184]

If the dried product has a high tendency for excipient recrystallization during storage, protein aggregation could be potentially accelerated. Such a tendency depends not only on the type and level of excipients as mentioned above but also on the storage temperature. It has been shown that the time to recrystallization of amorphous lactose decreases with increasing T–Tg.[416]

It would be greatly beneficial if the storage stability of dried proteins can be predicted one way or another in the presence of excipients. Several parameters have been examined in the past, including glass transition temperature, retention of protein structure as monitored by a biophysical method during lyophilization, and molecular mobility of the protein formulation. Literature data have shown that a higher glass transition temperature of a protein formulation is not always associated with better protein stability with a less tendency for protein aggregation during storage.[337] The glass dynamics mechanism does not explain the IgG1 stabilization well, while the degree of retention of the native structure (by Fourier transform infrared [FTIR]) of the protein during drying could be a more reliable measure of storage stability.[395] On the contrary, the differences in the aggregation rate of the lyophilized β-galactosidase observed between formulations containing different excipients could not be attributed to differences in protein structure (by FTIR). Others used molecular mobility[366] or the thermal denaturation temperature of a solid protein/excipient mixture[417] as an indictor for long-term stability prediction. It was found that the aggregation rate of lyophilized β-galactosidase is unexpectedly correlated with the local mobility of the protein and that sucrose exhibited the most intense stabilizing effect due to the most intense ability to inhibit local protein mobility during storage.[366] It remains to be seen whether and how the storage stability of solid proteins can be predicted.

4.6.4 Moisture Content

Another critical factor in controlling the rate of protein aggregation in solid states is the moisture content of the formulation.[418] Generally, increasing the moisture content increases the rate of protein aggregation during storage, such as hGH,[397] mAbs,[419,420] insulin,[50] and bovine pancreatic RNase.[2] For example, incubation of lyophilized hGH with a sucrose/GH ratio of 6 at 40°C for 6 months at moisture levels of 0.5%, 3.1%, and 4.9% led to an increase in aggregates of 0.26%, 0.35%, and 1.11%, respectively.[397] This is likely due to a reduction in glass transition temperature and an increase in the mobility of proteins at a higher moisture content.

In many cases, however, the effect of moisture on protein aggregation is a complex function. For example, maximum aggregation (78%) of lyophilized tetanus toxoid occurred at a water content of about 36% during storage at 37°C for 10 days, and less aggregation was observed at water contents either below or above that level.[52] Aggregation of both (lyophilized) rHA and BSA during storage at 37°C had a bell-shaped relationship as a function of water content with maximum aggregation at about 32% and 28% moisture, respectively.[407,408] The bell-shaped relationship was confirmed for BSA in a recent study.[398] Similar bell-shaped aggregation dependence on water content was also observed for several other proteins, including whey proteins,[262] a mAb,[418] ovalbumin (chicken egg albumin), glucose oxidase, bovine β-lactoglobulin,[407] and insulin.[411,421]

The moisture content of a solid protein formulation could dictate the aggregation pathway. The mechanism of aggregation of lyophilized tetanus toxoid was different under different relative humidities during storage.[422]

4.7 SUMMARY

A common phenomenon of protein instability is the formation of aggregates. Several aggregation pathways have been described (Fig. 4.2), leading to physical or chemically linked aggregates. A variety of factors have been identified affecting the rate and extent of protein aggregation. These factors can be divided into several categories: (1) general factors (temperature and light); (2) solution conditions and composition (pH, buffer type and concentration, ionic strength, excipients and level, protein concentration, metal ions, denaturing and reducing agents, impurities, organic solvents, containers/closures, and sources of proteins); (3) processing steps (fermentation/expression, unfolding/refolding, purification, freeze-thaw, shaking and shearing, pressurization, formulation/filling, drying, preparation of modified protein or delivery systems, analytical methodologies, etc.); and (4) solid-state conditions and composition (solid-state pH, excipients and level, physical state of the solid, and moisture content).

Among these factors, temperature is the most critical and commonly encountered variable that needs to be accurately controlled. The next critical factor is arguably the solution pH. All of these aggregation-influencing factors need to be kept in mind when dealing with proteins. Since the aggregation pathways and ultimate aggregate composition/structure are different under different experimental conditions, it is critical to identify the aggregation mechanism(s) before designing an effective way to inhibit/prevent protein aggregation. Inhibition of protein aggregation is frequently achieved by changing the immediate environment of the protein through the use of a compatible excipient(s). The often used protein stabilizing excipients include sugars, polyols, surfactants, salts, PEGs, polymers, metal ions, and amino acids.

ACKNOWLEDGMENT

The authors appreciate the critical review of the manuscript by Prof. Ted Randolph.

REFERENCES

1. Wang W. 2005. Protein aggregation and its inhibition in biopharmaceutics. Int J Pharm 289(1–2):1–30.
2. Townsend MW, Byron PR, DeLuca PP. 1990. The effects of formulation additives on the degradation of freeze-dried ribonuclease A. Pharm Res 7(10):1086–1091.
3. Cleland JL, Wang DI. 1990. Refolding and aggregation of bovine carbonic anhydrase B: Quasi-elastic light scattering analysis. Biochemistry 29(50):11072–11078.
4. Runkel L, Meier W, Pepinsky RB, Karpusas M, Whitty A, Kimball K, Brickelmaier M, Muldowney C, Jones W, Goelz SE. 1998. Structural and functional differences between glycosylated and non-glycosylated forms of human interferon-beta (IFN-beta). Pharm Res 15(4):641–649.
5. Nesta DP, Elliott JS, Warr JP. 2000. Supercritical fluid precipitation of recombinant human immunoglobulin from aqueous solutions. Biotechnol Bioeng 67(4): 457–464.
6. Wang W, Kelner DN. 2003. Correlation of rFVIII inactivation with aggregation in solution. Pharm Res 20(4):693–700.
7. Clark AC. 2008. Protein folding: Are we there yet? Arch Biochem Biophys 469(1):1–3.
8. Carpenter JF, Randolph TW, Jiskoot W, Crommelin DJ, Middaugh CR, Winter G, Fan YX, Kirshner S, Verthelyi D, Kozlowski S, Clouse KA, Swann PG, Rosenberg A, Cherney B. 2008. Overlooking subvisible particles in therapeutic protein products: Gaps that may compromise product quality. J Pharm Sci. doi: 10.1002/jps.21530.
9. Fink AL. 1998. Protein aggregation: Folding aggregates, inclusion bodies and amyloid. Fold Des 3(1):R9–R23.
10. Fields G, Alonso D, Stiger D, Dill KA. 1992. Theory for the aggregation of proteins and copolymers. J Phys Chem 96:3974–3981.

11. Murphy KP, Bhakuni V, Xie D, Freire E. 1992. Molecular basis of co-operativity in protein folding. III. Structural identification of folding units and folding intermediates. J Mol Biol 227(1):293–306.
12. Kendrick BS, Carpenter JF, Cleland JL, Randolph TW. 1998. A transient expansion of the native state precedes aggregation of recombinant human interferon-gamma. Proc Natl Acad Sci U S A 95(24):14142–14146.
13. Damodaran S, Song KB. 1988. Kinetics of adsorption of proteins at interfaces: Role of protein conformation in diffusional adsorption. Biochim Biophys Acta 954(3):253–264.
14. Uversky VN, Karnoup AS, Khurana R, Segel DJ, Doniach S, Fink AL. 1999. Association of partially-folded intermediates of staphylococcal nuclease induces structure and stability. Protein Sci 8(1):161–173.
15. Espargaro A, Castillo V, de Groot NS, Ventura S. 2008. The in vivo and in vitro aggregation properties of globular proteins correlate with their conformational stability: The SH3 case. J Mol Biol 378(5):1116–1131.
16. Zhang L, Lu D, Liu Z. 2008. How native proteins aggregate in solution: A dynamic Monte Carlo simulation. Biophys Chem 133(1–3):71–80.
17. Weiss WFt, Young TM, Roberts CJ. 2008. Principles, approaches, and challenges for predicting protein aggregation rates and shelf life. J Pharm Sci. doi: 10.1002/jps.21521.
18. Raso SW, Abel J, Barnes JM, Maloney KM, Pipes G, Treuheit MJ, King J, Brems DN. 2005. Aggregation of granulocyte-colony stimulating factor in vitro involves a conformationally altered monomeric state. Protein Sci 14(9):2246–2257.
19. Safar J, Roller PP, Gajdusek DC, Gibbs JCJ. 1994. Scrapie amyloid (prion) protien has the conformational characteristics of an aggregated molten globule folding intermediate. Biochemistry 33:8375–8383.
20. Cleland JL, Wang DI. 1992. Transient association of the first intermediate during the refolding of bovine carbonic anhydrase B. Biotechnol Prog 8(2):97–103.
21. Hammarstrom P, Persson M, Freskgard PO, Martensson LG, Andersson D, Jonsson BH, Carlsson U. 1999. Structural mapping of an aggregation nucleation site in a molten globule intermediate. J Biol Chem 274(46):32897–32903.
22. Bam NB, Cleland JL, Randolph TW. 1996. Molten globule intermediate of recombinant human growth hormone: Stabilization with surfactants. Biotechnol Prog 12(6):801–809.
23. Brange J, Andersen L, Laursen ED, Meyn G, Rasmussen E. 1997. Toward understanding insulin fibrillation. J Pharm Sci 86(5):517–525.
24. Booth DR, Sunde M, Bellotti V, Robinson CV, Hutchinson WL, Fraser PE, Hawkins PN, Dobson CM, Radford SE, Blake CC, Pepys MB. 1997. Instability, unfolding and aggregation of human lysozyme variants underlying amyloid fibrillogenesis. Nature 385(6619):787–793.
25. Speed MA, King J, Wang DIC. 1997. Polymerization mechanism of polypeptide chain aggregation. Biotechnol Bioeng 54:333–343.
26. Schuler B, Rachel R, Seckler R. 1999. Formation of fibrous aggregates from a non-native intermediate: The isolated P22 tailspike beta-helix domain. J Biol Chem 274(26):18589–18596.

27. Teschke CM. 1999. Aggregation and assembly of phage P22 temperature-sensitive coat protein mutants in vitro mimic the in vivo phenotype. Biochemistry 38(10):2873–2881.

28. Chi EY, Krishnan S, Kendrick BS, Chang BS, Carpenter JF, Randolph TW. 2003. Roles of conformational stability and colloidal stability in the aggregation of recombinant human granulocyte colony-stimulating factor. Protein Sci 12(5): 903–913.

29. Militello V, Casarino C, Emanuele A, Giostra A, Pullara F, Leone M. 2004. Aggregation kinetics of bovine serum albumin studied by FTIR spectroscopy and light scattering. Biophys Chem 107(2):175–187.

30. Vetri V, Militello V. 2005. Thermal induced conformational changes involved in the aggregation pathways of beta-lactoglobulin. Biophys Chem 113(1):83–91.

31. Katayama DS, Nayar R, Chou DK, Valente JJ, Cooper J, Henry CS, Vander Velde DG, Villarete L, Liu CP, Manning MC. 2006. Effect of buffer species on the thermally induced aggregation of interferon-tau. J Pharm Sci 95(6):1212–1226.

32. Ecroyd H, Carver JA. 2008. The effect of small molecules in modulating the chaperone activity of alpha B-crystallin against ordered and disordered protein aggregation. FEBS J 275(5):935–947.

33. Necula M, Kayed R, Milton S, Glabe CG. 2007. Small molecule inhibitors of aggregation indicate that amyloid beta oligomerization and fibrillization pathways are independent and distinct. J Biol Chem 282(14):10311–10324.

34. Liu J, Nguyen MD, Andya JD, Shire SJ. 2005. Reversible self-association increases the viscosity of a concentrated monoclonal antibody in aqueous solution. J Pharm Sci 94(9):1928–1940.

35. Alford JR, Kendrick BS, Carpenter JF, Randolph TW. 2008. High concentration formulations of recombinant human interleukin-1 receptor antagonist: II. Aggregation kinetics. J Pharm Sci 97(8):3005–3021.

36. Kanai S, Liu J, Patapoff TW, Shire SJ. 2008. Reversible self-association of a concentrated monoclonal antibody solution mediated by Fab-Fab interaction that impacts solution viscosity. J Pharm Sci. doi:10.1002/jps.21322.

37. Saluja A, Kalonia DS. 2008. Nature and consequences of protein-protein interactions in high protein concentration solutions. Int J Pharm 358(1–2):1–15.

38. Chi EY, Krishnan S, Randolph TW, Carpenter JF. 2003. Physical stability of proteins in aqueous solution: Mechanism and driving forces in nonnative protein aggregation. Pharm Res 20(9):1325–1336.

39. Richards JP, Stickelmeyer MP, Flora DB, Chance RE, Frank BH, DeFelippis MR. 1998. Self-association properties of monomeric insulin analogs under formulation conditions. Pharm Res 15(9):1434–1441.

40. Bajaj H, Sharma VK, Badkar A, Zeng D, Nema S, Kalonia DS. 2006. Protein structural conformation and not second virial coefficient relates to long-term irreversible aggregation of a monoclonal antibody and ovalbumin in solution. Pharm Res 23(6):1382–1394.

41. Broersen K, Weijers M, de Groot J, Hamer RJ, de Jongh HHJ. 2007. Effect of protein charge on the generation of aggregation-prone conformers. Biomacromolecules 8(5):1648–1656.

42. Krebs MR, Devlin GL, Donald AM. 2007. Protein particulates: Another generic form of protein aggregation? Biophys J 92(4):1336–1342.
43. Wang YJ, Shahrokh Z, Vemuri S, Eberlein G, Beylin I, Busch M. 1996. Characterization, stability, and formulations of basic fibroblast growth factor. Pharm Biotechnol 9:141–180.
44. Shahrokh Z, Eberlein G, Buckley D, Paranandi MV, Aswad DW, Stratton P, Mischak R, Wang YJ. 1994. Major degradation products of basic fibroblast growth factor: Detection of succinimide and iso-aspartate in place of aspartate. Pharm Res 11(7):936–944.
45. Yoshioka S, Aso Y, Izutsu K, Terao T. 1993. Aggregates formed during storage of beta-galactosidase in solution and in the freeze-dried state. Pharm Res 10(5):687–691.
46. Jordan GM, Yoshioka S, Terao T. 1994. The aggregation of bovine serum albumin in solution and in the solid state. J Pharm Pharmacol 46(3):182–185.
47. Livney YD, Dalgleish DG. 2004. Specificity of disulfide bond formation during thermal aggregation in solutions of beta-lactoglobulin b and kappa-casein A. J Agric Food Chem 52(17):5527–5532.
48. Livney YD, Verespej E, Dalgleish DG. 2003. Steric effects governing disulfide bond interchange during thermal aggregation in solutions of beta-lactoglobulin B and alpha- lactalbumin. J Agric Food Chem 51(27):8098–8106.
49. Cabra V, Vazquez-Contreras E, Moreno A, Arreguin-Espinosa R. 2008. The effect of sulfhydryl groups and disulphide linkage in the thermal aggregation of Z19 alpha-zein. Biochim Biophys Acta 1784(7–8):1028–1036.
50. Costantino HR, Langer R, Klibanov AM. 1994. Moisture-induced aggregation of lyophilized insulin. Pharm Res 11(1):21–29.
51. Malencik DA, Anderson SR. 2003. Dityrosine as a product of oxidative stress and fluorescent probe. Amino Acids 25(3–4):233–247.
52. Schwendeman SP, Costantino HR, Gupta RK, Siber GR, Klibanov AM, Langer R. 1995. Stabilization of tetanus and diphtheria toxoids against moisture-induced aggregation. Proc Natl Acad Sci U S A 92(24):11234–11238.
53. Malencik DA, Anderson SR. 2003. Dityrosine as a product of oxidative stress and fluorescent probe. Amino Acids 25(3–4):233–247.
54. Khossravi M, Shire SJ, Borchardt RT. 2000. Evidence for the involvement of histidine A(12) in the aggregation and precipitation of human relaxin induced by metal-catalyzed oxidation. Biochemistry 39(19):5876–5885.
55. Liang JN, Rossi MT. 1990. In vitro non-enzymatic glycation and formation of browning products in the bovine lens alpha-crystallin. Exp Eye Res 50(4):367–371.
56. Zarina S, Zhao HR, Abraham EC. 2000. Advanced glycation end products in human senile and diabetic cataractous lenses. Mol Cell Biochem 210(1–2):29–34.
57. Hu D, Qin Z, Xue B, Fink AL, Uversky VN. 2008. Effect of methionine oxidation on the structural properties, conformational stability, and aggregation of immunoglobulin light chain LEN. Biochemistry (Mosc) 47(33):8665–8677.
58. Li S, Nguyen TH, Schoneich C, Borchardt RT. 1995. Aggregation and precipitation of human relaxin induced by metal-catalyzed oxidation. Biochemistry 34(17):5762–5772.

59. Hawkins CL, Davies MJ. 2005. The role of aromatic amino acid oxidation, protein unfolding, and aggregation in the hypobromous acid-induced inactivation of trypsin inhibitor and lysozyme. Chem Res Toxicol 18(11):1669–1677.
60. Gu LC, Erdos EA, Chiang HS, Calderwood T, Tsai K, Visor GC, Duffy J, Hsu WC, Foster LC. 1991. Stability of interleukin 1 beta (IL-1 beta) in aqueous solution: Analytical methods, kinetics, products, and solution formulation implications. Pharm Res 8(4):485–490.
61. Van Buren N, Rehder D, Gadgil H, Matsumura M, Jacob J. 2008. Elucidation of two major aggregation pathways in an IgG2 antibody. J Pharm Sci. doi: 10.1002/jps.21514.
62. Takata T, Oxford JT, Demeler B, Lampi KJ. 2008. Deamidation destabilizes and triggers aggregation of a lens protein, {beta}A3-crystallin. Protein Sci. doi:10.1110/ps.035410.108.
63. Flaugh SL, Mills IA, King J. 2006. Glutamine deamidation destabilizes human gammaD-crystallin and lowers the kinetic barrier to unfolding. J Biol Chem 281(41):30782–30793.
64. von Bergen M, Barghorn S, Li L, Marx A, Biernat J, Mandelkow EM, Mandelkow E. 2001. Mutations of tau protein in frontotemporal dementia promote aggregation of paired helical filaments by enhancing local beta-structure. J Biol Chem 276(51):48165–48174.
65. Klucken J, Shin Y, Masliah E, Hyman BT, McLean PJ. 2004. Hsp70 reduces alpha-synuclein aggregation and toxicity. J Biol Chem 279(24):25497–25502.
66. Mizuguchi M, Hayashi A, Takeuchi M, Dobashi M, Mori Y, Shinoda H, Aizawa T, Demura M, Kawano K. 2008. Unfolding and aggregation of transthyretin by the truncation of 50 N-terminal amino acids. Proteins 72(1):261–269.
67. Hoiberg-Nielsen R, Fuglsang CC, Arleth L, Westh P. 2006. Interrelationships of glycosylation and aggregation kinetics for *Peniophora lycii* phytase. Biochemistry (Mosc) 45(15):5057–5066.
68. Baynes BM, Wang DI, Trout BL. 2005. Role of arginine in the stabilization of proteins against aggregation. Biochemistry 44(12):4919–4925.
69. Gsponer J, Vendruscolo M. 2006. Theoretical approaches to protein aggregation. Protein Pept Lett 13(3):287–293.
70. Jarrett JT, Lansbury PT Jr. 1993. Amyoid fibril formation requires a chemically discriminating nucleation event: Studies of an amyloidogenic sequence from the bacterial protein OsmB. Biochemistry 31:12345–12352.
71. Finke JM, Roy M, Zimm BH, Jennings PA. 2000. Aggregation events occur prior to stable intermediate formation during refolding of interleukin 1beta. Biochemistry 39(3):575–583.
72. Lomakin A, Teplow DB, Kirschner DA, Benedek GB. 1997. Kinetic theory of fibrillogenesis of amyloid beta-protein. Proc Natl Acad Sci U S A 94:7942–7947.
73. Szabo Z, Klement E, Jost K, Zarandi M, Soos K, Penke B. 1999. An FT-IR study of the beta-amyloid conformation: standardization of aggregation grade. Biochem Biophys Res Commun 265(2):297–300.
74. Weiss WFt, Hodgdon TK, Kaler EW, Lenhoff AM, Roberts CJ. 2007. Nonnative protein polymers: Structure, morphology, and relation to nucleation and growth. Biophys J 93(12):4392–4403.

75. Boeris V, Spelzini D, Salgado JP, Pico G, Romanini D, Farruggia B. 2008. Chymotrypsin-poly vinyl sulfonate interaction studied by dynamic light scattering and turbidimetric approaches. Biochim Biophys Acta 1780(9):1032–1037.
76. Won CM, Molnar TE, McKean RE, Spenlehauer GA. 1998. Stabilizers against heat-induced aggregation of RPR 114849, an acidic fibroblast growth factor (AFGF). Int J Pharm 167:25–36.
77. Raibekas AA, Bures EJ, Siska CC, Kohno T, Latypov RF, Kerwin BA. 2005. Anion binding and controlled aggregation of human interleukin-1 receptor antagonist. Biochemistry (Mosc) 44(29):9871–9879.
78. Rezaei-Ghaleh N, Ramshini H, Ebrahim HA, Moosavi MAA, Nemat GM. 2008. Thermal aggregation of alpha-chymotrypsin: Role of hydrophobic and electrostatic interactions. Biophys Chem 132(1):23–32.
79. De Young LR, Dill KA, Fink AL. 1993. Aggregation and denaturation of apomyoglobin in aqueous urea solutions. Biochemistry 32(15):3877–3886.
80. Wang K, Kurganov BI. 2003. Kinetics of heat- and acidification-induced aggregation of firefly luciferase. Biophys Chem 106(2):97–109.
81. Giese A, Bader B, Bieschke J, Schaffar G, Odoy S, Kahle PJ, Haass C, Kretzschmar H. 2005. Single particle detection and characterization of synuclein co-aggregation. Biochem Biophys Res Commun 333(4):1202–1210.
82. Cellmer T, Douma R, Huebner A, Prausnitz J, Blanch H. 2007. Kinetic studies of protein L aggregation and disaggregation. Biophys Chem 125(2–3):350–359.
83. Sun Y, Hayakawa S. 2002. Heat-induced gels of egg white/ovalbumins from five avian species: Thermal aggregation, molecular forces involved, and rheological properties. J Agric Food Chem 50(6):1636–1642.
84. Sluzky V, Tamada JA, Klibanov AM, Langer R. 1991. Kinetics of insulin aggregation in aqueous solutions upon agitation in the presence of hydrophobic surfaces. Proc Natl Acad Sci U S A 88:9377–9381.
85. Sluzky V, Klibanov AM, Langer R. 1992. Mechanism of insulin aggregation and stabilization in agitated aqueous solutions. Biotechnol Bioeng 40:895–903.
86. Darrington RT, Anderson BD. 1995. Evidence for a common intermediate in insulin deamidation and covalent dimer formation: Effects of pH and aniline trapping in dilute acidic solutions. J Pharm Sci 84:275–282.
87. Costantino HR, Langer R, Klibanov AM. 1994. Moisture-induced aggregation of lyophilized insulin. Pharm Res 11:21–29.
88. Brange J, Havelund S, Hougaard P. 1992. Chemical stability of insulin. 2. Formation of higher molecular weight transformation products during storage of pharmaceutical preparations. Pharm Res 9(6):727–734.
89. Shahrokh Z, Stratton PR, Eberlein GA, Wang YJ. 1994. Approaches to analysis of aggregates and demonstrating mass balance in pharmaceutical protein (basic fibroblast growth factor) formulations. J Pharm Sci 83(12):1645–1650.
90. Rousseau F, Schymkowitz J, Serrano L. 2006. Protein aggregation and amyloidosis: Confusion of the kinds? Curr Opin Struct Biol 16(1):118–126.
91. Wikipedia. 2009. Available at http://enwikipediaorg/wiki/Thermodynamic_temperature.

92. Macchioni A, Ciancaleoni G, Zuccaccia C, Zuccaccia D. 2008. Determining accurate molecular sizes in solution through NMR diffusion spectroscopy. Chem Soc Rev 37(3):479–489.

93. Dill KA. 1990. Dominant forces in protein folding. Biochemistry 29(31):7133–7155.

94. Jaenicke R. 2000. Stability and stabilization of globular proteins in solution. J Biotechnol 79(3):193–203.

95. Pace CN, Shirley BA, McNutt M, Gajiwala K. 1996. Forces contributing to the conformational stability of proteins. FASEB J 10(1):75–83.

96. Chen Y, Ding F, Nie H, Serohijos AW, Sharma S, Wilcox KC, Yin S, Dokholyan NV. 2008. Protein folding: Then and now. Arch Biochem Biophys 469(1):4–19.

97. Talla-Singh D, Stites WE. 2008. Refinement of noncalorimetric determination of the change in heat capacity, DeltaC(p), of protein unfolding and validation across a wide temperature range. Proteins 71(4):1607–1616.

98. Rees DC, Robertson AD. 2001. Some thermodynamic implications for the thermostability of proteins. Protein Sci 10(6):1187–1194.

99. Jaenicke R. 1991. Protein stability and molecular adaptation to extreme conditions. Eur J Biochem 202(3):715–728.

100. Giger K, Vanam RP, Seyrek E, Dubin PL. 2008. Suppression of insulin aggregation by heparin. Biomacromolecules 9(9):2338–2344.

101. Demeule B, Lawrence MJ, Drake AF, Gurny R, Arvinte T. 2007. Characterization of protein aggregation: The case of a therapeutic immunoglobulin. Biochim Biophys Acta 1774(1):146–153.

102. Kato A, Takagi T. 1988. Formation of intermolecular b-sheet structure during heat denaturation of ovalbumin. J Agric Food Chem 36:1156–1159.

103. Fabian H, Falber K, Gast K, Reinstadler D, Rogov VV, Naumann D, Zamyatkin DF, Filimonov VV. 1999. Secondary structure and oligomerization behavior of equilibrium unfolding intermediates of the lambda cro repressor. Biochemistry 38(17):5633–5642.

104. Boctor AM, Mehta SC. 1992. Enhancement of the stability of thrombin by polyols: Microcalorimetric studies. J Pharm Pharmacol 44(7):600–603.

105. Charman SA, Mason KL, Charman WN. 1993. Techniques for assessing the effects of pharmaceutical excipients on the aggregation of porcine growth hormone. Pharm Res 10(7):954–962.

106. Narhi LO, Philo JS, Sun B, Chang BS, Arakawa T. 1999. Reversibility of heat-induced denaturation of the recombinant human megakaryocyte growth and development factor. Pharm Res 16(6):799–807.

107. Calamai M, Canale C, Relini A, Stefani M, Chiti F, Dobson CM. 2005. Reversal of protein aggregation provides evidence for multiple aggregated States. J Mol Biol 346(2):603–616.

108. Krishnan S, Chi EY, Webb JN, Chang BS, Shan D, Goldenberg M, Manning MC, Randolph TW, Carpenter JF. 2002. Aggregation of granulocyte colony stimulating factor under physiological conditions: Characterization and thermodynamic inhibition. Biochemistry 41(20):6422–6431.

109. Chen BL, Arakawa T, Morris CF, Kenney WC, Wells CM, Pitt CG. 1994. Aggregation pathway of recombinant human keratinocyte growth factor and its stabilization. Pharm Res 11(11):1581–1587.

110. Brange J, Langkjaer L. 1997. Insulin formulation and delivery. Pharm Biotechnol 10:343–409.

111. Hall D, Minton AP. 2005. Turbidity as a probe of tubulin polymerization kinetics: A theoretical and experimental re-examination. Anal Biochem 345(2):198–213.

112. Chi EY, Weickmann J, Carpenter JF, Manning MC, Randolph TW. 2005. Heterogeneous nucleation-controlled particulate formation of recombinant human platelet-activating factor acetylhydrolase in pharmaceutical formulation. J Pharm Sci 94(2):256–274.

113. Chi EY, Kendrick BS, Carpenter JF, Randolph TW. 2005. Population balance modeling of aggregation kinetics of recombinant human interleukin-1 receptor antagonist. J Pharm Sci 94(12):2735–2748.

114. Roy S, Katayama D, Dong A, Kerwin BA, Randolph TW, Carpenter JF. 2006. Temperature dependence of benzyl alcohol- and 8-anilinonaphthalene-1- sulfonate-induced aggregation of recombinant human interleukin-1 receptor antagonist. Biochemistry (Mosc) 45(12):3898–3911.

115. Chen B, Bautista R, Yu K, Zapata GA, Chamow SM. 2003. Influence of histidine on the stability and physical properties of a fully human antibody in aqueous and solid forms. Pharm Res 20(12):1952–1960.

116. Jiskoot W, Beuvery EC, de Koning AAM, Herron JN, Crommelin DJA. 1990. Analytical approaches to the study of monoclonal antibody stability. Pharm Res 7:1234–1241.

117. Oliva A, SantoveZa A, Llabres M, Rarina JB. 1999. Stability study of human serum albumin pharmaceutical preparations. J Pharm Pharmacol 51:385–392.

118. Soenderkaer S, Carpenter JF, Van De Weert M, Hansen LL, Flink J, Frokjaer S. 2004. Effects of sucrose on rFVIIa aggregation and methionine oxidation. Eur J Pharm Sci 21(5):597–606.

119. Zlateva T, Boteva R, Salvato B, Tsanev R. 1999. Factors affecting the dissociation and aggregation of human interferon gamma. Int J Biol Macromol 26(5):357–362.

120. Vilasi S, Iannuzzi C, Portaccio M, Irace G, Sirangelo I. 2008. Effect of trehalose on W7FW14F apomyoglobin and insulin fibrillization: New insight into inhibition activity. Biochemistry (Mosc) 47(6):1789–1796.

121. McGuffey MK, Epting KL, Kelly RM, Foegeding EA. 2005. Denaturation and aggregation of three alpha-lactalbumin preparations at neutral pH. J Agric Food Chem 53(8):3182–3190.

122. Yan YB, Wang Q, He HW, Zhou HM. 2004. Protein thermal aggregation involves distinct regions: Sequential events in the heat-induced unfolding and aggregation of hemoglobin. Biophys J 86(3):1682–1690.

123. Gaikwad SM, Khan MI. 2003. pH-dependent aggregation of oligomeric *Artocarpus hirsuta* lectin on thermal denaturation. Biochem Biophys Res Commun 311(2): 254–257.

124. Barzegar A, Moosavi-Movahedi AA, Rezaei-Zarchi S, Saboury AA, Ganjali MR, Norouzi P, Hakimelahi GH, Tsai FY. 2008. The mechanisms underlying the effect

of alpha-cyclodextrin on the aggregation and stability of alcohol dehydrogenase. Biotechnol Appl Biochem 49(Pt 3):203–211.

125. Benjwal S, Verma S, Roehm KH, Gursky O. 2006. Monitoring protein aggregation during thermal unfolding in circular dichroism experiments. Protein Sci 15(3):635–639.

126. Golub NV, Markossian KA, Kasilovich NV, Sholukh MV, Orlov VN, Kurganov BI. 2008. Thermal inactivation, denaturation and aggregation of mitochondrial aspartate aminotransferase. Biophys Chem 135(1–3):125–131.

127. Petersen SB, Jonson V, Fojan P, Wimmer R, Pedersen S. 2004. Sorbitol prevents the self-aggregation of unfolded lysozyme leading to an up to 13°C stabilisation of the folded form. J Biotechnol 114(3):269–278.

128. Derrick T, Grillo AO, Vitharana SN, Jones L, Rexroad J, Shah A, Perkins M, Spitznagel TM, Middaugh CR. 2007. Effect of polyanions on the structure and stability of repifermin (keratinocyte growth factor-2). J Pharm Sci 96(4):761–776.

129. Arnaudov LN, de Vries R. 2005. Thermally induced fibrillar aggregation of hen egg white lysozyme. Biophys J 88(1):515–526.

130. Matsuoka T, Tomita S, Hamada H, Shiraki K. 2007. Amidated amino acids are prominent additives for preventing heat- induced aggregation of lysozyme. J Biosci Bioeng 103(5):440–443.

131. Holm NK, Jespersen SK, Thomassen LV, Wolff TY, Pankaj S, Thomsen LA, Christiansen G, Andersen CB, Knudsen AD, Otzen DE. 2007. Aggregation and fibrillation of bovine serum albumin. BBA-Proteins Proteom 1774(9):1128–1138.

132. Purohit VS, Middaugh CR, Balasubramanian SV. 2006. Influence of aggregation on immunogenicity of recombinant human factor VIII in hemophilia A mice. J Pharm Sci 95(2):358–371.

133. Andrews JM, Weiss WF, Roberts CJ. 2008. Nucleation, growth, and activation energies for seeded and unseeded aggregation of alpha-chymotrypsinogen A. Biochemistry (Mosc) 47(8):2397–2403.

134. Fedurkina NV, Belousova LV, Mitskevich LG, Zhou HM, Chang Z, Kurganov BI. 2006. Change in kinetic regime of protein aggregation with temperature increase. Thermal aggregation of rabbit muscle creatine kinase. Biochemistry (Mosc) 71(3):325–331.

135. Vermeer AW, Norde W. 2000. The thermal stability of immunoglobulin: Unfolding and aggregation of a multi-domain protein. Biophys J 78(1):394–404.

136. Eisenthal R, Peterson ME, Daniel RM, Danson MJ. 2006. The thermal behaviour of enzyme activity: Implications for biotechnology. Trends Biotechnol 24(7):289–292.

137. Fitzsimons SM, Mulvihill DM, Morris ER. 2007. Denaturation and aggregation processes in thermal gelation of whey proteins resolved by differential scanning calorimetry. Food Hydrocolloids 21(4):638–644.

138. Saluja A, Badkar AV, Zeng DL, Kalonia DS. 2007. Ultrasonic rheology of a monoclonal antibody (IgG2) solution: Implications for physical stability of proteins in high concentration formulations. J Pharm Sci 96(12):3181–3195.

139. Zhang J, Peng X, Jonas A, Jonas J. 1995. NMR study of the cold, heat, and pressure unfolding of ribonuclease A. Biochemistry 34(27):8631–8641.
140. Kosa T, Maruyama T, Otagiri M. 1998. Species differences of serum albumins: II. Chemical and thermal stability. Pharm Res 15(3):449–454.
141. Ferri C, Zignego AL, Pileri SA. 2002. Cryoglobulins. J Clin Pathol 55(1):4–13.
142. Ramsland PA, Farrugia W. 2002. Crystal structures of human antibodies: A detailed and unfinished tapestry of immunoglobulin gene products. J Mol Recognit 15(5):248–259.
143. Sukumar M, Doyle BL, Combs JL, Pekar AH. 2004. Opalescent appearance of an IgG1 antibody at high concentrations and its relationship to noncovalent association. Pharm Res 21(7):1087–1093.
144. Piedmonte DM, Summers C, McAuley A, Karamujic L, Ratnaswamy G. 2007. Sorbitol crystallization can lead to protein aggregation in frozen protein formulations. Pharm Res 24(1):136–146.
145. Heller MC, Carpenter JF, Randolph TW. 1997. Manipulation of lyophilization-induced phase separation: Implications for pharmaceutical proteins. Biotechnol Prog 13(5):590–596.
146. Heller MC, Carpenter JF, Randolph TW. 1999. Protein formulation and lyophilization cycle design: Prevention of damage due to freeze-concentration induced phase separation. Biotechnol Bioeng 63(2):166–174.
147. Schein CH. 1990. Solubility as a function of protein structure and solvent components. Biotechnology 8:308–317.
148. Kohn WD, Kay CM, Hodges RS. 1997. Salt effects on protein stability: Two-stranded alpha-helical coiled-coils containing inter- or intrahelical ion pairs. J Mol Biol 267(4):1039–1052.
149. Vrkljan M, Foster TM, Powers ME, Henkin J, Porter WR, Staack H, Carpenter JF, Manning MC. 1994. Thermal stability of low molecular weight urokinase during heat treatment. II. Effect of polymeric additives. Pharm Res 11(7):1004–1008.
150. Fatouros A, Osterberg T, Mikaelsson M. 1997. Recombinant factor VIII SQ—Influence of oxygen, metal ions, pH and ionic strength on its stability in aqueous solution. Int J Pharm 155:121–131.
151. Chiti F, Stefani M, Taddei N, Ramponi G, Dobson CM. 2003. Rationalization of the effects of mutations on peptide and protein aggregation rates. Nature 424(6950):805–808.
152. DuBay KF, Pawar AP, Chiti F, Zurdo J, Dobson CM, Vendruscolo M. 2004. Prediction of the absolute aggregation rates of amyloidogenic polypeptide chains. J Mol Biol 341(5):1317–1326.
153. Wang W, Antonsen K, Wang YJ, Wang DQ. 2008. pH dependent effect of glycosylation on protein stability. Eur J Pharm Sci 33(2):120–127.
154. Hartmann WK, Saptharishi N, Yang XY, Mitra G, Soman G. 2004. Characterization and analysis of thermal denaturation of antibodies by size exclusion high-performance liquid chromatography with quadruple detection. Anal Biochem 325(2):227–239.
155. Kerwin BA, Heller MC, Levin SH, Randolph TW. 1998. Effects of Tween 80 and sucrose on acute short-term stability and long-term storage at –20 degrees C of a recombinant hemoglobin. J Pharm Sci 87(9):1062–1068.

156. Advant SJ, Braswell EH, Kumar CV, Kalonia DS. 1995. The effect of pH and temperature on the self-association of recombinant human interleukin-2 as studied by equilibrium sedimentation. Pharm Res 12(5):637–641.

157. Tsai AM, van Zanten JH, Betenbaugh MJ. 1998. I. Study of protein aggregation due to heat denaturation: A structural approach using circular dichroism spectroscopy, nuclear magnetic resonance, and static light scattering. Biotechnol Bioeng 59(3):273–280.

158. Thirumangalathu R, Krishnan S, Brems DN, Randolph TW, Carpenter JF. 2006. Effects of pH, temperature, and sucrose on benzyl alcohol-induced aggregation of recombinant human granulocyte colony stimulating factor. J Pharm Sci 95(7):1480–1497.

159. Roy S, Henderson I, Nayar R, Randolph TW, Carpenter JF. 2008. Effect of pH on stability of recombinant botulinum scrotype A vaccine in aqueous solution and during storage of freeze-dried formulations. J Pharm Sci. doi:10.1002/jps.21409.

160. Shukla AA, Gupta P, Han X. 2007. Protein aggregation kinetics during protein A chromatography case study for an Fc fusion protein. J Chromatogr 1171(1–2):22–28.

161. Ejima D, Tsumoto K, Fukada H, Yumioka R, Nagase K, Arakawa T, Philo JS. 2007. Effects of acid exposure on the conformation, stability, and aggregation of monoclonal antibodies. Proteins 66(4):954–962.

162. Barteri M, Gaudiano MC, Rotella S, Benagiano G, Pala A. 2000. Effect of pH on the structure and aggregation of human glycodelin A. A comparison with beta-lactoglobulin A. BBA-Protein Struct M 1479(12):255–264.

163. Munishkina LA, Ahmad A, Fink AL, Uversky VN. 2008. Guiding protein aggregation with macromolecular crowding. Biochemistry 47(34):8993–9006.

164. Kurochkin IV, Procyk R, Bishop PD, Yee VC, Teller DC, Ingham KC, Medved LV. 1995. Domain structure, stability and domain-domain interactions in recombinant factor XIII. J Mol Biol 248(2):414–430.

165. Majhi PR, Ganta RR, Vanam RP, Seyrek E, Giger K, Dubin PL. 2006. Electrostatically driven protein aggregation: Beta-lactoglobulin at low ionic strength. Langmuir 22(22):9150–9159.

166. Chung K, Kim J, Cho BK, Ko BJ, Hwang BY, Kim BG. 2007. How does dextran sulfate prevent heat induced aggregation of protein? The mechanism and its limitation as aggregation inhibitor. Biochim Biophys Acta 1774(2):249–257.

167. Long X, Zhang C, Cheng J, Bi S. 2008. A novel method for study of the aggregation of protein induced by metal ion aluminum(III) using resonance Rayleigh scattering technique. Spectrochim Acta A Mol Biomol Spectrosc 69(1):71–77.

168. Rana A, Gupta TP, Bansal S, Kundu B. 2008. Formation of amyloid fibrils by bovine carbonic anhydrase. Biochim Biophys Acta 1784(6):930–935.

169. Kamat MS, Tolman GL, Brown JM. 1996. Formulation development of an antifibrin monoclonal antibody radiopharmaceutical. Pharm Biotechnol 9:343–364.

170. Pikal MJ, Dellerman KM, Roy ML, Riggin RM. 1991. The effects of formulation variables on the stability of freeze-dried human growth hormone. Pharm Res 8(4):427–436.

171. Eberlein GA, Stratton PR, Wang YJ. 1994. Stability of rhbFGF as determined by UV spectroscopic measurements of turbidity. PDA J Pharm Sci Technol 48(5):224–230.

172. Paborji M, Pochopin NL, Coppola WP, Bogardus JB. 1994. Chemical and physical stability of chimeric L6, a mouse-human monoclonal antibody. Pharm Res 11(May):764–771.

173. Kita Y, Arakawa T. 2002. Salts and glycine increase reversibility and decrease aggregation during thermal unfolding of ribonuclease-A. Biosci Biotechnol Biochem 66(4):880–882.

174. Baussay K, Bon CL, Nicolai T, Durand D, Busnel JP. 2004. Influence of the ionic strength on the heat-induced aggregation of the globular protein beta-lactoglobulin at pH 7. Int J Biol Macromol 34(1–2):21–28.

175. Arnaudov LN, de Vries R. 2006. Strong impact of ionic strength on the kinetics of fibrilar aggregation of bovine beta-lactoglobulin. Biomacromolecules 7(12): 3490–3498.

176. Kerstens S, Murray BS, Dickinson E. 2005. Confocal microscopy of heat-induced aggregation and gelation of beta- lactoglobulin in presence of non-ionic surfactant. Food Hydrocolloids 19(3):625–633.

177. Trevino SR, Scholtz JM, Pace CN. 2008. Measuring and increasing protein solubility. J Pharm Sci 97(10):4155–4166.

178. Bagger HL, Ogendal LH, Westh P. 2007. Solute effects on the irreversible aggregation of serum albumin. Biophys Chem 130(1–2):17–25.

179. Rezaei-Ghaleh N, Ebrahim-Habibi A, Moosavi-Movahedi AA, Nemat-Gorgani M. 2007. Role of electrostatic interactions in 2,2,2-trifluoroethanol-induced structural changes and aggregation of alpha-chymotrypsin. Arch Biochem Biophys 457(2):160–169.

180. Grudzielanek S, Smirnovas V, Winter R. 2007. The effects of various membrane physical-chemical properties on the aggregation kinetics of insulin. Chem Phys Lipids 149(1–2):28–39.

181. Xia Y, Park YD, Mu H, Zhou HM, Wang XY, Meng FG. 2007. The protective effects of osmolytes on arginine kinase unfolding and aggregation. Int J Biol Macromol 40(5):437–443.

182. Goyal K, Walton LJ, Tunnacliffe A. 2005. LEA proteins prevent protein aggregation due to water stress. Biochem J 388(1):151–157.

183. Beranger F, Crozet C, Goldsborough A, Lehmann S. 2008. Trehalose impairs aggregation of PrPSc molecules and protects prion-infected cells against oxidative damage. Biochem Biophys Res Commun 374(1):44–48.

184. Schule S, Schulz-Fademrecht T, Garidel P, Bechtold-Peters K, Friess W. 2008. Stabilization of IgG1 in spray-dried powders for inhalation. Eur J Pharm Biopharm 69(3):793–807.

185. Rondeau P, Armenta S, Caillens H, Chesne S, Bourdon E. 2007. Assessment of temperature effects on beta-aggregation of native and glycated albumin by FTIR spectroscopy and PAGE: Relations between structural changes and antioxidant properties. Arch Biochem Biophys 460(1):141–150.

186. Maclean DS, Qian Q, Middaugh CR. 2002. Stabilization of proteins by low molecular weight multi-ions. J Pharm Sci 91(10):2220–2229.

187. Unterhaslberger G, Schmitt C, Sanchez C, Appolonia NC, Raemy A. 2006. Heat denaturation and aggregation of beta-lactoglobulin enriched WPI in the presence of arginine HCl, NaCl and guanidinium HCl at pH 4.0 and 7.0. Food Hydrocolloids 20(7):1006–1019.

188. Ignatova Z, Gierasch LM. 2006. Inhibition of protein aggregation in vitro and in vivo by a natural osmoprotectant. Proc Natl Acad Sci U S A 103(36):13357–13361.

189. Tsumoto K, Ejima D, Kita Y, Arakawa T. 2005. Review: Why is arginine effective in suppressing aggregation? Protein Pept Lett 12(7):613–619.

190. Hamada H, Takahashi R, Noguchi T, Shiraki K. 2008. Differences in the effects of solution additives on heat- and refolding-induced aggregation. Biotechnol Prog 24(2):436–443.

191. Lyutova EM, Kasakov AS, Gurvits BY. 2007. Effects of arginine on kinetics of protein aggregation studied by dynamic laser light scattering and tubidimetry techniques. Biotechnol Prog 23(6):1411–1416.

192. Das U, Hariprasad G, Ethayathulla AS, Manral P, Das TK, Pasha S, Mann A, Ganguli M, Verma AK, Bhat R, Chandrayan SK, Ahmed S, Sharma S, Kaur P, Singh TP, Srinivasan A. 2007. Inhibition of protein aggregation: Supramolecular assemblies of arginine hold the key. PLoS One 2(11):e1176.

193. Ignatova Z, Gierasch LM. 2007. Effects of osmolytes on protein folding and aggregation in cells. Methods Enzymol 428:355–372.

194. Kar K, Kishore N. 2007. Enhancement of thermal stability and inhibition of protein aggregation by osmolytic effect of hydroxyproline. Biopolymers 87(5–6):339–351.

195. Rezaei-Ghaleh N, Ebrahim HA, Moosavi MAA, Nemat GM. 2007. Effect of polyamines on the structure, thermal stability and 2,2,2- trifluoroethanol-induced aggregation of alpha-chymotrypsin. Int J Biol Macromol 41(5):597–604.

196. Schlieben NH, Niefind K, Schomburg D. 2004. Expression, purification, and aggregation studies of His-tagged thermoalkalophilic lipase from *Bacillus thermocatenulatus*. Protein Expr Purif 34(1):103–110.

197. Lam XM, Oeswein JQ, Ongpipattanakul B, Shahrokh Z, Wang SX, Weissburg RP, Wong RL. 2000. Stabilized antibody formulation. US6991790 B1.

198. Eronina TB, Chebotareva NA, Kurganov BI. 2005. Influence of osmolytes on inactivation and aggregation of muscle glycogen phosphorylase b by guanidine hydrochloride. Stimulation of protein aggregation under crowding conditions. Biochemistry (Mosc) 70(9):1020–1026.

199. Moussa CE, Mahmoodian F, Tomita Y, Sidhu A. 2008. Dopamine differentially induces aggregation of A53T mutant and wild type alpha-synuclein: Insights into the protein chemistry of Parkinson's disease. Biochem Biophys Res Commun 365(4):833–839.

200. Joshi SB, Kamerzell TJ, McNown C, Middaugh CR. 2008. The interaction of heparin/polyanions with bovine, porcine, and human growth hormone. J Pharm Sci 97(4):1368–1385.

201. Arakawa T, Wen J, Philo JS. 1993. Densimetric determination of equilibrium binding of sucrose octasulfate with basic fibroblast growth factor. J Protein Chem 12(6):689–693.

202. Considine T, Patel HA, Singh H, Creamer LK. 2007. Influence of binding conjugated linoleic acid and myristic acid on the heat- and high-pressure-induced unfolding and aggregation of beta- lactoglobulin B. Food Chem 102(4):1270–1280.

203. Baynes BM, Trout BL 2004. Rational design of solution additives for the prevention of protein aggregation. Biophys J 87(3):1631–1639.

204. Zhang MZ, Wen J, Arakawa T, Prestrelski SJ. 1995. A new strategy for enhancing the stability of lyophilized protein: The effect of the reconstitution medium on keratinocyte growth factor. Pharm Res 12(10):1447–1452.

205. Vemuri S, Yu CT, Roosdorp N. 1993. Formulation and stability of recombinant alpha 1-antitrypsin. Pharm Biotechnol 5:263–286.

206. Calamai M, Taddei N, Stefani M, Ramponi G, Chiti F. 2003. Relative influence of hydrophobicity and net charge in the aggregation of two homologous proteins. Biochemistry (Mosc) 42(51):15078–15083.

207. Arakawa T, Wen J, Philo JS. 1994. Stoichiometry of heparin binding to basic fibroblast growth factor. Arch Biochem Biophys 308(1):267–273.

208. Volkin DB, Tsai PK, Dabora JM, Gress JO, Burke CJ, Linhardt RJ, Middaugh CR. 1993. Physical stabilization of acidic fibroblast growth factor by polyanions. Arch Biochem Biophys 300(1):30–41.

209. Tsai AM, van Zanten JH, Betenbaugh MJ 1998. II. Electrostatic effect in the aggregation of heat-denatured RNase A and implications for protein additive design. Biotechnol Bioeng 59(3):281–285.

210. Song C, Wang Q, Li CC. 2007. Characterization of the aggregation-prevention activity of p97/valosin-containing protein. Biochemistry 46(51):14889–14898.

211. Manning MC, Matsuura JE, Kendrick BS, Meyer JD, Dormish JJ, Vrkljian M, Ruth JR, Carpenter JF, Shefter E. 1995. Approaches for increasing the solution stability of proteins. Biotechnol Bioeng 48:506–512.

212. Kehoe JJ, Morris ER, Brodkorb A. 2007. The influence of bovine serum albumin on beta-lactoglobulin denaturation, aggregation and gelation. Food Hydrocolloids 21(56):747–755.

213. Kubota S, Kubota H, Nagata K. 2006. Cytosolic chaperonin protects folding intermediates of G ss from aggregation by recognizing hydrophobic ss-strands. Proc Natl Acad Sci U S A 103(22):8360–8365.

214. Pivovarova AV, Chebotareva NA, Chernik IS, Gusev NB, Levitsky DI. 2007. Small heat shock protein Hsp27 prevents heat-induced aggregation of F-actin by forming soluble complexes with denatured actin. FEBS J 274(22):5937–5948.

215. Pivovarova AV, Mikhailova VV, Chernik IS, Chebotareva NA, Levitsky DI, Gusev NB. 2005. Effects of small heat shock proteins on the thermal denaturation and aggregation of F-actin. Biochem Biophys Res Commun 331(4):1548–1553.

216. Yamashita H, Kawamata J, Okawa K, Kanki R, Nakamizo T, Hatayama T, Yamanaka K, Takahashi R, Shimohama S. 2007. Heat-shock protein 105 interacts with and suppresses aggregation of mutant Cu/Zn superoxide dismutase: Clues to a possible strategy for treating ALS. J Neurochem 102(5):1497–1505.

217. Lee S, Carson K, Rice FA, Good T. 2006. Small heat shock proteins differentially affect A beta aggregation and toxicity. Biochem Biophys Res Commun 347(2):527–533.

218. Linetsky M, Shipova E, Cheng R, Ortwerth BJ. 2008. Glycation by ascorbic acid oxidation products leads to the aggregation of lens proteins. Biochim Biophys Acta 1782(1):22–34.

219. Estrada LD, Soto C. 2006. Inhibition of protein misfolding and aggregation by small rationally- designed peptides. Curr Pharm Des 12(20):2557–2567.

220. Etienne MA, Aucoin JP, Fu Y, McCarley RL, Hammer RP. 2006. Stoichiometric inhibition of amyloid beta-protein aggregation with peptides containing alternating alpha, alpha-disubstituted amino acids. J Am Chem Soc 128(11):3522–3523.

221. Yazdanparast R, Esmaeili MA, Khodagholi F. 2007. Control of aggregation in protein refolding: Cooperative effects of artificial chaperone and cold temperature. Int J Biol Macromol 40(2):126–133.

222. Tokihiro K, Irie T, Uekama K. 1997. Varying effects of cyclodextrin derivatives on aggregation and thermal behavior of insulin in aqueous solution. Chem Pharm Bull (Tokyo) 45(3):525–531.

223. Wang W, Wang YJ, Wang DQ. 2008. Dual effects of Tween 80 on protein stability. Int J Pharm 347(1–2):31–38.

224. Randolph TW, Jones LS. 2002. Surfactant-protein interactions. Pharm Biotechnol 13:159–175.

225. Bam NB, Randolph TW, Cleland JL. 1995. Stability of protein formulations: Investigation of surfactant effects by a novel EPR spectroscopic technique. Pharm Res 12(1):2–11.

226. Chou DK, Krishnamurthy R, Randolph TW, Carpenter JF, Manning MC. 2005. Effects of Tween 20 and Tween 80 on the stability of Albutropin during agitation. J Pharm Sci 94(6):1368–1381.

227. Kerwin BA. 2008. Polysorbates 20 and 80 used in the formulation of protein biotherapeutics: Structure and degradation pathways. J Pharm Sci 97(8): 2924–2935.

228. Bam NB, Cleland JL, Yang J, Manning MC, Carpenter JF, Kelley RF, Randolph TW. 1998. Tween protects recombinant human growth hormone against agitation-induced damage via hydrophobic interactions. J Pharm Sci 87(12):1554–1559.

229. Arakawa T, Kita Y. 2000. Protection of bovine serum albumin from aggregation by Tween 80. J Pharm Sci 89(5):646–651.

230. Katakam M, Banga AK. 1997. Use of poloxamer polymers to stabilize recombinant human growth hormone against various processing stresses. Pharm Dev Technol 2(2):143–149.

231. Rozema D, Gellman SH. 1996. Artificial chaperone-assisted refolding of carbonic anhydrase B. J Biol Chem 271(7):3478–3487.

232. Treuheit MJ, Kosky AA, Brems DN. 2002. Inverse relationship of protein concentration and aggregation. Pharm Res 19(4):511–516.

233. Panyukov YV, Nemykh MA, Dobrov EN, Drachev VA. 2008. Surfactant-induced amorphous aggregation of tobacco mosaic virus coat protein: A physical methods approach. Macromol Biosci 8(2):199–209.

234. Panyukov YV, Nemykh MA, Rafikova ER, Kurganov BI, Yaguzhinsky LS, Arutyunyan AM, Drachev VA, Dobrov EN. 2006. Low cetyltrimethylammonium bromide concentrations induce reversible amorphous aggregation of tobacco

mosaic virus and its coat protein at room temperature. Int J Biochem Cell Biol 38(4):533–543.

235. Collins T, D'Amico S, Georlette D, Marx JC, Huston AL, Feller G. 2006. A nondetergent sulfobetaine prevents protein aggregation in microcalorimetric studies. Anal Biochem 352(2):299–301.
236. Zhang Y, Roy S, Jones LS, Krishnan S, Kerwin BA, Chang BS, Manning MC, Randolph TW, Carpenter JF. 2004. Mechanism for benzyl alcohol-induced aggregation of recombinant human interleukin-1 receptor antagonist in aqueous solution. J Pharm Sci 93(12):3076–3089.
237. Roy S, Jung R, Kerwin BA, Randolph TW, Carpenter JF. 2005. Effects of benzyl alcohol on aggregation of recombinant human interleukin-1-receptor antagonist in reconstituted lyophilized formulations. J Pharm Sci 94(2):382–396.
238. Harn N, Allan C, Oliver C, Middaugh CR. 2007. Highly concentrated monoclonal antibody solutions: Direct analysis of physical structure and thermal stability. J Pharm Sci 96(3):532–546.
239. Ellis RJ, Minton AP. 2006. Protein aggregation in crowded environments. Biol Chem 387(5):485–497.
240. Minton AP. 2006. Macromolecular crowding. Curr Biol 16(8):R269–R271.
241. Guo J, Harn N, Robbins A, Dougherty R, Middaugh CR. 2006. Stability of helix-rich proteins at high concentrations. Biochemistry 45(28):8686–8696.
242. Jimenez M, Rivas G, Minton AP. 2007. Quantitative characterization of weak self-association in concentrated solutions of immunoglobulin G via the measurement of sedimentation equilibrium and osmotic pressure. Biochemistry (Mosc) 46(28):8373–8378.
243. Elshereef R, Budman H, Moresoli C, Legge RL. 2008. Fluorescence-based soft-sensor for monitoring beta-lactoglobulin and alpha-lactalbumin solubility during thermal aggregation. Biotechnol Bioeng 99(3):567–577.
244. Sukumar M, Doyle BL, Combs JL, Pekar AH. 2004. Opalescent appearance of an IgG1 antibody at high concentrations and its relationship to noncovalent association. Pharm Res 21(7):1087–1093.
245. Pease LF III, Elliott JT, Tsai DH, Zachariah MR, Tarlov MJ. 2008. Determination of protein aggregation with differential mobility analysis: application to IgG antibody. Biotechnol Bioeng 101(6):1214–1222.
246. Ramani K, Purohit V, Middaugh CR, Balasubramanian SV. 2005. Aggregation kinetics of recombinant human FVIII (rFVIII). J Pharm Sci 94(9):2023–2029.
247. Roefs SP, De Kruif KG. 1994. A model for the denaturation and aggregation of beta-lactoglobulin. Eur J Biochem 226(3):883–889.
248. Alexandrescu AT, Rathgeb-Szabo K. 1999. An NMR investigation of solution aggregation reactions preceding the misassembly of acid-denatured cold shock protein A into fibrils. J Mol Biol 291(5):1191–1206.
249. Golub N, Meremyanin A, Markossian K, Eronina T, Chebotareva N, Asryants R, Muronets V, Kurganov B. 2007. Evidence for the formation of start aggregates as an initial stage of protein aggregation. FEBS Lett 581(22):4223–4227.
250. Mounsey JS, O'Kennedy BT. 2007. Conditions limiting the influence of thiol-disulphide interchange reactions on the heat-induced aggregation kinetics of beta- lactoglobulin. Int Dairy J 17(9):1034–1042.

251. Powers ET, Powers DL. 2008. Mechanisms of protein fibril formation: Nucleated polymerization with competing off-pathway aggregation. Biophys J 94(2):379–391.
252. Huus K, Havelund S, Olsen HB, van de Weert M, Frokjaer S. 2005. Thermal dissociation and unfolding of insulin. Biochemistry (Mosc) 44(33):11171–11177.
253. Wang W, Wang YJ, Kelner DN. 2003. Coagulation factor VIII: Structure and stability. Int J Pharm 259(1–2):1–15.
254. Alcaraz LA, Donaire A. 2004. Unfolding process of rusticyanin—Evidence of protein aggregation. Eur J Biochem 271(21):4284–4292.
255. Giese A, Levin J, Bertsch U, Kretzschmar H. 2004. Effect of metal ions on de novo aggregation of full-length prion protein. Biochem Biophys Res Commun 320(4):1240–1246.
256. Derrick TS, Kashi RS, Durrani M, Jhingan A, Middaugh CR. 2004. Effect of metal cations on the conformation and inactivation of recombinant human factor VIII. J Pharm Sci 93(10):2549–2557.
257. Vessely CR, Carpenter JF, Schwartz DK. 2005. Calcium-induced changes to the molecular conformation and aggregate structure of beta-casein at the air-water interface. Biomacromolecules 6(6):3334–3344.
258. Capanni C, Taddei N, Gabrielli S, Messori L, Orioli P, Chiti F, Stefania M, Ramponia G. 2004. Investigation of the effects of copper ions on protein aggregation using a model system. Cell Mol Life Sci 61(7–8):982–991.
259. Simmons MJH, Jayaraman P, Fryer PJ. 2007. The effect of temperature and shear rate upon the aggregation of whey protein and its implications for milk fouling. J Food Eng 79(2):517–528.
260. Bermudez O, Forciniti D. 2004. Aggregation and denaturation of antibodies: A capillary electrophoresis, dynamic light scattering, and aqueous two-phase partitioning study. J Chromatogr B 807(1):17–24.
261. Katakam M, Bell LN, Banga AK. 1995. Effect of surfactants on the physical stability of recombinant human growth hormone. J Pharm Sci 84(6):713–716.
262. Zhou P, Liu X, Labuza TP. 2008. Moisture-induced aggregation of whey proteins in a protein/buffer model system. J Agric Food Chem 56(6):2048–2054.
263. Lee YN, Chen LK, Ma HC, Yang HH, Li HP, Lo SY. 2005. Thermal aggregation of SARS-CoV membrane protein. J Virol Methods 129(2):152–161.
264. Patro SY, Przybycien TM. 1994. Simulations of kinetically irreversible protein aggregate structure. Biophys J 66:1274–1289.
265. Gotte G, Libonati M. 2008. Oligomerization of ribonuclease A under reducing conditions. Biochim Biophys Acta 1784(4):638–650.
266. Maeda R, Ado K, Takeda N, Taniguchi Y. 2007. Promotion of insulin aggregation by protein disulfide isomerase. Biochim Biophys Acta 1774(12):1619–1627.
267. Schokker EP, Singh H, Creamer LK. 2000. Heat-induced aggregation of beta-lactoglobulin A and B with alpha- lactalbumin. Int Dairy J 10(12):843–853.
268. Gorbenko GP, Ioffe VM, Kinnunen PK. 2007. Binding of lysozyme to phospholipid bilayers: Evidence for protein aggregation upon membrane association. Biophys J 93(1):140–153.
269. Creusot N, Gruppen H. 2007. Enzyme-induced aggregation and gelation of proteins. Biotechnol Adv 25(6):597–601.

270. Pace CN, Trevino S, Prabhakaran E, Scholtz JM. 2004. Protein structure, stability and solubility in water and other solvents. Philos Trans R Soc Lond B Biol Sci 359(1448):1225–1234; discussion 1234–1235.

271. Feng S, Yan YB. 2008. Effects of glycerol on the compaction and stability of the wild type and mutated rabbit muscle creatine kinase. Proteins 71(2):844–854.

272. Amani M, Yousefi R, Moosavi-Movahedi AA, Pintus F, Mura A, Floris G, Kurganov BI, Saboury AA. 2008. Structural changes and aggregation process of Cu/containing amine oxidase in the presence of 2,2,2′-trifluoroethanol. Protein Pept Lett 15(5):521–527.

273. Sukumar M, Storms SM, De Felippis MR. 2005. Non-native intermediate conformational states of human growth hormone in the presence of organic solvents. Pharm Res 22(5):789–796.

274. Cholewinski M, Luckel B, Horn H. 1996. Degradation pathways, analytical characterization and formulation strategies of a peptide and a protein. Calcitonine and human growth hormone in comparison. Pharm Acta Helv 71(6):405–419.

275. Joshi O, McGuire J, Wang DQ. 2008. Adsorption and function of recombinant factor VIII at solid-water interfaces in the presence of Tween-80. J Pharm Sci. doi:10.1002/jps.21333.

276. Holmberg M, Stibius KB, Ndoni S, Larsen NB, Kingshott P, Hou XL. 2007. Protein aggregation and degradation during iodine labelling and its consequences for protein adsorption to biomaterials. Anal Biochem 361(1):120–125.

277. Santos O, Nylander T, Paulsson M, Trägårdh C. 2006. Whey protein adsorption onto steel surfaces—Effect of temperature, flow rate, residence time and aggregation. J Food Eng 74(4):468–483.

278. Joshi O, McGuire J. 2008. Adsorption behavior of lysozyme and Tween 80 at hydrophilic and hydrophobic silica-water interfaces. Appl Biochem Biotechnol. doi:10.1007/s12010-008-8246-8.

279. Kueltzo LA, Wang W, Randolph TW, Carpenter JF. 2008. Effects of solution conditions, processing parameters, and container materials on aggregation of a monoclonal antibody during freeze-thawing. J Pharm Sci 97(5):1801–1812.

280. Gryniewicz CM, Kauffman JF. 2008. Multivariate calibration of covalent aggregate fraction to the Raman spectrum of regular human insulin. J Pharm Sci 97(9):3727–3734.

281. Jones LS, Kaufmann A, Middaugh CR. 2005. Silicone oil induced aggregation of proteins. J Pharm Sci 94(4):918–927.

282. McGuffey MK, Otter DE, Zanten JHv, Foegeding EA. 2007. Solubility and aggregation of commercial alpha-lactalbumin at neutral pH. Int Dairy J 17(10): 1168–1178.

283. Kerwin BA, Remmele RL Jr. 2007. Protect from light: Photodegradation and protein biologics. J Pharm Sci 96(6):1468–1479.

284. Li DY, Borkman RF, Wang RH, Dillon J. 1990. Mechanisms of photochemically produced turbidity in lens protein solutions. Exp Eye Res 51(6):663–669.

285. Vanhooren A, Devreese B, Vanhee K, Van Beeumen J, Hanssens I. 2002. Photoexcitation of tryptophan groups induces reduction of two disulfide bonds in goat alpha-lactalbumin. Biochemistry (Mosc) 41(36):11035–11043.

286. Davies MJ. 2003. Singlet oxygen-mediated damage to proteins and its consequences. Biochem Biophys Res Commun 305(3):761–770.

287. Georgiou G, Valax P. 1999. Isolating inclusion bodies from bacteria. Methods Enzymol 309:48–58.

288. de Groot NS, Ventura S. 2006. Protein activity in bacterial inclusion bodies correlates with predicted aggregation rates. J Biotechnol 125(1):110–113.

289. Hao Y, Chu J, Wang Y, Zhuang Y, Zhang S. 2007. The inhibition of aggregation of recombinant human consensus interferon-alpha mutant during *Pichia pastoris* fermentation. Appl Microbiol Biotechnol 74(3):578–584.

290. Wu D, Hao YY, Chu J, Zhuang YP, Zhang SL. 2008. Inhibition of degradation and aggregation of recombinant human consensus interferon-alpha mutant expressed in *Pichia pastoris* with complex medium in bioreactor. Appl Microbiol Biotechnol. doi:10.1007/s00253-008-1629-5.

291. Bahrami A, Shojaosadati SA, Khalilzadeh R, Mohammadian J, Vashghani Farahani E, Masoumian MR. 2008. Prevention of human G-CSF protein aggregation in recombinant *Pichia pastoris* fed-batch fermentation using additives. Biotechnol Appl Biochem. doi:10.1042/BA20070267.

292. Chaderjian WB, Chin ET, Harris RJ, Etcheverry TM. 2005. Effect of copper sulfate on performance of a serum-free CHO cell culture process and the level of free thiol in the recombinant antibody expressed. Biotechnol Prog 21(2): 550–553.

293. Sola RJ, Griebenow K. 2008. Effects of glycosylation on the stability of protein pharmaceuticals. J Pharm Sci. doi:10.1002/jps.21504.

294. Lim KH, Chang HI. 2006. O-linked N-acetylglucosamine suppresses thermal aggregation of Sp1. FEBS Lett 580(19):4645–4652.

295. Srour MA, Grupp J, Aburubaiha Z, Albert T, Brondke H, Oldenburg J, Schwaab R. 2008. Modified expression of coagulation factor VIII by addition of a glycosylation site at the N terminus of the protein. Ann Hematol 87(2): 107–112.

296. Singleton KD, Wischmeyer PE. 2008. Glutamine induces heat shock protein expression via O-glycosylation and phosphorylation of HSF-1 and Sp1. JPEN J Parenter Enteral Nutr 32(4):371–376.

297. Castle AM, Castle JD. 1998. Enhanced glycosylation and sulfation of secretory proteoglycans is coupled to the expression of a basic secretory protein. Mol Biol Cell 9(3):575–583.

298. Weintraub BD, Stannard BS, Meyers L. 1983. Glycosylation of thyroid-stimulating hormone in pituitary tumor cells: Influence of high mannose oligosaccharide units on subunit aggregation, combination, and intracellular degradation. Endocrinology 112(4):1331–1345.

299. Mishra R, Seckler R, Bhat R. 2005. Efficient refolding of aggregation-prone citrate synthase by polyol osmolytes: How well are protein folding and stability aspects coupled? J Biol Chem 280(16):15553–15560.

300. De Bernardez Clark E, Schwarz E, Rudolph R. 1999. Inhibition of aggregation side reactions during in vitro protein folding. Methods Enzymol 309:217–236.

301. Clark EDB. 1998. Refolding of recombinant proteins. Curr Opin Biotechnol 9(2):157–163.

302. Klein J, Dhurjati P. 1995. Protein aggregation kinetics in an *Escherichia coli* strain overexpressing a Salmonella typhimurium CheY mutant gene. Appl Environ Microbiol 61(4):1220–1225.
303. Georgiou G, Valax P, Ostermeier M, Horowitz PM. 1994. Folding and aggregation of TEM beta-lactamase: Analogies with the formation of inclusion bodies in *Escherichia coli*. Protein Sci 3(11):1953–1960.
304. Karuppiah N, Sharma A. 1995. Cyclodextrins as protein folding aids. Biochem Biophys Res Commun 211(1):60–66.
305. Xie Y, Wetlaufer DB. 1996. Control of aggregation in protein refolding: The temperature-leap tactic. Protein Sci 5(3):517–523.
306. Stempfer G, Holl-Neugebauer B, Rudolph R. 1996. Improved refolding of an immobilized fusion protein. Nat Biotechnol 14(3):329–334.
307. Gupta P, Hall CK, Voegler AC. 1998. Effect of denaturant and protein concentrations upon protein refolding and aggregation: A simple lattice model. Protein Sci 7(12):2642–2652.
308. Cardamone M, Puri NK, Brandon MR. 1995. Comparing the refolding and reoxidation of recombinant porcine growth hormone from a urea denatured state and from *Escherichia coli* inclusion bodies. Biochemistry 34(17):5773–5794.
309. Kim D, Yu MH. 1996. Folding pathway of human alpha 1-antitrypsin: Characterization of an intermediate that is active but prone to aggregation. Biochem Biophys Res Commun 226(2):378–384.
310. Rariy RV, Klibanov AM. 1997. Correct protein folding in glycerol. Proc Natl Acad Sci U S A 94:13520–13523.
311. Yasuda M, Murakami Y, Sowa A, Ogino H, Ishikawa H. 1998. Effect of additives on refolding of a denatured protein. Biotechnol Prog 14(4):601–606.
312. Maeda Y, Ueda T, Imoto T. 1996. Effective renaturation of denatured and reduced immunoglobulin G in vitro without assistance of chaperone. Protein Eng 9:95–100.
313. Cleland JL, Builder SE, Swartz JR, Winkler M, Chang JY, Wang DI. 1992. Polyethylene glycol enhanced protein refolding. Biotechnology (N Y) 10(9): 1013–1019.
314. Wetlaufer DB, Xie Y. 1995. Control of aggregation in protein refolding: A variety of surfactants promote renaturation of carbonic anhydrase II. Protein Sci 4(8):1535–1543.
315. Monterroso B, Minton AP. 2007. Effect of high concentration of inert cosolutes on the refolding of an enzyme: Carbonic anhydrase B in sucrose and Ficoll 70. J Biol Chem 282(46):33452–33458.
316. van den Berg B, Ellis RJ, Dobson CM. 1999. Effects of macromolecular crowding on protein folding and aggregation. EMBO J 18(24):6927–6933.
317. Hevehan D, De Bernardez-Clark E. 1997. Oxidative renaturation of lysozyme at high concentrations. Biotechnol Bioeng 54:221–230.
318. Stigter D, Dill KD. 1993. Theory for protein solubilities. Fluid Phase Equilib 82:237–249.
319. Hamada H, Shiraki K. 2007. L-argininamide improves the refolding more effectively than L-arginine. J Biotechnol 130(2):153–160.

320. Kuhlman B, Luisi DL, Evans PA, Raleigh DP. 1998. Global analysis of the effects of temperature and denaturant on the folding and unfolding kinetics of the N-terminal domain of the protein L9. J Mol Biol 284(5):1661–1670.

321. Scalley ML, Baker D. 1997. Protein folding kinetics exhibit an Arrhenius temperature dependence when corrected for the temperature dependence of protein stability. Proc Natl Acad Sci U S A 94(20):10636–10640.

322. Clarke AR, Waltho JP. 1997. Protein folding and intermediates. Curr Opin Biotechnol 8(4):400–410.

323. King J. 1997. Refolding with a piece of the ring. Nat Biotechnol 15(6):514–515.

324. Golbik R, Zahn R, Harding SE, Fersht AR. 1998. Thermodynamic stability and folding of GroEL minichaperones. J Mol Biol 276(2):505–515.

325. Thomas JG, Ayling A, Baneyx F. 1997. Molecular chaperones, folding catalysts, and the recovery of active recombinant proteins from *E. coli*. To fold or to refold. Appl Biochem Biotechnol 66(3):197–238.

326. Rozema D, Gellman SH. 1996. Artificial chaperone-assisted refolding of denatured-reduced lysozyme: Modulation of the competition between renaturation and aggregation. Biochemistry 35(49):15760–15771.

327. Stockel J, Doring K, Malotka J, Jahnig F, Dornmair K. 1997. Pathway of detergent-mediated and peptide ligand-mediated refolding of heterodimeric class II major histocompatibility complex (MHC) molecules. Eur J Biochem 248(3):684–691.

328. Yamaguchi S, Yamamoto E, Tsukiji S, Nagamune T. 2008. Successful control of aggregation and folding rates during refolding of denatured lysozyme by adding N-methylimidazolium cations with various N′-substituents. Biotechnol Prog 24(2):402–408.

329. Maeda Y, Yamada H, Ueda T, Imoto T. 1996. Effect of additives on the renaturation of reduced lysozyme in the presence of 4m urea. Protein Eng 9:461–465.

330. Arakawa T, Tsumoto K. 2003. The effects of arginine on refolding of aggregated proteins: Not facilitate refolding, but suppress aggregation. Biochem Biophys Res Commun 304(1):148–152.

331. Tsumoto K, Umetsu M, Kumagai I, Ejima D, Arakawa T. 2003. Solubilization of active green fluorescent protein from insoluble particles by guanidine and arginine. Biochem Biophys Res Commun 312(4):1383–1386.

332. Phillips J, Drumm A, Harrison P, Bird P, Bhamra K, Berrie E, Hale G. 2001. Manufacture and quality control of CAMPATH-1 antibodies for clinical trials. Cytotherapy 3(3):233–242.

333. Heavner GA, Arakawa T, Philo JS, Calmann MA, Labrenz S. 2007. Protein isolated from biopharmaceutical formulations cannot be used for comparative studies: Follow-up to "a case study using Epoetin Alfa from Epogen and EPREX." J Pharm Sci 96(12):3214–3225.

334. Cromwell ME, Hilario E, Jacobson F. 2006. Protein aggregation and bioprocessing. AAPS J 8(3):E572–E579.

335. Chang BS, Kendrick BS, Carpenter JF. 1996. Surface-induced denaturation of proteins during freezing and its inhibition by surfactants. J Pharm Sci 85(12):1325–1330.

336. Lee LS. 1997. Stabilized monomeric protein compositions. US5656730, Enzon, Inc.
337. Wang W. 2000. Lyophilization and development of solid protein pharmaceuticals. Int J Pharm 203(1–2):1–60.
338. Nakagawa K, Hottot A, Vessot S, Andrieu J. 2006. Influence of controoled nucleation by ultrasounds on ice morphology of froaen formulations for pharmaceutical proteins freeze-drying. Chem Eng Proc 45:783–791.
339. Eckhardt BM, Oeswein JQ, Bewley TA. 1991. Effect of freezing on aggregation of human growth hormone. Pharm Res 8(11):1360–1364.
340. Anchordoquy TJ, Carpenter JF. 1996. Polymers protect lactate dehydrogenase during freeze-drying by inhibiting dissociation in the frozen state. Arch Biochem Biophys 332(2):231–238.
341. Nema S, Avis KE. 1992. Freeze-thaw studies of a model protein, lactate dehydrogenase, in the presence of cryoprotectants. J Parenter Sci Technol 47:76–83.
342. Lashmar UT, Vanderburgh M, Little SJ. 2007. Bulk freeze-thawing of macromolecules. Bioprocess Int 5:44–54.
343. Farruggia B, Garcia G, D'Angelo C, Pico G. 1997. Destabilization of human serum albumin by polyethylene glycols studied by thermodynamical equilibrium and kinetic approaches. Int J Biol Macromol 20(1):43–51.
344. Lee LS. 1997. Stabilized monomeric protein compositions. US5656730.
345. Bush L, Webb C, Bartlett L, Burnett B. 1998. The formulation of recombinant factor IX: Stability, robustness, and convenience. Semin Hematol 35(2 Suppl. 2):18–21.
346. Kreilgaard L, Jones LS, Randolph TW, Frokjaer S, Flink JM, Manning MC, Carpenter JF. 1998. Effect of Tween 20 on freeze-thawing-and agitation-induced aggregation of recombinant human factor XIII. J Pharm Sci 87(12):1597–1603.
347. Volkin DB, Klibanov AM. 1989. Minimizing protein inactivation. In Creighton TE, ed., *Protein Function a Practical Approach*. Oxford, England: Information Press, pp. 1–24.
348. Kiese S, Papppenberger A, Friess W, Mahler HC. 2008. Shaken, not stirred: Mechanical stress testing of an IgG1 antibody. J Pharm Sci. doi:10.1002/jps.21328.
349. Levine HL, Ransohoff TC, Kawahata RT, McGregor WC. 1991. The use of surface tension measurements in the design of antibody-based product formulations. J Parenter Sci Technol 45(3):160–165.
350. Kueltzo LA. 2004. *Formulation Strategies for Protein Therapeutics*. Boston: IBC Life Sciences.
351. Biddlecombe JG, Craig AV, Zhang H, Uddin S, Mulot S, Fish BC, Bracewell DG. 2007. Determining antibody stability: Creation of solid-liquid interfacial effects within a high shear environment. Biotechnol Prog 23(5):1218–1222.
352. Kwon YM, Baudys M, Knutson K, Kim SW. 2001. In situ study of insulin aggregation induced by water-organic solvent interface. Pharm Res 18(12):1754–1759.
353. Kerwin BA, Akers MJ, Apostol I, Moore-Einsel C, Etter JE, Hess E, Lippincott J, Levine J, Mathews AJ, Revilla-Sharp P, Schubert R, Looker DL. 1999. Acute and long-term stability studies of deoxy hemoglobin and characterization of ascorbate-induced modifications. J Pharm Sci 88(1):79–88.
354. Grácia-Juliá A, René M, Cortés MM, Picart L, López PT, Chevalier D, Dumay E. 2008. Effect of dynamic high pressure on whey protein aggregation: a

comparison with the effect of continuous short-time thermal treatments. Food Hydrocolloids 22(6):1014–1032.

355. Patel HA, Singh H, Havea P, Considine T, Creamer LK. 2005. Pressure-induced unfolding and aggregation of the proteins in whey protein concentrate solutions. J Agric Food Chem 53(24):9590–9601.

356. Seefeldt MB, Kim YS, Tolley KP, Seely J, Carpenter JF, Randolph TW. 2005. High-pressure studies of aggregation of recombinant human interleukin-1 receptor antagonist: Thermodynamics, kinetics, and application to accelerated formulation studies. Protein Sci 14(9):2258–2266.

357. Huppertz T, Fox PF, de Kruif KG, Kelly AL. 2006. High pressure-induced changes in bovine milk proteins: a review. Biochim Biophys Acta 1764(3):593–598.

358. Kolakowski P, Dumay E, Cheftel JC. 2001. Effects of high pressure and low temperature on beta-lactoglobulin unfolding and aggregation. Food Hydrocolloids 15(3):215–232.

359. St John RJ, Carpenter JF, Balny C, Randolph TW. 2001. High pressure refolding of recombinant human growth hormone from insoluble aggregates. Structural transformations, kinetic barriers, and energetics. J Biol Chem 276(50):46856–46863.

360. St John RJ, Carpenter JF, Randolph TW. 1999. High pressure fosters protein refolding from aggregates at high concentrations. Proc Natl Acad Sci U S A 96(23):13029–13033.

361. St John RJ, Carpenter JF, Randolph TW. 2002. High-pressure refolding of disulfide-cross-linked lysozyme aggregates: Thermodynamics and optimization. Biotechnol Prog 18(3):565–571.

362. Kuhlman B, Yang HY, Boice JA, Fairman R, Raleigh DP. 1997. An exceptionally stable helix from the ribosomal protein L9: Implications for protein folding and stability. J Mol Biol 270(5):640–647.

363. Rupley JA, Careri G. 1991. Protein hydration and function. Adv Protein Chem 41:37–172.

364. Abdul-Fattah AM, Lechuga-Ballesteros D, Kalonia DS, Pikal MJ. 2008. The impact of drying method and formulation on the physical properties and stability of methionyl human growth hormone in the amorphous solid state. J Pharm Sci 97(1):163–184.

365. Salnikova MS, Middaugh CR, Rytting JH. 2008. Stability of lyophilized human growth hormone. Int J Pharm 358(1–2):108–113.

366. Yoshioka S, Miyazaki T, Aso Y, Kawanishi T. 2007. Significance of local mobility in aggregation of beta-galactosidase lyophilized with trehalose, sucrose or stachyose. Pharm Res 24(9):1660–1667.

367. Yu Z, Johnston KP, Williams RO III. 2006. Spray freezing into liquid versus spray-freeze drying: Influence of atomization on protein aggregation and biological activity. Eur J Pharm Sci 27(1):9–18.

368. Sarciaux JM, Mansour S, Hageman MJ, Nail SL. 1999. Effects of buffer composition and processing conditions on aggregation of bovine IgG during freeze-drying. J Pharm Sci 88(12):1354–1361.

369. Cleland JL, Lam X, Kendrick B, Yang J, Yang TH, Overcashier D, Brooks D, Hsu C, Carpenter JF. 2001. A specific molar ratio of stabilizer to protein is

required for storage stability of a lyophilized monoclonal antibody. J Pharm Sci 90(3):310–321.

370. Hagiwara H, Yuasa H, Yamamoto Y. 2000. Stabilized human monoclonal antibody preparation. US6165467.

371. Ressing ME, Jiskoot W, Talsma H, Van Ingen CW, Crommelin DJ. 1992. Influence of sucrose, dextran, and hydroxypropyl-beta-cyclodextrin as lyoprotectants for a freeze-dried mouse IgG2a monoclonal antibody (MN12). Pharm Res 9(Feb): 266–270.

372. Andya JD, Hsu CC, Shire SJ. 2003. Mechanisms of aggregate formation and carbohydrate excipient stabilization of lyophilized humanized monoclonal antibody formulations. AAPS PharmSci 5(2):E10.

373. Morales AA, Nunez-Gandolff G, Perez NP, Veliz BC, Caballero-Torres I, Duconge J, Fernandez E, Crespo FZ, Veloso A, Iznaga-Escobar N. 1999. Freeze-dried formulation for direct 99mTc-labeling ior-egf/r3 MAb: Additives, biodistribution, and stability. Nucl Med Biol 26(6):717–723.

374. Lueckel B, Helk B, Bodmer D, Leuenberger H. 1998. Effects of formulation and process variables on the aggregation of freeze-dried interleukin-6 (IL-6) after lyophilization and on storage. Pharm Dev Technol 3(3):337–346.

375. Kreilgaard L, Frokjaer S, Flink JM, Randolph TW, Carpenter JF. 1998. Effects of additives on the stability of recombinant human factor XIII during freeze-drying and storage in the dried solid. Arch Biochem Biophys 360(1):121–134.

376. Volkin DB, Middaugh CR. 1996. The characterization, stabilization, and formulation of acidic fibroblast growth factor. Pharm Biotechnol 9:181–217.

377. Webb SD, Cleland JL, Carpenter JF, Randolph TW. 2003. Effects of annealing lyophilized and spray-lyophilized formulations of recombinant human interferon-gamma. J Pharm Sci 92(4):715–729.

378. Maury M, Murphy K, Kumar S, Mauerer A, Lee G. 2005. Spray-drying of proteins: Effects of sorbitol and trehalose on aggregation and FT-IR amide I spectrum of an immunoglobulin G. Eur J Pharm Biopharm 59(2):251–261.

379. Andya JD, Maa YF, Costantino HR, Nguyen PA, Dasovich N, Sweeney TD, Hsu CC, Shire SJ. 1999. The effect of formulation excipients on protein stability and aerosol performance of spray-dried powders of a recombinant humanized anti-IgE monoclonal antibody. Pharm Res 16(3):350–358.

380. Maa YF, Nguyen PA, Hsu SW. 1998. Spray-drying of air-liquid interface sensitive recombinant human growth hormone. J Pharm Sci 87(2):152–159.

381. Webb SD, Golledge SL, Cleland JL, Carpenter JF, Randolph TW. 2002. Surface adsorption of recombinant human interferon-gamma in lyophilized and spray-lyophilized formulations. J Pharm Sci 91(6):1474–1487.

382. Burke PA, Klumb LA, Herberger JD, Nguyen XC, Harrell RA, Zordich M. 2004. Poly(lactide-co-glycolide) microsphere formulations of darbepoetin alfa: Spray drying is an alternative to encapsulation by spray-freeze drying. Pharm Res 21(3):500–506.

383. Nguyen XC, Herberger JD, Burke PA. 2004. Protein powders for encapsulation: A comparison of spray-freeze drying and spray drying of darbepoetin alfa. Pharm Res 21(3):507–514.

384. Costantino HR, Firouzabadian L, Hogeland K, Wu C, Beganski C, Carrasquillo KG, Cordova M, Griebenow K, Zale SE, Tracy MA. 2000. Protein spray-freeze drying. Effect of atomization conditions on particle size and stability. Pharm Res 17(11):1374–1383.
385. Jovanovic N, Bouchard A, Sutter M, Van Speybroeck M, Hofland GW, Witkamp GJ, Crommelin DJ, Jiskoot W. 2008. Stable sugar-based protein formulations by supercritical fluid drying. Int J Pharm 346(1–2):102–108.
386. Castellanos IJ, Crespo R, Griebenow K. 2003. Poly(ethylene glycol) as stabilizer and emulsifying agent: A novel stabilization approach preventing aggregation and inactivation of proteins upon encapsulation in bioerodible polyester microspheres. J Control Release 88(1):135–145.
387. Wei G, Lu LF, Lu WY. 2007. Stabilization of recombinant human growth hormone against emulsification-induced aggregation by pluronic surfactants during microencapsulation. Int J Pharm 338(1–2):125–132.
388. Geng Y, Yuan W, Wu F, Chen J, He M, Jin T. 2008. Formulating erythropoietin-loaded sustained-release PLGA microspheres without protein aggregation. J Control Release 130(3):259–265.
389. Sah H. 1999. Stabilization of proteins against methylene chloride/water interface-induced denaturation and aggregation. J Control Release 58(2):143–151.
390. Kanaoka E, Nagata S, Hirano K. 1999. Stabilization of aerosolized IFN-gamma by liposomes. Int J Pharm 188(2):165–172.
391. Ip AY, Arakawa T, Silvers H, Ransone CM, Niven RW. 1995. Stability of recombinant consensus interferon to air-jet and ultrasonic nebulization. J Pharm Sci 84(10):1210–1214.
392. Assemand E, Lacroix M, Mateescu MA. 2003. L-tyrosine prevents aggregation of therapeutic proteins by gamma-irradiation. Biotechnol Appl Biochem 38(Pt 2):151–156.
393. Kang J, Lambert O, Ausborn M, Schwendeman SP. 2008. Stability of proteins encapsulated in injectable and biodegradable poly(lactide-co-glycolide)-glucose millicylinders. Int J Pharm 357(1–2):235–243.
394. Costantino HR, Griebenow K, Mishra P, Langer R, Klibanov AM. 1995. Fourier-transform infrared spectroscopic investigation of protein stability in the lyophilized form. Biochim Biophys Acta 1253(1):69–74.
395. Chang LL, Shepherd D, Sun J, Ouellette D, Grant KL, Tang XC, Pikal MJ. 2005. Mechanism of protein stabilization by sugars during freeze-drying and storage: Native structure preservation, specific interaction, and/or immobilization in a glassy matrix? J Pharm Sci 94(7):1427–1444.
396. Roy ML, Pikal MJ, Rickard EC, Maloney AM. 1992. The effects of formulation and moisture on the stability of a freeze-dried monoclonal antibody-vinca conjugate: A test of the WLF glass transition theory. Dev Biol Stand 74:323–339; discussion 340.
397. Pikal MJ, Rigsbee D, Roy ML, Galreath D, Kovach KJ, Wang BS, Carpenter JF, Cicerone MT. 2008. Solid state chemistry of proteins: II. The correlation of storage stability of freeze-dried human growth hormone (hGH) with structure and dynamics in the glassy solid. J Pharm Sci 97(12):5106–5121.
398. Jain NK, Roy I. 2008. Role of trehalose in moisture-induced aggregation of bovine serum albumin. Eur J Pharm Biopharm 69(3):824–834.

399. Chang BS, Reeder G, Carpenter JF. 1996. Development of a stable freeze-dried formulation of recombinant human interleukin-1 receptor antagonist. Pharm Res 13(2):243–249.

400. Hora MS, Rana RK, Smith FW. 1992. Lyophilized formulations of recombinant tumor necrosis factor. Pharm Res 9(1):33–36.

401. Garzon-Rodriguez W, Koval RL, Chongprasert S, Krishnan S, Randolph TW, Warne NW, Carpenter JF. 2004. Optimizing storage stability of lyophilized recombinant human interleukin-11 with disaccharide/hydroxyethyl starch mixtures. J Pharm Sci 93(3):684–696.

402. Strickley RG, Anderson BD. 1997. Solid-state stability of human insulin. II. Effect of water on reactive intermediate partitioning in lyophiles from pH 2–5 solutions: Stabilization against covalent dimer formation. J Pharm Sci 86(6):645–653.

403. Kreilgaard L, Frokjaer S, Flink JM, Randolph TW, Carpenter JF. 1999. Effects of additives on the stability of *Humicola lanuginosa* lipase during freeze-drying and storage in the dried solid. J Pharm Sci 88(3):281–290.

404. Costantino HR, Schwendeman SP, Griebenow K, Klibanov AM, Langer R. 1996. The secondary structure and aggregation of lyophilized tetanus toxoid. J Pharm Sci 85(12):1290–1293.

405. Sane SU, Wong R, Hsu CC. 2004. Raman spectroscopic characterization of drying-induced structural changes in a therapeutic antibody: Correlating structural changes with long-term stability. J Pharm Sci 93(4):1005–1018.

406. Costantino HR, Andya JD, Nguyen PA, Dasovich N, Sweeney TD, Shire SJ, Hsu CC, Maa YF. 1998. Effect of mannitol crystallization on the stability and aerosol performance of a spray-dried pharmaceutical protein, recombinant humanized anti-IgE monoclonal antibody. J Pharm Sci 87(11):1406–1411.

407. Liu WR, Langer R, Klibanov AM. 1990. Moisture-induced aggregation of lyophilized proteins in the solid state. Biotechnol Bioeng 37:177–184.

408. Costantino HR, Langer R, Klibanov AM. 1995. Aggregation of a lyophilized pharmaceutical protein, recombinant human albumin: Effect of moisture and stabilization by excipients. Biotechnology (N Y) 13(5):493–496.

409. Hora MS, Rana RK, Wilcox CL, Katre NV, Hirtzer P, Wolfe SN, Thomson JW. 1992. Development of a lyophilized formulation of interleukin-2. Dev Biol Stand 74:295–303.

410. Chen B, Costantino HR, Liu J, Hsu CC, Shire SJ. 1999. Influence of calcium ions on the structure and stability of recombinant human deoxyribonuclease I in the aqueous and lyophilized states. J Pharm Sci 88(4):477–482.

411. Katakam M, Banga AK. 1995. Aggregation of insulin and its prevention by carbohydrate excipients. PDA J Pharm Sci Technol 49(4):160–165.

412. Breen ED, Curley JG, Overcashier DE, Hsu CC, Shire SJ. 2001. Effect of moisture on the stability of a lyophilized humanized monoclonal antibody formulation. Pharm Res 18(9):1345–1353.

413. Andya J, Cleland JL, Hsu CC, Lam XM, Overcashier DE, Shire SJ, Yang JY-F, Wu SS-Y. 2004. *Protein Formulation*. Genentech, Inc. US6685940 B2.

414. Chang LL, Shepherd D, Sun J, Ouellette D, Grant KL, Tang XC, Pikal MJ. 2005. Mechanism of protein stabilization by sugars during freeze-drying and storage:

Native structure preservation, specific interaction, and/or immobilization in a glassy matrix? J Pharm Sci 94(7):1427–1444.

415. Broersen K, Elshof M, de Groot J, Voragen AGJ, Hamer RJ, de Jongh HHJ. 2007. Aggregation of beta-lactoglobulin regulated by glucosylation. J Agric Food Chem 55(6):2431–2437.
416. Schenz TW. 1995. Glass transition and product stability—An overview. Food Hydrocolloids 9:307–315.
417. Pikal MJ, Rigsbee D, Akers MJ. 2009. Solid state chemistry of proteins IV. What is the meaning of thermal denaturation in freeze dried proteins? J Pharm Sci 98(4):1387–1399.
418. Breen ED, Curley JG, Overcashier DE, Hsu CC, Shire SJ. 2001. Effect of moisture on the stability of a lyophilized humanized monoclonal antibody formulation. Pharm Res 18(9):1345–1353.
419. Maa YF, Nguyen PA, Andya JD, Dasovich N, Hsu CC, et al. 1998. Effect of spray drying and subsequent processing conditions on residual moisture content and physical/biochemical stability of protein inhalation powders. Pharm Res 15(May):768–775.
420. Ma X, Wang DQ, Bouffard R, MacKenzie A. 2001. Characterization of murine monoclonal antibody to tumor necrosis factor (TNF-MAb) formulation for freeze drying cycle development. Pharm Res 18(Feb):196–202.
421. Separovic F, Lam YH, Ke X, Chan HK. 1998. A solid-state NMR study of protein hydration and stability. Pharm Res 15(12):1816–1821.
422. Schwendeman SP, Costantino HR, Gupta RK, Tobio M, Chang AC, Alonso MJ, Siber GR, Langer R. 1996. Strategies for stabilising tetanus toxoid towards the development of a single-dose tetanus vaccine. Dev Biol Stand 87:293–306.
423. Renard D, Lefebvre J, Robert P, Llamas G, Dufour E, Dufour E. 1999. Structural investigation of beta-lactoglobulin gelation in ethanol/water solutions. Int J Biol Macromol 26(1):35–44.
424. Maa Y-F, Hsu CC. 1997. Protein denaturation by combined effect of shear and air-liquid interface. Biotechnol Bioeng 54(6):503–512.
425. Tsai YR, Yang SJ, Jiang SS, Ko SJ, Hung SH, Kuo SY, Pan RL. 1998. High-pressure effects on vacuolar H^+-ATPase from etiolated mung bean seedlings. J Protein Chem 17(2):161–172.
426. Tsai PK, Volkin DB, Dabora JM, Thompson KC, Bruner MW, Gress JO, Matuszewska B, Keogan M, Bondi JV, Middaugh CR. 1993. Formulation design of acidic fibroblast growth factor. Pharm Res 10(5):649–659.
427. Navarra G, Leone M, Militello V. 2007. Thermal aggregation of beta-lactoglobulin in presence of metal ions. Biophys Chem 131(13):52–61.

CHAPTER 5

Experimental Detection and Characterization of Protein Aggregates

VIKAS K. SHARMA and DEVENDRA S. KALONIA

5.1 INTRODUCTION

Aggregation of protein-based biotherapeutics continues to present a challenge during various stages of processing and product development, including cell culture/fermentation, purification, formulation, shipping, and storage.[1–4] The concern around the presence of aggregates in the final finished product as a consequence of processing steps or upon storage is profound, given the resulting potential for undesirable physiological consequences upon systemic administration. Aggregates could reduce altered potency and toxicity, and could elicit an immunological response.[5–12] The potential mechanisms involved for linking the aggregate size, structure, and shape to immunogenicity have been reviewed previously (see also Chapter 10).[13–15] Therefore, significant effort and resources are employed to understand, characterize, and quantify the type and level of aggregates during product development of biotherapeutics, with an ultimate goal of minimizing the level of aggregates in the final product.

In this chapter, we focus on the characterization, detection, and analyses of different types of aggregates commonly observed during the development of protein-based therapeutics and discuss the analytical tools employed to achieve such understanding. It is well realized that no single technique exists to analyze all of the various forms of commonly observed aggregates, and often the choice of a given technique is governed by the size and the characteristics of aggregates. In the absence of any a priori knowledge about the characteristic of the aggregate species formed, it typically becomes necessary

Aggregation of Therapeutic Proteins, Edited by Wei Wang and Christopher J. Roberts

to employ multiple techniques to obtain as much information as possible about the type and levels of various forms of aggregates. To be able to use a given technique, it is important to first understand the types of aggregates that could be commonly present in protein products. The underlying mechanisms involved toward the formation of aggregates and approaches to control the formation of aggregates during processing, shipping, and storage are discussed elsewhere in this book and will not be elaborated in this chapter.

5.2 AGGREGATE CLASSIFICATION

Protein aggregates can be categorized into several different types and are usually defined in the context of the mechanism, interactions involved in the formation of aggregates, and size.[2,16,17] Currently, there is no unified system of nomenclature to define an aggregate, and often an aggregate can be defined using a variety of terms.

Protein aggregates can be classified based on the following four categories:

1. conformation: native or non-native,
2. linkage: covalent or non-covalent,
3. reversibility: reversible or irreversible, and
4. size: soluble or subvisible or visible.

Table 5.1 lists various terms commonly used in the literature to describe protein aggregates, along with a brief definition of each term. Almost all aggregates reported in the literature for therapeutic proteins can be defined by a combination of terms from the above-mentioned four categories. Therefore, it becomes necessary that an aggregate must be defined using all of the four attributes/categories, that is, conformation, linkage, reversibility, and size.

To facilitate this further, we propose an abbreviated nomenclature to define aggregates in a comprehensive manner.[18] We propose a "lettered" approach to represent each category and the type of aggregate within each category. We represent conformation as "C," linkage as "L," reversibility as "R," and size as "S." The classification within each category is represented as a subscript with a lowercase letter. Table 5.1 lists the proposed abbreviated nomenclature for each type of aggregate. Commonly accepted definitions of the various terms to define various types of aggregates and the abbreviated nomenclature are described below.

1. Conformation, C

Native aggregates, C_n, are those in which the higher-order structure (tertiary and secondary) of the monomeric unit is largely retained in the aggregated state.

TABLE 5.1. Classification, Definitions, and Proposed Nomenclature of Protein Aggregates

Aggregate Category	Aggregate Type	Nomenclature	Definition
Conformation	Native	C_n	Aggregates where the protein largely retains its higher-order tertiary and secondary structure
	Non-native	C_u	Aggregates where the protein exhibits significantly altered structure compared with the native state
Linkage	Covalent or non-dissociable	L_c	Aggregates held together through covalent links, for example, disulfide linked, ester linked, and so on, and will not typically dissociate in the presence of a chemical denaturant
	Physical (non-covalent or dissociable)	L_p	Aggregates held together through physical interactions, for example, electrostatic, hydrophobic, hydrogen bonding, and so on, and would typically dissociate in the presence of a chemical denaturant, such as SDS, urea, or guanidine hydrochloride
Reversibility	Reversible	R_r	Aggregates that dissociate upon reversal of the solution condition that initially resulted in their very formation
	Irreversible	R_i	Aggregates that remain associated in their higher-order state upon reversal of the solution conditions that initially resulted in their very formation
Size	Soluble	S_s	Typically oligomeric aggregates such as dimers, trimers, and so on, which are not visible to the naked eye or under a laboratory microscope and would typically pass through a 0.2-µm filter

(*Continued*)

TABLE 5.1. *Continued*

Aggregate Category	Aggregate Type	Nomenclature	Definition
	Soluble oligomers with known number of monomers	S_{sn} (n could be 2,3 … for dimer, trimer, etc.)	A general term to describe a *soluble* aggregate with a defined number of monomeric units present such as in dimer, trimer, tetramer, and so on
	Subvisible	S_{sv}	Typically between 0.1 and 100 µm in size. May induce visible or measurable turbidity if present in sufficient concentration, however, may render the solution clear at low concentrations
	Visible aggregates	S_v	Aggregates typically larger than 100 µm in size that are visible to a trained eye or observed under a laboratory microscope

Non-native aggregates, C_u, are those in which the higher-order structure of the monomeric unit is significantly altered in the aggregated state.

2. Linkage, L

Covalent aggregates, L_c, also referred to as "non-dissociable" aggregates, form as a result of a chemical bond. Common examples of such aggregates are those formed through the disulfide linkage.[19–21] Such aggregates do not typically dissociate in the presence of a chemical denaturant such as urea, guanidinium hydrochloride, or sodium dodecyl sulfate (SDS).

Physical or non-covalent aggregates, L_p, also referred to as "dissociable" aggregates, are those where the monomeric units adhere through physical forces such as hydrophobic, electrostatic, van der Waals or hydrogen-bond-mediated interactions.[22–24] Such aggregates would typically dissociate in the presence of a chemical denaturant such as urea, guanidinium hydrochloride, or SDS.

3. Reversibility, R

Reversible aggregates, R_r, are those that "reverse" to the lower-molecular-weight state (oligomer or monomer) upon removal of the solution condition that, in the first place, resulted in the formation of the aggregate. The solution condition promoting reversible aggregates could be protein concentration, pH, temperature, or presence of a cosolute such as polyethylene glycol or salt. Examples of reversible aggregates are oligomeric protein assemblies formed through self-association[25–28] or larger visible particulates (crystals or amorphous particles)[29–32] that result when the intrinsic solubility of a protein is exceeded under a given solution condition or through addition of a cosolute.

In reversible aggregates, the protein is largely believed to retain its native structure except for the site on a protein that is involved in making the contact points for association or crystallization/precipitation. Furthermore, reversible aggregates typically exist in some state of equilibrium with each other or with the monomeric protein (cf. Chapter 2).

Irreversible aggregates, R_i, are those that do not reverse into the lower-molecular-weight state (oligomer or monomer) upon removal of the condition that resulted in the formation of the aggregate.[22,33,34] Such aggregates typically do not participate in an equilibrium reaction.

4. Size, S

Soluble aggregates, S_s, are typically oligomeric aggregates, for example, dimers, trimers, and tetramers and, in some cases, could be composed of hundreds of monomers per aggregate present in a polymer-like state.[35,36] These aggregates in most cases would typically pass through a 0.22-μm filter or through a smaller pore size-exclusion chromatography (SEC) column and are not visible to the naked eye or through a laboratory optical microscope.[1,37,38] If the number of monomeric units present in a soluble aggregate is known, then the oligomer can be represented as S_{s2}, S_{s3}, S_{s4}, and so forth for a dimer, trimer, and tetramer, respectively.

Subvisible aggregates, S_{sv}, also commonly referred to as subvisible particles, are defined as those that typically fall in the particle size range of ~100 nm to 100 μm.[2,39] It should be noted that this size typically refers to the hydrodynamic radius, R_h, of an equivalent sphere. The exact morphology of the aggregate may differ, however, and therefore, the measured hydrodynamic radius may underrepresent or may overrepresent the true dimensions of the aggregate, depending on whether the molecule is compact/spherical, extended/polymeric, or, in general, irregularly shaped. Therefore, caution should be exerted when presenting the size of a subvisible particle and, if possible, the morphology of the particle should be discussed as well. The upper size range of subvisible particles may be visible to a trained eye or through an optical microscope. However, for the larger, more general definitions, this should still be considered subvisible. It should be noted that subvisible particles, when present in sufficiently high concentration, may give rise to solutions that are visibly turbid.

Visible aggregates or *visible particles*, S_v, are those where discrete particles are visible to the naked eye and typically are larger than 50 μm in size.[2,39] The term "insoluble" to define protein particles or aggregates should be used with caution since this term does not unambiguously define any characteristic of the aggregate, except that it probably represents large visible aggregates. As defined above, based on size, it is more appropriate to classify aggregates as soluble, subvisible, and visible aggregates/particulates.

5.2.1 Aggregate Nomenclature

All aggregates could be defined using at least one term from all of the above-mentioned four categories. Therefore, we propose an approach to completely define a given type of aggregate by combining the letter representation for each category of aggregate and represent a given aggregate using the four letters "CLRS" (pronounced as "clears"), along with their respective defined subscripts. For example, a native, covalent, irreversible, soluble dimer could be described as $C_nL_cR_iS_{s2}$. Various potential combinations of terms from different categories as described above to define an aggregate are listed in Table 5.2. The potential outcomes listed in Table 5.2 are based on the assumption that covalent aggregates are always irreversible in nature and non-covalent aggregates that involve non-native or unfolded protein conformations are typically not reversible in nature. To further demonstrate the usage of the above nomenclature, we cite a few examples of aggregates reported in literature for several therapeutic proteins.

One of the earliest examples of aggregates reported for a biotherapeutic was that of insulin.[40–45] It was well recognized early on that insulin is stored in the pancreas in the hexameric form in complexation with zinc, whereas only the monomeric form was the active form.[46] Based on the detailed characterization of insulin "self-association" over several years, various forms of insulin aggregates or the self-associated species have been identified to exist in the hexameric, tetrameric, and the dimeric forms. The number of units in a soluble aggregate is always in reference to the monomer. Based on our proposed nomenclature, the soluble, reversible, native, non-covalent aggregates of insulin would be classified as, for example, $C_nL_pR_rS_{s6}$, $C_nL_pR_rS_{s4}$, and $C_nL_pR_rS_{s2}$ for the hexamer, tetramer, and dimer forms, respectively. These aggregates have been considered native since the monomer insulin retains its higher-order structure in the hexameric form to a large extent (except for local changes in the regions involved at the site of association). The formation of higher oligomeric states does not involve any covalent linkages in this case. Furthermore, the distribution of sizes is modulated by the protein concentration and the presence of zinc, and hence, aggregates are classified as reversible in nature.

On the other hand, shaking or stirring typically results in the formation of larger aggregates that span from soluble to visible aggregates and are often irreversible in nature involving non-native protein conformations. These aggregates typically involve non-covalent interactions; however, covalent linkages may also be present in the final aggregated form. Such aggregates have been observed with a wide variety of proteins including human growth hormone[22] and more recently with monoclonal antibodies (mAbs).[47,48] Such aggregates would be classified into several different categories and would be represented as $C_uL_pR_iS_s$, $C_uL_pR_iS_{sv}$, and $C_uL_pR_iS_v$ for non-covalent aggregates and similarly as $C_uL_cR_iS_s$, $C_uL_cR_iS_{sv}$, and $C_uL_cR_iS_v$ for covalently linked aggregates.

An example of reversible visible aggregates would be those formed as a consequence of limited solubility of certain proteins under given formulation/solution conditions. For example, insulin suspension, which is available

TABLE 5.2. Various Potential Combinations of Aggregate Definitions from Table 5.1 to Completely Define an Aggregate

	Type of Aggregate (Conformation, Linkage, Reversibility, Size)	Nomenclature	Example
1	Native, physical (non-covalent), reversible, soluble	$C_nL_pR_rS_s$	Reversible self-association of monomers to dimers, trimers, and so on, for example, insulin,[147] β-lactoglobulin[148]
2	Native, physical (non-covalent), irreversible, soluble	$C_nL_pR_iS_s$	Fibrillar aggregates with native-like conformation[149]
3	Native, covalent, irreversible, soluble	$C_nL_cR_iS_s$	Disulfide-linked dimers, trimers, and so on[20]
4	Unfolded, physical (non-covalent), irreversible, soluble	$C_uL_pR_iS_s$	Commonly observed aggregates in many pharmaceutical proteins[22–24,50]
5	Unfolded, covalent, irreversible, soluble	$C_uL_cR_iS_s$	Aggregates formed in solid state[21]
6	Native, physical (non-covalent), reversible, subvisible	$C_nL_pR_rS_{sv}$	Precipitates formed due to a lower protein solubility[31,32]
7	Native, physical (non-covalent), irreversible, subvisible	$C_nL_pR_iS_{sv}$	Fibrillar aggregation involving native conformation[150]
8	Native, covalent, irreversible, subvisible	$C_nL_cR_iS_{sv}$	Disulfide-linked larger aggregates
9	Unfolded, physical (non-covalent), irreversible, subvisible	$C_uL_pR_iS_{sv}$	Agitation-induced aggregates, silicon oil-induced aggregates[77,151]
10	Unfolded, covalent, irreversible, subvisible	$C_uL_cR_iS_{sv}$	Aggregates formed in solid state[21]
11	Native, physical (non-covalent), reversible, visible	$C_nL_pR_rS_v$	Protein precipitates/crystals resulting due to lack of solubility
12	Native, physical (non-covalent), irreversible, visible	$C_nL_pR_iS_v$	Not commonly observed
13	Native, covalent, irreversible, visible	$C_nL_cR_iS_v$	Not commonly observed
14	Unfolded, physical (non-covalent), irreversible, visible	$C_uL_pR_iS_v$	Large amyloids, protein fibrils, visible aggregates upon aggregation[47]
15	Unfolded, covalent, irreversible, visible	$C_uL_cR_iS_v$	Visible aggregates upon aggregation with potential covalent cross-links

commercially for subcutaneous administration, is comprised of larger, visible, reversible crystals of insulin that slowly dissolve after injection to release insulin.[25] In contrast, reversible visible aggregates of certain non-glycosylated proteins may present a formulation challenge if the desired concentration lies above the solubility limit of a given protein. These types of larger aggregates would be classified as $C_nL_pR_rS_{sv}$ and $C_nL_pR_rS_v$, thus capturing both subvisible and visible aggregates.

The most common forms of aggregates observed in biotherapeutics are the soluble irreversible aggregates formed upon processing and storage. These aggregates are typically retained in the final product since they are not filtered by a 0.2-μm filter, and therefore, their concentrations in the final finished product must be minimized. These aggregates could be covalent or non-covalent in nature and may or may not retain their native structure. Therefore, these could be categorized as $C_nL_cR_iS_s$, $C_nL_pR_iS_s$, $C_uL_cR_iS_s$, or $C_uL_cR_iS_s$. Several studies exist in the literature that report the formation of these aggregates. Some of the noted literature examples include recombinant human factor VIII (rhFVIII),[24] human growth hormone,[22] granulocyte colony-stimulating factor,[23,34] interferons and interleukins,[11,49,50] and mAbs.[47,51–53] In general, these types of aggregates are present at some level for almost all therapeutic proteins.

As evident from the above discussion, the complexity and the diversity involved in various types of aggregates often requires multiple analytical techniques for their detection and characterization. Often, it is necessary to make use of multiple techniques, also commonly referred to as "orthogonal" methods for complete characterization of aggregated species present in a solution. In the section below, we discuss various analytical technologies currently used for the detection, characterization, and quantitation of aggregates/particulates in biotherapeutics products.

5.3 ANALYTICAL TOOLS FOR THE CHARACTERIZATION OF AGGREGATES

At the outset, most analytical techniques can be divided into two main categories: those used for the detection and quantitation in terms of size and shape of aggregates and those used for the characterization in terms of aggregate composition, that is, proteinaceous versus non-proteinaceous, and protein structure. Table 5.3 provides a comprehensive list of a variety of analytical techniques used for the detection and characterization of aggregates and summarizes the approximate cost associated with the technique, the type, and size of aggregates that can be analyzed, and the advantages and disadvantages of each technique. The choice of an analytical technique for the detection and quantitation is often determined by the size range of aggregates that a particular technique can measure in a consistent and reproducible manner, by the detection and quantitation limits of the technique,

TABLE 5.3. Summary of Analytical Tools Utilized for the Characterization of Protein Aggregates

Technique/ Instrument Cost	Cost	Type of Aggregates Characterized[a]	Size Range/ Molecular Weight Range	Advantages	Limitations
Size-exclusion chromatography (SEC)	~$60K	Soluble aggregates $C_nL_pR_iS_s$, $C_uL_pR_iS_s$, $C_nL_cR_iS_s$, $C_uL_cR_iS_s$, $C_nL_pR_rS_s$	5–1000kDa	Easy to use and validate, robust, high throughput, high sensitivity, information about protein hydrodynamics, absolute molecular weight information with online light scattering detector, high throughput	Protein–column interactions, sample dilution, filtration of larger soluble aggregates, different solution conditions following injection, limited mobile phase options, inaccurate molecular weight based on SEC analysis only
Static light scattering	Batch mode: ~$100K, with SEC-HPLC and refractive index detector: ~$120K	Soluble aggregates $C_nL_pR_iS_s$, $C_uL_pR_iS_s$, $C_nL_cR_iS_s$, $C_uL_cR_iS_s$, $C_nL_pR_rS_s$	5kDa to several million daltons	Batch mode allows characterization under native solution conditions, provides information about radius of gyration, combined with SEC for absolute molecular weight analysis of eluting species, characterization of self-associating species, high throughput	Tedious procedure with batch mode, interference from dust and large impurities, shares disadvantages of SEC in flow mode, column may shed particles and contribute to noise in the eluting peaks

(Continued)

TABLE 5.3. *Continued*

Technique/ Instrument Cost	Cost	Type of Aggregates Characterized[a]	Size Range/ Molecular Weight Range	Advantages	Limitations
Dynamic light scattering	~$70 K	Soluble aggregates, subvisible particles $C_nL_pR_iS_s$, $C_uL_pR_iS_s$, $C_nL_cR_iS_s$, $C_uL_cR_iS_s$, $C_nL_pR_rS_s$, $C_nL_pR_iS_{sv}$, $C_uL_pR_iS_{sv}$, $C_nL_cR_iS_{sv}$, $C_uL_cR_iS_{sv}$, $C_nL_pR_rS_{sv}$	1 nm–10 μm	Allows characterization in native solution conditions, wide range of size determination, high throughput	Unable to differentiate between monomer and smaller aggregates, interference from dust and larger impurities, measurement biased toward larger particles, difficult to quantitate size fractions and validate data
Analytical ultracentrifugation (AUC)	~$250 K	Soluble aggregates $C_nL_pR_iS_s$, $C_uL_pR_iS_s$, $C_nL_cR_iS_s$, $C_uL_cR_iS_s$, $C_nL_pR_rS_s$	5–1000 kDa	Analysis under native solution conditions, well suited to study self-association, fluorescence detection system allows analysis of high concentration solutions, estimation of sedimentation coefficient related to shape and size	Data collection and analysis is complex, requires advanced training, sample dilution is often required, difficult to validate, time-consuming

Asymmetric flow field flow fractionation (AF4)	~$70 K	Soluble aggregates $C_nL_pR_iS_s$, $C_uL_pR_iS_s$, $C_nL_cR_iS_s$, $C_uL_cR_iS_s$, $C_nL_pR_rS_s$	5 kDa to several million daltons	Absence of protein–matrix interactions such as those present in SEC, detection of aggregates that are retained by SEC column, high throughput	Potential of protein–membrane interactions, limited to soluble aggregates, sample dilution
Sodium dodecyl sulfate–polyacrylamide gel electrophoresis (SDS-PAGE)	~$30 K	Soluble aggregates $C_nL_cR_iS_s$, $C_uL_cR_iS_s$	5–1000 kDa	Analysis of covalent aggregates, easy to use, high throughput, orthogonal technique to characterize whether aggregates are covalent or non-covalent, reduced SDS-PAGE differentiates between disulfide and other covalent aggregates	Analysis under denaturing conditions, non-covalent aggregates dissociate and cannot be analyzed, indirect molecular weight analysis, sample preparation can form aggregates, quantitation less accurate
Capillary electrophoresis–sodium dodecyl sulfate (CE-SDS)	~$50 K	Soluble aggregates $C_nL_cR_iS_s$, $C_uL_cR_iS_s$	5–1000 kDa	High throughput analysis of covalent aggregates, quantitation is accurate, orthogonal technique to characterize whether aggregates are covalent or non-covalent, reducing conditions differentiate between disulfide and other covalent aggregates	Analysis under denaturing conditions, non-covalent aggregates dissociate and cannot be analyzed, indirect molecular weight analysis, sample preparation can form aggregates

(*Continued*)

TABLE 5.3. *Continued*

Technique/ Instrument Cost	Cost	Type of Aggregates Characterized[a]	Size Range/ Molecular Weight Range	Advantages	Limitations
Microfluidics chip-based electrophoresis	~$30 K	Soluble aggregates $C_nL_cR_iS_s$, $C_uL_cR_iS_s$	5–1000 kDa	Very high throughput analysis of covalent aggregates, quantitation is accurate, orthogonal technique to characterize whether aggregates are covalent or non-covalent, reducing conditions differentiate between disulfide and other covalent aggregates	Analysis under denaturing conditions, non-covalent aggregates dissociate and cannot be analyzed, indirect molecular weight analysis, sample preparation can form aggregates

Turbidimetry/nephelometry	~30 K	Subvisible particles and visible particles $C_nL_pR_iS_{sv}$, $C_uL_pR_iS_{sv}$, $C_nL_cR_iS_{sv}$, $C_uL_cR_iS_{sv}$, $C_nL_pR_rS_{sv}$, $C_nL_pR_iS_v$, $C_uL_pR_iS_v$, $C_nL_cR_iS_v$, $C_uL_cR_iS_v$, $C_nL_pR_rS_v$	~100 nm to several microns	Simple and easy to use, relative and qualitative assessment of aggregates/particles growth	Provides only a qualitative assessment of presence of aggregates/particles, does not provide information about shape, size nor nature of aggregates, background turbidity from high-concentration protein solutions may confound results, turbidity could change due to effect of solution conditions on Rayleigh scattering and not necessarily change in aggregate population
Light obscuration	~$50 K	Subvisible particles $C_nL_pR_iS_{sv}$, $C_uL_pR_iS_{sv}$, $C_nL_cR_iS_{sv}$, $C_uL_cR_iS_{sv}$, $C_nL_pR_rS_{sv}$	~1 µm to >100 µm	Reliable estimation of particles >10 µm, automated measurement	Results affected by bubbles and translucent particles, cannot differentiate between protein- and non-protein-related particles, requires large sample volume (~25 mL), not suitable for particles below 2 µm

(Continued)

TABLE 5.3. *Continued*

Technique/ Instrument Cost	Cost	Type of Aggregates Characterized[a]	Size Range/ Molecular Weight Range	Advantages	Limitations
Microscopy	~$20 K	Subvisible particles $C_nL_pR_iS_{sv}$, $C_uL_pR_iS_{sv}$, $C_nL_cR_iS_{sv}$, $C_uL_cR_iS_{sv}$, $C_nL_pR_rS_{sv}$	~1 µm to >100 µm	Reliable estimation of particles >10 µm	Tedious, manual counting of particles, not suitable for particles below 10 µm
Flow microscopy imaging	~$50 K	Subvisible particles $C_nL_pR_iS_{sv}$, $C_uL_pR_iS_{sv}$, $C_nL_cR_iS_{sv}$, $C_uL_cR_iS_{sv}$, $C_nL_pR_rS_{sv}$	~1 µm to >100 µm	Reliable estimation of particles >2 µm, automated measurement, allows image analysis, differentiate between air bubbles, oil droplets, and other particles	Difficult to validate, not suitable for particles below 1 µm
Coulter counter	~$50 K	Subvisible particles $C_nL_pR_iS_{sv}$, $C_uL_pR_iS_{sv}$, $C_nL_cR_iS_{sv}$, $C_uL_cR_iS_{sv}$, $C_nL_pR_rS_{sv}$	~1 µm to >100 µm	Reliable estimation of particles >2 µm, automated measurement	High conductivity solution required, not suitable for particles below 2 µm, underestimation of smaller particles
Electrospray differential mobility analyzer	~$90 K	Soluble aggregates $C_nL_pR_iS_s$, $C_uL_pR_iS_s$, $C_nL_cR_iS_s$, $C_uL_cR_iS_s$, $C_nL_pR_rS_s$	3–250 nm	Absence of protein–matrix or protein–membrane interactions, quantitation of larger soluble aggregates not quantifiable by SEC	Requires low protein concentration, low salt concentration

Polarization-induced differential scattering	~$50 K	Soluble aggregates, subvisible particles $C_nL_pR_iS_s$, $C_uL_pR_iS_s$, $C_nL_cR_iS_s$, $C_uL_cR_iS_s$, $C_nL_pR_rS_s$, $C_nL_pR_iS_{sv}$, $C_uL_pR_iS_{sv}$, $C_nL_cR_iS_{sv}$, $C_uL_cR_iS_{sv}$, $C_nL_pR_rS_{sv}$	40 nm–2000 μm	Allows characterization in native solution conditions, wide range of size determination, high throughput	New technique, lack of sufficient data to assess precision and accuracy
Scanning electron microscope/transmission electron microscope	~$200 K	Subvisible aggregates, visible aggregates $C_nL_pR_iS_{sv}$, $C_uL_pR_iS_{sv}$, $C_nL_cR_iS_{sv}$, $C_uL_cR_iS_{sv}$, $C_nL_pR_rS_{sv}$, $C_nL_pR_iS_v$, $C_uL_pR_iS_v$, $C_nL_cR_iS_v$, $C_uL_cR_iS_v$, $C_nL_pR_rS_v$	Few nanometers to several microns	Size and shape analysis, visualization of aggregates	Not suitable for quantitation, sample handling and preparation could form or destroy particles or aggregates
Fluorescence spectrometer	~$70 K	Structural characterization, aggregate detection	NA	Tertiary structure analysis, visualization of aggregates, applicable to soluble, subvisible and visible aggregates, detection of small amount of aggregates	Extrinsic fluorescence dyes could lead to formation of aggregates, qualitative assessment of aggregates, significant aggregate population to be present for structural analysis

(Continued)

TABLE 5.3. *Continued*

Technique/ Instrument Cost	Cost	Type of Aggregates Characterized[a]	Size Range/ Molecular Weight Range	Advantages	Limitations
Circular dichroism spectropolarimeter	~$100K	Structural characterization	NA	Tertiary and secondary structure analysis, applicable to soluble aggregates	Significant population of soluble aggregate to be present for analysis, difficult to analyze larger/visible aggregates
FTIR spectrometer	~$100K	Structural characterization	NA	Secondary structure analysis in liquid and solid state	Interference from water, significant population of aggregate to be present for analysis
Raman spectrometer	~$200K	Structural characterization	NA	Tertiary and secondary structure analysis, wide variety of samples could be tested, minimal interference from water	Significant population of aggregate to be present for analysis

[a]This column describes the aggregates that can be potentially detected/analyzed by a given technique. The actual determination would depend on each specific case and could be affected by several factors such as aggregates concentration, solution conditions, technique, and instrument operation.
NA, not applicable.

and by the potential effects on aggregates during sample preparation. This assumes that instrument availability and cost are not limiting. When selecting an analytical technique, one should be aware of the underlying theory of the measurement, and conclusions should be carefully drawn, keeping in consideration the sample preparation and the scope and limitations of a given technique. These aspects of the various analytical techniques utilized for detection and characterization of aggregates are discussed in the remainder of this chapter.

5.3.1 SEC

SEC, also referred to as gel filtration chromatography (when aqueous mobile phase is used) or molecular sieve chromatography, is considered the industrial workhorse for the detection and quantitation of most types of soluble aggregates in protein solutions. SEC is often used as a "lot release" method as it is often relatively simple to validate with high resolution, precision, and accuracy.

Detection of protein-based aggregates in SEC relies on the separation of macromolecules in a chromatographic column typically packed with silica-based microparticles of 5–10 μm in size.[54] The silica-based packed column acts as a molecular sieve toward various oligomeric states of protein, wherein, as the protein solution flows through the column, the lower-molecular-weight species diffuse into the pores of the particles as well as into the void space between the particles of the packed column and elute later in time. Larger-molecular-weight species, however, being unable to enter smaller pores, elute earlier. The concentration of the eluting protein is monitored using a spectrophotometric detector or a refractive index detector. In more recent years, the use of an in-line light scattering detector has increased in popularity to obtain an estimate of the absolute molecular weight of the eluting species (discussed later in text). In the absence of an absolute molecular weight determination, relative information about the molecular weight of the eluting species can be obtained by injecting proteins of known molecular weight as standards and by comparing the elution time of the sample protein with standard proteins under similar experimental conditions of mobile phase and column.

The principal advantage of SEC is its ease of use and high throughput in combination with high pressure liquid chromatography (HPLC) to monitor irreversible soluble aggregates. In addition, the mobile phase can be varied to characterize and monitor reversibility of the aggregates. Through selection of an appropriate column in terms of size of the packing particles and pore size, separation and detection of aggregates in the range of 5 to ~10^3 kDa can be achieved using SEC. Through the use of automated HPLC methods and quantitative integration of peaks of interest in the eluting chromatogram, SEC allows accurate and precise quantitation of soluble aggregates with high throughput. The method also requires minimal sample preparation, and often, the samples can be injected directly without any sample modification, except for occasional dilution. SEC is robust and easy to validate and is commonly used for lot release as well as for in-process testing toward quantification of soluble aggregates.

Studies reporting the use of SEC to monitor soluble irreversible aggregates of protein therapeutics are abundant in literature. A few examples include detection of aggregates in rhFVIII,[24,55] granulocyte colony-stimulating factor (G-CSF),[34] human growth hormone,[22] interferon gamma[56,57] and interferon alpha,[58] and, more recently, mAbs.[59] Although SEC is commonly used for the detection of irreversible soluble aggregates, in principle, SEC can be utilized to characterize protein self-association. In SEC, while irreversible aggregates can be resolved by choosing appropriate column and solution conditions, the behavior of reversibly associating species would depend on the equilibrium and the interaction of these species with column, thus affecting the profile of the eluting peaks.[60,61] Three cases could exist: slow interconversion between the two species, intermediate rates of interconversion, or rapid interconversion. In the case of slow interconversion, the column separates the associated species present in a solution on the timescale of the experiment, and these species elute as distinct peaks. In this scenario, the ratio of the monomer-to-dimer peak area changes as a function of concentration due to the different proportions of the two species present in the injection solution as defined by the dissociation constant, K_a (or K_d in the nomenclature of Chapter 2). In the case of rapidly equilibrating associating species, the associating species rapidly re-equilibrates following elution from the column. The dimer and the monomer are not resolved, the elution time will vary with concentration, and the peak contains a polydisperse molecular weight distribution. For the intermediate case, one would expect some separation as well as asymmetry in the eluting peak depending on K_a and initial concentration. The three cases could be distinguished by injecting solutions of different initial protein concentration and at various rates. For example, for rapid interconversion, flow rate does not affect the distribution of the associating species, whereas, for intermediate or slow interconversion, the flow rate will have a distinct effect.

In one example, short-column SEC has been utilized to characterize the association behavior of recombinant human growth hormone.[62] The method separated the monomer and dimer of pre-equilibrated solutions of different protein concentrations at fast flow rates through a short column and allowed estimation of the dissociation constant through consideration of the dilution of protein and subsequent dissociation upon elution. Through computer simulation of the moving boundaries of eluting peaks, the dissociation constant, the rate of dissociation, and the rate of association were determined. In another example, Moore et al. used SEC as the primary method to understand the kinetics of dissociation of a self-associating mAb, recombinant human monoclonal antibody-vascular endothelial growth factor (rhuMAb-VEGF), under various solution conditions of pH, salt, and temperature.[26] This antibody reversibly formed dimer under certain conditions, and the monomer–dimer equilibrium association constant was derived through assessment of the relative proportions of the dimer and the monomer. In this particular example, the antibody also formed an irreversible dimer, the levels of which typically increased upon storage under accelerated conditions. Since the reversible dimer coeluted with the irreversible dimer, and the implications of the reversible dimer on the *in vivo* efficacy and safety were unknown, a more rigorous SEC method was developed for the characterization of rhuMAb VEGF size heterogeneity during its product development phase.[16]

While easy to use and amenable to high throughput, SEC has its limitations and these have been discussed in the recent literature.[63,64] Often, the mobile phase needs to be altered compared with the formulation in terms of its composition, pH, temperature, and so on, to facilitate the elution of the solute and to obtain good resolution among different eluting peaks. This modification may alter the concentration and state of the aggregates and overall size distribution of protein that might be present in the original solution. The sample, once injected, is diluted into the mobile phase and typically elutes at a lower concentration than that of the original sample. Therefore, aggregates that can dissociate upon dilution but are present in the original sample will not be accurately quantified or even detected upon elution.[65] One of the major drawbacks of SEC is the potential interaction of protein or protein aggregates with the column packing.[66,67] Aggregates may adsorb to the column through electrostatic/hydrophobic interactions and may not elute with 100% recovery, leading to underestimation of their levels. Large enough aggregates could be retained by the column and would not be detectable following elution. To overcome these protein–column interactions, the mobile phase composition is typically modified extensively and is often substantially different from the protein sample composition.[67] Although this approach may minimize the protein–column interaction, it introduces another risk of modifying protein aggregates, either by dissociation or formation, as a result of exposure to the modified mobile phase composition. SEC is typically unable to differentiate between the covalent-linked and non-covalent-linked aggregates. However, in some cases, introduction of a denaturant in the mobile phase such as SDS, or pretreatment of the sample with reducing agents may help in differentiating between the levels of disulfide-linked covalent aggregates and other non-covalently linked aggregates.

5.3.2 Light Scattering

The technique of light scattering can be categorized into "static light scattering" (SLS) and "dynamic light scattering" (DLS). The two techniques provide very different characteristics of the aggregates in solution and hence are discussed separately.

5.3.2.1 SLS The technique of SLS involves measurement of the excess time-averaged intensity of the scattered monochromatic light from particles in a solution at various angles. The time-averaged intensity of the scattered light depends on the molecular weight, concentration, shape, second virial coefficient, B, of the interacting particles, wavelength of light, and the angle of measurement. The SLS data are processed by the classical method of Zimm[68] using the equation

$$\frac{Kc}{R_\theta} = \left(\frac{1}{M} + 2Bc\right)\left[1 + \frac{16\pi^2 R_g^2 \sin^2\left(\theta/2\right)}{3\lambda^2}\right], \tag{5.1}$$

where R_θ is the measured excess Rayleigh's ratio of protein solution and can be measured as a function of concentration, c, at multiple angles, θ. M is the weight-average molecular weight, R_g is the radius of gyration, and λ is the wavelength of the incident light.

The constant K is defined as

$$K = \frac{4\pi^2 (n\, dn/dc)^2}{\lambda^4 N_A}, \tag{5.2}$$

where n is the refractive index of the solvent, dn/dc is the refractive index increment of the solute in a given solution, and N_A is the Avogadro's number. A "Zimm plot" is obtained by plotting Kc/R_θ versus $[\sin^2(\theta/2) + kc]$, where k is an arbitrary constant. The molecular weight is obtained from the intercept following extrapolation to zero concentration, R_g is obtained from the slope of the line extrapolated to zero concentration, and B is determined from the slope of the concentration data, when extrapolated to zero angle. When the data are collected at multiple angles, the technique is commonly referred to as multi-angle laser light scattering or "MALLS."

While MALLS provides detailed and more accurate representation of higher molecular aggregates, in most cases of unaggregated proteins or smaller protein aggregates (dimers, trimer, etc.), isotropic scattering can be assumed. This is because proteins do not typically exhibit an angular scattering dependence in the wavelength region commonly employed for light scattering measurements (400–800 nm), that is, when the R_g of the scatterer is less than 1/20th of the wavelength of the incident light. With this assumption, Equation 5.1 can be simplified and reduced to measurement at a single angle, such as 90°. This simplified equation is represented as

$$\frac{Kc}{R_\theta} = \frac{1}{M_w} + Bc. \tag{5.3}$$

For generating Kc/R_θ versus c data, the measurements are usually carried out at 90° to minimize scattering contributions from dust particles. The inverse of the intercept provides the molecular weight, and the slope is related to the second virial coefficient, B.

SLS can be used to measure molecular weights in the range of 5 kDa to several million daltons and provides information about the radius of gyration of the molecules. While SLS was first developed to be used in a "batch" mode, this type of measurement provides little information about aggregates in a solution containing both lower-molecular-weight species and high-molecular-weight species. This is primarily because of the fact that the intensity recorded by SLS measurements is a weight-average intensity from all species present in a solution. A more useful and widely used mode of SLS for characterization of protein aggregates is in flow mode, wherein the SLS detector is combined online with SEC-HPLC along with ultraviolet/refractive index (UV/RI) detection.[69] This type of setting allows the separation of aggregates through SEC, and the SLS detector allows the accurate and absolute determination of the average molecular weight of the eluting species. The data can be collected in either "MALLS" mode or single-angle mode. The size limit of aggregates that can be detected using this method is limited by those that elute from the SEC column and is typically in the molecular weight range of 5–10^3 kDa.

Typically, the type of aggregates characterized by online SLS combined with SEC are similar to those characterized by SEC alone, that is, irreversible soluble aggregates. More recently, the two methodologies have also been utilized to characterize reversible self-association. In such techniques, the protein solution is introduced in the detector either through a syringe pump or through an HPLC system employing an SEC column. The former technique has been developed by Attri and Minton and has been used to characterize homo- as well as hetero-association of proteins.[44,70] The latter technique has been reported by us for the characterization of self-association using β-lactoglobulin as the model associating protein.[71,72] In the technique of Attri and Minton, a dual syringe pump is utilized to generate different concentrations of protein by varying the flow rate of the two syringes followed by in-flow mixing of the two solutions. The solution is allowed to equilibrate and then analyzed through a concentration and a light scattering detector. The concentration and scattering data thus generated are used to characterize the self-association behavior of a given protein. This method, if used appropriately, can be used to characterize both the association and dissociation behavior of a self-associating species and can also be applied to study association of two different protein species. However, caution must be exerted for association/dissociation kinetics that are slow enough compared with the duration of experiment and mixing allowed between each dilution. Furthermore, the sample must be pure and free from macromolecular impurities to prevent interference by irreversible aggregates and non-proteinaceous material.

Recently, we reported the design and use of a dual-source, dual-detector system combining the concentration detector and the light scattering detector in a single cell that can be used in conjunction with SEC-HPLC to characterize protein–protein interactions.[71] A concentration range required to generate the Debye plot can be obtained from a single injection and through the utilization of the concentration distribution in the eluting peak. Each data point on the eluting peak can be converted to concentration and light scattering intensity using signals from the appropriate detectors, and these data can be utilized to investigate the self-association behavior as well as non-ideality of a given protein in a solution.

The advantages of this technique include the removal of macromolecular impurities and separation of irreversible aggregates by the SEC column, short duration of analysis, and availability of data from a single injection. While this technique is well suited for rapidly equilibrating systems, in a slowly equilibrating system, the monomer and oligomer peaks could appear as separate peaks. A dependence of the chromatographic profile and the distribution of different species as a function of flow rate and concentration will help distinguish between slow and fast association equilibria.

Although the technique of light scattering in "batch mode" involves direct measurement of scattered light intensity and is non-invasive, the use of this technique in aqueous solutions is often challenging and requires tedious cleaning procedures to collect reproducible data. Due to enhanced variation and limited reproducibility, experiments are both labor and time intensive in order to minimize contribution from dust particles from the samples and sample tubes. While SLS in batch mode may be challenging for low-molecular-weight proteins and at low concentrations, the technique typically provides more reproducible data for high-molecular-weight proteins and at relatively high concentrations. As mentioned earlier, batch-mode SLS measurements cannot resolve aggregates from the native protein in a solution. The limitations of online SLS with SEC are similar to those of SEC itself. In addition, the light scattering detector may not provide a reproducible and accurate mass determination at low concentrations of low-molecular-weight species due to higher background scattering. The samples and the mobile phase should also be maintained dust and particulate free, requiring special sample preparation protocols.

5.3.2.2 DLS The technique of DLS, also known as quasi-elastic light scattering (QELS) or photon correlation spectroscopy (PCS), involves the measurement of the fluctuations in the intensity of the scattered light resulting due to continuous random motion of molecules/particles in solution over a period of time.[73] Since the intensity fluctuations contain information about the random motion of molecules/particles, DLS measurements provide information about the diffusion coefficient of the particles. For non-interacting dilute solutions and assuming spherical particles, the diffusion coefficient, D, thus obtained is related to the hydrodynamic radius, R_H, of the particles in solution through the Stokes–Einstein equation

$$D = \frac{k_B T}{6\pi\eta R_H}, \tag{5.4}$$

where k_B is the Boltzmann constant, T is the absolute temperature, and η is the solution viscosity.

DLS can be used to measure aggregates in the size range of 1 nm to 10 μm.[74] A principle advantage of DLS is minimal sample preparation. Samples typically can be placed in a clear glass tube or cuvette, and direct, non-invasive measurements can be made to assess the size distribution. Since the data are based on the collection of scattered light intensity, the size distribution typically obtained is intensity weighted. Most software allows the conversion of an intensity-weighted size distribution to a weight-based, volume-based, or number-based distribution through the use of appropriate equations, and making assumptions about particle shape. Therefore, a broad range of information could be obtained from a single experiment. In addition, since the measurement is done directly on the sample, DLS is capable of following aggregation kinetics as the reaction proceeds.[75–77] If the size distributions of various species in solution are well separated, DLS can provide information about the relative percentage of species in terms of weight-based, volume distribution or number-based distribution.[66,74] With further advances in data collection technology, instruments are now available that can perform DLS measurements in a high-throughput mode using a 96-well format. While, traditionally, DLS measurements are carried out in the configuration where the laser enters the sample and the scattered light is collected at a 90° angle, instruments are now available that allow analysis of the scattered light at higher angles, or in the "back-scattering" mode.[78] This configuration allows DLS measurements to be performed on relatively high-concentration protein samples or turbid samples without internal or secondary scattering effects.

DLS studies carried out in "batch mode" suffer from similar limitations as those of SLS, that is, interference from dust particles, air bubbles, and other larger impurities. Since the data are intensity weighted, larger particles tend to affect size distribution analysis of the aggregates/particulates. Although this can be avoided using sample filtration, one also runs into the risk of removing particulates during filtration. In a heterogeneous mixture, the capability of DLS to resolve various size distributions is dependent on the size range of interest. For example, in the case of proteins, where the monomer radius typically has a range between 1 and 10 nm, DLS may not be able to resolve smaller oligomers such as dimers and trimers from the monomer.[66] Turbid samples or samples containing a relatively high concentration of protein may cause a loss in the intensity of scattered light due to internal or secondary scattering effects or divergence of the incoming laser beam, thereby providing erratic results. Furthermore, the models typically used for data quantitation in DLS studies assume spherical shape; therefore, caution should be exerted in the data interpretation for non-spherical particles. In addition, the models used to convert intensity-based distributions to other types of distributions have assumptions built in, and therefore, such data should be interpreted with caution. For accurate size measurements, the viscosity of the solvent should be known and the temperature should be controlled. Due to these reasons, DLS is more commonly used as a characterization method and as a research tool.

5.3.3 Analytical Ultracentrifugation (AUC)

AUC provides information on the size of a macromolecule or its aggregated state through studying its sedimentation behavior in a solution when exposed to a high centrifugal speed/force. The movement of the molecule of interest under centrifugal force is monitored by interference optics, absorption spectroscopy, or fluorescence spectroscopy. AUC is typically run in two modes: sedimentation velocity (SV-AUC), which utilizes relatively high speeds, and sedimentation equilibrium (SE-AUC), which utilizes a somewhat lower centrifugal speed.[79,80]

5.3.3.1 Sedimentation Velocity In sedimentation velocity mode (SV-AUC), the molecules sediment toward the bottom of the cell due to the relatively high centrifuge speeds. They rapidly reach a terminal, constant velocity, which reflects a balance of the centrifugal force on the molecule (the buoyant mass) and the frictional forces. The net sedimentation behavior of the molecule can be described by the Svedberg equation[80,81]

$$s = \frac{u}{\omega^2 r} = \frac{M(1-\bar{v}\rho)}{Nf} = \frac{MD(1-\bar{v}\rho)}{RT}, \tag{5.5}$$

where s is the apparent sedimentation coefficient, u is the observed radial velocity of the molecule, ω is the centrifugation speed, r is the radial position, M is the molecular mass, $\bar{v}$ is the partial specific volume of the solute, ρ is the solvent density, N is the Avogadro's number, f is the friction coefficient, and R is the universal gas constant. The sedimentation coefficient (s) is also related to the diffusion in the ultracentrifuge cell through the following equation

$$J = s\omega^2 rc - D\frac{\partial c}{\partial r}, \tag{5.6}$$

where J is the flux of the solute, c is the solute concentration, D is the diffusion coefficient, and $\frac{\partial c}{\partial r}$ is the concentration gradient. In a typical SV-AUC experiment, a protein solution is subjected to sedimentation forces at high rotational speeds, and the movement of the protein boundary, relative to the position of the meniscus, is monitored by detection methods such as interference optics, absorption, or fluorescence spectroscopy.[82] The molecules move during centrifugation at a rate proportional to their s value, which can be determined by measuring the rate of movement of the boundary midpoint. The design of the typical sample compartment (sector) leads to radial dilution, which is taken into consideration for the complete data analysis. SV-AUC measurements therefore provide information about s and D and, in some cases, M of the molecule under study.

During a typical SV-AUC run, a large number of data points are collected and subject to advanced fitting algorithms to extract information about the s values of various species and their respective amounts/percentages. The experiments are typically carried out under well-controlled conditions of temperature, rotor type, and rotor speed. Advanced data analysis tools are now available to deconvolute the s values for a mixture of various molecular weight species present in a solution. Some of the commonly used methods include the determination of the integral sedimentation coefficient distribution $G(s)$, the time derivative method of obtaining dc/dt, and the recently developed method by Schuck et al., which models the moving boundaries using the underlying transport equation, known as the Lamm equation.[65,80,81,83] The latter method utilizes the SEDFIT software, which has an improved resolution to cover a broad range of size distribution since it explicitly corrects the broadening from diffusion by the numerical methods, and many of the scans can be used in the analysis.

5.3.3.2 Sedimentation Equilibrium In a typical sedimentation equilibrium study, relatively slow centrifugation speeds are employed, and the protein concentration is measured as a function of the radial distance within the centrifugal cell. At equilibrium, there is a balance in the flux of sedimenting molecules (driven by the centrifugal force) and diffusing molecules (driven by the concentration gradient). For an ideal, non-interacting, single-component system, the concentration gradient as a function of radial distance can be represented as[79,84]

$$\frac{d\ln c}{d(r^2)} = \frac{\omega^2(1-\bar{v}\rho)}{2RT} M_{\mathrm{WT}}, \tag{5.7}$$

or upon integration as

$$c(r) = c(a)\exp\left\lfloor AM_{\mathrm{wr}}(r^2 - a^2)\right\rfloor \tag{5.8}$$

where

$$A = \left[(1-\bar{v}\rho)\omega^2/2RT\right] \tag{5.9}$$

and $c(r)$ represents the concentration of the protein at a distance r, $c(a)$ is the concentration at a reference position, a is the radial distance of the reference, ω is the angular velocity of the rotor (radians per second), R is the gas constant, T is the temperature, ρ is the solution density, $\bar{v}$ is the partial specific volume of the solute, and M_{wr} is the weight-average molecular weight of the species. The above equation describes the concentration gradient at equilibrium over a distance r in a centrifugation cell. For a single, non-associating solute, assuming that partial specific volume of the species does not vary with c, the molecular weight can be determined from the slope of ln c versus r^2.

One of the principle applications of SE-AUC is the characterization of associating–dissociating systems and, therefore, the characterization of participating oligomeric species. If the sample is ideal and monodisperse, a plot of ln c versus r^2 will be linear (or M_{app} vs. c or r^2 will be constant). However, these profiles become non-linear if the system is more complex, that is, exhibits non-ideality and/or self-association. For an associating species, a series of association constants can be defined as

$$K_i = c_i/(c_1)^{L(i)}, \tag{5.10}$$

where K_i is the association constant between the monomer and the ith aggregate, $L(i)$ is the degree of association for the ith aggregate, and $K_1 = 1$. Assuming dilute solutions such that the higher virial coefficients can be ignored, the total concentration at a radial position r can be written as[84]

$$c_r = \sum_{i=1}^{n+1} c_{i0} e^{L(i)\left[AM_1\left(r^2 - r_0^2\right) - BM_1\left(c_r - c_{r0}\right)\right]}, \tag{5.11}$$

where r_0 is the radial reference position, c_{r0} is the protein concentration at reference position, and B is the second virial coefficient. At the reference radial position, r_0, the total concentration can be written as

$$c_{r0} = \sum_{i=1}^{n+1} e^{\ln K_i + L(i)\ln c_{10}}, \tag{5.12}$$

where c_{10} is the concentration of the monomer at the reference radial position. Equations 5.10–5.12 represent generalized forms of the sedimentation equilibrium equation to analyze a wide variety of discrete self-association systems and can be modified for each specific system.

As mentioned earlier, SV-AUC is primarily useful to characterize the sedimentation coefficients, s, of various species in a solution, whereas SE-AUC can be used to obtain information about reversibly associating species, their respective molecular weights, the stoichiometry of association, and the association constants. AUC, in general, offers advantages over other methods, especially SEC and field-flow fractionation (FFF, discussed later) such as minimal sample modification, the wide range of solution conditions that can be studied, often including the formulation buffer, and direct measurement of protein aggregates in a sample.[59,65,74] The technique of AUC is suitable for soluble aggregates and, depending on the mode of analysis, can be used to distinguish between reversible and irreversible aggregates. Specially, the use of SV-AUC has recently gained interest as an orthogonal method to confirm results obtained from SEC or FFF, and a few reports have been published recently comparing results from the three techniques. For example, Liu et al. reviewed and compared the techniques of AUC and FFF for the detection and quantitation of aggregates in protein samples.[65] In another example, Gabrielson et al. used SV-AUC, SEC, FFF, and DLS to characterize aggregation of a recombinant mAb.[59] The authors demonstrated that SV-AUC and FFF provided better estimates and recovery of the aggregates compared with SEC and recommended the former techniques as orthogonal methods for characterization of aggregates. With the use of an appropriate detection system, a wide range of concentrations can be studied using SV-AUC. The recent development of a fluorescence detection system has further enabled AUC studies of highly concentrated protein solutions.[85] In such cases, a tracer protein molecule, which is labeled with a fluorescence probe, is incorporated into the protein solution, and the sedimentation behavior of the labeled molecules is followed using the fluorescence detection system. The underlying assumption is that the labeled protein behaves similar to the unlabeled molecules, and the labeling procedure itself has minimal impact on the protein's behavior. This methodology utilizes special AUC equipment that has a fluorescence detection system. If such equipment is not accessible, protein association at high concentration could also be characterized by utilizing preparative analytical centrifugation as reported by Liu et al.[86]

While the use of SV-AUC has increased over the past few years for the detection of aggregates in biopharmaceuticals, the technique still requires special training and expertise to perform experiments and analyze the complex data output. In addition, the data could be influenced by the presence of cosolutes, high buffer concentration, and non-ideality of the system.[87] Furthermore, the reproducibility and precision of the data can be affected by inappropriate instrument configuration and/or by artifacts introduced during data collection and data interpretation.[88,89] Unlike SEC or FFF, the technique of SV-AUC is not considered a high-throughput methodology and only a limited number of samples can be analyzed within a given experimental time frame.

5.3.4 FFF

The technique of FFF utilizes the effect of an external field on molecules under flow to cause a size-based separation. The external field could be flow, electrical, thermal, sedimentation, or magnetic.[2,65,90] The field is typically applied perpendicular to the direction of flow in thin channels (50–300 μm), and the separation occurs due to the movement of solution components into different stream laminae within these thin channels. Different laminar velocities of the laminae within the channel (typically parabolic, similar to the laminar flow of a liquid in a thin tube) then drive the separation of components as the solvent elutes from the channels. Due to the versatility in the types of fields that can be applied and the thickness of the channels, FFF can be applied for a wide range of size of molecules/aggregates from 1 nm to as high as several microns.

Among the various fields applied, flow FFF (F4) has gained considerable usage in the field of biopharmaceuticals for separation of protein aggregates.[59,65,91] In F4, a crossflow stream is applied perpendicular to the flow inside the channel. The channel flow drives the components to the other end of the channel, whereas the crossflow drives the components toward the wall of the channel. When one wall of the channel is made of a semipermeable ultrafiltration type of membrane allowing only solvent and small molecules to pass, the technique is also known as asymmetric F4, or AF4. In this type of assembly, the applied crossflow drives molecules toward the semipermeable membrane. Low-molecular-weight molecules, however, diffuse more readily away from the membrane, whereas larger-molecular-weight molecules tend to be closer to the membrane. Due to unequal velocities of the laminae in the channel, the velocity being fastest in the center of the channel, low-molecular-weight species elute first and the larger-molecular-weight species elute later. As in SEC, the eluted species are detected by UV, fluorescence, or a refractive index detector. The crossflow stream velocity tends to reduce the velocity of the liquid flowing perpendicular inside the channel. This effect is typically minimized by the use of various geometries of the channel such as the very thin trapezoidal-shaped channels. However, thin channels also result in the reduction of velocity gradient between the center and the side of the channel thereby affecting the resolution of the technique.

Since the AF4 technique uses a simple hollow channel, where the liquid sample in the longitudinal flow direction is subject to a crossflow stream, protein–matrix interactions are greatly reduced compared with those present in SEC. This further allows detection and quantitation of larger-size aggregates, which otherwise may be retained and therefore not elute through an SEC column. A few recent reports compare the technique of AF4 to SEC and SV-AUC for the detection and quantitation of aggregates in samples of mAbs.[59,65] These studies demonstrate the usefulness of AF4 for the detection of larger protein aggregates; however, there is a need for more studies to be able to validate this technique for routine analysis of aggregates in protein samples.

Although the usage of AF4 is becoming more popular, there are certain issues that need to be considered while employing this technique for accurate and precise determination of protein aggregates. While the separation is carried out in a hollow channel with a semipermeable membrane on one side of the wall, the separation may be affected by protein–membrane interactions, especially under high crossflow conditions. To prevent such interactions, modifications in the composition of the crossflow fluid may be necessary, which may then alter the size distribution of injected samples, either via inducing aggregation or through dissociation of aggregates. This issue can be alleviated by choice of a non-interacting membrane such as regenerated cellulose. The sample may also experience a concentration change following injection.

Dilution may occur when the sample mixes with the fluid medium, and concentration occurs as various sized species are focused into the lamellar streamlines inside the channel. Such concentration changes may affect quantitation of reversible aggregates. Although the technique of FFF can be used for a wide range of sizes, the technique of AF4 is still limited to soluble aggregates.

5.3.5 Electrophoresis

The technique of electrophoresis utilizes an electric field to separate macromolecules based on their charge and molecular weight. The overall separation, in fact, depends on several factors including the strength of the field; the net charge, size, and shape of the macromolecule; the ionic strength and viscosity of the separation medium; and the composition of the separation medium. Two common electrophoresis-based techniques used for the detection and characterization of protein aggregates are sodium dodecyl sulfate–polyacrylamide gel electrophoresis (SDS-PAGE) and capillary electrophoresis utilizing sodium dodecyl sulfate gel (CE-SDS).

5.3.5.1 SDS-PAGE Gel electrophoresis utilizing polyacrylamide-based gels has been one of the most popular techniques for several decades to study molecular weight-based separation of protein molecules.[92–94] This technique typically utilizes a cast gel in the shape of a slab made of a defined concentration of polyacrylamide in a buffer. While, traditionally, the gels are created in the lab using primarily acrylamide and a small concentration of a cross-linker such as bisacrylamide in a buffer, precast ready-to-use gels including gradient gels are now available from various manufacturers, making the technique more efficient and reproducible. The protein sample is added to one end of the gel, and protein migration is initiated between a cathode and an anode through the use of an electric field. In the absence of any denaturant, the migration depends on protein charge and protein size/shape. Under these conditions, also referred to as "native gel electrophoresis," interpretation of the migration pattern may become complex since the effect of net charge and shape and size need to be deconvoluted to draw conclusions from the data. To overcome these limitations, gel electrophoresis is often carried out on protein samples pretreated with SDS, and this method is commonly known as SDS-PAGE. SDS is an anionic denaturant and binds to proteins with typically 1.4 g of SDS bound to 1 g of the protein. In the presence of excess binding of SDS to denatured protein, often facilitated with the help of heating, the protein acquires a net negative charge arising from SDS binding and typically assumes a somewhat elongated shape because of intramolecular charge repulsion. When such sample is applied to a polyacrylamide gel, the separation of various molecular weight species in a sample under a given electrical field depends largely on the molecular weight of each species.

Following separation of various molecular weight species, the protein bands are stained to provide a visual mode of detection and, more recently, quantitation. The most common method is to stain with Coomassie Brilliant Blue R-250 dye,[95] which can detect protein levels as low as 1–10 μg/band. For lower protein levels, cupric-silver staining in commonly employed that can detect protein levels as low as 10 ng/band.[96,97] The use of fluorescence staining using Sypro® Ruby dye has also been reported for sensitive detection of proteins in gels.[98] In general, accurate quantitation using Coomassie blue staining is more reliable than by silver staining. It is common to run a set of standards of proteins of known molecular weight to obtain estimates of the molecular weight of the sample protein. Instruments are now available that are capable of quantifying protein bands in an SDS-PAGE gel using densitometry and appropriate software analysis.

SDS-PAGE is commonly applied to study the covalent versus non-covalent nature of the aggregates. This technique can provide separation between various protein-related species in the molecular weight range of 5–500 kDa. Since treatment with SDS would eliminate any non-covalent-linked species, that is, those held together only through physical interactions, the bands that appear of greater molecular weight than the monomeric protein are considered to be linked through covalent linkages. When dithiothreitol (DTT) or 2-mercaptoethanol is used along with SDS during sample preparation, the method is often referred to as "reducing" SDS-PAGE gel and is used to disrupt disulfide-linked species in a sample. This method is typically used to further confirm the nature of the covalent-linked aggregates. Therefore, by using the non-reducing and reducing conditions, SDS-PAGE offers an attractive method to differentiate between the disulfide-linked and other covalently linked aggregates.

While native gel electrophoresis in certain cases can provide information about all of the soluble aggregate species in a sample, it is often cumbersome to interpret the data and achieve reasonable separation, since the separation is affected by charge, size, and shape. On the other hand, while SDS-PAGE overcomes some of these limitations, it does not allow the estimation of the non-covalent aggregates. Furthermore, the dissociation and denaturation of certain β-sheet proteins and disulfide-linked proteins may not be complete and depends on the SDS concentration and temperature employed during sample preparation. Also, during heating itself, there is a potential of either breaking up of the aggregates or even forming aggregates. Such considerations should be kept in mind during analysis and interpretation of the data obtained from SDS-PAGE.

5.3.5.2 CE-SDS In principle, the technique of CE-SDS differs from SDS-PAGE in that the separation occurs in a fine capillary. The sample preparation is similar to that of SDS-PAGE, either under non-reducing or reducing conditions. The separation takes place in a gel-filled capillary, which is typically cleaned and rinsed between each injection. The separated species are monitored by UV detection, which provides sensitivity similar to that of Coomassie stain SDS-PAGE or by laser-induced fluorescence (LIF) detection, where the sample is treated with a fluorescent dye, such as 5-carboxytetramethyl rhodamine succinimidyl ester (5-TAMRA, SE), during sample preparation.[99–101] The levels of detection obtained with LIF are similar to those obtained by using silver stain SDS-PAGE.

Because of its defined diameter capillary, the lack of a slab gel, and the automated reuse of the capillary, CE-SDS allows one to easily analyze multiple samples in a chromatographic-like manner and provides reproducibility, resolution, robustness, and speed. The CE-SDS method has been most widely used for the characterization of aggregates and fragments in mAbs and is now routinely used for in-process testing and for lot release testing. Certain manufacturers have converted the concept of CE-SDS to "lab on a chip" by utilizing microfluidics. These chips utilize microfabricated channels etched onto glass or plastic, and separation is achieved electrophoretically.[102,103] Detection is accomplished by incorporating a fluorescent dye in the polymer solution, which fills the microchannels and binds protein and SDS, both protein bound and free. The background fluorescence from dye associated with non-protein-bound SDS micelles is lowered by a controlled backflow of destain solution, migrating electrophoretically, removing free SDS micelles, and keeping the SDS concentration below its critical micelle concentration. Some systems may in fact utilize prelabeled protein samples. Chip-based CE-SDS offers advantages such as speed, relatively short sample preparation times, and ease of application and use.

The primary limitations of CE-SDS are the same as those noted for SDS-PAGE; that is, it disrupts the non-covalent aggregates and may affect the level of aggregates during sample preparation itself. The use of this technique is also limited to soluble aggregates, and the technique is also sample invasive, thus preventing the collection of separated bands for further characterization.

5.3.6 Turbidimetry and Nephelometry

Both turbidimetry and nephelometry are scattering-based techniques and directly or indirectly measure the extent of the light scattered by a solution.[104] Typically, turbidimetry refers to measuring the loss in the intensity of incident light, as a result of scattering, as estimated by recording the optical density (OD) or absorbance using a standard spectrophotometer using wavelengths at which the protein does not absorb. Therefore, a positive OD value or an absorbance value is indicative of the presence of scattering species in the sample. Nephelometry typically refers to measuring the scattered light intensity directly, most commonly at right angles to the incident light. Both techniques only allude to the presence of a scattering species when compared against the vehicle or control and do not provide information about the size or shape of the scattering species.

The term "turbidity" is generally used to describe the presence of cloudiness or haze in a solution, and the mild form of turbidity is often referred to as "opalescence." In its strict definition, opalescence simply arises from scattering, often multiple scattering events, of visible light by elements present in a solution. The scattering elements could be protein molecules, which intrinsically scatter visible light and may give rise to opalescence when present at moderate to high concentrations.[105,106] In certain cases, the opalescence is termed as "critical opalescence," which is a result of scattering of light from substantial fluctuations in concentration or density (and hence in refractive index) for a system that approaches equilibrium or its critical point, but has not yet completely achieved phase separation or complete immiscibility.[107–109] This phenomenon is typically observed for binary mixtures such as liquid–liquid or liquid–vapor systems. Recently, the phenomenon of critical opalescence has been reported for a mAb solution.[110]

Since turbidity arises due to the net scattered light, it typically depends on the sample path length, protein concentration, and size of the scatterer (protein/aggregates/particles). An increase in turbidity for the same path length can occur due to an increase in the size of the scatterers or an increase in the concentration of scatterers or both. In protein solutions, turbidity is typically measured using a UV-visible spectrophotometer at wavelengths in the range of 320–800 nm (protein non-absorbing). Since turbidity is a result of scattering, the sensitivity of turbidity measurements increases at lower wavelengths due to inverse dependence of intensity of scattered light on the fourth power of the wavelength of the incident light. However, certain protein solutions may exhibit color-induced absorbance at lower wavelengths or intrinsically higher turbidity, therefore necessitating turbidity measurements at higher wavelengths.

Relative increases in the turbidity upon storage or upon solution perturbation is often cited as an indication of aggregation in protein biopharmaceuticals, particularly since methods such as SEC of AF4 may not detect larger aggregates that contribute toward turbidity. However, during formulation screening and preparation, the initial turbidity of different formulations should be cautiously compared, especially for highly concentrated antibody solutions. This is because different solution conditions may affect the intrinsic Rayleigh scattering from protein molecules and therefore could significantly contribute and alter turbidity values even at similar protein concentrations, when measured by OD or absorbance.[105] Opalescence or turbidity could also be a precursor to liquid–liquid phase separation.[110] Caution should be taken when comparing turbidity from two different spectrophotometers since the design of the cuvette holding cell with respect to its distance from the detector could give rise to different OD/absorbance values for the same sample.

While there is no standard definition to describe turbidity or opalescence of a protein-based parenteral drug product in the United States Pharmacopeia (USP), the European Pharmacopoeia (Ph. Eur.) Section 2.2.1 defines visual clarity and describes standard levels of turbidity relative to water.[111] This

TABLE 5.4. Reference Suspensions for Clarity and Degree of Opalescence of Liquids Test as per Ph. Eur. 2.2.1[111]

Reference Suspensions	Opalescent Values, NTU	Description
I	3	≤Ref I = clear
II	6	≤Ref II = slightly opalescent
III	18	≤Ref III = opalescent
IV	30	≤Ref IV = highly opalescent

allows a standardized and reproducible assignment of turbidity to parenterals compared with that obtained from a visual inspection. The method described in Ph. Eur. 2.2.1 compares the turbidity of a protein sample against formazin reference suspensions of defined turbidity. The level of turbidity is then assigned based on the formazin turbidity unit (FTU) reading as per Table 5.4. The turbidity assessment of the sample against reference suspensions could be achieved either visually or by the use of a turbidimeter or nephelometer. A typical turbidimeter/nephelometer consists of a light source such as a tungsten filament lamp and a light detector at 90° angle. The turbidity measurement output is often recorded as nephelometric turbidity units (NTUs), if measurements are performed using a calibrated nephelometer or as FTU if formazin suspensions are used as a reference.

As mentioned earlier, turbidity measurements are advantageous for comparative assessment of samples, especially during prolonged storage as an indicator of protein aggregation. When present in freshly prepared solutions, turbidity may suggest insolubility, but could arise intrinsically as a result of high protein concentration and may or may not indicate that protein aggregation has occurred. Turbidity measurements, in general, do not provide any information on aggregate size and shape.

5.3.7 Analytical Techniques for Subvisible Particles

Subvisible particles are typically defined in the size range of 0.1–100 µm. The lower end of this size range of particles is not typically quantifiable by techniques used to detect soluble aggregates, such as SEC and AF4, and on the other hand, the upper end of the range may not be detectable by visual inspection. The techniques that have been traditionally utilized for the detection of such particles are based on light obscuration and microscopy.[112] These techniques are in fact listed in the pharmacopoeias with defined limits around particles greater than 10 µm in size. It should be noted that the historical relevance of these limits is primarily driven through contamination by non-product-related foreign contaminants that might also lead to *in vivo* capillary occlusion. Additionally, a non-pharmacopoeial technique that can be used for detection of subvisible particles is the Coulter Counter. All these techniques are discussed below.

5.3.7.1 Light Obscuration Particle Count Test The light obscuration method is based on the simple principle that when a particle passes through the path of a light beam (or a laser beam), the extent of the loss in the light intensity is proportional to the area and hence size of the particle.[112] Through proper design of the flow cell, which typically includes capillaries, the particles are assumed to enter in a single file arrangement as the sample passes through the flow channel. The decrease in the intensity as a result of scattering/reflection of the incoming light leads to a decreased magnitude of the pulse on the detector, which is then converted to the particle diameter. Newer instruments typically use lasers as the light source for precise measurement of the particle size. The collected data are then processed using sizing algorithms to generate the size histogram and other parameters. Theoretically, the particle size that can be measured using light obscuration lies in the range of 1–600 μm; however, the data should be used with caution at the extremes of the size range with respect to the precision and accuracy of the instrument.

The light obscuration method is one of the most widely used methods for quantitation of subvisible particles in parenterals. This method is one of the two methods described in USP, European Pharmacopoeia (EP), and Japanese Pharmacopoeia (JP) to fulfill the particulate testing requirements for large volume (≥100 mL) as well as small volume injectables (<100 mL). Such requirements and acceptance criteria form part of the USP chapter <788>, Ph. Eur. 2.9.19, and JP and are described in Table 5.5.[113–115] Similar requirements for ophthalmic solutions form part of the USP <789> and are also included in Table 5.5.[116]

While the light obscuration method works well with aggregates having a distinct refractive index compared with the bulk solution, the method may not accurately detect aggregates, which are somewhat translucent. Additionally, air bubbles may generate false positives toward particle counting. Furthermore, with concentrated protein samples, dilution may be necessary prior to analysis, either because of background arising from high concentration of proteins or too many particles could enter the capillary causing blockage. With high concentrations of particles, detector saturation may occur, thereby leading to an underestimation of particle count and distribution. Also, the pharmacopoeial method describes using 25 mL of the sample volume for analysis, which could be a limitation for high-concentration protein solutions, especially during the early stages of the product development. Lastly, the method cannot differentiate between particles of protein origin and those that are non-proteinaceous in nature.

5.3.7.2 Microscopy The second method described in the pharmacopoeias (USP, Ph. Eur., JP) is based on microscopy for the measurement of particle size and counting particles.[113–115] The test involves filtration of the sample onto a membrane filter, followed by analysis of the collected particles by a binocular microscope. A circular diameter graticule is used to count particles and to

TABLE 5.5. Criteria for Passing the Particulate Matter in Injections Test Listed in Pharmacopoeias (USP, Ph. Eur., JP)[113–115] and Particulate Matter in Ophthalmic Solutions Test Listed in USP[116]

Container Volume	Method	Limits
≤100 mL (small volume parenterals)	Light obscuration	Does not exceed 6000 particles/container of ≥10 μm
		Does not exceed 600 particles/container ≥25 μm
	Microscopy	Does not exceed 3000 particles/container of ≥10 μm
		Does not exceed 300 particles/container of ≥25 μm
>100 mL (large volume parenterals)	Light obscuration	Does not exceed 25 particles/container of ≥10 μm
		Does not exceed 3 particles/container ≥25 μm
	Microscopy	Does not exceed 12 particles/container of ≥10 μm
		Does not exceed 2 particles/container ≥25 μm
Ophthalmic solutions	Light obscuration	Does not exceed 50 particles/mL of ≥10 μm
		Does not exceed 5 particles/mL of ≥25 μm
	Microscopy	Does not exceed 50 particles/mL of ≥10 μm
		Does not exceed 5 particles/mL of ≥25 μm
		Does not exceed 2 particles/mL of ≥50 μm

obtain the size of the particles through comparison with standards. The data are then converted to size distributions. Table 5.5 lists the acceptance criteria with respect to subvisible particle counts in parenterals using the microscopy-based counting method.

The microscopy-based method described in the pharmacopoeias is often tedious, and the analytical outcome could be operator dependent. Newer instruments are now available that could provide automated detection of particles through a microscope in a flow mode.[117–119] Such a technique, also referred to as micro-flow imaging (MFI), combines a flow-based cell with the microscope. The total particle count and size distribution is obtained as the sample flows in front of the microscope objective. This system has the advantage that it is completely automated and hence efficient. Besides counting particles, the instrument can store data electronically and analyze images for the attributes of shape and transparency, allowing the differentiation of particles such as oil droplets, air bubbles, and solid particles. Although the technique needs to be further assessed and validated before it is included into pharmacopoeias, its use during early development and for research and development is increasingly becoming popular.

5.3.7.3 Coulter Counter Particle counting and sizing of the Coulter Counter technique is based on the effect a particle has on the conductance between two electrodes placed in a conducting solution (Beckman, Application Note BCPCT0001_CE).[120] Typically, the counter consists of an aperture, also known as the "sensing zone," placed in between two electrodes. As a particle passes through the orifice, it induces resistance and gives rise to a voltage impulse, the height of which is proportional to the particle size, and the width is proportional to the transit time. Therefore, the resulting distribution provides information on the number and size of the particles. The technique provides fast data collection using small sample volumes. Some of the limitations include the underestimation of the number of smaller-size particles, as smaller particles may multiply and enter the aperture at the same time, and the requirement to use an electrolyte with sufficient conductivity, which may affect samples initially formulated in a low electrolyte solution.

5.3.7.4 Subvisible Particles Summary As described above, while techniques exist to detect and quantify subvisible aggregates/particles in protein formulations in the size range of 10–100 μm, there is a need to develop analytical tools for quantitative detection of these particles in the size range on 0.1–10 μm, especially in the size range of 0.1–1 μm. This has been discussed recently and has been cited as a gap since aggregates in this size range often go undetected and, therefore, unaccounted for.[39] The extent to which these particles are present and their clinical consequences remain largely unanswered at this point. Yet, there seems to be considerable interest in developing analytical tools to quantify aggregates/particles in this size range. While DLS provides information in cases where aggregates are present in large enough concentration, at low concentrations, aggregates may not be detected above the noise and go undetected. Another area of development that is required for the characterization of subvisible particles is the ability to differentiate between proteinaceous and non-proteinaceous aggregates/particles.

5.3.8 Visible Particles

As defined earlier in this text, visible particles or particulates are defined as those which are typically greater than 50 μm in diameter and are visible to the naked eye. It should be noted that by its definition, visibility of a given particle may depend on several factors such as the refractive index of the particle, contrast from its surrounding solution and the background, and training and experience toward visualization of particles. For example, a high refractive index particle of size as low as 50 μm could be easily visible and on the other hand, a 200-μm particle of low contrast may not be visible through a semitransparent plastic container. Currently, there are no instruments available that can accurately and precisely provide automated quantitation of visible particles. Various pharmacopoeias (USP, Ph. Eur., and JP) require that the injectables be "essentially free" (USP) or "practically free" (Ph. Eur.) from

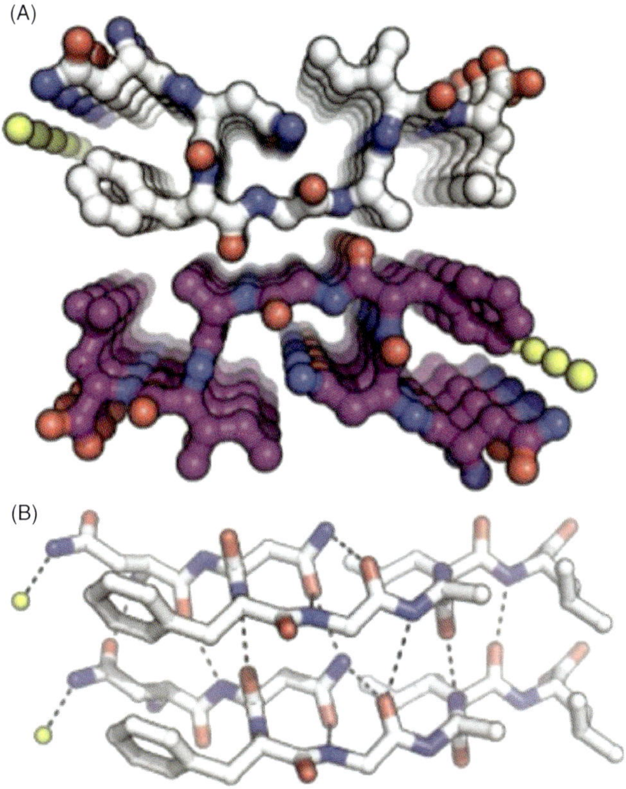

Figure 1.2. MOLECULAR model of cross-beta structure showing interdigitation of side chains that form the tight packing arrangement in the IAPP peptide sequence NNFGAIL (A) and the backbone hydrogen bonding between strands (B).

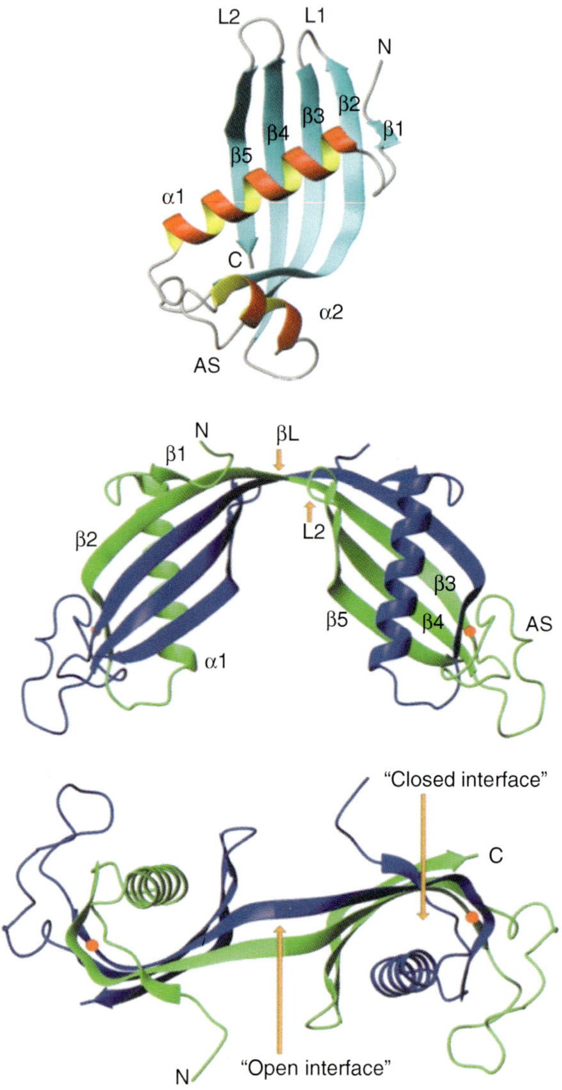

Figure 1.6. Domain-swapped dimer structure of cystatin C. The fold of chicken cystatin (PDB entry 1CEW), defining the topology of this class of proteins (top). Domain-swapped dimer of HCC in a view similar to (middle) and in a perpendicular orientation (bottom) emphasizing the beta sheet in the domain switch region (open interface) and the site of the L68Q mutation (red dot). From Reference 64.

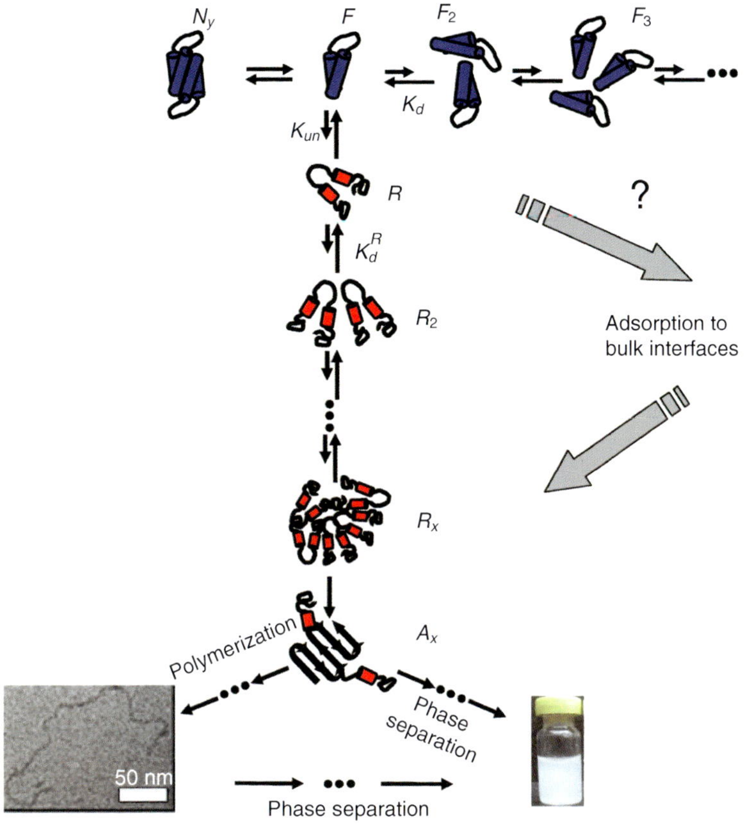

Figure 2.1. Schematic overview of multistage pathways for protein aggregation. Double arrows indicate steps that are reversible, with associated equilibrium constants defined for selected stages. Single arrows indicate steps that are effectively irreversible. The cryo-TEM image in the lower left corner is a high-molecular-weight soluble aggregate of bovine G-CSF (modified from Reference 54).

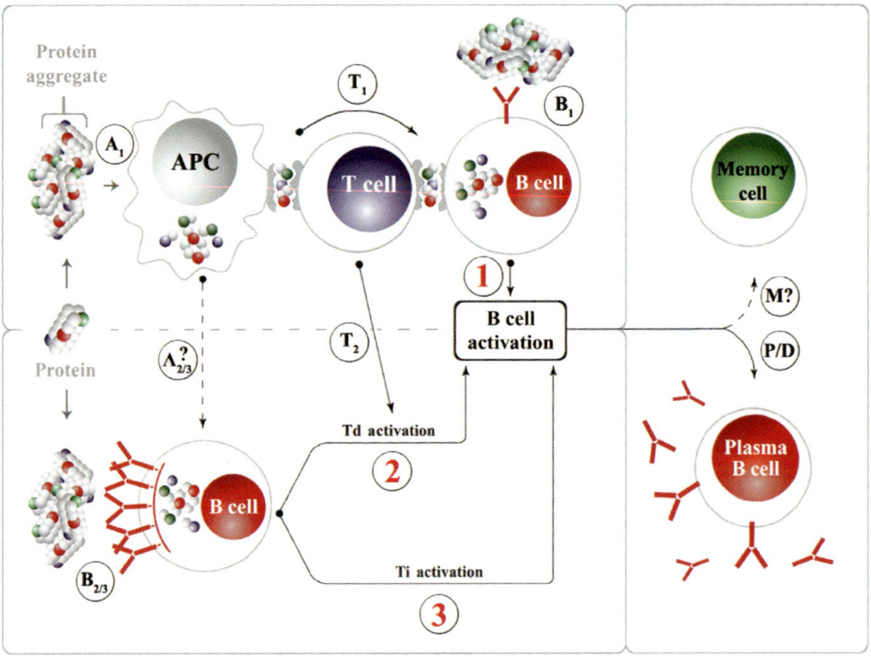

Figure 10.6. Proposed mechanisms of antibody induction by protein aggregates. This mechanism follows a typical Td activation (see Fig. 9.2 for details). It involves the uptake and processing of protein aggregates by professional APCs (A_1), followed by T-cell activation (T_1), accompanied by aggregate uptake and processing by B cells (B_1), which eventually leads to B-cell proliferation and differentiation (P/D) and to the induction of memory cells (M?). (1) This mechanism occurs when aggregates trigger multivalent cross-linking of B-cell receptors ($B_{2/3}$). B cells may then receive activating signals from accessory cells ($A_{2/3}$). T cells are then recruited to activate B cells (T_2), which leads to their proliferation and differentiation (P/D) and possibly to the induction of memory cells (M?). (2) This mechanism also occurs when aggregates trigger multivalent cross-linking of B-cell receptors ($B_{2/3}$), which triggers the production of several activation signals. B cells may then receive additional signals from accessory cells ($A_{2/3}$). T-cell help is not required for this type of B-cell activation. B cells are then activated, which leads to their proliferation and differentiation (P/D). The induction of memory cells via this mechanism remains unclear (M?).

particulates that could be detected visually.[114,115,121] These pharmacopoeias define visible particulates as extraneous, mobile undissolved particulates, other than gas bubbles, unintentionally present in the solutions.

In the absence of any validated, automated methods, visual inspection of containers (vials, etc.) is carried out by manual means to fulfill the pharmacopoeial requirement of injectables to be free from visible particulates. While no detailed procedure is outlined in the USP, Ph. Eur. and JP outline a procedure for the visual inspection of injectables against a defined black and white background, given light intensity for a defined duration of time. Ph. Eur. describes the use of a black/white background with a light intensity of 2–3.75 klux and an inspection time of 5 s,[121] whereas JP describes the use of a black background with a light intensity of 1.0 klux and an inspection time of 15 s for visual inspection of injectables.[114] The detection of visible particulates remains subjective and depends on the training and experience of the inspector along with other factors such as particle size, shape, and contrast against the background. Automated analysis of visible particles is highly desirable; however, it currently remains a gap.

5.3.9 Miscellaneous Technologies for Characterization of Protein Aggregates

5.3.9.1 Fluorescence Spectroscopy Using Extrinsically Fluorescent Dyes The characterization of protein aggregates using extrinsically fluorescent dyes is primarily based on the shift in their fluorescence spectral properties upon interaction with the aggregates as compared with the protein in the native state.[122] The interaction is primarily non-covalent in nature, and the shift is observed because of the altered dielectric environment around the dyes, as it moves from a polar environment of the bulk solution to a more complex environment on the protein surface or within protein aggregates. Therefore, the underlying presumption of this technique is that upon formation of aggregates, there is a change in the nature of the solvent-accessible surface of the protein, for example, an altered surface hydrophobicity resulting from partial unfolding of the protein or due to the formation of misfolded species. Since the dye fluorescence properties are different when bound to protein aggregates, information about aggregates can be obtained through fluorescence spectroscopy or a fluorescence microscope.

A number of extrinsic fluorescence dyes have been used for the characterization of protein aggregates. Typical examples include 1-anilinonaphthalene 8-sulfonate (ANS), 4,4′-bis-1-anilinonaphthalene 8-sulfonate (bis-ANS), Nile red, thioflavin T, and Congo red. The solution properties and mechanisms governing the use of these dyes have been reviewed recently.[122] ANS and its dimeric analog bis-ANS exhibit a blue shift and an increase in the fluorescence intensity as its environment changes from polar (water) to relatively low dielectric constant medium (ethanol, dimethyl sulfoxide [DMSO]). Therefore, an increase in the ANS (or bis-ANS) fluorescence intensity along with a blue shift in the dye fluorescence in a protein-containing solution is indicative of an increase in the accessible hydrophobic surface suggesting changes in protein tertiary structure. When used in combination with other techniques such as SEC, DLS, studies with ANS, or bis-ANS could provide insights into the mechanism of aggregation. For example, bis-ANS was used to demonstrate formation of structurally altered species of rhFVIII with exposed hydrophobic residues that contributed toward the formation of soluble aggregates.[24] In this respect, ANS and bis-ANS may provide information toward early events in protein structure change that may subsequently lead to protein aggregation. One needs to be cautious using ANS or bis-ANS when studying phenomena related to electrostatic interactions since ANS and bis-ANS are charged dyes and their binding could be pH and/or salt dependent.

Nile red (9-diethylamino-5*H*-benzo[α]phenoxazine-5-one) overcomes these limitations as it is an uncharged dye. The fluorescence of Nile red is sensitive to the polarity of its surroundings and is similar to ANS/bis-ANS, whereby a decrease in dielectric constant leads to a blue shift and an increase in its fluorescence intensity. Nile red binds to hydrophobic protein surface, and a change in the solvent-accessible hydrophobic surface area would result in a change in the properties of its fluorescence spectrum.[123] This information is then utilized to gain insights into protein conformation change and associated aggregation. Sutter et al. utilized Nile red to study heat-induced aggregation of β-galactosidase.[124] The authors were able to quantify aggregates at levels as low as 1 nM of heat-denatured aggregates that were not quantifiable by SEC. In addition, protein aggregates bound to Nile red can be visualized using fluorescence microscopy. This approach was used by Demeule et al. to visualize antibody aggregates following staining of antibody solution with Nile red.[53] The authors were able to detect aggregates in high-concentration antibody solutions (193 mg/mL) without dilution. Using a particle counting approach similar to microscopy, a particle size distribution histogram was generated for the antibody from micrographs obtained from the fluorescence microscope. The use of Nile red has been proposed for the early detection of small amounts of aggregates due to its sensitivity to changes in accessible hydrophobic surface of proteins upon aggregation. The early detection of calcitonin fibrils was facilitated by using Nile red and fluorescence microscopy.[125] Advantages of Nile red include its photostability, broad wavelength range, and stable fluorescence with high quantum yield over a range of pH. Other fluorescence dyes such as thioflavin T and Congo red have been used primarily for characterization of amyloid fibrils.[126–130] The changes in fluorescence properties of thioflavin T are used to characterize amyloid fibrils, whereas with Congo red (in contrast to thioflavin T), it is more common to use its absorbance, birefringence, and dichroism properties.

While the use of extrinsic dyes facilitates the early detection of aggregates arising from the perturbation of protein structure, these dyes may not detect aggregates that involve protein in a largely native conformation. The use of the dye itself may induce aggregate formation, especially soluble aggregates, because of its tendency to perturb protein structure.[131,132] The presence of surfactant micelles may further confound fluorescence data and give false positives since these dyes may be entrapped in the hydrophobic interior of micelles.

5.3.9.2 Electrospray-Differential Mobility Analysis (ES-DMA) ES-DMA, also known as gas-phase electrophoretic mobility molecular analysis (GEMMA), has been recently used to characterize protein aggregates in the size range of 3–250 nm.[133,134] The technique of ES-DMA is similar in concept to mass spectrometry and is based on the electrophoretic mobility of singly charged molecules in air. Briefly, the technique utilizes electrospray nebulizers to generate charged droplets, and the high charge on the droplets is rapidly removed by inducing ionization of the surrounding air using a radiation source. The ions generated in air rapidly diffuse and reach the charged droplets and prevent disintegration of the droplet. In conventional electrospray ionization mass spectrometry (ESI-MS) experiments, the charged droplets actually disintegrate under low vacuum to produce multiply charged ions. In ES-DMA, the primary droplets lose their liquid by evaporation leaving free analyte molecules. The molecules have either a +1 or a −1 charge and enter the differential mobility analyzer (DMA) under atmospheric pressure and experience an atmospheric drag. The molecules separate based on their charge-to-aerodynamic size ratio.

DMA typically consists of a coaxial electrode pair through which a laminar axial flow of particle-free air is established. Once a potential is applied, it establishes a radial electric field, which further causes charged particles to migrate across the flow and enter the central rod. The particles that fall within a defined narrow range of mobilities, that is, whose radial drag force is balanced by the radial voltage, exit through the gap and enter the particle detector. The size band of ions that pass to the detector depends on the flow rates of the sample and the gas, the size of the electrodes, and the applied voltage. Typically, a step increase in the voltage allows sampling the particle size distribution. The voltage applied is then used to obtain the electrophoretic diameter. The molecules are then counted using a condensation particle counter (CPC) in which particles/molecules are subjected to saturated vapor and act as nucleation sites. The droplets are allowed to grow on each particle until they can be detected by an optical system and are then counted to provide particle concentration per milliliter. Finally, a particle distribution is obtained, plotting the particle counts against the diameter.

The utility of ES-DMA has been demonstrated in the separation and analysis of several globular proteins ranging from 3 to 17 nm corresponding to a molecular weight of 6 kDa to over 1000 kDa.[134] Recently, the technique was used to separate and characterize IgG antibodies and its multimers.[133] Information about the molecule size of any given peak is obtained through the known diameter or crystal structure dimension of the molecules.

The advantages of ES-DMA include a direct readout of soluble protein aggregates up to 50 nm in size and extended even to 250 nm. The technique may thus provide quantitation of larger soluble aggregates that may not be detected or resolved by SEC. The limitations of this technique include low protein concentrations (nanogram per milliliter to microgram per milliliter) and solutions with low salt content since a higher salt concentration may lead to salt deposition on particles or detector walls. In addition, the technique is applicable for only soluble aggregates, especially irreversible aggregates, since soluble reversible aggregates may be affected by initial dilution step and the technique itself.

5.3.9.3 Polarization Intensity Differential Scattering (PIDS) The technique of PIDS has been developed by Beckman to provide size measurements in the range of 40 nm to 2000 μm including non-spherical particles (Beckman, Application Note A1995-A).[135] The technique is based on the measurement of the difference between the intensity of the horizontal and vertical scattered light when a particle-containing solution is illuminated separately with a horizontally and vertically polarized light. When a dipole in a particle is illuminated by polarized light, the resulting oscillating dipole radiates light in all directions in the same plane of polarization as the incident light. While the scattered light in a given plane has a distinct profile with maxima and minima for larger particles (>1 μm) in the scattering pattern at low angles, this profile vanishes for smaller particles (<1 μm) and is detectable only at very high angles making the detection and resolution of the scattering pattern difficult. Therefore, an alternate approach of PIDS was developed to measure particle size in the submicron range.

PIDS utilizes the fact that when submicron-size particles are illuminated by the two polarized lights independently, a small contrast is observed between the horizontal and the vertical scattered light. This difference is used to obtain information about the particle fine structure and is translated to particle size. The data are typically collected for three different wavelengths (450, 600, and 900 nm) for the horizontal and vertical polarized light over a range of angles, and the scattering intensity difference versus scattering angle data is used to generate the continuous particle size distribution data. The interference from very large particles is eliminated as these do not exhibit differences between the vertical and horizontal scattered light. This is a new instrument and technique; as such, data need to be generated to assess its precision, accuracy, and validation as well as a comparison with other techniques such as DLS for the quantitation and sizing of submicron-size particles.

5.3.9.4 Electron Microscopy Scanning electron microscopy (SEM) and transmission electron microscopy (TEM) are often utilized to image aggregates to obtain information about their shape and size.[54] In TEM, an electron beam is produced using a hot filament and focused on the sample. To eliminate the poor contrast resulting from small differences in atomic structure and thickness of the sample compared with its surrounding support, the samples are often dried under vacuum, coated with gold, and then bombarded with an electron beam to produce an image. This technique is also known as "shadow casting" and the resulting shadow is analyzed based on beam angle and a reference shadow of a sphere. In the case of SEM, the surface of the sample is scanned using an electron beam with the help of deflection coils and focusing lenses. The reflected electron signal is used synchronously to generate a high magnification image.

TEM has been used to characterize amyloid fibrils, insulin fibrils, and shape and flexibility of mAbs.[136,137] SEM has been used to visualize protein aggregates on filtration membranes, especially in the case of fouled membranes.[138,139] Both techniques, SEM and TEM, are well suited for image analysis of aggregates/particles and can be used in the size range from a few nanometers to several microns. The techniques, however, are not suitable for aggregate quantification except for the size of the aggregates. Furthermore, the sample treatment must be taken into consideration, that is, drying and coating, when analyzing the data as these could affect the original state of the sample and even give rise to the formation of aggregates. Furthermore, the sample area viewed may not be representative of the entire sample.

5.3.10 Protein Structural Characterization in Aggregates: Spectroscopy-Based Techniques

Besides the characterization and quantitation of number, size, and shape of aggregates in protein-based biopharmaceuticals, it is often essential to obtain information about the protein structure in aggregates. This information could provide insights as to the mechanism of aggregate formation as well as help to determine whether aggregates involve protein in native or non-native states. Several techniques can be utilized to obtain structural information such as fluorescence spectroscopy, circular dichroism (CD) spectroscopy, Fourier transform infrared (FTIR) and Raman spectroscopy, and nuclear magnetic resonance spectroscopy. Intrinsic tryptophan fluorescence spectroscopy provides information about the tryptophan microenvironment within a protein and hence, the protein tertiary structure. CD spectroscopy can be used in the far-UV region (180–260 nm) to provide information on the secondary structure conformation or in the near-UV region (240–340 nm) to provide information on the tertiary protein structure. FTIR spectroscopy is primarily used to obtain information about the secondary structure, whereas Raman spectroscopy can be used to monitor changes in secondary as well as tertiary structure conformation.

Several examples exist in literature where various spectroscopic techniques have been used to characterize aggregates or understand the mechanism of aggregation. CD spectroscopy in the near-UV region has been shown to be sensitive to the association state of insulin and relaxin.[46,84,140,141] FTIR has been widely used to characterize protein structure in precipitates (e.g., IL-2), in amyloids and fibrils, and in solid/lyophilized state.[52,137,142–145] FTIR has also been useful to understand the mechanism of non-native aggregation as it can point to conformational changes that may take place prior to aggregation. Raman spectroscopy has been increasingly popular in the field of protein-based pharmaceuticals and provides advantages over FTIR as the relative signal contribution from water compared with protein is more favorable. In addition, Raman spectroscopy can be used for any physical state of the protein (solution, suspension, gels, solid, etc.) without much sample manipulation. Although there are limited reports of the use of Raman for the characterization of aggregates, this technique is promising to study protein conformation in aggregates in a wide variety of samples.[146]

It is evident from the literature that spectroscopic techniques provide adequate information about protein structure and structural perturbations in a sample containing a single species or a sample that is rich in aggregated protein. However, when aggregates are present in the solution at low concentration, they may need to be separated prior to analysis as the signal from these species is overwhelmed by that of the native protein. There are limitations in the methods to separate low levels of protein aggregates for further spectroscopic analysis. When using spectroscopic techniques for the analysis of protein structure in aggregates, in a solution containing a heterogeneous mix of species (in terms of structure and aggregate molecular weight/size), one should take into consideration the effect of sample preparation (e.g., dilution) and methodology employed to separate aggregates (e.g., filtration). Regardless, spectroscopic techniques are used widely to gain mechanistic insights on the formation of aggregates in protein samples.

5.4 SUMMARY

Proteins have an intrinsic tendency to aggregate; therefore, it is expected that aggregates could potentially form during the fermentation, purification, processing, formulation, shipping, and storage of protein-based biotherapeutics. Because of the potential undesirable biological consequences of protein aggregates, attempts are made to minimize and control the levels of various types of aggregates in any clinical product during its intended shelf life. Therefore, it is essential that appropriate analytical techniques are utilized to characterize and quantitate the various types of aggregates, which may occur during various stages of protein pharmaceutical development. The diversity in the types of aggregates that could be formed requires the use of multiple techniques for their quantitative analyses and characterization. In this chapter, we have proposed a novel nomenclature to define the complex nature of various aggregates to define the four essential attributes of all aggregates, that is, reversibility, linkage, structure, and size. The use of our proposed nomenclature should harmonize the use of various terms currently used to define aggregates as well as facilitate simpler recognition of the various types of aggregates.

We have presented a comprehensive overview of the various technologies currently used for quantitation and characterization of aggregates in terms of the four attributes as defined above. Many techniques can be used in the high-throughput mode to facilitate rapid and efficient analyses of protein aggregates such as SEC, SEC-MALLS, DLS, AF4, CE-SDS, and flow-based microscopy. When using various techniques, the protein chemist should carefully consider the size limitations, the detection limits, sensitivity, resolution, and reproducibility in the quantitation and characterization of aggregates. Furthermore, "orthogonal" methods must be used to compare and confirm results between different techniques. While several techniques are not ready to be used in a validated manner (except for SEC and CE-SDS), they do provide useful information in our understanding of the nature and size of aggregates. Lastly, while techniques exist to analyze soluble aggregates (up to ~100 nm) and larger subvisible and visible aggregates (>1 μm), currently there exists a gap to reliably and quantitatively analyze moderately sized subvisible particles (100 nm to 1 μm). Even though the characterization of such aggregates is desired, the clinical consequence of such aggregates remains largely unknown since these aggregates should not present the concern of *in vivo* capillary occlusion. In the near future, we anticipate increased activity toward characterization of these subvisible aggregates.

REFERENCES

1. Cromwell MEM, Hilario E, Jacobson F. 2006. Protein aggregation and bioprocessing. AAPS J 8(3):E572–E579.
2. Mahler HC, Friess W, Grauschopf U, Kiese S. 2009. Protein aggregation: Pathways, induction factors and analysis. J Pharm Sci. 98(9):2909–2934.
3. Shire SJ, Cromwell M, Liu J. 2006. Concluding summary: Proceedings of the AAPS biotec open forum on "aggregation of protein therapeutics." AAPS J 8(4):E729–E730.
4. Wang W. 2005. Protein aggregation and its inhibition in biopharmaceutics. Int J Pharm 289(1-2):1–30.
5. Bucciantini M, Giannoni E, Chiti F, Baroni F, Formigli L, Zurdo JS, Taddei N, Ramponi G, Dobson CM, Stefani M. 2002. Inherent toxicity of aggregates implies a common mechanism for protein misfolding diseases. Nature 416(6880): 507–511.
6. Hermeling S, Aranha L, Damen JMA, Slijper M, Schellekens H, Crommelin DJA, Jiskoot W. 2005. Structural characterization and immunogenicity in wild-type and immune tolerant mice of degraded recombinant human interferon alpha2b. Pharm Res 22(12):1997–2006.
7. Hermeling S, Crommelin DJA, Schellekens H, Jiskoot W. 2004. Structure-immunogenicity relationships of therapeutic proteins. Pharm Res 21(6):897–903.
8. Schellekens H. 2002. Bioequivalence and the immunogenicity of biopharmaceuticals. Nat Rev Drug Discov 1(6):457–462.
9. Schellekens H. 2005. Factors influencing the immunogenicity of therapeutic proteins. Nephrol Dial Transplant 20:3–9.

10. Stefani M, Dobson CM. 2003. Protein aggregation and aggregate toxicity: New insights into protein folding, misfolding diseases and biological evolution. J Mol Med 81(11):678–699.
11. Braun A, Kwee L, Labow MA, Alsenz J. 1997. Protein aggregates seem to play a key role among the parameters influencing the antigenicity of interferon alpha (IFN-alpha) in normal and transgenic mice. Pharm Res 14(10):1472–1478.
12. Moore WV, Leppert P. 1980. Role of aggregated human growth-hormone (hGH) in development of antibodies to hGH. J Clin Endocrinol Metab 51(4):691–697.
13. Rosenberg AS. 2005. AAPS Biotechnology Conference Open Forum on Aggregation of Protein Therapeutics, San Francisco, CA, June 5–8, pp. E501–E507.
14. Bachmann MF, Zinkernagel RM. 1997. Neutralizing antiviral B cell responses. Annu Rev Immunol 15:235–270.
15. Dintzis RZ, Okajima M, Middleton MH, Greene G, Dintzis HM. 1989. The immunogenicity of soluble haptenated polymers is determined by molecular mass and hapten valence. J Immunol 143(4):1239–1244.
16. Cromwell MEM, Felten C, Flores H, Liu J, Shire SJ. 2007. Self-association of therapeutic proteins. In Murphy RM, Tsai AM, eds., *Misbehaving Proteins*. New York: Springer, pp. 313–330.
17. Cordoba-Rodriguez RV. 2008. Aggregates in MAbs and recombinant therapeutic proteins: A regulatory perspective. Biopharm Int 21(11):44–53.
18. Sharma VK, Patapoff TW, Kalonia DS. 2010. A harmonized classification and nomenclature of protein aggregates. J Pharm Sci, in revision.
19. Trivedi M, Davis RA, Shabaik Y, Roy A, Verkhivker G, Laurence JS, Middaugh CR, Siahaan TJ. 2009. The role of covalent dimerization on the physical and chemical stability of the EC1 domain of human E-cadherin. J Pharm Sci. 98(10): 3562–3574.
20. Buren NV, Rehder D, Gadgil H, Matsumura M, Jacob J. 2009. Elucidation of two major aggregation pathways in an IgG2 antibody. J Pharm Sci 98(9):3013–3030.
21. Klibanov AM, Schefiliti JA. 2004. On the relationship between conformation and stability in solid pharmaceutical protein formulations. Biotechnol Lett 26(14): 1103–1106.
22. Bam NB, Cleland JL, Yang J, Manning MC, Carpenter JF, Kelley RF, Randolph JW. 1998. Tween protects recombinant human growth hormone against agitation-induced damage via hydrophobic interactions. J Pharm Sci 87(12):1554–1559.
23. Chi EY, Krishnan S, Kendrick BS, Chang BS, Carpenter JF, Randolph TW. 2003. Roles of conformational stability and colloidal stability in the aggregation of recombinant human granulocyte colony-stimulating factor. Protein Sci 12(5): 903–913.
24. Grillo AO, Edwards KLT, Kashi RS, Shipley KM, Hu L, Besman MJ, Middaugh CR. 2001. Conformational origin of the aggregation of recombinant human factor VIII. Biochemistry (Mosc) 40(2):586–595.
25. Hansen JF. 1991. The self-association of zinc-free human insulin and insulin analog B13-glutamine. Biophys Chem 39(1):107–110.
26. Moore JMR, Patapoff TW, Cromwell MEM. 1999. Kinetics and thermodynamics of dimer formation and dissociation for a recombinant humanized monoclonal

antibody to vascular endothelial growth factor. Biochemistry 38(42): 13960–13967.

27. Timasheff SN, Aune KC, Goldsmith LC. 1971. Dimerization of a-chymotrypsin. II. Ionic strength and temperature dependence. Biochemistry 10:1617–1622.
28. Verheul M, Pedersen JS, Roefs S, de Kruif KG. 1999. Association behavior of native beta-lactoglobulin. Biopolymers 49(1):11–20.
29. Schwartz AM, Berglund KA. 2001. Comparison of supersaturation profiles employed on lysozyme crystallization from a hanging drop. Cryst Growth Des 1(1):81–85.
30. St Clair N, Shenoy B, Jacob LD, Margolin AL. 1999. Cross-linked protein crystals for vaccine delivery. Proc Natl Acad Sci U S A 96:9469–9474.
31. Sharma VK, Kalonia DS. 2004. Polyethylene glycol-induced precipitation of interferon alpha-2a followed by vacuum drying: Development of a novel process for obtaining a dry, stable powder. AAPS J 6(1):31–44.
32. Matheus S, Friess W, Schwartz D, Mahler HC. 2008. Liquid high concentration IgG1 antibody formulations by precipitation. J Pharm Sci. 98(9):3043–3057.
33. Brange J, Andersen L, Laursen ED, Meyn G, Rasmussen E. 1997. Toward understanding insulin fibrillation. J Pharm Sci 86(5):517–525.
34. Krishnan S, Chi EY, Webb JN, Chang BS, Shan DX, Goldenberg M, Manning MC, Randolph TW, Carpenter JF. 2002. Aggregation of granulocyte colony stimulating factor under physiological conditions: Characterization and thermodynamic inhibition. Biochemistry (Mosc) 41(20):6422–6431.
35. Li Y, Weiss WF, Roberts CJ. 2009. Characterization of high-molecular-weight nonnative aggregates and aggregation kinetics by size exclusion chromatography with inline multi-angle laser light scattering. J Pharm Sci. 98(11):3997–4016.
36. Weiss WF, Hodgdon TK, Kaler EW, Lenhoff AM, Roberts CJ. 2007. Nonnative protein polymers: Structure, morphology, and relation to nucleation and growth. Biophys J 93(12):4392–4403.
37. Brnjaskraljevic J, Maricic S. 1978. Hydration and self-association of hemoglobin in solution. Biochem Biophys Res Commun 83(3):1048–1054.
38. Milthorpe BK, Nichol LW, Jeffrey PD. 1977. The polymerization pattern of zinc(II)-insulin at pH 7.0. Biochim Biophys Acta 495(2):195–202.
39. Carpenter JF, Randolph TW, Jiskoot W, Crommelin DJA, Middaugh CR, Winter G, Fan Y-X, Kirshner S, Verthelyi D, Kozlowski S, Clouse KA, Swann PG, Rosenberg A, Cherney B. 2009. Overlooking subvisible particles in therapeutic protein products: Gaps that may compromise product quality. J Pharm Sci 98:1201–1205.
40. Helmerhorst E, Stokes GB. 1987. Self-association of insulin. Its pH dependence and effect of plasma. Diabetes 36(3):261–264.
41. Jeffrey PD. 1986. Self-association of des-(B26-B30)-insulin. The effect of calcium and some other divalent cations. Biol Chem Hoppe Seyler 367(5):363–369.
42. Jeffrey PD, Coates JH. 1966. An equilibrium ultracentrifuge study of the self-association of bovine insulin. Biochemistry (Moscow, Russian Federation) 5(2):489–498.
43. Lord RS, Gubensek F, Rupley JA. 1973. Insulin self-association. Spectrum changes and thermodynamics. Biochemistry 12(22):4385–4391.

44. Doty P, Gellert M, Rabinovitch B. 1952. The association of insulin. I. Preliminary investigations. J Am Chem Soc 74:2065–2069.

45. Gutfreund H. 1948. Molecular weight of insulin and its dependence on pH, concentration, and temperature. Biochem J 42:544–548.

46. Uversky VN, Garriques LN, Millett IS, Frokjaer S, Brange J, Doniach S, Fink AL. 2003. Prediction of the association state of insulin using spectral parameters. J Pharm Sci 92(4):847–858.

47. Kiese S, Papppenberger A, Friess W, Mahler HC. 2008. Shaken, not stirred: Mechanical stress testing of an IgG1 antibody. J Pharm Sci 97(10):4347–4366.

48. Mahler HC, Muller R, Friess W, Delille A, Matheus S. 2005. Induction and analysis of aggregates in a liquid IgG1-antibody formulation. Eur J Pharm Biopharm 59(3):407–417.

49. Advant SJ, Braswell EH, Kumar CV, Kalonia DS. 1995. The effect of pH and temperature on the self-association of recombinant human interleukin-2 as studied by equilibrium sedimentation. Pharm Res 12(5):637–641.

50. Finke JM, Jennings PA. 2001. Early aggregated states in the folding of interleukin-1 beta. J Biol Phys 27(2-3):119–131.

51. Ahrer K, Buchacher A, Iberer G, Josic D, Jungbauer A. 2002. 22nd International Symposium on the Separation of Proteins, Peptides and Polynucleotides (ISPPP 02), Heidelberg, Germany, November 10–13, pp. 89–96.

52. Breen ED, Curley JG, Overcashier DE, Hsu CC, Shire SJ. 2001. Effect of moisture on the stability of a lyophilized humanized monoclonal antibody formulation. Pharm Res 18(9):1345–1353.

53. Demeule B, Lawrence MJ, Drake AF, Gurny R, Arvinte T. 2007. Characterization of protein aggregation: The case of a therapeutic immunoglobulin. Biochim Biophys Acta 1774(1):146–153.

54. Hiemenz PC, Rajagopalan R. 1997. Colloid and surface chemistry: Scope and variables. In Hiemenz PC, Rajagopalan R, eds., *Principles of Colloid and Surface Chemistry*. New York: Marcel Dekker.

55. Ramani K, Purohit V, Middaugh CR, Balasubramanian SV. 2005. Aggregation kinetics of recombinant human FVIII (rFVIII). J Pharm Sci 94(9):2023–2029.

56. Kendrick BS, Carpenter JF, Cleland JL, Randolph TW. 1998. A transient expansion of the native state precedes aggregation of recombinant human interferon-gamma. Proc Natl Acad Sci U S A 95(24):14142–14146.

57. Kendrick BS, Cleland JL, Lam X, Nguyen T, Randolph TW, Manning MC, Carpenter JF. 1998. Aggregation of recombinant human interferon gamma: Kinetics and structural transitions. J Pharm Sci 87(9):1069–1076.

58. Knepp VM, Muchnik A, Oldmark S, Kalashnikova L. 1998. Stability of nonaqueous suspension formulations of plasma derived factor IX and recombinant human alpha interferon at elevated temperatures. Pharm Res 15(7):1090–1095.

59. Gabrielson JP, Brader ML, Pekar AH, Mathis KB, Winter G, Carpenter JF, Randolph TW. 2007. Quantitation of aggregate levels in a recombinant humanized monoclonal antibody formulation by size-exclusion chromatography, asymmetrical flow field flow fractionation, and sedimentation velocity. J Pharm Sci 96(2):268–279.

60. Ackers GK, Thompson TE. 1965. Determination of stoichiometry and equilibrium constants for reversibly associating systems by molecular sieve chromatography. Proc Natl Acad Sci U S A 53:342–349.
61. Prochazka K, Mandak T, Bednar B, Trnena J, Tuzar Z. 1990. Behavior of reversibly associating systems in size exclusion chromatography. Interpretation of experimental data based on theoretical model. J Liq Chromatogr 13:1765–1783.
62. Patapoff TW, Mrsny RJ, Lee WA. 1993. The application of size exclusion chromatography and computer simulation to study the thermodynamic and kinetic parameters for short-lived dissociable protein aggregates. Anal Biochem 212:71–78.
63. Arakawa T, Ejima D, Li T, Philo JS. 2010. The critical role of mobile phase composition in size exclusion chromatography of protein pharmaceuticals. J Pharm Sci 99(4):1674–1692.
64. Carpenter JF, Randolph TW, Jiskoot W, Crommelin DJA, Middaugh CR, Winter G. 2010. Potential inaccurate quantitation and sizing of protein aggregates by size exclusion chromatography: Essential need to use orthogonal methods to assure the quality of therapeutic protein products. J Pharm Sci 99(5):2200–2208.
65. Liu J, Andya JD, Shire SJ. 2005. AAPS Biotechnology Conference Open Forum on Aggregation of Protein Therapeutics, San Francisco, CA, June 5–8, pp. E580–E589.
66. Philo JS. 2005. AAPS Biotechnology Conference Open Forum on Aggregation of Protein Therapeutics, San Francisco, CA, June 5–8, pp. E564–E571.
67. Arakawa T, Philo JS, Ejima D, Tsumoto K, Arisaka F. 2006. Aggregation analysis of therapeutic proteins, part 1: General aspects and techniques for assessment. Bioprocess Int 4:42–49.
68. Zimm BH. 1946. Application of the methods of molecular distribution to solutions of large molecules. J Chem Phys 14:164–179.
69. Wen J, Arakawa T, Philo JS. 1996. Size-exclusion chromatography with on-line light-scattering, absorbance, and refractive index detectors for studying proteins and their interactions. Anal Biochem 240(2):155–166.
70. Attri AK, Minton AP. 2005. Composition gradient static light scattering: A new technique for rapid detection and quantitative characterization of reversible macromolecular hetero-associations in solution. Anal Biochem 346:132–138.
71. Bajaj H, Sharma VK, Kalonia DS. 2004. Determination of second virial coefficient of proteins using a dual-detector cell for simultaneous measurement of scattered light intensity and concentration in SEC-HPLC. Biophys J 87:4048–4055.
72. Bajaj H, Sharma VK, Kalonia DS. 2007. A high-throughput method for detection of protein self-association and second virial coefficient using size-exclusion chromatography through simultaneous measurement of concentration and scattered light intensity. Pharm Res 24:2071–2083.
73. Lomakin A, Benedek GB, Teplow DB. 1999. Monitoring protein assembly using quasielastic light scattering spectroscopy. In Sigurdsson EM, ed., *Amyloid, Prions, and Other Protein Aggregates*. San Diego, CA: Academic Press, pp. 429–459.
74. Arakawa T, Philo JS, Ejima D, Tsumoto K, Arisaka F. 2007. Aggregation analysis of therapeutic proteins, part 2: Analytical ultracentrifugation and dynamic light scattering. Bioprocess Int 5:36–47.

75. Aymard P, Nicolai T, Durand D, Clark A. 1999. Static and dynamic scattering of beta-lactoglobulin aggregates formed after heat-induced denaturation at pH 2. Macromolecules 32(8):2542–2552.
76. Bloomfield VA. 2000. Static and dynamic light scattering from aggregating particles. Biopolymers 54(3):168–172.
77. Sluzky V, Klibanov AM, Langer R. 1992. Mechanism of insulin aggregation and stabilization in agitated aqueous-solutions. Biotechnol Bioeng 40(8):895–903.
78. Mintova S, Petkov N, Karaghiosoff K, Bein T. 2002. Crystallization of nanosized MEL-type zeolite from colloidal precursors. Mat Sci Eng C-Biomim 19(1–2): 111–114.
79. Hansen JC, Lebowitz J, Demeler B. 1994. Analytical ultracentrifugation of complex macromolecular systems. Biochemistry 33(45):13155–13163.
80. Lebowitz J, Lewis MS, Schuck P. 2002. Modern analytical ultracentrifugation in protein science: A tutorial review. Protein Sci 11(9):2067–2079.
81. Schuck P. 2000. Size-distribution analysis of macromolecules by sedimentation velocity ultracentrifugation and Lamm equation modeling. Biophys J 78(3):1606–1619.
82. Laue TM, Stafford WF. 1999. Modern applications of analytical ultracentrifugation. Annu Rev Biophys Biomol Struct 28:75–100.
83. Schuck P. 2003. On the analysis of protein self-association by sedimentation velocity analytical ultracentrifugation. Anal Biochem 320(1):104–124.
84. Shire SJ, Holladay LA, Rinderknecht E. 1991. Self-association of human and porcine relaxin as assessed by analytical ultracentrifugation and circular dichroism. Biochemistry 30(31):7703–7711.
85. Alday PH, Correia JJ, Laue T, Stafford WF. 2007. 51st Annual Meeting of the Biophysical Society, Baltimore, MD, March 3–7, p. 307A.
86. Liu J, Nguyen MDH, Andya JD, Shire SJ. 2005. Reversible self-association increases the viscosity of a concentrated monoclonal antibody in aqueous solution. J Pharm Sci 94(9):1928–1940.
87. Gabrielson JP, Arthur KK, Kendrick BS, Randolph TW, Stoner MR. 2009. Common excipients impair detection of protein aggregates during sedimentation velocity analytical ultracentrifugation. J Pharm Sci 98(1):50–62.
88. Gabrielson JP, Randolph TW, Kendrick BS, Stoner MR. 2007. Sedimentation velocity analytical ultracentrifugation and SEDFIT/c(s): Limits of quantitation for a monoclonal antibody system. Anal Biochem 361(1):24–30.
89. Arthur KK, Gabrielson JP, Kendrick BS, Stoner MR. 2008. Detection of protein aggregates by sedimentation velocity analytical ultracentrifugation (SV-AUC): Sources of variability and their relative importance. J Pharm Sci. 98(10):3522–3539.
90. Giddings JC. 1993. Field-flow fractionation—Analysis of macromolecular, colloidal, and particulate materials. Science 260(5113):1456–1465.
91. Arakawa T, Philo JS, Ejima D, Sato H, Tsumoto K. 2007. Aggregation analysis of therapeutic proteins, part 3: Principles and optimization of field-flow fractionation (FFF). Bioprocess Int 5:52–70.
92. Shapiro AL, Viñuela E, Maizel JV. 1967. Molecular weight estimation of polypeptide chains by electrophoresis in SDS-polyacrylamide gels. Biochem Biophys Res Commun 28(5):815–820.

93. Raymond S, Weintraub L. 1959. Acrylamide gel as a supporting medium for zone electrophoresis. Science 130(3377):711.
94. Bier M. 1959. *Electrophoresis: Theory, Methods, and Applications*. New York: Academic Press.
95. Merril CR. 1990. Gel-staining techniques. Methods Enzymol 182:477–488.
96. Merril CR. 1990. Silver staining of proteins and DNA. Nature 343(6260): 779–780.
97. Switzer RC, Merril CR, Shifrin S. 1979. A highly sensitive silver stain for detecting proteins and peptides in polyacrylamide gels. Anal Biochem 98(1):231–237.
98. Schulenberg B, Ahnert N, Patton W. 2005. Sensitive quantitative fluorescence detection of proteins in gels using SYPRO® Ruby Protein Gel Stain. In Walker JM, ed., *The Proteomics Protocols Handbook*. Totowa, NJ: Humana Press, pp. 209–214.
99. Han M, Phan D, Nightlinger N, Taylor L, Jankhah S, Woodruff B, Yates Z, Freeman S, Guo A, Balland A, Pettit D. 2006. Optimization of CE-SDS method for antibody separation based on multi-users experimental practices. Chromatographia 64(5-6):335–342.
100. Rustandi RR, Washabaugh MW, Wang Y. 2008. Applications of CE SDS gel in development of biopharmaceutical antibody-based products. Electrophoresis 29(17):3612–3620.
101. Wehr T. 2005. Capillary electrophoresis in the biopharmaceutical industry: Part II. Available at http://chromatographyonline.findanalytichem.com/lcgc/Misc/Capillary-Electrophoresis-in-the-Biopharmaceutical/ArticleStandard/Article/detail/189943.
102. Yao S, Anex DS, Caldwell WB, Arnold DW, Smith KB, Schultz PG. 1999. SDS capillary gel electrophoresis of proteins in microfabricated channels. Proc Natl Acad Sci U S A 96(10):5372–5377.
103. Kuschel M. 2000. *Protein Sizing and Analysis Using the Agilent 2100 Bioanalyzer and Protein 200 LabChip® Kit*. Waldbronn, Germany: Agilent.
104. Blirup-Jensen SR. 2001. Protein standardization III: Method optimization. Basic principles for quantitative determination of human serum proteins on automated instruments based on turbidimetry or nephelometry. Clin Chem Lab Med 39(11):1098–1109.
105. Sukumar M, Doyle BL, Combs JL, Pekar AH. 2004. Opalescent appearance of an IgG1 antibody at high concentrations and its relationship to noncovalent association. Pharm Res 21(7):1087–1093.
106. Wang N, Hu B, Ionescu R, Mach H, Sweeney J, Hamm C, Kirchmeier MJ, Meyer BK. 2009. Opalescence of an IgG1 MAb formulation is mediated by ionic strength and excipients. Biopharm Int 22(4):36–47.
107. Zubkov LA, Romanov VP. 1988. Critical opalescence. Sov Phys Uspekhi 31(4): 328.
108. Gopal E. 2000. Critical opalescence. Resonance 5(4):37–45.
109. Kao WP, Chu B. 1969. Light-scattering studies of critical opalescence in binary liquid mixtures. IV paraffins in beta,beta [prime]-dichlorodiethyl ether. J Chem Phys 50(9):3986–4007.
110. Cromwell M. 2008. BioProcess International Conference, Vienna, Austria.

111. Eur. P. 2008. *2.2.1. Clarity and Degree of Opalescense of Liquids*, 6th edition. European Directorate for the Quality of Medicine (EDQM).
112. Akers MJ, Larrimore DS, Guazzo DM. 2003. Particulate matter testing. In Akers MJ, Larrimore DS, Guazzo DM, eds., *Parenteral Quality Control: Sterility, Pyrogen, Particulate, and Package Integrity Testing*. New York: Marcel Dekker, pp. 197–280.
113. Eur. P. 2008. *2.9.19. Particulate Contamination: Sub-Visible Particles*, 6th edition. European Directorate for the Quality of Medicine (EDQM).
114. JP. 2001. *Japanese Pharmacopoeia*, 14th edition. Tokyo: Yakuji Nippo Ltd.
115. USP. 2008. *USP/NF General Chapter <788> Particulate Matter in Injections*. Rockville, MD: United States Pharmacopeial Convention.
116. USP. 2008. *USP/NF General Chapter <789> Particulate Matter in Ophthalmic Solutions*. Rockville, MD: United States Pharmacopeial Convention.
117. Sharma D, King D, Thomas D, Oma P. 2007. Flow microscopy for particulate analysis in parenteral and pharmaceutical fluids. Eur J Paranteral Pharm Sci 12:97–101.
118. Micro Flow Imaging. Ottawa, Canada. Available at http://www.brightwelltech.com/.
119. Huang C-T, Sharma D, Oma P, Krishnamurthy R. 2008. Quantitation of protein particles in parenteral solutions using micro-flow imaging. J Pharm Sci. 98(9):3058–3071.
120. Beckman technical report. Publication AN-12763A. http://www.beckmancoulter.com/coultercounter/showDoc/doc.do?filename=&fileid=11746.
121. Eur. P. 2008. *2.9.20. Particulate Contamination: Visible Particles*, 6th edition. European Directorate for the Quality of Medicine (EDQM).
122. Hawe A, Sutter M, Jiskoot W. 2008. Extrinsic fluorescent dyes as tools for protein characterization. Pharm Res 25(7):1487–1499.
123. Sackett DL, Wolff J. 1987. Nile red as a polarity-sensitive fluorescent-probe of hydrophobic protein surfaces. Anal Biochem 167(2):228–234.
124. Sutter M, Oliveira S, Sanders NN, Lucas B, van Hoek A, Hink MA, Visser AJWG, De Smedt SC, Hennink WE, Jiskoot W. 2007. Sensitive spectroscopic detection of large and denatured protein aggregates in solution by use of the fluorescent dye Nile red. J Fluoresc 17:181–192.
125. Demeule B, Gurny R, Avinte T. 2007. Detection and characterization of protein aggregates by fluorescence microscopy. Int J Pharm 329:37–45.
126. Khurana R, Coleman C, Ionescu-Zanetti C, Carter SA, Krishna V, Grover RK, Roy R, Singh S. 2005. Mechanism of thioflavin T binding to amyloid fibrils. J Struct Biol 151(3):229–238.
127. Khurana R, Uversky VN, Nielsen L, Fink AL. 2001. Is Congo red an amyloid-specific dye? J Biol Chem 276(25):22715–22721.
128. Kim YS, Randolph TW, Manning MC, Stevens FJ, Carpenter JF. 2003. Congo red populates partially unfolded states of an amyloidogenic protein to enhance aggregation and amyloid fibril formation. J Biol Chem 278(12):10842–10850.
129. Klunk WE, Jacob RF, Mason RP. 1999. Quantifying amyloid by Congo red spectral shift assay. In Kheterpal I, Wetzel R, eds., *Amyloid, Prions, and Other Protein Aggregates*. San Diego, CA: Academic Press, pp. 285–305.

130. Krebs MRH, Bromley EHC, Donald AM. 2005. The binding of thioflavin-T to amyloid fibrils: Localisation and implications. J Struct Biol 149(1):30–37.
131. Engelhard M, Evans PA. 1995. Kinetics of interaction of partially folded proteins with a hydrophobic dye: Evidence that molten globule character is maximal in early folding intermediates. Protein Sci 4:1553–1562.
132. Kundu B, Guptasarma P. 2002. Use of a hydrophobic dye to indirectly probe the structural organization and conformational plasticity of molecules in amorphous aggregates of carbonic anhydrase. Biochem Biophys Res Commun 293(1):572–577.
133. Pease LF III, Elliott JT, Tsai DH, Zachariah MR, Tarlov MJ. 2008. Determination of protein aggregation with differential mobility analysis: Application to IgG antibody. Biotechnol Bioeng 101(6):1214–1222.
134. Kaufman SL, Skogen JW, Dorman FD, Zarrin F, Lewis KC. 1996. Macromolecule analysis based on electrophoretic mobility in air: Globular proteins. Anal Chem 68(11):1895–1904.
135. Beckman Coulter, Inc. 2005. The Sizing of Non-Spherical, Sub-Micron Particles Using Polarization Intensity Differential Scattering (PIDS™). Application Note: A1995-A. https://www.beckmancoulter.com/literature/Bioresearch/A-1995A.pdf.
136. Taschner N, Muller SA, Alumella VR, Goldie KN, Drake A, Aebi U, Arvinte T. 2001. Modulation of antigenicity related to changes in antibody flexibility upon lyophilization. J Mol Biol 310:169–179.
137. Nielsen L, Frokjaer S, Carpenter JF, Brange J. 2001. Studies of the structure of insulin fibrils by Fourier transform infrared (FTIR) spectroscopy and electron microscopy. J Pharm Sci 90(1):29–37.
138. Güell C, Davis RH. 1996. Membrane fouling during microfiltration of protein mixtures. J Memb Sci 119(2):269–284.
139. Salgin S, Takac S, Ozdamar TH. 2005. A parametric study on protein-membrane-ionic environment interactions for membrane fouling. Sep Sci Technol 40(6):1191–1212.
140. Goldman J, Carpenter FH. 1974. Zinc binding, circular dichroism, and equilibrium sedimentation studies on insulin (bovine) and several of its derivatives. Biochemistry 13:4566–4574.
141. Hughes H, Morgan C, Brunyak E, Barranco K, Cohen E, Edmunds T, Lee K. 2009. A multi-tiered analytical approach for the analysis and quantitation of high-molecular-weight aggregates in a recombinant therapeutic glycoprotein. AAPS J 11(2):335–341.
142. Allison SD, Dong AC, Carpenter JF. 1996. Counteracting effects of thiocyanate and sucrose on chymotrypsinogen secondary structure and aggregation during freezing, drying, and rehydration. Biophys J 71(4):2022–2032.
143. Szabo Z, Klement E, Jost K, Zarandi M, Soos K, Penke B. 1999. An FT-IR study of the beta-amyloid conformation: Standardization of aggregation grade. Biochem Biophys Res Commun 265(2):297–300.
144. Militello V, Casarino C, Emanuele A, Giostra A, Pullara F, Leone M. 2004. Aggregation kinetics of bovine serum albumin studied by FTIR spectroscopy and light scattering. Biophys Chem 107(2):175–187.

145. Dong A, Prestrelski SJ, Allison SD, Carpenter JF. 1995. Infrared spectroscopic studies of lyophilization- and temperature-induced protein aggregation. J Pharm Sci 84(4):415–424.

146. Sane SU, Wong R, Hsu CC. 2004. Raman spectroscopic characterization of drying-induced structural changes in a therapeutic antibody: Correlating structural changes with long-term stability. J Pharm Sci 93(4):1005–1018.

147. Richards JP, Stickelmeyer MP, Flora DB, Chance RE, Frank BH, DeFelippis MR. 1998. Self-association properties of monomeric insulin analogs under formulation conditions. Pharm Res 15(9):1434–1441.

148. Sakurai K, Oobatake M, Goto Y. 2001. Salt-dependent monomer-dimer equilibrium of bovine β-lactoglobulin at pH 3. Protein Sci 10:2325–2335.

149. Ho M-R, Lou Y-C, Lin W-C, Lyu P-C, Huang W-N, Chen C. 2006. Human pancreatitis-associated protein forms fibrillar aggregates with a native-like conformation. J Biol Chem 281:33566–33576.

150. Grondelle WV, Iglesias CL, Coll E, Artzner F, Paternostre M, Lacombe F, Cardus M, Martinez G, Montes M, Cherif-Cheikh R, Valéry C. 2007. Spontaneous fibrillation of the native neuropeptide hormone Somatostatin-14. J Struct Biol 160(2):211–223.

151. Thirumangalathu R, Krishnan S, Ricci MS, Brems DN, Randolph TW, Carpenter JF. 2009. Silicone oil- and agitation-induced aggregation of a monoclonal antibody in aqueous solution. J Pharm Sci. 98(9):3167–3181.

CHAPTER 6

Approaches to Control Protein Aggregation during Bulk Production

LINDA O. NARHI, YIJIA JIANG, ROHINI DESHPANDE, SOHYE KANG, and JOSEPH SHULTZ

6.1 INTRODUCTION

In recent years, biological/protein pharmaceutics have become an important class of therapeutics due to their high specificity, unique mechanism of action, novel targets, increased half-life, and improved safety profiles. Bulk production is a critical process during biopharmaceutical manufacturing since a biological product is often defined by its manufacturing process.[1] Aggregation is a critical product quality attribute that can affect the efficacy and safety of a biological product and recently has become the focus of increased regulatory concerns.[2,3] Avoiding (or at the minimum, controlling) protein aggregation during processing can result in a drug product that is stable during long-term storage, with minimal safety issues. In this chapter, we discuss the approaches to control aggregation during three different stages of bulk production: candidate selection, cell culture, and purification.

6.2 CANDIDATE SELECTION

With the advent of high-throughput screening methods, phage display, and so on, it is now possible to generate multiple candidates against the same target. Selection of which candidates to move forward is frequently a part of the discovery phase of the product life cycle. The quality of the candidates moving forward can be favorably influenced by considering manufacturability criteria during the candidate screening stage, where candidates that perform poorly against those criteria are eliminated. Candidate selection should take into

Aggregation of Therapeutic Proteins, Edited by Wei Wang and Christopher J. Roberts

consideration characteristics that increase the chance of a protein therapeutic successfully surviving manufacturing, storage, and delivery. Assessment of a candidate at the early development stage as part of the assessment of product quality should be an integral part of quality by design (QbD) for the development of biotherapeutics.

The criteria established for candidate selection should include things that affect biological product quality and stability during processing, storage, and delivery. This subsection will focus on sequence analysis, and the assessment of stability to process conditions and long-term storage conditions, with the goal of choosing the candidate with the least tendency to aggregate under these conditions.

6.2.1 Sequence Analysis

Protein sequence analysis involves the comparison of the sequences of product candidates to those of standard proteins to find homologies, to determine basic product characteristics such as molecular weight and isoelectric point (pI), to identify potential sites of chemical modification and so on, and to identify other sequences that could result in aggregation (see Chapter 3 for further discussion). To take a QbD approach, sequence analysis can be done early during candidate screening, as soon as the primary sequence information becomes available. The purpose of the sequence analysis is to identify labile residues that may result in decreased protein stability, activity, and so on. Residues susceptible to clipping, oxidation, deamidation, and isomerization that may lead to protein instability and aggregation are identified, and the results are used to guide the storage stability assessment and the handling of the protein therapeutic as it advances through the development life cycle. In some cases, protein engineering may be required to replace the problematic amino acids and increase the efficacy, stability, and homogeneity of the protein. These efforts should be based on knowledge of similar products and processes. Protein engineering can potentially reduce the number of critical quality attributes and the complexity of process development, control, and validation, resulting in a product that can survive processing, storage, and delivery with a minimum of modification and aggregation. There have been many publications focused on using protein engineering/mutations to improve the stability, yield, and activity of protein molecules.[4,5]

The work on IL-18 is a good example of how one can use protein engineering to increase the yield and improve the homogeneity of the protein, in addition to increasing bioactivity. IL-18 is secreted *in vivo* in a pro form, with the pro sequence then clipped by caspase 1 in the bacterial cell to yield the mature molecule. Coexpressing the protease and the IL-18 gene is not very feasible, as the caspase itself can be lethal. Expressing IL-18 with the signal sequence of an antibody light chain resulted in a protein that could be expressed in mammalian cells at high levels, and that was secreted with the appropriate N-terminus of the mature IL-18 following clipping of the light chain signal sequence in the endoplasmic reticulum. The bioactivity of this protein was also enhanced by mutations to increase interactions with the receptor, arrived at using X-ray structures of the IL-18/IL-18 receptor complex to identify opportunities for enhancement of salt bridges, and so on. This engineering was necessary in order to develop this protein into a potentially viable commercial product.[4]

Subtilisin is another example of successful sequence analysis and protein engineering, resulting in a protein that is more stable to auto-proteolysis and storage. Comparison of the sequence of six homologous proteases identified a few sites where subtilisin BPN′ diverged from the others. Mutating BPN′ subtilisin to contain the consensus sequence increased the stability of the protein. Other opportunities for increasing stability that were successful for subtilisin include adding salt bridges, adding a disulfide bond, or even mutating residues in the interior of the protein to supply methyl groups, or other hydrophobic moities, to fill cavities in the interior of the molecule. Subtilisin is a metal-binding protein, with a high- and low-affinity Ca binding site. Increasing the affinity of these binding sites for calcium also increased the conformational stability of the molecule.[6,7] All of these efforts resulted in a product that was more stable, producing a more homogeneous product with less clipping and aggregation.

6.2.2 Stability to Process Conditions

For biopharmaceutical manufacturing, the purification process is critical in eliminating product- and process-related impurities, including aggregates. However, the process conditions themselves, including pH, buffer components, and denaturants, may induce unfolding of proteins. The unfolded species can lead to protein aggregation, which may cause filter clogging, loss of product, and poor yield during the purification process. If the unfolded species are not completely removed, they can also serve as nucleation sites resulting in aggregation of the protein with time; this may cause specification failure, poor appearance (visible particulates), loss of storage stability, loss of efficacy, and safety concerns in the product, including potential for immunogenicity. Therefore, comparing the stability of the candidates under potential process conditions can help eliminate less stable molecules. It can also provide information on the stability of the selected molecule, identifying potential problems that may be encountered.[8,9]

This is an area that is fairly new but is being actively pursued within the industry. The examples below illustrate the analysis of the conformational stability and propensity for self-association of the candidate molecules under process-related conditions.

6.2.2.1 Specialized Techniques for Stability Analysis In addition to the techniques described in the previous chapters for measuring oligomers and aggregates, traditional conformation-sensitive techniques can be employed to assess the effects of process and formulation conditions on the conformations of the product candidates. Particular attention should be paid to any irreversible changes in the molecule. Secondary structure changes can be assessed by Fourier transform infrared (FTIR), circular dichroism (CD), and Raman spectroscopies, tertiary structure by fluorescence, CD and Raman spectroscopies, changes in the surface properties by dye binding, and thermal stability by differential scanning calorimeter (DSC). For the secondary structure analyses, CD is particularly sensitive to changes in alpha-helical structure, while FTIR and Raman spectroscopies are more sensitive to beta-sheet structures as well as typically being performed at higher protein concentrations (about 10 mg/mL and higher for FTIR and >30 mg/mL for FT-Raman). FTIR in particular can identify the presence of intermolecular beta-sheet structures, transitions which occur during protein aggregation. Near-UV CD spectra are sensitive to changes in the environment of the aromatic amino acids and disulfide bonds, while fluorescence monitors changes in the environment of Trp, and distances between fluorescent Tyr and tryptophan. The most commonly used dye for probing changes in the surface of the proteins is 8-anilinonaphthalene 1-sulfonate (ANS); this probe fluoresces upon binding to exposed hydrophobic patches on a protein surface while it has almost no fluorescence when free in solution. For proteins such as antibodies that are very hydrophilic, there is very little background fluorescence in the presence of the native protein. Thus, ANS fluorescence occurs only when the protein structure is perturbed, exposing residues that are not normally accessible, making it a very useful assay for these types of proteins.[9] All of the analyses described below are performed at ambient temperature, unless otherwise indicated, at protein concentrations optimized for the technique. Reversibility of the changes is probed by incubation of the protein for 2 h under the conditions indicated, followed by dialysis overnight at 4°C into phosphate buffered saline. These are included to illustrate how this approach can be applied to optimize candidate selection; the exact protocols used should be optimized for the system being studied.

6.2.2.2 Example 1: Analysis of Fc Fusion Protein Candidates This candidate screening approach was successfully applied to assess six Fc-conjugated constructs of molecules, or peptibodies, where either two or four copies (2x or 4x) of the non-Fc moieties were attached to the Fc molecule (2xP, 2xQ, 2xR and 4xP, 4xQ, and 4xR). The conformational and thermal stability of the constructs were compared at low and neutral pH values using FTIR and fluorescence spectroscopies and DSC.[10] From these analyses, the 2x constructs were more stable to low-pH and thermal-induced unfolding than the corresponding 4x constructs. The 4x constructs also showed decreased solubility at low pH and precipitated before reaching the concentrations required for FTIR spectroscopy.

Changes in the surface of the proteins resulting from unfolding were also compared, using ANS binding as a probe for increasing exposure of hydrophobic residues. The numbers of hydrophobic residues in the sequence of the 4x constructs are very similar, and yet the ANS binding is quite different. At neutral pH, ANS fluorescence in the presence of the 4xR is at least twofold higher than that of the other 4x constructs, indicating that this construct has more of its hydrophobic residues exposed to the surface. The 4xR construct had the highest ANS fluorescence, while the 4xP had the least amount of ANS fluorescence. This trend was repeated for the 2x analogs. The results suggested that the P analogs should be the most stable to process conditions, and result in decreased aggregate formation relative to the other candidates.

DSC can be used to identify the changes in thermal stability of a protein (or a protein domain) and the changes in solubility of the unfolded forms of a protein.[11,12] The thermal stability as determined by the first (CH2) transition, and solubility information (indicated by where the exotherm from aggregation occurs) can help to evaluate the relative stability of protein constructs in the same buffer and predict the possibility of aggregation and precipitation. DSC results suggested that the 4x constructs at both neutral and low pH values are not as stable as their corresponding 2x constructs, and that the 2xP appears to be slightly more stable than the other molecules. This is consistent with the ANS and other results described above. These results can be used to help select which candidate to advance to the next stage of development.

6.2.2.3 Example 2: Monoclonal Antibody (mAb) Candidates In another example of candidate screening, four mAbs (mAbs A, B, C, and D) generated against the same target were compared. For mAbs, the downstream process steps for protein A column elution and viral clearance involve a pH of around 3.5, and the process also includes exposures to pH 5.[13] Therefore, the effect of pH 3, 3.5, and 5 on the conformation, thermal stability, and self-association propensity of the candidates and the reversibility (into PBS) of any changes that occurred were compared to rank their relative stability under the process-related conditions using CD, DSC, and dynamic light scattering (DLS). The reversibility of any changes induced by these conditions can be assessed by incubation in the desired buffer followed by exchange into physiologically relevant conditions.

Figures 6.1 and 6.2 show the near- and far-UV CD spectra of the four candidates in PBS after incubation at pH 3, 3.5, and 5, as well as the original pH 3 spectra. At low pH, both the near- and far-UV CD spectra of the four candidates showed changes (data not shown) indicating tertiary and secondary structural changes in the proteins. Significant structure is still retained at pH 3 for all four candidates; if the proteins were totally unfolded, the near-UV CD signal would be a flat line at about zero across the spectrum. The near- and far-UV CD spectra of the proteins in PBS after incubation in pH 3.5 are identical to those after the pH 5 incubation; the changes induced at pH 3.5 appear to be reversible. The spectra of candidates B, C, and D in PBS after incubation at pH 3 and 5 are different, suggesting that the changes in the

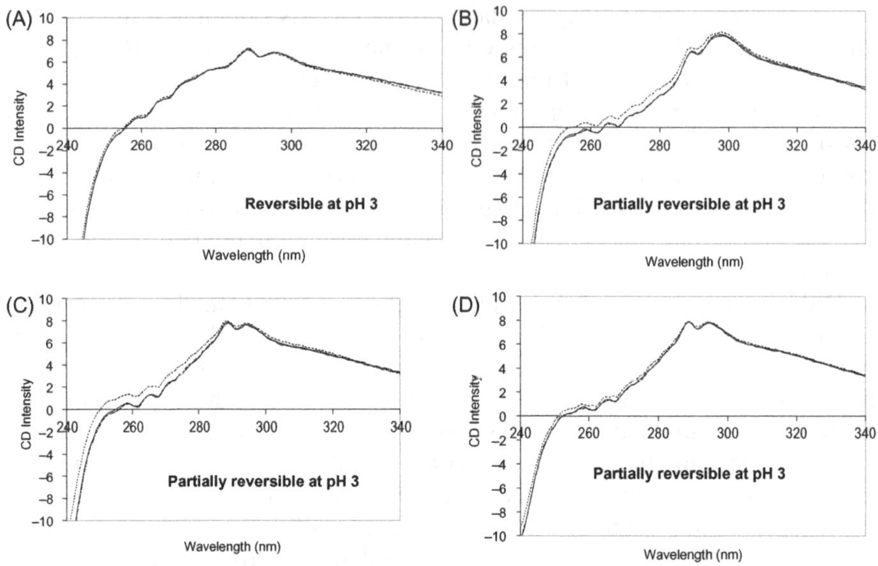

Figure 6.1. The near-UV CD spectra of the four mAb candidates (A, B, C, D) in PBS after incubation at pH 3 (-----), 3.5 (-··-·-), and 5 (—).

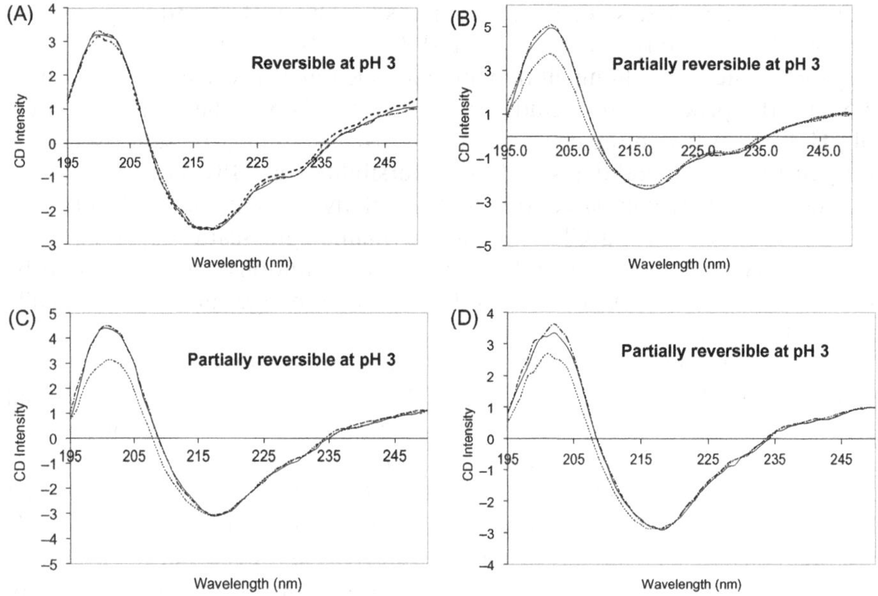

Figure 6.2. The far-UV CD spectra of the four mAb candidates (A, B, C, D) in PBS after incubation at pH 3 (-----), 3.5 (-··-·-), and 5 (—).

overall conformation of those three candidates at pH 3 are not fully reversible. In contrast, the spectra of mAb A in PBS after incubation at both pH 3 and 5 are identical; thus, the changes in the conformation of candidate A induced at pH 3 appear to be reversible.

Figure 6.3 shows the DSC scans of the four candidates at pH 5, 3.5, and 3. At pH 5, there are two endothermic thermal transitions, which represent the unfolding of the C_H2 domain at ~64°C and the unfolding of the Fab and C_H3 domains at ~74°C. The thermal stability of all candidates decreased with decreasing pH. The unfolded proteins precipitate at pH 5 but become very soluble at pH 3.5 and pH 3. Candidates A and B showed slightly better thermal stability than C and D at all pH values. The solubilities of the unfolded forms of the candidates are very similar. Changes that were observed in the thermal stability of the four mAbs at lower pH values were fully reversible upon dialysis into PBS within detection limits (data not shown). Overall, the DSC results indicate that the candidates are most stable at pH 5, and the unfolded forms are soluble at low pH.

DLS measures the time correlation of the intensity of the scattered light to determine the diffusion coefficient/size and the size distribution of the particles. As an example of this type of analysis, Fig. 6.4 shows the size distribution by intensity profiles of the four candidates at pH 3.5 and room temperature. Candidate B appeared to be the most heterogeneous with the highest *Z*-averages at all three pH values, and contained oligomers and large particles. The heterogeneity increased with decreasing pH for all four candidates. Candidate B also showed a greater tendency for aggregation at pH 3 and 3.5; the increased heterogeneity of the initial sample might contribute to this. The extents of pH-induced aggregation of the other candidates are similar to each other; they become heterogeneous at pH 3.5 and contain even more of the high-molecular-weight species (HMWS) at pH 3.

The size distribution by intensity profiles of the samples at neutral pH, following incubation in low pH (reversibility studies), was also studied. The self-associations that occurred at pH 3.5 and below were not fully reversible for any of the four candidates, with more HMWS remaining after exposure to pH 3. Candidate B appears to be the most heterogeneous and exhibits the least reversibility among all four candidates after the pH 3 treatment. This candidate appears to have an increased tendency to aggregate under conditions that might be encountered during processing of mammalian-derived antibodies. The relative stability of the candidates was similar for all of these analyses; this will not always be the case.

These results also indicate that even though the changes in secondary and tertiary structure as analyzed by changes in spectra are reversible, at least some aggregation was not. This could be due to differences in the sensitivity of the analytical methods, or could indicate that the HMWS detected by the DLS have native conformation.

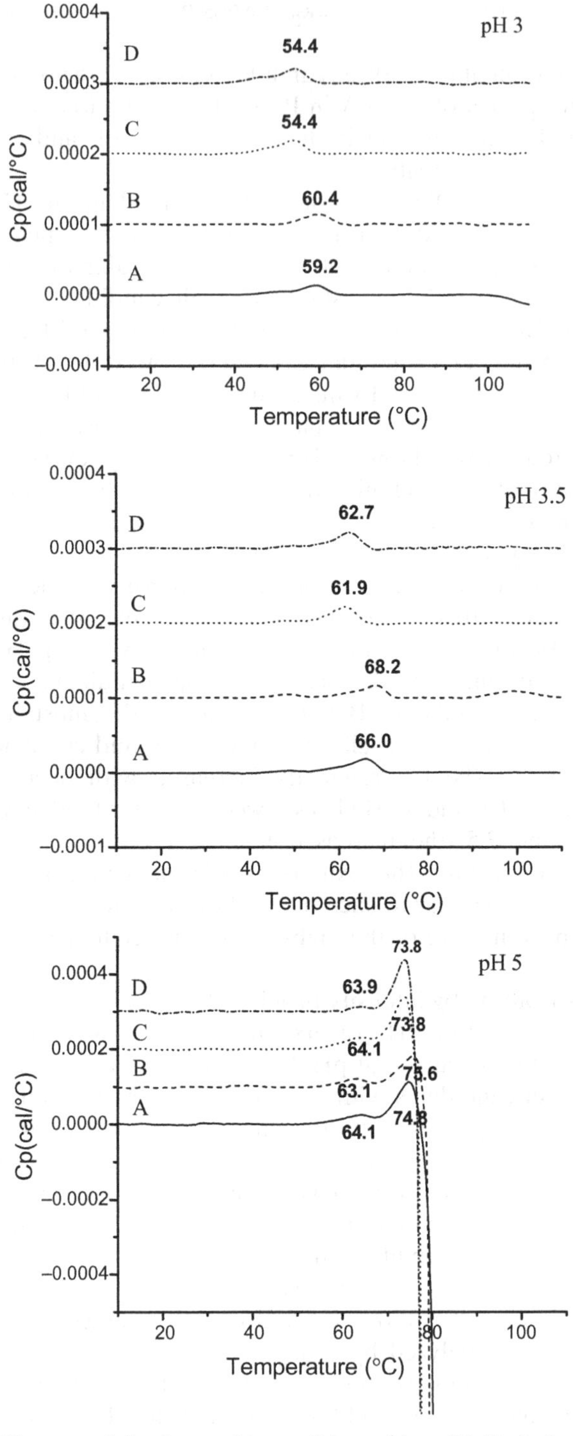

Figure 6.3. DSC scans of the four mAb candidates (A, solid; B, dash; C, dot; D, dash dot) at pH 3, 3.5, and 5.

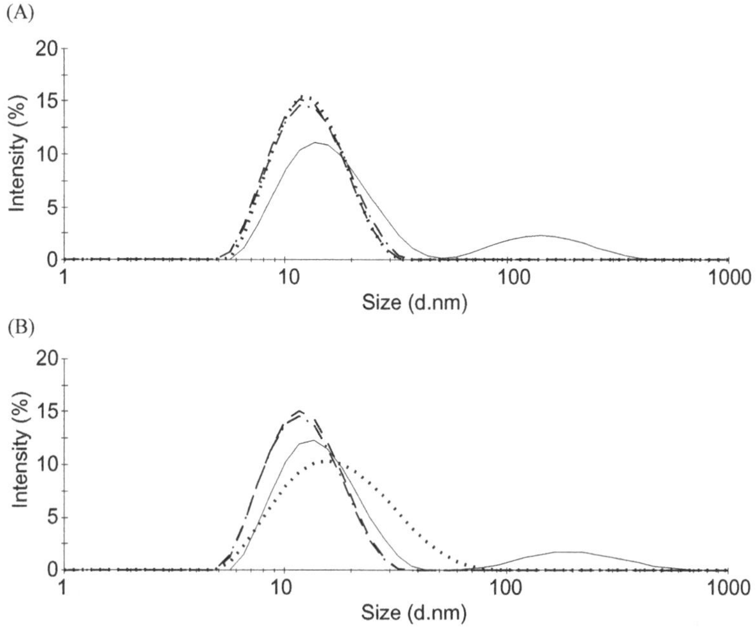

Figure 6.4. Size distribution by intensity profiles of the four mAb candidates (A, dot; B, solid; C, dash; D, dash dot) at pH 3.5 (a) and in PBS following incubation at pH 3.5 (b) as examples.

6.2.3 Focus on Stability to Formulation and Long-Term Storage Conditions

Protein stability during storage and delivery is the primary focus of formulation development. Aggregation during storage and delivery is one of the major factors that affect protein stability and is a product quality attribute that must be monitored and controlled. Therefore, the propensity for aggregation of the different candidates under storage and delivery conditions should be considered in selecting the candidate to advance to the next stage in the development cycle. Various studies, including forced degradation, can be applied to proteins to assess their tendency to aggregate, including aggregate formation during storage, light exposure, agitation, and freeze-thaw in potential formulation buffer(s). Studies also need to assess the amount of aggregation that occurs following the transition from formulation buffer into neutral pH to simulate administration into physiological conditions. The stability of a protein to different formulation buffers, devices, and bulk processing steps can limit the delivery options available and affect the potential uses of the product. Several studies to assess the stability of a protein to formulation conditions that illustrate this approach are given below. More extensive general discussions and review of the underlying physical principles are provided in Chapters 7 and 8, with additional case study examples in Chapter 9.

6.2.3.1 Assessing Stability after Storage and Freeze-Thaw Hold storage and freeze-thaw are important bulk production steps, which are known to stress proteins and cause aggregation.[14] Unstable candidates incur a greater resource burden during bulk process and formulation development. The clinical and marketed dosage forms and drug product shelf life will be affected by suboptimal stability of the therapeutic protein. During candidate selection, the stabilities of the candidates can be compared in potential formulation buffer(s), after incubation at different temperatures, and after freeze-thaw cycles. Visual inspection and size exclusion–high performance liquid chromatography (SE-HPLC), sodium dodecyl sulfate–polyacrylamide gel electrophoresis (SDS-PAGE), or equivalent assays should be used to ensure product stability under storage conditions, and to rank the relative stability of product candidates when stressed at elevated temperature. Examples of studies to assess the stability of candidates to formulation conditions are presented in the subsections below.

6.2.3.1.1 Example: mAb Candidates Candidates M and N are two different IgG2 antibodies generated to the same target. To help inform the choice of which one to advance to the next stage of development, their stability to storage conditions was compared. No instability was detected in either candidate following freeze-thaw cycling, storage at –20°C or storage at 4°C (data not shown). A difference in the degradation patterns of the two candidates was observed by size-exclusion chromatography (SEC) for samples incubated at 37°C (data not shown). Incubation of candidate M induced the growth of one early eluting peak and two late eluting peaks; incubation of candidate N induced growth of multiple early eluting peaks in the SEC chromatograms. For candidate M, the early eluting SEC species are aggregated forms of the protein, and the late eluting peaks represent low-molecular-weight protein fragments. For candidate N, one of the early eluting peaks has been shown to be a chemical degradation product; other early eluting isoforms are aggregated and/or self-associated forms of the protein.

The SEC chromatograms were integrated, and the percent main peak for each candidate was plotted as a function of time (data not shown). Loss of main peak for candidate N occurred at a rate that was approximately five times faster than that of candidate M.

6.2.3.2 Assessing Protein Stability upon Transitioning into Neutral pH To mimic the pH shift encountered by the protein following administration to patients, the solubility and aggregation of the candidates following a change from formulation buffer to PBS (at 37°C) can be assessed. The samples are analyzed for formation of visible particles and smaller oligomers.

The two antibody candidates (X and Y) were also subjected to this analysis, and the results of the change from potential formulation buffer to PBS, fol-

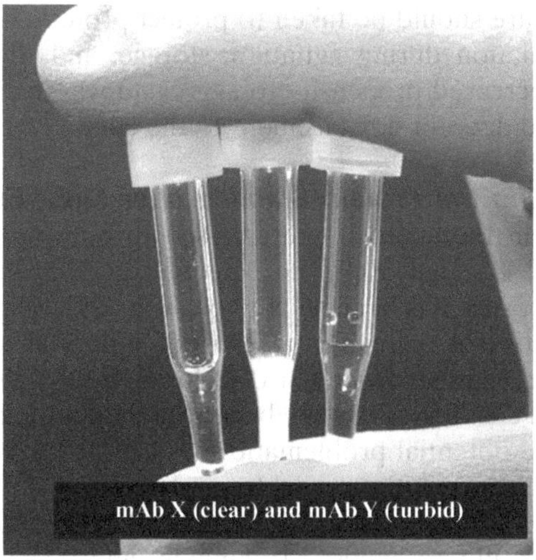

Figure 6.5. Picture of mAb candidate X and Y and buffer control in PBS following a change from potential formulation buffer and after a subsequent incubation of 6 h at 37°C.

lowed by a subsequent incubation of 6 h at 37°C of two mAb candidates and a buffer control, are shown in Fig. 6.5. mAb X appears most stable by visual inspection. SEC showed no significant increase in the amount of soluble aggregates for both candidates at time zero after the treatment (data not shown). Assessment of particles with sizes between those of the oligomers measured with SEC and those detected by visual inspection could also be included as part of this assessment.

6.2.3.3 Assessing the Stability after Agitation Agitation can be encountered frequently during transportation, sample handling, and filling of the bulk production process. To assess the propensity of a candidate to form particles as a result of agitation and exposure to the air/water interface in potential formulation buffer(s), the protein can be agitated in the buffer(s) for a period of time, followed by visual observations, DLS, and SE-HPLC analysis. The agitation protocol used should be one that has demonstrated predictability for the system being analyzed.

The SEC analysis of the protein samples before and after agitation demonstrated that the agitation resulted in a significant increase (>50%) in the amount of soluble aggregates in protein Z. DLS analysis was also performed, and the results were consistent with those of the SEC, clearly indicating that agitation resulted in an increase in the amounts of larger-sized species, an increase in sample heterogeneity, and a concurrent decrease in the amount of monomer (data not shown). The results from all the aggregation-sensitive assays are consistent and suggest that this therapeutic candidate is labile to agitation stress. Care should be taken to protect protein Z from aggregation and, thus, particulation during agitation, storage, and shipping. If another candidate with decreased propensity to aggregate was available, it would be the preferred candidate for advancement to the clinic.

6.2.3.4 Assessing the Protein Stability after Light Exposure During production, protein therapeutics are exposed to both ambient light in the facilities used and UV light from the detectors used during the purification process. Light exposure has been shown to cause oxidation and cross-linking of proteins, formation of aggregates, and loss of biological activity.[15] Assessing the propensity for aggregation and/or discoloring upon light exposure of different therapeutic candidates may be beneficial in identifying potential problems early on. The potential problematic sites can be engineered out, or the results can be used to rank the available candidates.

To evaluate the susceptibility of light-induced discoloration and aggregation, candidates should be exposed to ambient and intense sources of light under controlled conditions for various times, up to several days. Following this exposure, visual inspection can be used to follow the formation of large visible aggregates. Yellowing can be determined by visual inspection, or by the use of colorimetric measurements. Assessment for aggregates including DLS, SEC, or subvisible particle analysis could also be included in this assessment, and perhaps most importantly, bioactivity needs to be assessed. If this candidate is advanced through the pipeline, then this knowledge can be used to ensure that it is protected from light as much as possible.

6.2.4 Specific Considerations for Mammalian Cell Culture-Derived Protein

Proteins derived from mammalian cell cultures are generally folded and glycosylated.[16] The presence of carbohydrate on a protein has been shown to increase protein stability and solubility.[17–19] Carbohydrate also often plays a crucial role in the bioactivity of the protein.[20,21] The purification of mammalian cell culture-derived protein includes a viral inactivation step at a low pH (pH 3–4).[22] Since low pH may induce unfolding of the proteins and lead to protein aggregation, assessing the aggregation propensity and reversibility of the candidates following low pH as described above is strongly recommended to ensure that the most stable candidates are selected for further development to avoid potential downstream problems.

6.2.5 Specific Considerations for Bacteria-Derived Protein

Proteins derived from bacterial cells, especially *Escherichia coli*, are generally unfolded/aggregated (e.g., inclusion body) and non-glycosylated.[23] The purification of these proteins typically involves refolding processes to ensure the correct higher-order structure, disulfide linkages, and activity. Comparing the structural changes, aggregation, and reversibility of the candidates under potential refolding conditions may provide value for rank ordering the candidates and selecting the most viable molecule for further development.

6.3 PROTEIN AGGREGATION AND CELL CULTURE

Generally, after candidate selection to minimize particle formation, as discussed in the previous section, protein aggregation is a product quality issue addressed more intensively at the purification and formulation steps as compared with the upstream cell culture stage. However, several published studies, albeit limited, describe various efforts to minimize protein aggregation during the cell culture process.

6.3.1 Expression Strategies

One of the possible mechanisms for aggregation at the cellular level involves high levels of recombinant protein expression overloading the endoplasmic reticulum (ER) machinery, resulting in an increased number of misfolded or aggregated proteins improperly exiting.[24,25] Recognizing the role of chaperones in regulating the folding and secretion of the proteins, several groups have attempted coexpressing chaperone proteins in recombinant cell lines.[26,27] However, this coexpression strategy has not been universally successful, with negative impact on expression levels observed in some cases.[27]

An alternative approach that has been shown to be successful in reducing protein aggregation levels is coexpressing the soluble receptor with its cognate ligand. For example, the levels of HMWS were significantly reduced when a soluble 1L-13 receptor (sIL13R), an Fc fusion protein, was coexpressed with its cognate cytokine, IL-13.[24] This improvement in product quality was accompanied by a significant increase in the sIL13R expression level. A possible explanation for the observed improvement due to this coexpression is that the cytokine functions as a scaffold, which guides the recombinant protein to adopt and maintain a stable conformation as it exits the ER and traverses through the secretory pathway.

Studies have shown that excess heavy chain has a propensity to aggregate, and efficient mAb production requires excess amounts of light-chain polypeptide.[28–30] The heavy-chain polypeptide has been shown to associate with ER stress proteins such as BiP, until it can be properly folded with the light chain.[28,30] In line with these observations, an inverse correlation between the antibody aggregation levels and the light chain-to-heavy chain mRNA ratio was detected during the early-stage clone screening process.[31] Thus, high-throughput measurement of heavy- and light-chain transcript levels will allow identification of clones with the propensity to form antibody aggregates during early cell line development. This information can be combined with the information obtained during the candidate screening process, to select a clone that produces the greatest percent of properly folded protein that has the best stability to process and storage conditions. However, since not all clones with high levels of antibody aggregation were identified by a low ratio of light-chain mRNA to heavy-chain mRNA, it is likely that there are multiple mechanisms of aggregation.

6.3.2 Bioreactor Conditions

In addition to expression strategy, certain bioreactor conditions have been shown to favor reduced aggregation. In Chinese hamster ovary (CHO) cells, decreased protein aggregation and misfolding have been observed by adjusting the temperature and/or the pH of the cell culture condition.[24,32,33] Culture temperature was also observed to affect the amount of protein produced in the aggregate form in the *E. coli* microbial system, in which shifting the temperature to 18°C allowed expression of largely soluble proteins, in contrast to completely insoluble protein expressed in inclusion bodies at 37°C.[34]

Components in the cell culture medium can also affect the levels of aggregation. The addition of copper sulfate ($CuSO_4^+$) to the culture medium of a mAb expressed in CHO cells was shown to result in a threefold decrease in the level of free thiol.[24,35] This reduction in free thiol level can reduce the level of covalent dimer formation, which results from the interaction between the two free thiols each present on individual Ab monomers. Another component that has been reported to reduce high-molecular-weight (HMW) aggregation is carnosine.[36] Cells grown in a culture medium containing carnosine and other carnosine-like antioxidant/anti-senescence compounds were shown to produce decreased levels of HMW aggregates. In addition to reduction in aggregate levels, enhanced overall production of recombinant protein and cell viability were observed upon addition of these compounds.[36] A possible explanation for the observed improvement in product quality is that carnosine provides a favorable milieu (both inside and outside of the cells) in which the recombinant proteins can fold properly and maintain structural integrity. It is difficult to pinpoint the exact mechanism through which carnosine provides a "favorable milieu" because carnosine is a multifunctional dipeptide. The known beneficial properties of carnosine include antioxidant activity, buffering activity, membrane protection properties, anti-senescent activity, and metal-chelating activity. It is not known whether one, or several, or all of these properties contribute to providing a favorable milieu. Another study demonstrated that by minimizing the exposure time to cell culture fluid (i.e., time in the conditioned media after the cells were harvested), the extent of mAb aggregation can be minimized.[24] These observations suggest that certain components in the cell culture media, whether secreted from the cells themselves or in combination with the media components, can induce or expedite the protein aggregation process.

Careful control of the culture conditions can minimize the amount of aggregate in the initial media used as the source for purification of the drug product, resulting in less aggregate for the process to remove, and potentially a more stable, aggregate free product.

6.4 PROTEIN AGGREGATION AND PURIFICATION

The downstream purification process is the critical place where the drug product purity profile is defined. There are a number of impurities and contaminants that must be removed or controlled by the purification process, including host proteins, DNA, and product variants. Perhaps the most prevalent and important impurities are product aggregates.

During processing, aggregates are hard to visualize and track because while simple online measurements, like turbidity, may be able to quantify relative levels of aggregation, they cannot distinguish different particle sizes and they do not adequately track soluble aggregates.[37] Typical conditions present during the production, purification, storage, and handling of proteins can affect protein structure such that aggregation occurs. This aggregation can result in reduced efficacy, yields, and stability.[38] There is a seemingly endless list of triggers for aggregation, including pH, temperature, ionic concentration, protein concentration, interfacial stresses, and cavitation, all of which can result in perturbation of the protein structure or unfolding, which may in turn lead to aggregation.

C. B. Anfinsen proposed that proteins are in equilibrium between the native and unfolded states and that the native state will tend to be at the lowest Gibbs free energy that can be achieved without breaking covalent bonds;[39] this has been described in more detail in Chapter 1. This assumption is termed the thermodynamic hypothesis for protein folding. The thermodynamic hypothesis assumes both that protein is in equilibrium and that the native state is at the global minimum Gibbs energy.[39] This is important because environmental conditions and chemical or physical stressors can affect systems in equilibrium, which can either be used to our advantage or work against us. The thermodynamic hypothesis was developed to explain the folding equilibrium of proteins in isolation, where self-association does not occur; application to systems where self-association can also occur is relatively recent (cf. Chapters 1 and 2). In order to minimize both aggregation and other non-native protein conformations, the environment proteins are exposed to during purification can be actively manipulated either physically or chemically, to shift the equilibrium away from the aggregated and unfolded states and toward the native, monomeric protein.

While the thermodynamic theory maintains that the native protein state is at or near the Gibbs energy minimum, in cases where aggregates form through unfolded intermediates or even covalent linkages, it is possible that the local free energy minimum for some aggregates may be stable enough to effectively sequester protein in a stable aggregate form and essentially preclude those molecules from participation in the equilibrium balance (see Fig. 6.6). Couple

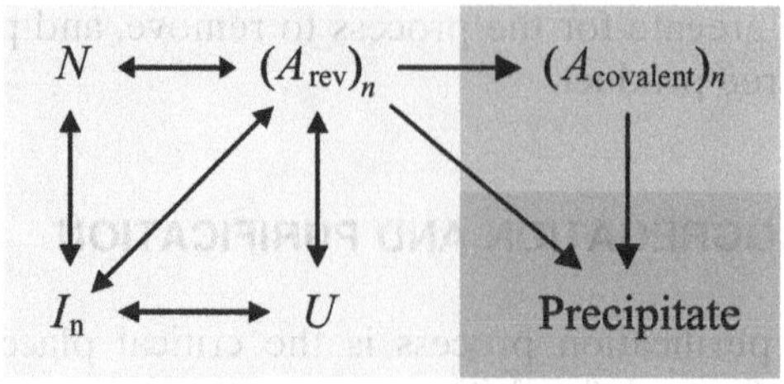

Figure 6.6. Equilibrium state diagram. The native state (N) is in equilibrium with the unfolded state (U) and the reversible, non-covalent aggregated state (A_{rev}) through a series of intermediates (I_n). States that are in a reversible equilibrium are represented in light gray. A_{rev} can also form covalent intermolecular bonds but remain soluble ($A_{covalent}$). This protein is no longer recoverable in the equilibrium equation (medium gray). Aggregated forms that become insoluble (precipitate) can no longer be recovered and represent a process yield loss (dark gray).

this with the fact that under some conditions aggregation kinetics can be at least 10 times faster than folding kinetics, and the situation seems hopeless.[40] Thus, intervention on the part of the protein chemist is necessary to reduce the rate of aggregation and prevent or limit shifts to irreversible aggregate.

On the other hand, not all aggregates are large irreversible aggregates that lead to precipitation or high-order covalent aggregates; many in-process aggregates do not form irreversibly. These reversible, self-associated proteins should adhere most closely to the thermodynamic hypothesis and can be driven to predictable proportions of monomer and aggregate via manipulations of the solution environment.

Hall and Abraham used equilibrium sedimentation studies to determine that IgG1-X myeloma protein (IgG-MIT), which is associated with the hyperviscosity syndrome, formed reversible aggregates. In this work, they demonstrated an effect of ionic strength and temperature on IgG-MIT self-association rates, which suggested that the weak protein interactions were mediated by dispersion forces, hydrophobic interactions, and potential ion shielding of local repulsive charge effects.[41]

Shultz and colleagues[22] presented work that highlighted how molecular structure and small peptide sequence differences can lead to strikingly different propensities for self-association, between two Fc peptide fusion molecules (peptibodies). Early evaluation of the stability of a highly hydrophobic peptibody molecule showed a very limited range of pH and salt levels where the monomeric form could be maintained. Further analytical work showed that two forms of aggregates were present (non-reversible/covalent and reversible/non-covalent species). The DSC analysis is shown in Fig. 6.7A,B. pH 5 is quite unique, in that "peptibody A" remains soluble even after the second thermal transition (C_H3 domain). Addition of salts to the buffer greatly destabilizes the molecule, lowering the onset temperature for protein unfolding in the C_H2

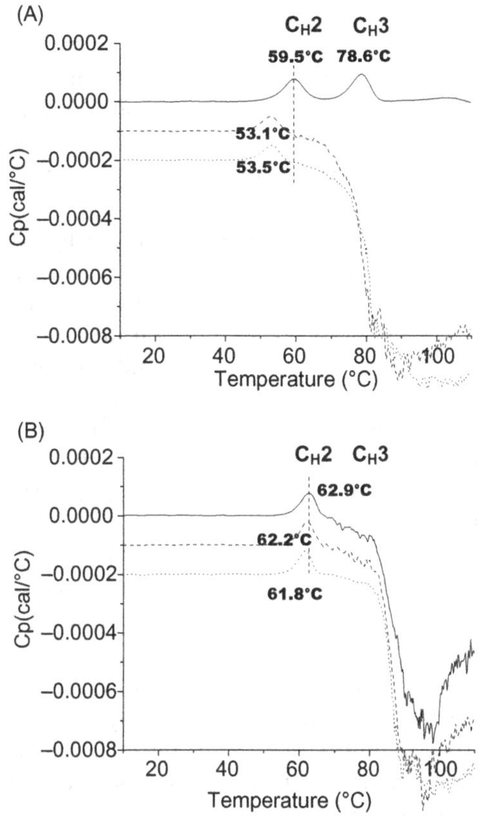

Figure 6.7. (A) "Peptibody A" stability at pH 5.0 by differential scanning calorimetry (Shultz et al.[22]) (solid: 20 mM sodium acetate, pH 5.0; dash: 20 mM sodium acetate, 0.3 M NaCl, pH 5.0; dot: 20 mM sodium acetate, 0.5 M ammonium sulfate, pH 5.0). (B) "Peptibody A" stability at pH 6.5 by differential scanning calorimetry (Shultz et al.[22]) (solid: 20 mM sodium phosphate, pH 6.5; dash: 20 mM sodium phosphate, 0.3 M NaCl, pH 6.5; dot: 20 mM sodium phosphate, 0.5 M ammonium sulfate, pH 6.5).

domain of the Fc region and precipitation of the molecule before the unfolding of the C_H3 domain can be visualized. At pH 6.5, "peptibody A" precipitates prior to visualization of the unfolding in the C_H3 domain. Addition of salts to the buffer shows slightly lower unfolding temperatures for the C_H2 domain, and similar precipitation of the molecule before the unfolding of the C_H3 domain can be visualized. Combining in-depth thermal stability measurements with DSC, ANS binding studies, tracking of molecular size by analytical ultracentrifugation (AUC), and techniques such as those shown in the candidate selection section of this chapter (for instance, Figs. 6.1 and 6.2) showed that the molecule was inherently unstable and that the mechanism of aggregation was via an anion-mediated, hydrophobic interaction driven by protein concentration, ion concentration, and pH.

The degree and form of aggregation was dependent on the protein concentration. In the presence of isotonic salt levels below 10 g/L protein concentrations, aggregates spanning all forms from monomer to heptamers were visualized. At concentrations above 10 g/L, the protein formed monodisperse heptameric clusters. It was further demonstrated that the self-association could be minimized by purifying the protein at pH 5.0 and very low salt concentrations.[22]

Regardless of the type of aggregation present during downstream processing, there are methods available to control them. In fact, it is not always necessary to prevent aggregation during processing. But rather, the requirement is to have minimal levels of aggregate in the final drug product and to control aggregate growth during drug substance (DS) and drug product storage. It has been suggested that not all aggregates are the same, in that non-dissociable HMW aggregates are likely to have a higher potential for immunogenicity in humans (cf. Chapter 10), while aggregates that readily dissociate in serum are of significantly less concern from a clinical perspective.[42] The stabilizing effects of hydrogen bonding and hydrophobic interactions are counterbalanced to a degree by configurational entropy (cf. Chapter 1). Therefore, any environmental effector that modulates hydrogen bonding, hydrophobic effects, or electrostatic interactions can shift the balance toward protein unfolding. Examples of these effectors include pH, temperature, and the presence of cosolvents that either stabilize or disrupt intramolecular interactions.[39] The rest of this chapter will summarize a number of aggregate prevention/control strategies for in-process intermediates and bulk DS processing. We will also discuss strategies for removal/reduction and considerations for controlling aggregation in the formulated DS.

6.4.1 Preventing Aggregation with Appropriate Solvent Conditions

Prevention of aggregation covers two categories: (1) *in-process*, where removal and control are both options and (2) *during DS storage*, where removal is no longer an option and the focus has to be on control. This section will provide an overview of common chemical, physical, and molecular means to prevent and/or control aggregates in in-process protein streams.

A number of elements must be considered when selecting the best approach to use, such as the stage of processing, interference with analytical assays, effect on equipment (e.g., steel corrosion with chlorides or guanidine hydrochloride), and the regulatory implications. Any additives must be cleared to non-toxic levels by the process. This can make it difficult to utilize certain agents in the final purification operations. It is also important to understand the appropriate concentrations of additives to use: too strong and the agent will denature or precipitate the protein; too low and no useful effect will be gained. Each additive has the potential to alter the protein in a deleterious manner and should be investigated with caution and balance. Also, some of the approaches discussed may be useful to the researcher but are not appropriate for large-scale

TABLE 6.1. Useful Additives to Increase Protein Solubility and Reduce Aggregation[38,47,56,57]

Additive Class	Additive	Recommended Concentration Range
Kosmotropes (non-precipitating levels)	$MgSO_4$	0–0.4 M
	$(NH_4)_2SO_4$	0–0.3 M
	K_2SO_4	0.4–0.6 M
	Na_2SO_4	0–0.2 M
	Cs_2SO_4	0–0.2 M
Weak kosmotropes (non-precipitating levels)	NaCl	0–1 M
	KCl	0–1 M
Chaotropes (non-denaturing levels)	$CaCl_2$	0–0.2 M
	$MgCl_2$	0–0.2 M
	LiCl	0–0.8 M
	RbCl	0–0.8 M
	NaSCN	0–0.2 M
	NaI	0–0.4 M
	$NaClO_4$	0–0.4 M
	NaBr	0–0.4 M
	Urea	0–2 M
Amino acids	Glycine	0.5–2% w/v
	L-arginine hydrochloride	0–5 M
Sugars	Sucrose	0–1 M
	Glucose	0–2 M
	Lactose	0.1–0.5 M
	Trehalos	0–1 M
	Raffinose	0–1 M
Polyhydric alcohols	Ethylene glycol	0–60% v/v
	Xylitol	0–30% w/v
	Mannitol	0–15% w/v
	Sorbitol	0–40% w/v
	Glycerol	5–40% v/v
Surfactants	Tween 80	0–0.2% w/v
	Tween 20	0–120 mM
	Triton X-100	0–10 mM
Polymers	Polyethylene glycol 4000	5–20% w/v
	Hydroxyethyl (heta) starch	2–20% w/v
	Gelatin	0–1% w/v
	β-Cyclodextrin	20–100 mM

biopharmaceutical manufacture (i.e., antibody-inhibitor binding or chaperones). Nonetheless, these approaches are included for completeness.

6.4.1.1 Additives Chemical additives can be utilized to modify the equilibrium away from aggregation and intermediates known to lead to aggregation. Table 6.1 summarizes many common additives that have been shown to be useful in controlling protein aggregate formation; a more detailed discussion of these is included in Chapter 4.

There are many types of solutes that may be added to a process stream to prevent aggregation. These agents may work via a number of mechanisms such as preferential interaction with the protein, preferential hydration of proteins, preferential exclusion of solutes from the protein surface, and destabilization of protein–protein interactions. Regardless of the mode of action, the unifying attribute of each of the described additives is their usefulness in driving the thermodynamic equilibrium in a way that favors the native protein state over the intermediate, unfolded, or aggregated states (cf. Chapter 2).

A wide variety of buffer cosolvents can be utilized to modulate protein folding and solubility. Additives can be used to directly destabilize aggregates while enhancing native protein stability. For example, polyhydric alcohols and sugars, like glycerol and sucrose, have been used as protein stabilizers for many years.[43] Sugars have been shown to effectively decrease the rate of aggregation as a formulation excipient and low concentrations of salts can disrupt electrostatic interactions that may lead to protein–protein interactions.[38,43] Kosmotropes raise the energy required to hydrate intermediate or denatured states relative to the native state, and chaotropes can disrupt protein aggregates.[38] Proteins themselves are charged molecules, which can result in specific ion interactions between the small counterions present in the buffers typically used for processing. Recent studies have further complicated this picture by demonstrating that the effect of the Hofmeister series of salts is concentration dependent. For anti-streptavidin, the relative stabilizing effect of ions below physiological concentration was the reverse of their relative effect above 150 mM.[44]

Biopharmaceutical process streams are complex mixtures containing many protein species, each with a different potential for unfolding and aggregation. By understanding how each additive class affects aggregation, it becomes possible to proactively adjust specific protein or system conditions toward the prevention of aggregation. The following sections discuss different additive classes and their mechanism for affecting aggregation.

6.4.1.1.1 Kosmotropes Kosmotropes are small molecules that tend to have high charge densities. Examples of ionic kosmotropes are sulfate, phosphate, magnesium, and lithium. These salts tend to bind water and thus have strongly negative hydration free energies (Chapter 4, this volume). The more negative the $\Delta G_{\text{hydration}}$, the more kosmotropic the salt; these are the ions found on the left of the Hofmeister series. The kosmotropes typically increase air–water surface tension and decrease the solubility of non-polar molecules.[45] This is often interpreted as leading to an increase in the strength of hydrophobic protein interactions, but has yet to be shown conclusively.

In moderation, kosmotropes can be used to stabilize the native states of proteins. This stability has been proposed to be gained, at least in part, because the entropic cost of hydrating intermediate and denatured states is greater than the native state.[38] Kosmotropes make water a poorer solvent, and this favors the protein minimizing its surface area. The charged state of the protein in the solutions being used, and the counterions present at this surface, can also play a role in interactions with kosmotropes and chaotropic agents. Despite the open questions regarding the complete molecular basis for kosmotrope and chaotrope effects, there are numerous experimental examples of their stabilizing and destabilizing effects for proteins.

An example of kosmotropic stabilization was reported by Foster et al., where they showed that the aggregation levels of heat-treated low-molecular-weight urokinase (LMW-UK) is reduced in the presence of up to 0.19 M ammonium sulfate. Little additional benefit was observed above 0.19 M, and at concentrations above 1.7 M, the LMW-UK protein precipitates.[37] Magnesium chloride was also shown to stabilize LMW-UK during heat treatment, at levels as low as 0.103 M. In this case, both of these agents were most likely stabilizing the folded monomer and thus decreasing aggregation of non-native protein. Arakawa and Timasheff noted the stabilizing effects of $MgCl_2$ and further demonstrated that the effectiveness of $MgCl_2$ is dependent on the pH and the nature of the protein, because magnesium ions can also bind to proteins.[37]

6.4.1.1.2 Chaotropes Chaotropes have the opposite effect of kosmotropes. These large monovalent ions of low charge density can increase the solvation layer around proteins and increase the solubility of non-polar molecules. Hofmeister ions, like I^- or SCN^-, which "salt in" peptide groups and interact more strongly with proteins in the unfolded state, are considered to be strong denaturants.[45] Large Hofmeister salts with low charge densities can act as chaotropes. Examples of such ions are bromide, iodine, potassium, and guanidinium. These ions shift the protein equilibrium toward unfolding by "salting in" the non-polar peptide groups.

Certain denaturants, such as urea or guanidine, can be used to either stabilize the unfolded protein state or aid in solubilization of proteins. These molecules can have strong direct interactions with the protein via the formation of hydrogen bonds with the backbone.[43] Other chaotropes may disrupt stable protein structure by interfering with hydrogen bonding or affecting the hydration shell.[46] Denaturants such as guanidine hydrochloride or urea are bound preferentially to proteins, and the free energy of the denatured state is reduced to a greater degree than that of the native state because the unfolded state presents a greater surface area to the solvent for binding.[14] Therefore, at high concentrations, chaotropes will shift the equilibrium of unfolding toward unfolded states.

However, at non-denaturing concentrations, chaotropes can inhibit intermolecular interactions that may lead to the formation of aggregates. Also, drawing from well-known protein refolding strategies, low levels of chaotropes can be used to lower the energy of activation such that the protein system can shift between aggregated, unfolded, and native state more freely. The intent of this strategy is to optimize the system at an equilibrium state that favors the native protein, and allows enough time for equilibrium to be reached, overcoming potential kinetic control. As will be illustrated later, chaotropes may be used to modify the proteins in a way that benefits the chromatographic separation of similar species.

6.4.1.1.3 Sugars and Polyhydric Alcohols Sugars are well-known protein stabilizers. The mechanism of action has been well described, by a number of groups, as preferential exclusion;[14,37,43] see also Chapter 2. Simple saccharide stabilizers are preferentially excluded from the surface of the protein and thus the protein becomes preferentially hydrated.[43] This results in an increase in chemical potential that is directly proportional to the surface area of the protein.[37,38] Therefore, the degree of exclusion is greater for unfolded or intermediate states, which have a larger surface area, increasing the free energy barrier between the more tightly folded native state and larger unfolded states. The resulting thermodynamic equilibrium will favor the native protein state and make it more structurally stable.[37]

In developing this understanding, sucrose was most often used as the model. However, Arakawa and Timasheff proposed that preferential exclusion is a common element in aqueous sugar systems at the concentrations typically utilized, and therefore, it is reasonable to assume that other simple sugars would have similar stabilizing effects as sucrose.[43] In fact, Foster and coworkers demonstrated this to be true for trehalose and raffinose on a mole for mole basis, and also showed that while glucose had stabilizing effects, it was less effective than sucrose.[37] They also suggested that sugars can increase the melting temperature (T_m) of many proteins. As a side note, fructose has been demonstrated to precipitate LMW-UK at levels as low as 5%. It is unclear why fructose acts as a precipitant, and there is little, if any, literature describing fructose use in preferential solute interactions.[37]

Another class of additives that can be used to stabilize native structure in proteins, and act by a similar mechanism to the sugars, is the polyhydric alcohols. Glycerol is a model polyol that has been repeatedly demonstrated to have utility in stabilizing proteins and has demonstrated its use as a cryoprotectant.[47,48] Glycerol is an osmolyte that does not interfere with most chromatography operations, other than increased viscosity. This and other similar low-molecular-weight polyols usually require concentrations of greater than several hundred millimolars to be effective.[14] The mechanism of action is thought to be similar to sugars and can provide non-specific protection to freeze-thaw damage by the preferential solute exclusion mechanism described earlier.[14] As an example of polyol stabilization, glycerol has been shown to protect maltodextrin glucosidase (MALZ) from denaturation and deactivation by urea.[23] Alternative polyols, with similar function, include xylitol, sorbitol, or mannitol.

6.4.1.1.4 Surfactants Surfactants are commonly used to inhibit protein aggregation at very low concentrations.[14] They inhibit aggregation by two mechanisms: by binding to hydrophobic surfaces on the protein and by competing with the protein for adsorption at hydrophilic–hydrophobic surface interfaces.[49] These compounds have been used to prevent aggregation during the purification of enzymes.[50]

The most common surfactants used for protein stabilization in the pharmaceutical industry are polysorbate 20 (Tween 20) and polysorbate 80 (Tween 80). These non-ionic detergents have been used extensively as cryoprotectants in therapeutic DS and drug product formulations. At very low levels (<0.01% w/v), polysorbate 20 can stabilize a wide variety of proteins to freezing damage without affecting the activity of the protein. Polysorbate has been effective in preventing protein aggregation in a number of process operations where interfacial stresses may occur, including protein folding, agitation, and freezing.[14] The surfactants are typically used at low concentration to avoid denaturing the proteins, and also to maintain a solution viscosity and density that will not interfere with the operation of the equipment being used, rates of filtration, and so on. In general, the surfactant concentration during processing is below that of the critical micelle concentration (CMC), particularly for chromatography where it could decrease interactions with the column matrix if used at high concentrations.

6.4.1.1.5 Polymers Hydrophilic polymers, at high concentrations, typically stabilize proteins via a mechanism of solute preferential exclusion from the protein surface, similar to that described above for sugars and other small hydrophilic solutes. At lower concentrations, polymers may decrease protein aggregation by increasing the molecular density and viscosity of the solvent.[46] Examples relevant to the pharmaceutical industry include dextran and polyethylene glycols (PEGs). Both are commonly used protein precipitants, because precipitation with these agents is less likely to denature the protein than other methods such as the use of isoelectric or kosmotropic precipitation.

Many other polymers have been reported to be effective as protein stabilizers. Vrkljan and coworkers demonstrated the effectiveness of gelatin, hydroxyethyl (heta) starch, PEG 4000 in preventing aggregation in heat-treated LMW-UK, while lower-order PEG forms were ineffective.[47] PEGs exhibit both hydrophobic and hydrophilic characteristics and HMW forms have shown protective abilities dependent on the PEG concentration, temperature, and the hydrophobicity of the protein.[47] PEGs may also be used in protein refolding to prevent aggregation.

6.4.1.1.6 pH and Ionic Strength during Processing Proteins are often least soluble near their pI. This is the pH at which a protein has an overall neutral charge, and will thus have decreased electrostatic repulsion with itself. Even though the reaction is reversible, precipitation during processing can result in complications during processing, including decreased yield, and should be avoided whenever possible. This leaves the protein chemist with two options when developing a process: avoid pH conditions at or near the pI, or adjust the ionic strength of the system to modulate the association of molecules in the system.[51] Data collected using LMW-UK suggest that at pH 6.5 in phosphate buffer, minimal ionic strength was required to inhibit aggregation. However, at pH 7.5, the benefit of increased ionic strength was severely muted.[51] Furthermore, large classes of molecules, such as mAbs, have been shown to unfold at low pH leading to aggregation.[13]

6.4.1.2 Protein Concentration during Processing Protein aggregation is greatly favored at relatively high concentrations (e.g., >10 mg/mL). These concentrations are common during the processing of protein therapeutic products.[14] Since protein unfolding is a unimolecular event for monomeric proteins, it should not be concentration dependent. Aggregation on the other hand, by basic principles of chemical equilibria and mass action kinetics, is likely to be concentration dependent, with higher levels of aggregate expected to form at faster rates with increasing protein concentrations.[51] During processing, aggregation can result from multiple pathways, both reversible and irreversible, including association of native monomers, and association of different unfolded intermediates. The different intermediates can each have their own reaction rate for oligomerization, and pH, salt concentration, and so on, can have different effects on the various equilibria involved. Thus, the conditions that favor the isolation of the native monomer can be highly protein specific.

Experimental work on LMW-UK observed a lag time prior to the formation of measurable aggregates. This work demonstrated that the duration of the lag was not dependent on protein concentration and was correlated with the rate of denaturation prior to aggregation.[51]

In another example of the effect of concentration on self-association during processing, Gokarn and colleagues demonstrated concentration dependence on the self-association of a hydrophobic peptibody protein.[52] Figure 6.8 provides an example of how protein concentration can affect the size of self-associated species in the presence of salt. In this example, not only is aggregation concentration dependent, but the protein forms monodisperse aggregates of specific size in relation to concentration. At pH 5.0, this protein does not self-associate unless salt is present at which point the protein associates through hydrophobic interactions. At concentrations between 1 and 10 g/L, the protein aggregate size increases with protein concentration until a final heptameric state is achieved and maintained at all higher concentrations. It is believed that the protein forms stable heptameric clusters at these higher concentrations. For this protein, the mechanism of self-association is via reversible hydrophobic or ionic interactions and does not involve unfolded intermediates, resulting in a rate of aggregation that is concentration dependent.[52]

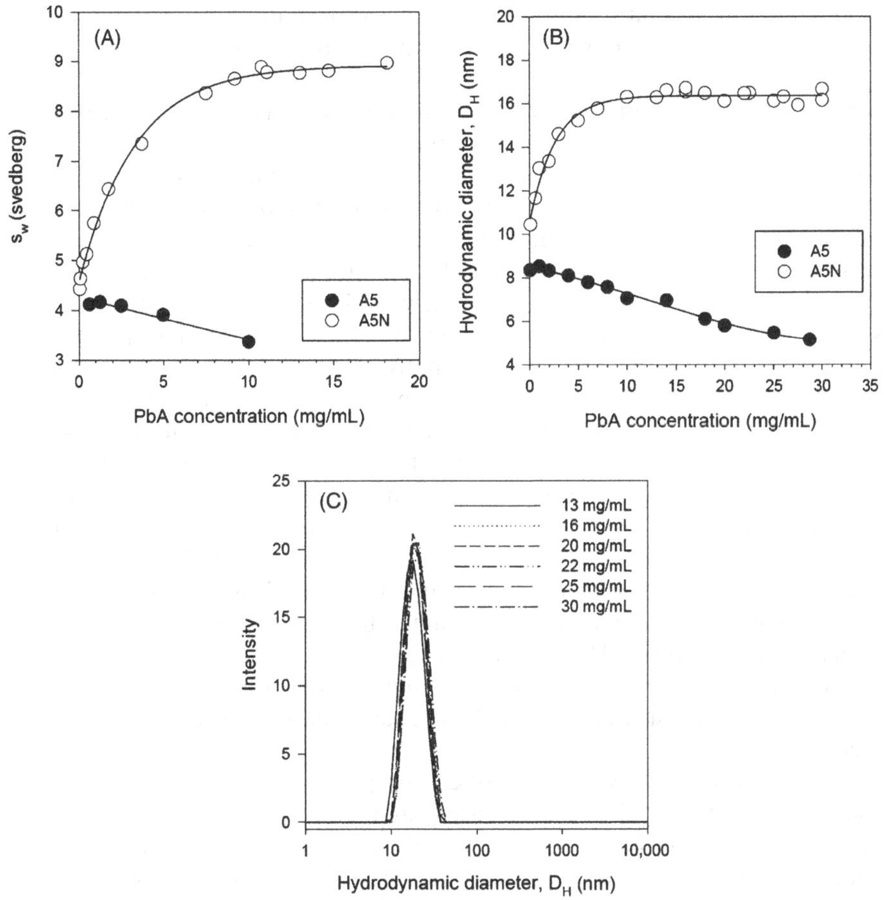

Figure 6.8. Example of concentration-dependent, reversible self-association in a peptibody. The protein self-associates, at pH 5.0, only in the presence of salt and forms monodisperse oligomers that are dependent on protein concentration. The sedimentation velocity (SV) as determined by AUC, dynamic light scattering (DLS), and static light scattering (SLS) are shown versus protein concentration in both pH 5.0 acetate buffered solution (A5) and the same acetate buffered solution with isotonic sodium chloride (A5N).

6.4.2 Removal of Aggregates during Processing

The ultimate goal of protein purification is to have a protein, or in the case of biotherapeutics, a drug product, that is as homogeneous and free of impurities as possible. Aggregates fall into the impurities category. Some level of aggregate will inevitably be present in the cell culture, and could also be generated during processing. Thus, one of the functions of the final steps in the process is to remove as much aggregate as possible, finishing with the protein under storage conditions that stabilize the monomeric protein. This section describes some techniques for removing aggregate during processing.

6.4.2.1 Precipitation A crude separation method that may be useful in the removal of undesirable aggregates is precipitation. This is a very old concept, and this technique can be applied to large volume bulk. As such, it is often the first step in a large-scale purification process. Precipitation can be affected in a number of ways. For example, kosmotropes at higher concentrations can be used to "salt out" proteins. These salts enhance protein association and can result in precipitation of aggregates that would otherwise have stayed in the solution. A careful balance of salt concentration may preferentially precipitate unfolded or aggregated protein versus native monomer.[45]

Other well-practiced precipitation techniques include isoelectric precipitation and polymer precipitation. Isoelectric or "acid" precipitation works by transitioning the pH of the solution to or through the pI of the protein. At the pI, there is no charge repulsion, and the net neutrally charged protein can interact and self-associate. The aggregated protein often becomes insoluble and forms large particles. It is possible to affect a significant degree of aggregate removal with this method. This has been used effectively to reduce aggregation in early crude process streams. Hart et al. presented work on the effects of precipitation pH, salt, urea, and protein concentration, and temperature on the effectiveness of aggregate removal across 10 microbial molecules. This work demonstrated up to 30% reduction of aggregate using this technique.[53]

At high enough concentrations of certain polymers, such as PEG, proteins can be precipitated. This may be used, under the right conditions, for the separation of monomer and aggregates based on the premise that molecules with larger surface (e.g., unfolded or aggregated proteins) will be most affected by the polymer exclusion and will preferentially precipitate.

6.4.2.2 Chromatography The most common and selective method for the removal of aggregate from a process stream involves the use of various chromatographic separations. Many modalities have proven useful for this purpose, including hydrophobic interaction chromatography (HIC), ion exchange chromatography (IEX), and, in some cases, hydroxyapatite chromatography, depending on the protein.

6.4.2.1.1 HIC HIC can be very effective in separating aggregate species from monomeric (native) proteins. This technique is useful across a wide range of protein types. HIC often exploits the differences in the hydrophobicities of the respective species to affect a separation.[54] Since the unfolded and aggregated protein states frequently have increased degrees of solvent-exposed hydrophobic patches, they should be retained to a greater extent in HIC media. In fact, in typically large biotherapeutic molecule classes, such as mAbs and Fc-protein conjugates, aggregates are usually resolved as late eluting species. However, McCue et al. also showed that irreversible adsorption of aggregates to the resin was a major contributor to the separation of monomer and aggregate species, as opposed to differences in the binding properties of the two species to the hydrophobic adsorbent.[54]

6.4.2.2.2 IEX IEX has been used to effectively separate aggregate from monomer, across a wide range of protein types. In typical large biotherapeutic molecule classes, such as mAbs or Fc-protein conjugates, aggregates are typically separated as late eluting species on both anion and cation exchange media. Regardless of the pI of the molecule, aggregate usually binds tighter than the desired protein product. This suggests that the aggregated protein has a larger surface interfacing with the resin and is more difficult for salt to dissociate.

There are, however, instances when fast-reversing self-association can hinder separations on ion exchange media. When the kinetics of aggregation and subsequent dissociation are rapid, the protein may transition between native and aggregated state many times during the chromatographic experience. Figure 6.9A depicts the effect of fast-acting transient aggregation on the separation of protein over an ion exchange operation. If the product-related impurities (I) normally elute earlier than the desired product (P) species and the aggregated species tend to elute even later, then in a fast-reversing dynamic equilibrium, the impurities would not be as well resolved from the product species as was projected. In this example, the separation was improved by adding 0.5 M urea to the chromatography solutions and the protein load. Figure 6.9B depicts the improved yield of product species and concurrent reduction of impurity species in the elution window. The urea acted to reduce the tendency of the molecule to self-associate in the presence of salt and allowed for significant improvement in the separation and yield.[22]

6.4.2.3 Dissociation As mentioned earlier, while separation of process impurities (e.g., host cell proteins, DNA, viruses, HMW covalent aggregates, and product-related impurities) is a must, it may not be necessary to remove all aggregate types. Reversible self-associated species may be disassociated during the final DS formulation or other operations during processing. Many separation operations are unable to differentiate between covalent and reversible aggregates, resulting in a potentially unnecessary yield loss. If reversible aggregate can be dissociated, this would increase the amount of folded monomer that could be isolated.

Most covalent aggregates are the result of disulfide cross-linking, and typically require treatment with strong reductants, such as dithiothreitol (DTT) or β-mercaptoethanol, to dissociate. These conditions also reduce disulfides that can be crucial for maintaining native structure, and therefore, this discussion will focus on reversibly associated aggregates. For these species, the use of any additive that competes with the mechanisms of association (such as

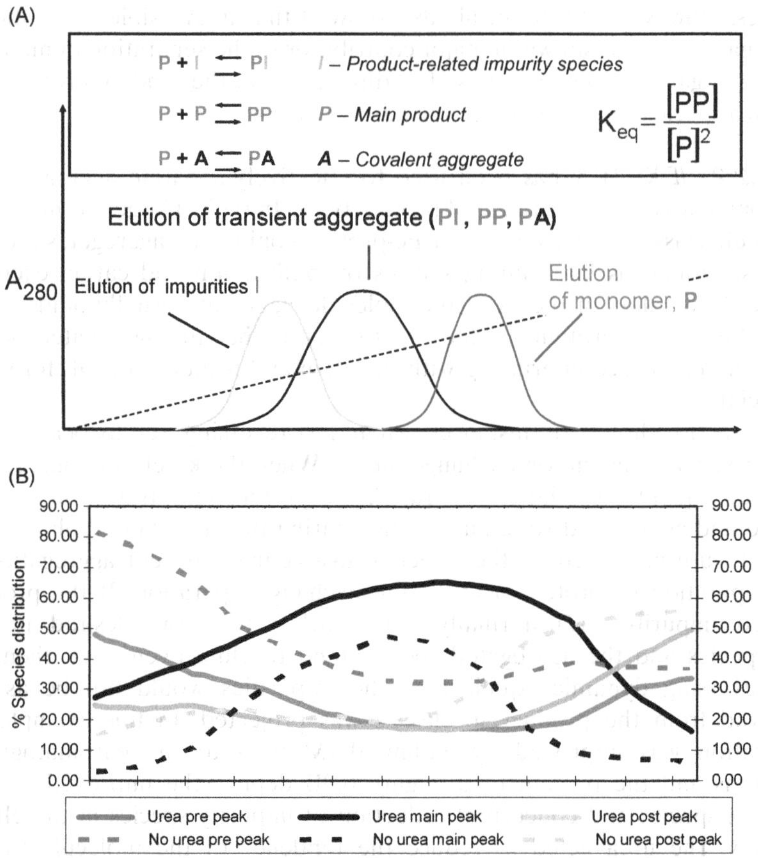

Figure 6.9. (A) Depiction of the effects of transient aggregation on separation during an ion exchange chromatography operation. The net effect of rapid transitioning between native and self-associated states is a loss in resolution between the species. (B) Improvement in the separation of a self-associating peptibody by addition of urea (Shultz et al.[22]). An ion exchange separation, with a protein exhibiting a rapid, reversible transitioning between native and aggregated states, was improved via addition of urea to the load and running buffers. The dashed lines represent the untreated separation, and the solid lines represent the improved separation achieved in the presence of 0.5 M urea.

chaotropes, surfactants, etc.) may be useful as long as the excipient can be removed or is acceptable in the final formulation solution. As described above, urea has been used to affect improved separations.[22] Polysorbate 20 has also been used to decrease aggregate formation.[50]

An additional approach has been the dissociation of self-associated proteins by formulation of the protein into a stabilizing solution. The DS and drug product formulation solutions are specifically crafted to maximize long-term stability and/or resilience to freezing and thawing cycles. During the purification process, it is very difficult to operate all of the separation steps under conditions that will prevent self-association. However, the exchange of the purified protein pool into the final formulation solution is an opportunity to recover the self-associated species into the native monomeric state. This approach has been utilized in the purification of an unstable microbial-derived protein that is associated under all conditions other than pH 5, no salt, and high levels of sucrose.[22] In this case, it was shown that the self-associated protein would readily and rapidly disassemble into the monomeric form upon exchange out of the final chromatography pool excipients and into the formulation buffer.[22,52]

6.4.3 Specific Considerations for Mammalian Cell-Derived Protein

The most common mammalian proteins that are being used as biopharmaceuticals today are mAbs. Due to the number of antibodies that have been examined, certain specific considerations can be summarized, based on experience. The lessons learned from the mAbs can be extrapolated to other mammalian-derived proteins, as they are all typically expressed in the folded form, avoiding the need for a refolding step, and are often glycosylated.

mAbs are known to be prone to aggregation at lower pH. Selection of candidates that are stable to these conditions, as discussed in Section 6.2, can minimize aggregation under these process conditions. Care should be taken to control the pH during viral inactivation and isolation such that it stays within acceptable ranges for the particular protein being processed, as well. Specific interactions with certain host-derived impurities (e.g., host cell proteins, DNA, etc.) have been linked to precipitation during processing. The ever-increasing push to higher protein concentration in the in-process pools and drug product formulations results in conditions where aggregation may be thermodynamically favored.

Proteins are exposed to fairly high concentrations of salt during the purification process; the known sensitivities to aggregation are all controllable with careful consideration during process development. For example, very low pH conditions can be avoided with minimal manufacturing consequence, and additives and chromatographic washes will often reduce the effects of host impurities.[55] Also, mammalian proteins are usually stabilized by the presence of their carbohydrate structures.

6.4.4 Specific Considerations for Bacterial-Derived Proteins

Microbial-derived therapeutic proteins represent a wider range of diversity; from small molecules like insulin to large and fairly complex molecules like Fc-protein conjugates, which may contain up to 24 intramolecular disulfide bonds and multiple peptide chains. Therefore, certain considerations must be made for microbial-derived proteins.

Proteins expressed in inclusion bodies must be denatured and refolded into the correct structure. Typical refolding conditions will yield a high level of protein misfolds that typically form aggregates. Other common attributes of these proteins include a propensity to aggregate at low pH, lack of a stabilizing carbohydrate structure, and the potential for N-terminal substitutions. In particular, certain classes of Fc conjugates are known to be generally pH sensitive, have peptide sequences that are prone to autolysis or modification, can be sensitive to aggregation at high concentrations, and have a higher sensitivity to high salt concentration than mAbs. This is often due to the unnatural structures and sequences that can be developed in microbial systems. On the other hand, specific mutations and genetic engineering can more readily be utilized as a tool to minimize aggregation and structural effectors and should be considered frequently during the development of a protein product (see Section 6.2).

To overcome the less desirable characteristics commonly observed in microbial-derived proteins, a strategic approach must be implemented in developing a purification process. There is significant variability in the behavior of different protein classes, and there is not yet an extensive knowledge base like that acquired for mAbs. For example, the conditions necessary for the removal of specific host proteins include extreme pH or high salt levels. These conditions, in turn, can lead to aggregation. Many of the strategies to control the equilibrium between native and unfolded states, described above, can be utilized directly during the refolding of proteins to achieve stable products with minimal aggregation.

6.4.4.1 Strategies to Control Aggregation during Protein Refolding Aggregation in microbial production systems poses an interesting challenge. On the one hand, high production titers can be obtained from microbial fermentation, because the expressed peptide chain is deposited into dense protein aggregates that form particles, commonly called inclusion bodies. While this concentrated aggregation provides a beneficial storage mechanism, the inclusion bodies are not in a usable state and must be solubilized (denatured and reduced) and renatured. This renaturation or refolding process is cumbersome and often occurs at low yields. For HMW proteins, misfolding and aggregation are common refolding problems.[23] In fact, an overwhelming majority of the aggregate species found in microbial processes are derived in the refold step. Therefore, the efficient refolding of proteins and prevention of aggregate formation is critical to biotherapeutic production.

The folding of native proteins is in kinetic competition with unproductive side reactions such as misfolding and aggregation.[56] Not only must the secondary structure be correctly formed, but often the process must facilitate concurrent formation of disulfide bond linkages and native structure. This is an added challenge, in that incorrect formation of the disulfide bonds can lead to covalent aggregate (via intermolecular linkages) or improper conformation that is held in place by a covalent bond. Goldberg et al. showed that incorrect disulfide bonding may not be the major cause of aggregation;[40] any non-native structure can propagate further aggregation during processing, and small aggregates may grow via intermolecular disulfide bond formation between unpaired cysteines.[57]

Protein folding involves a complex series of equilibria that form a balance between unfolded, intermediate, native, and aggregated states. This folding path is guided through a series of folding intermediates by the intramolecular association of hydrophobic surfaces within the peptide chain.[58] These intermediates have significant native-like secondary structure, but hydrophobic surfaces that would eventually form the internal core of the native protein can be exposed to the solvent.[57] Aggregation can occur when these hydrophobic surfaces on separate peptide chains interact in an intermolecular manner.[57,58] Consequently, it is possible to use structural and primary sequence data to identify hydrophobic patches, which participate in aggregation of folding intermediates.[57] It is then possible to develop strategies to decrease these detrimental intermolecular associations whether by molecular alterations, addition of excipients, or physical intervention via chaperones, fusions, or antibody assistants.

6.4.4.1.1 General Approaches to Prevent Aggregation during Refolding Aggregates come in multiple forms, and different types of aggregate may exist simultaneously in a refolding reaction. Large aggregates, or precipitates, fall out of solution and manifest as a loss in total protein yield. Smaller non-precipitating aggregates fall into two categories: (1) irreversible and (2) reversible. The reversible forms are the most easily recoverable, although they more often than not lead to significant protein yield loss during processing. Accordingly, the focus of this section will be on techniques aimed at preventing aggregation of all forms.

Multiple techniques have been practiced and discussed in the literature. The most common approaches involve chemical additives that aim to stabilize the native protein structure, while making the unfolded or denatured states thermodynamically unfavorable. The specific classes of additives will be discussed in greater detail below. Furthermore, other factors, such as the protein concentration effects, and various methods of physical inhibition of aggregate formation will be outlined.

6.4.4.1.2 Concentration If possible, industrial protein refolding should be performed at high concentrations. However, often the yield of native protein decreases with increased concentration in the refold solution, as large inactive aggregates readily form.[23,40,57] Literature reports suggest that the optimum refolding yields are obtained in the range of 10–50 mg/L.[23,57] At these concentrations, the production scale of microbial therapeutics is extremely limited due to the large refolding volumes that are required, and the economics are unfavorable. Refolding at higher concentrations is being actively explored.

Aggregate formation involves the reaction of multiple molecules, and thus, it is expected that the rate and degree of formation will increase rapidly with concentration. This is in contrast to the folding reaction of monomeric proteins, which involves a single molecule and thus should not be concentration dependent.[40] Thus, the rate of aggregate formation is expected to increase faster than the folding reaction as concentration is increased.

It has been suggested that there is a thermodynamically stable point in each reaction pathway where the reversibility is minimal. This implies that once the native protein is formed, it can be concentrated without aggregating and that once aggregate is formed, further dilution will not result in dissociation, without the use of denaturants.[40] Given the many possible protein states that can be achieved in a refolding reaction (see Fig. 6.7), it is necessary to either limit the concentration of the folding reaction to decrease aggregate formation or adjust solution conditions to inhibit the association of folding intermediates that are prone to aggregation. The latter approach is often the focus of refold buffer development. Other alternatives, such as the "temperature-leap tactic," in which the protein is refolded at low temperatures to minimize aggregation and is then rapidly heated to promote folding to the native state, have been explored.[57] However, these techniques are not useful at large scale as the heat transfer rates are usually impractical.

6.4.4.1.3 Use of Additives/Stabilizers during Refolding The development of a refolding process is often an empirical exercise involving dilution of denatured and reduced inclusion bodies into a refolding buffer, which contains stabilizing excipients. The selection of these additives is guided by the desire to minimize the formation of non-native species, especially aggregates.

A variety of additives have proven useful, such as osmolytes, low levels of kosmotropes and chaotropes, certain amino acid salts, polymers, surfactants, and sugars (see Table 6.2). Higher productivities may be achieved by selective application of a number of these agents. While certain general rules can be applied, it is difficult to predict the degree of effect a given additive will have on the prevention of aggregation and the formation of native protein. It is often necessary to experimentally determine the optimal excipients and solution concentrations required. Therefore, a range of demonstrated concentrations is provided as well as a brief discussion of the potential stabilizing mechanisms for some of the additive classes.

6.4.4.1.3.1 AMINO ACIDS AND POLYMERS AS STABILIZERS Amino acid salts have been used extensively as stabilizers in refolding reactions. L-arginine is the most commonly used refold additive.[56] Depending on the molecule and protein concentration in the refold, arginine levels ranging from 0 to 0.8M have been shown to increase native protein yields. This additive class enhances the solubility of partially structured folding intermediates, reducing aggregate formation.[56] Proline has also been investigated and shown to inhibit aggregation during refolding.[23] Proline has been shown to bind to and trap folding inter-

TABLE 6.2. Additives Used to Increase Yield of Correct Folding and Prevent Aggregation[23,56,57]

Additive Class	Additive	Recommended Concentration Range
Amino acids	L-arginine hydrochloride	0.4–0.8 M
	Glycine	0.5–2% w/v
	Proline	0–2 M
Chaotropes (non-denaturing levels)	Urea	<2.0 M
	Gdm/C1	<1.0 M
	Methylurea	1.5–2.5 M
	Ethylurea	1.0–2.0 M
	Acetamide	1.5–2.5 M
Alcohols	Ethanol	Up to 25% v/v
	n-Pentanol	1.0–10.0 mM
	n-Hexanol	0.1–10.0 mM
	Cyclohexanol	0.01–10.0 mM
Salts	Tris	>0.4 M
	Na_2SO_4 or K_2SO_4	0.4–0.6 M
Osmolyte	Dimethylsulfoxide	0–4 M
	Trimethylamine N-oxide	0–6 M
Sugars	Sucrose	0–20%
	Trehalose	0–3 M
Polyhydric alcohols	Glycerol	20–40% v/v
	Sorbitol	20–30% v/v
Surfactants	Lauryl maltoside	ca. 0.06 mg/mL
	Cetyltrimethylammonium bromide	ca. 0.6 mg/mL
	CHAPS	10.0–60.0 mM
	Triton X-100	~10.0 mM
	Dodecyl maltoside	2.0–5.0 mM
Polymers	Polyethylene glycol (PEG, 3350 MW)	0.1–0.4 g/L
	β-Cyclodextrin	20.0–100.0 mM
Others	Mixed micelles	Depending on surfactant used

mediates into "aggregation-insensitive" states. High concentrations of proline (>1.5 M) are required in order to function as a refolding aid.[23]

High concentrations of amino acids may increase the ionic strength of the solution to a point that subsequent use of chromatography operations is affected. This needs to be taken into consideration when including them in the refolding step.

Polymers, such as PEGs, may act by a mechanism similar to that of the amino acid stabilizers. It has been suggested that PEG interacts preferentially with the denatured state of the protein, but is excluded from the native state.[56] PEG, which is partially hydrophobic, may interact with the hydrophobic surfaces that are exposed in the intermediate and unfolded protein states.[56] This preferential interaction may assist in aggregate prevention.

6.4.4.1.3.2 DENATURANTS Non-denaturing concentrations of chaotropes, such as urea or guanidine, are common refolding additives. These agents can both solubilize the inclusion body and strongly decelerate the formation of aggregates.[56] However, a balance must be struck where inhibition of aggregation is achieved without hindering the formation of the native state. Other denaturing additives, such as alkylureas, carbonic acid amides, or alcohols have also been reported as improving refolding reactions.[56]

6.4.4.1.3.3 SUGARS AND POLYHYDRIC ALCOHOLS As described extensively above, sugars and polyols influence the preferential interaction of proteins and tend to stabilize proteins in a solution with a smaller surface area. If the unfolded state has a larger surface area than the native state, then the native state will be preferred. As the unfolded state is a pathway to aggregation, minimization of the unfolded state will also reduce aggregation. Sucrose, glycerol, glucose, lactose, and trehalose have all been shown to reduce levels of aggregation in protein refolds.[37,56,57] Glycerol provides a benefit via a mechanism comparable to that of sugars.

In most cases, the levels of sugars or polyols must be balanced in proportion to the chaotrope levels in the system. There appears to be an interactive relationship between these two different agents that should be optimized. Higher levels of these agents may be required for a given protein, but this usually must be accompanied by a subsequent increase in chaotrope level for full benefit.

Sugars and polyols can significantly increase the surface tension and viscosity of the fluid. This can have an effect on the movement of the fluid at large scale through subsequent steps and can limit the scale-up of the process.

6.4.4.1.3.4 SURFACTANTS AND MIXED MICELLES Surfactants or mixed micelles have been shown to bind to folding intermediates and inhibit self-association. Surfactant binding to hydrophobic patches on the intermediates dampens amino acid side-chain fluctuations.[56] These additives can also be used to solubilize the inclusion bodies. This binding inhibits the overall folding reaction, until the surfactant is removed from the system. This can be achieved in a number of ways, including extraction with cyclodextrin or removal over a resin matrix (e.g., DOWEX™ media). The optimal concentration of the surfactant depends on the protein system and detergent being used. In fact, some detergents are most effective above their CMC.[56] One limitation for the use of surfactants is the potential effect that the additive may have on subsequent processing, such as chromatography, where it can result in decreased binding to the column.

6.4.4.1.4 High-Pressure Refolding of Aggregates Another developing strategy to increase refolding efficiency is the use of high hydrostatic pressure to refold proteins from aggregates. Under mild conditions, disaggregation is typically so slow that aggregates are often considered to be irreversible.[37]

However, refolding has been recently demonstrated at high levels of hydrostatic pressure with high efficiency.

Typical ambient refolding processes require high levels of denaturant to disrupt the hydrophobic interactions and hydrogen bonding within the aggregates, prior to dilution into a refolding buffer. The dilution to lower denaturant levels simultaneously reduces the inhibition of hydrophobic interactions and hydrogen bonding, increasing the potential for aggregation by intermolecular hydrophobic interactions. In contrast, under high pressure, hydrophobic interactions are inhibited and aggregation is suppressed.[59] High-pressure refolding is typically performed in the presence of non-denaturing concentrations of guanidine or urea, at pressures exceeding 1000 bar.[59] The method uses increased pressure to alter the thermodynamics of the system by shifting the non-covalent intramolecular distances between secondary and tertiary structure in a protein to minimize the total protein–solvent system volume. The result is increased hydration of the protein and denaturation.[37] The use of this technique thermodynamically favors the native state and has been shown to be useful in limiting aggregate formation during the folding process. St John et al. demonstrated that under conditions of high hydrostatic pressure and temperature, aggregates of human growth hormone could be refolded to near 100% native state.[59] This strategy has not exhibited a dependence on protein concentration and typically does not require high levels of chemical additives to be effective.[59,60] However, there are no published reports demonstrating large-scale application of this method in the production of therapeutics.

6.4.4.1.5 Use of Protein Partners to Decrease Aggregate Formation during Refolding An additional, yet seldom utilized, technique for aggregate prevention during refolding is the physical inhibition of aggregation. This approach utilizes large external molecules to assist in the prevention of self-association.

6.4.4.1.5.1 ANTIBODIES AND FUSION CONSTRUCTS Since the exposed hydrophobic surfaces on folding intermediates are prone to aggregation, approaches to prevent aggregation by "blocking" these surfaces with mAbs or fusion constructs have been researched. mAbs have been utilized to specifically bind to hydrophobic patches on the refolding protein, protecting it from intermolecular aggregation reactions. This technique has been used to stabilize the reduced S-protein fragment of ribonuclease A during refolding.[56] In another example, staphylococcal protein A ligand has been used to substantially improve the folding of insulin-like growth factor (IGF)-1, and granulocyte colony-stimulating factor (GCSF) folding was improved using small hydrophilic peptide fusion partners.[56]

However, these approaches are not likely to be useful for large-scale therapeutic manufacture, as a number of issues surface. For example, tightly bound antibodies may inhibit proper formation of the protein structure, and addition of an additional protein provides an extra purification challenge, not to mention the additional cost of the antibody or fusion reagents analytical methods to demonstrate removal of the additional protein reagent would also be required.

6.4.4.1.5.2 CHAPERONES *In vivo* protein folding is often assisted by protein chaperones.[58] These molecules can also effectively modulate the competition between folding and aggregation, *in vitro*.[57] Chaperones assist folding via a mechanism similar to that of the antibodies as described above. They interact with the hydrophobic residues that are exposed in folding intermediates to suppress aggregation pathways. These molecules are also not likely to be useful in large-scale refolding as they are costly to produce, must be removed in the purification process, and have not been demonstrated in the production of large quantities of protein.[56,57] So far, molecular chaperones have only been used as stabilizers during protein storage.[56]

6.4.4.1.5.3 IMMOBILIZATION A final approach, which may eventually be useful at production scale, is immobilization of the unfolded protein on a solid support (e.g., chromatographic resin). This method binds each denatured molecule to a surface prior to refolding, which effectively separates the molecules, physically decreasing local concentration of proteins, limiting the intermolecular interactions that can lead to the formation of aggregates.[58] Immobilized protein folding using a number of strategies, including IEX, HIC, histidine tags, and polyionic fusion constructs, has been successfully explored.[56] While these approaches may reduce aggregate formation, they have not yet been shown to significantly improve refolding yields. This is likely because binding to the matrix, especially HIC, prevents the formation of the proper three-dimensional structure of the immobilized protein.[56]

Specific binding of the unfolded protein to a matrix has been achieved using histidine (His) tags or polyionic fusion constructs. This approach provides maximal freedom for folding by distancing the protein from the resin and decreasing steric hindrance. Aggregation can be completely prevented by this specific, non-covalent immobilization of the folding polypeptides via the affinity tags at their N- or C-terminus. This method must, however, be further optimized to allow its routine application for *in vitro* folding.[56]

6.4.5 Bulk DS Stability and Storage

Once a protein therapeutic is purified, it is critical that its stability be maintained and no further damage or aggregation occurs, as there are no longer any opportunities to remove these impurities. The only options left at this point in the process are prevention and control. The chemistry of the formulation solution can be used, to some degree, to control the kinetics of degradation. Every protein is susceptible to damage during the processing of the DS and interim storage. The most common indicator of physical instability is aggregation. The aggregates are usually non-native in structure and can be as large as visible particulates. In some cases, the aggregates may form intermolecular linkages, via disulfide bonding.[14]

Common stresses that may be experienced during DS processing may include transient exposure to high temperature, freezing and thawing events, and lyophilization.[14] There are also various opportunities for the protein to be denatured at surface interfaces, such as the filter surface and air–liquid interfaces.[14] It is, therefore, critical that potential sources of aggregation be minimized during all stages of DS processing. This is addressed in detail in the following two chapters.

6.4.6 Scale-Up

There are a number of considerations when scaling up a process. These factors mostly fall in the category of physical stressors (e.g., cavitation, shear, liquid–solid interface, etc.). However, there are also physical factors specific to manufacturing plant design that can contribute to local chemical extremes, which result in conditions that are prone to cause aggregation. An example of such a situation would be the efficiency of mixing upon scale-up. Mixing in large tanks is much less efficient than mixing at the bench scale, which can affect many operations, such as refolding (aeration, addition of the solubilized inclusion bodies to the refolding buffer, temperature control, etc.). Each of these factors can be controlled to differing degrees across scales. It is important to understand the effects each of the factors can have and how they may change as a purification process is scaled to manufacturing scale.

6.4.6.1 Mechanical Effectors Large-scale equipment and flow rates must be designed and balanced to limit physical stress on proteins. The primary sources of stress come from valves and pumps, which can create zones of shear and cavitation, and increased exposure to air/liquid interfaces.

Cavitation is common in pumps and occurs when an extremely low-pressure zone (vacuum) on the inlet side of the pump causes liquid to vaporize and quickly collapse at the high-pressure discharge side of the pump. The collapse of the vapor bubbles can be so violent that the temperature of the vapor within the bubble can reach very high temperatures and pressures. These local extremes in pressure and temperature can denature proteins and cause aggregation. Cavitation is most extreme in low-efficiency pumps. Increased air–liquid interface surface can also result in increased aggregate formation. Each of these factors can be controlled by carefully designing and selecting appropriate pumps for use during processing and testing for protein damage during scale-up. It may also be possible to limit damage by adjusting the solution conditions (e.g., pH, salt concentration, addition of stabilizers, etc.) (J. E. Shultz, unpublished data) and by designing the system to limit gas–liquid interfaces. It has also been reported that nanoparticles of steel shed from a pump may act as heterogeneous nuclei for protein microparticle formation.[61]

6.4.6.2 Mixing One of the most difficult parameters to scale is mixing. At large scale, mixing is much less efficient than what occurs at bench scale. Therefore, if caution is not taken, process parameters developed at small scale may not be repeated at large scale. Poor mixing can indirectly contribute to aggregation in a number of ways, by allowing zones of high protein concentration, areas of local pH or temperature extremes, inefficient diafiltration, and so on. All of these example conditions may be conducive to aggregate formation. Extensive literature has been dedicated to the concepts of mixing at large-scale and will not be replicated here, but suffice it to say that there are practical limits to the level of control that can be ultimately achieved.[62] Therefore, it is necessary to understand the sensitivity of the protein system to each of these factors and address them during the development of the initial process.

6.4.6.3 Freeze/Thaw Freeze-thaw rates can be very difficult to duplicate upon scaling. Freezing is a nucleation-dependent process. The number of nuclei is much greater at large volumes than for small volumes. Supercooling processes are consequently quite different as scale changes. As process batch sizes increase, so do the volumes of frozen intermediates. Practical partitioning of the fluids requires that the individual aliquots increase in volume as well. This directly affects the thermal transfer rates and, thus, freezing and thawing rates. Slow freezing and thawing increases the exposure of the protein to localized zones of high protein concentrations, high solute concentrations, and increased ice–liquid surface exposure.

The resultant aggregation can be controlled to some degree by engineering the formulation buffer to ameliorate these effects. However, there are also various equipment solutions that allow for rapid, controlled rate, freezing and thawing of very large volumes of liquid. These techniques are still being perfected but have already been applied to commercial-stage biopharmaceuticals. For example, specific vessels have been developed that allow for highly controlled heat transfer, resulting in more rapid freezing and/or thawing rates. There are also systems that allow liquid products to be frozen in plastic bags placed between fin-like heat exchangers.

6.4.6.4 Temperature Large-scale heat transfer is slower and less efficient than at small scale. At large scale, heating and cooling is attained through jackets on the sides of tanks and through heat exchangers. As tank volumes get larger, the surface area to volume ratio decreases rapidly. To affect temperature changes as rapidly as at small scale, the temperature required at the surface of tanks and heat exchangers must be significantly higher and the mixing must be better. This creates situations where the local temperatures at the heating surface can be significant enough to cause thermal denaturation and aggregation. To avoid these scenarios, attempts must be made to design systems for better heat transfer at safe surface temperatures. One way to ensure this is to develop processes with the largest scale in mind, so as to avoid the necessity for high surface temperatures/gradients. Small-scale laboratory equipment can be designed to mimic the slower heat transfer rate seen in large-scale manufacturing facilities. By doing so, processes are developed within the limitations of future scales and a key source of physical denaturation can be avoided.

6.4.6.5 pH There are many titration or conditioning events in a process. During batch titration operations, the presence of less efficient mixing can lead to local pH extremes that may not be experienced at smaller scales. As described above, these local extremes can lead to an aggregation cascade. To mitigate the risks of titration scaling, careful selection of titrant, strength, and addition rate are important. Moreover, avoidance of batch titration in favor of in-line conditioning is preferred. This operational strategy has been implemented in pharmaceutical manufacturing utilizing in-line mixing/dilution skids to continuously mix product pools with a buffering titrant that is defined in terms of solution ratio and mixing residence time. Once such a system is defined, it can be readily scaled within a defined confidence of reproducibility across scales.

6.5 SUMMARY

One of the goals in developing a commercial process for the manufacturing of a protein therapeutic is to obtain a high yield of properly folded protein while minimizing aggregation. This involves a concerted effort at every stage of the development process. This begins with the assessment of the stability of product candidates to process and storage conditions. Cell culture conditions and cell lines can also be chosen to maximize folded protein production and minimize aggregate present in the original material used as the source for purification. And finally, strategies to mitigate aggregate formation while favoring the formation of native protein should be explored and implemented.

REFERENCES

1. FDA guidance concerning demonstration of comparability of human biological products, including therapeutic biotechnology-derived products, Center for Biologics Evaluation and Research and Center for Drug Evaluation and Research. April, 1996.
2. Carpenter JF, Randolph TW, Jiskoot WIM, Crommelin DJA, Middaugh CR, Winter G, Fan YX, Kirshner S, Verthelyi D, Kozlowski S, Clouse KA, Swann PG, Rosenberg AS, Cherney B. 2008. Overlooking subvisible particles in therapeutic protein products: Gaps that may compromise product quality. J Pharm Sci 98(4):1201–1205.
3. Rosenberg AS. 2006. Effects of protein aggregates: An immunologic perspective. AAPS J 8:501–507.

4. Swencki-Underwood B, Cunningham MR, Heavner GA, Blasie C, McCarthy SG, Dougherty T, Brigham-Burke M, Gunn GR, Goletz TJ, Snyder LA. 2006. Engineering human IL-18 with increased bioactivity and bioavailability. Cytokine 34(1–2):114–124.
5. Alber T. 1989. Mutational effects on protein stability. Annu Rev Biochem 58:765–798.
6. Narhi LO, Stabinsky Y, Levitt M, Miller L, Sachdev R, Finley S, Park S, Kolvenbach C, Arakawa T, Zukowski M. 1991. Enhanced stability of subtilisin by three point mutations. Biotechnol Appl Biochem 13(1):12–24.
7. Bryan PN. 1992. Engineering dramatic increases in the stability of subtilisin. In Ahern TJ, Manning MC, eds., *Stability of Protein Pharmaceuticals*, Vol. B. New York: Plenum Press.
8. Hathway K, Gao S, Salmeron L, Chang J, Buko A. 2008. Assessment of manufacturability of protein drug candidates by ranking of antibody candidates based on accelerated stability studies. 236th ACS National Meeting, Division of Biochemical Technology, BIOT120, Philadelphia, PA, p. 52.
9. Ramachander R. 2008. Manufacturability assessments for early stage therapeutic candidate screenings. Presented at the IBC BPI Meeting, Anaheim, CA.
10. Jiang Y, Ramachander R, Wen J, Li C, Li J, Angell N, Bondarenco P, Narhi L. 2008. Manufacturability assessment for a successful therapeutic product. 236th ACS National Meeting, Division of Biochemical Technology, BIOT110, Philadelphia, PA, p. 49.
11. Wen J, Jiang Y, Narhi L. 2007. Applications of DSC for antibodies and Fc-conjugated proteins. Am Pharm Rev 10:10–15.
12. Wen J, Jiang Y, Hymes K, Gong K, Narhi L. 2007. Using differential scanning calorimetry in understanding the correlation between thermal stability and protein stability. MicroCal Application Note.
13. Shukla AA, Hubbard B, Tressel T, Guhan S, Low D. 2007. Downstream processing of monoclonal antibodies—Application of platform approaches. J Chromatogr B Analyt Technol Biomed Life Sci 848(1):28–39.
14. Carpenter JF, Kendrick BS, Chang BS, Manning MC, Randolph TW. 1999. Inhibition of stress-induced aggregation of protein therapeutics. Methods Enzymol 309:236–255.
15. Espinosa L, Schebor C, Nudelman NS, Chirife J. 2004. Stability of enzymes and proteins in dried glassy systems: Effect of simulated sunlight conditions. Biotechnol Prog 20(4):1220–1224.
16. Wurm FM. 2004. Production of recombinant proteins in cultivated mammalian cells. Nat Biotechnol 22(11):1393–1398.
17. Narhi LO, Arakawa T, Aoki KH, Elmore R, Rohde MF, Boone T, Strickland TW. 1991. The effect of carbohydrate on the structure and stability of erythropoietin. J Biol Chem 266(34):23022–23026.
18. Wen J, Jiang Y, Narhi L. 2008. DSC for studying carbohydrate effect on antibodies. Am Pharm Rev 11(6):98–104.
19. Sol RJ, Greibenow K. 2009. Effects of glycosylation on the stability of protein pharmaceuticals. J Pharm Sci 98:1223–1245.

20. Dissing-Olesen L, Thaysen-Andersen M, Meldgaard M, Hojrup P, Finsen B. 2008. The function of the human interferon-beta 1a glycan determined in vivo. J Pharmacol Exp Ther 326(1):338–347.
21. Makanji Y, Harrison CA, Stanton PG, Krishna R, Robertson DM. 2007. Inhibin A and B in vitro bioactivities are modified by their degrees of glycosylation and affinities to betaglycan. Endocrinology 148(5):2309–2316.
22. Shultz JE, Jiang Y, Wen J, Li C, Ramachander R, Stenson R, Pampel L, Gokarn Y, Narhi LO. 2008. Not all aggregates are the same: So, should they all be removed? Presented at the American Chemical Society National Meeting, Philadelphia, PA, August.
23. Paul S, Punam S, Chaudhuri TK. 2007. Chaperone-assisted refolding of *Escherichia coli* maltodextrin glycosidase. FEBS J 274(22):6000–6010.
24. Cromwell ME, Hilario E, Jacobson F. 2006. Protein aggregation and bioprocessing. AAPS J 8(3):E572–579.
25. Schroder M, Schafer R, Friedl P. 2002. Induction of protein aggregation in an early secretory compartment by elevation of expression level. Biotechnol Bioeng 78(2):131–140.
26. Bertolotti A, Zhang Y, Hendershot LM, Harding HP, Ron D. 2000. Dynamic interaction of BiP and ER stress transducers in the unfolded-protein response. Nat Cell Biol 2(6):326–332.
27. Lee GW. 2008. Stabilizing protein-folding conformations. Gen Eng Biotechnol News 28(5):60–63.
28. Lenny N, Green M. 1991. Regulation of endoplasmic reticulum stress proteins in COS cells transfected with immunoglobulin mu heavy chain cDNA. J Biol Chem 266(30):20532–20537.
29. Schlatter S, Stansfield SH, Dinnis DM, Racher AJ, Birch JR, James DC. 2005. On the optimal ratio of heavy to light chain genes for efficient recombinant antibody production by CHO cells. Biotechnol Prog 21(1):122–133.
30. Vanhove M, Usherwood YK, Hendershot LM. 2001. Unassembled Ig heavy chains do not cycle from BiP in vivo but require light chains to trigger their release. Immunity 15(1):105–114.
31. Lee CJ, Seth G, Tsukuda J, Hamilton RW. 2009. A clone screening method using mRNA levels to determine specific productivity and product quality for monoclonal antibodies. Biotechnol Bioeng 102(4):1107–1118.
32. Franco R, Daniela G, Fabrizio M, Ilaria G, Detlev H. 1999. Influence of osmolarity and pH increase to achieve a reduction of monoclonal antibodies aggregates in a production process. Cytotechnology 29(1):11–25.
33. Gomes JM, Hiller GW. 2008. Use of low temperature and/or low pH in cell culture. US Patent Application #US20080269132, Wyeth.
34. Freimuth P, Springer K, Berard C, Hainfeld J, Bewley M, Flanagan J. 1999. Coxsackievirus and adenovirus receptor amino-terminal immunoglobulin V-related domain binds adenovirus type 2 and fiber knob from adenovirus type 12. J Virol 73(2):1392–1398.
35. Chaderjian WB, Chin ET, Harris RJ, Etcheverry TM. 2005. Effect of copper sulfate on performance of a serum-free CHO cell culture process and the level of free thiol in the recombinant antibody expressed. Biotechnol Prog 21(2):550–553.

36. Gomes JM, Hiller GW, Luan Y-T, Wang W. 2008. Methods of protein production using anti-senescence compounds. US Patent Application #US20080274507, Wyeth.
37. Foster TM, Dormish JJ, Narahari U, Meyer JD, Vrkljan M, Henkin J, Porter WR, Staack H, Carpenter JF, Manning MC. 1996. Thermal stability of low molecular weight urokinase during heat treatment. III. Effect of salts, sugars and Tween 80. J Pharm 134(1–2):193–201.
38. Bondos SE, Bicknell A. 2003. Detection and prevention of protein aggregation before, during, and after purification. Anal Biochem 316(2):223–231.
39. Murphy KP. 2001. Stabilization of protein structure. Methods Mol Biol 168:1–16.
40. Goldberg ME, Rudolph R, Jaenicke R. 1991. A kinetic study of the competition between renaturation and aggregation during the refolding of denatured-reduced egg white lysozyme. Biochemistry 30(11):2790–2797.
41. Hall CG, Abraham GN. 1984. Reversible self-association of a human myeloma protein. Thermodynamics and relevance to viscosity effects and solubility. Biochemistry 23(22):5123–5129.
42. Mire-Sluis A. 2008. Purification processes—Sorting out the good, the bad and the ugly. Presented at the Recovery of Biological Products XIII Conference, Philadelphia, PA, June.
43. Arakawa T, Timasheff SN. 1982. Stabilization of protein structure by sugars. Biochemistry 21(25):6536–6544.
44. Saluja A, Crampton S, Kras E, Fesinmeyer RM, Remmele RL Jr., Narhi LO, Brems DN, Gokarn YR. 2009. Anion binding mediated precipitation of a peptibody. Pharm Res 26(1):152–160.
45. Baldwin RL. 1996. How Hofmeister ion interactions affect protein stability. Biophys J 71(4):2056–2063.
46. Schein CH. 1990. Solubility as a function of protein structure and solvent components. Biotechnology 8(4):308–317.
47. Vrkljan M, Foster TM, Powers ME, Henkin J, Porter WR, Staack H, Carpenter JF, Manning MC. 1994. Thermal stability of low molecular weight urokinase during heat treatment. II. Effect of polymeric additives. Pharm Res 11(7):1004–1008.
48. Strambini GB, Gabellieri E. 1996. Proteins in frozen solutions—Evidence of ice-induced partial unfolding. Biophys J 70(2):971–976.
49. Kerwin BA, Heller M, Levin SH, Randolph TW. 1998. Effects of Tween 80 and sucrose on acute short-term stability and long-term storage at -20 degrees C of a recombinant hemoglobin. J Pharm Sci 87(9):1062–1068.
50. Watabe T, Ogura K, Satsukawa M, Okuda H, Hiratsuka A. 1994. Molecular cloning and functions of rat liver hydroxysteroid sulfotransferases catalysing covalent binding of carcinogenic polycyclic arylmethanols to DNA. Chem Biol Interact 92(1–3):87–105.
51. Porter WR, Staack H, Brandt K, Manning MC. 1993. Thermal stability of low molecular weight urokinase during heat treatment. I. Effects of protein concentration, pH and ionic strength. Thromb Res 71(4):265–279.
52. Gokarn Y, Fesinmeyer RM, Saluja A, Cao S, Dankberg J, Goetze A, Remmele RL Jr., Narhi LO, Brems DN. 2009. Ion specific modulation of protein interactions:

Anion-induced, reversible oligomerization of a fusion protein. Protein Sci 18(1):169–179.

53. Hart RA, Smith SG, Shultz JE. 2005. Post-refolding harvest of *E. coli* expressed Fc-fusion proteins. Presented at the BioProcess International World Conference, Boston, September.
54. McCue JT, Engel P, Ng A, Macniven R, Thommes J. 2008. Modeling of protein monomer/aggregate purification and separation using hydrophobic interaction chromatography. Bioprocess Biosyst Eng 31(3):261–275.
55. Yigsaw Y. 2008. Improving upstream feed stock to downstream operations. Presented at the Recovery of Biological Products XIII, Quebec City, Quebec, Canada, June 22–27.
56. Clark EDB, Schwarz E, Rudolph R. 1999. Inhibition of aggregation side reactions during in vitro protein folding. Methods Enzymol 309:217–236.
57. Clark EDB. 1998. Refolding of recombinant proteins. Curr Opin Biotechnol 9:157–163.
58. Fink AL. 1998. Protein aggregation: Folding aggregates, inclusion bodies and amyloid. Fold Des 3:R9–R23.
59. St John RJ, Carpenter JF, Balny C, Randolph TW. 2001. High pressure refolding of recombinant human growth hormone from insoluble aggregates. Structural transformations, kinetic barriers, and energetics. J Biol Chem 276(50): 46856–46863.
60. Bondos SE, Sligar S, Jonas J. 2000. High-pressure denaturation of apomyoglobin. Biochim Biophys Acta 1480(1–2):353–364.
61. Tyagi AK, Randolph TW, Dong A, Maloney KM, Hitscherich C Jr., Carpenter JF. 2009. IgG particle formation during filling pump operation: A case study of heterogeneous nucleation on stainless steel nanoparticles. J Pharm Sci 98:94–104.
62. Amanullah A, McFarlane CM, Emery AN, Nienow AW. 2001. Scale-down model to simulate spatial pH variations in large-scale bioreactors. Biotechnol Bioeng 73(5):390–399.

CHAPTER 7

Protein Aggregation and Particle Formation: Effects of Formulation, Interfaces, and Drug Product Manufacturing Operations

HANNS-CHRISTIAN MAHLER, STEFAN FISCHER, THEODORE W. RANDOLPH, and JOHN F. CARPENTER

7.1 INTRODUCTION

In therapeutic protein products, aggregates are considered process-related or product-related impurities (degradation products), and great effort is made to minimize their levels.[1] In this context, aggregates are typically defined as assemblies of protein molecules that increase in level with time of storage and/or in response to stress. This behavior is in contrast to the native, equilibrium self-assembly that is a fundamental biological characteristic of many proteins. Therapeutic proteins may form native multimers, as well as form aggregated degradation products. For the purposes of this chapter, only the latter aggregates will be considered.

The aggregates that are classified in pharmaceutical protein products as "soluble" are oligomers, ranging from dimers to multimers with molecular weights of millions of daltons. These aggregates are routinely characterized and quantified by methods such as size exclusion chromatography and analytical ultracentrifugation. As an operational definition, aggregates that form large enough assemblies to sediment at moderate g-forces (e.g., in a microcentrifuge) are termed "insoluble aggregates." A special case of large aggregates (e.g., 1–125 μm) that do not readily sediment are "subvisible particles." These protein aggregates are usually quantified with particle counting equipment. Particles that are large enough to be observed by the eye are called "visible particles." Clearly, the boundaries between types of aggregates and

Aggregation of Therapeutic Proteins, Edited by Wei Wang and Christopher J. Roberts

particles are not absolutely defined, and classification may depend on the analytical method(s) used for detection (see also Chapter 5).

The type of aggregates formed by a given protein can vary greatly depending on the stresses causing aggregation (e.g., freeze-thawing vs. agitation) and the solution conditions. In some instances, under a given set of conditions, there may be a progression from monomer to dimer to higher-order soluble oligomers and then insoluble aggregates. In other cases, it may appear that monomers are directly assembling into insoluble aggregates (for a review of protein aggregation kinetics, see, e.g., Reference 2).

Subvisible and visible particles can be detected even at concentrations that represent minute amounts of protein mass. In practice, this can mean that a therapeutic protein product in which loss of monomer is virtually immeasurable can still contain hundreds of thousands of subvisible particles and perhaps visible particles. In some cases, this is because the particles are formed by adsorption of protein molecules to nano- or microparticles of foreign materials.[3,4] In other cases, the causes of protein particles are not known.

7.2 ROLES OF CONFORMATIONAL AND COLLOIDAL STABILITY IN REDUCING RATES OF AGGREGATION

A critical goal of therapeutic protein product development is minimizing the rate of aggregation during manufacturing and in the final drug product. These are daunting tasks because protein molecules may be exposed to aggregation-causing stresses (e.g., liquid–solid interfaces, liquid–air interfaces, high temperature, freeze-thawing, and pH extremes) during upstream and downstream processing, storage of bulk drug substance, the fill/finish operation, and during shipping, storage, and administration of the drug product.[5,6] Furthermore, even in the absence of obvious stress, such as during storage in solution at 2–8°C, significant protein aggregation can occur on timescales (e.g., 18–24 months) of pharmaceutical storage.

Minimizing protein aggregation is a challenge because protein molecules that form aggregates have only minimally perturbed conformation compared to the most compact, structured native molecules.[7,8] Conditions (e.g., change in pH or addition of bacteriostatic levels of benzyl alcohol[9–15]) that cause barely detectable perturbations of tertiary structure in a protein sample can greatly accelerate protein aggregation. Furthermore, it has been estimated that the degree of structural expansion association with the transition of a non-aggregating native protein molecule into a partially unfolded, aggregation-prone species is only on the order of 10–20%.[8] The aggregation-prone species are present even in protein samples under conditions that are known to greatly favor the native state. That is, within the native state ensemble of substates, there is a small fraction of partially unfolded protein molecules that are aggregation prone. However, it is also speculated that native proteins also may undergo aggregation.[16]

In the absence of stresses (e.g., exposure to interfaces), the aggregation rate may be slow. Yet, it also only takes a small increase in the absolute level of aggregation-prone species to greatly accelerate the rate of aggregation. Optical spectra used to assess protein structure in solution represent an average of the contributions of all protein molecules in a sample, and those of the rare, aggregation-prone protein molecules are greatly overcome by that of native molecules without alterations in tertiary structure. This is why changing from conditions under which the aggregation rate is low to those that accelerate aggregation often results in a minimally detectable change in the optical spectra of the protein indicative of tertiary structure.

To minimize aggregation in aqueous solution, conditions must be found that maximize two physical stability properties of the protein.[17,18] First, the conformational stability of the native state must be increased, which is concomitant with a reduction in the level of unfolded or partially unfolded aggregation-protein protein molecules in the molecular population. Second, the energetics of the intermolecular interactions between protein molecules must be manipulated to maximize intermolecular repulsion, that is, to maximize the so-called colloidal stability. It is important to point out that changes in solution conditions that affect conformational stability are likely also to affect colloidal stability and vice versa. Also, for a given protein, in some instances, the critical control of the aggregation rate will be due to conformational stability, and in the others, colloidal stability will dominate.

Solution conditions that are found to minimize aggregation during formulation development will probably also maximize conformational and colloidal stability. But the practical process of developing a formulation usually does not need to focus on measuring directly these physical parameters for a protein as a function of solution conditions. These measurements can be very time-consuming and unnecessary, compared to the speed and utility of measurements of relative rates of aggregation during formulation screening studies. For example, a formal measurement of conformational stability could involve obtaining equilibrium unfolding curves using urea or guanidine HCl as a denaturant.[19] This is a very labor-intensive and time-consuming process, which in the end will only help explain the mechanism by which conditions, shown in screening studies to reduce aggregation rate, operate. Similarly, the colloidal stability of a protein can be assessed by determining the osmotic second virial coefficient,[17] a process that also can be relatively impractical. However, a general mechanistic understanding of how solution conditions affect conformational and colloidal stabilities can help to guide formulation screening studies and to provide insight into why some conditions work and others fail at reducing protein aggregation rate.

7.2.1 Conformational Stability

Conformational stability can greatly be affected by solution pH; thus, it is always a critical variable to investigate during formulation studies. With many proteins, for example, at a certain level of acidity, the tertiary structure can be perturbed substantially, but the secondary structure may not be altered. Often under these conditions, aggregation is greatly favored. Even without observable (e.g., detectable via optical spectroscopy) alterations in the overall average conformation of a population of protein molecules, changes in pH may cause an increase in the levels of partially unfolded molecules and, hence, a substantial increase in the aggregation rate.

The conformational stability of a protein can be greatly increased by a ligand that binds specifically and with relatively high affinity to the native state of the molecule but not to unfolded states. Typically, stoichiometric amounts of ligand are effective, and in a formulation, the concentration of the ligand may be relatively low, for example, <10 μM. By the Wyman linkage function,[20,21] ligand binding to the native state will shift the molecular population toward the species with the most compact native conformations and away from unfolded species. As a result, the rate of aggregation can be substantially reduced. For example, the binding of heparin and other polyanions to fibroblast growth factor greatly increases conformational stability of the native state and thereby reduces the aggregation rate.[22–25] High-throughput screening studies testing the effects of a broad range of excipients (e.g., selections from the generally recognized as safe [GRAS] list) can lead to discovery of safe and highly effective stabilizing agents that function as strong binding ligands.[26,27] Usually, such results are not predictable a priori because the ligands identified may not be known from previous studies of the protein's biology and may not be normally found *in vivo*.

Nonspecific stabilizing compounds also can be used to increase the conformational stability of proteins. For example, sucrose is generally effective at stabilizing the native state of most proteins. Typically, relatively high concentrations (e.g., >300 mM) of such excipients are needed to obtain substantial increases in protein conformational stability, and the magnitude of the effect often is much less than that seen with a specific ligand. Nonspecific stabilizers are preferentially excluded from the surface of proteins; there is a lower concentration of stabilizer around the protein than in the bulk solution. This interaction increases the chemical potential of the protein. The magnitudes of preferential exclusion and the increase in chemical potential vary directly with the surface area of the protein molecule exposed to solvent. Thus, the chemical potential increase is greater for unfolded species than for the native state. As a result, the equilibrium between species is shifted toward those with the lowest surface area. This effect manifests itself as a reduction in the level of partially unfolded aggregation-prone molecules and as a resultant decrease in the rate of protein aggregation. In practice, however, nonspecific stabilizers alone, at concentrations that are acceptable for injection into patients, usually do not sufficiently reduce aggregation rate for pharmaceutically acceptable storage stability. But if optimized pH is used to greatly reduce aggregation rate, the incremental benefit from nonspecific stabilizers might provide the extra stability needed to achieve pharmaceutically acceptable rates of aggregation.[5]

7.2.2 Colloidal Stability

Colloidal stability also can contribute to decreased aggregation rates.[17] Protein surface charge may be a key parameter affecting colloidal stability. Charge–charge repulsion between protein molecules can be the dominant factor controlling the rate of aggregation. Thus, alterations in pH and the resulting change in protein surface charge can greatly influence aggregation rate, independent of the effects of pH on protein conformational stability. An interesting example has been observed with recombinant human granulocyte colony-stimulating factor (GCSF) for which aggregation is extremely rapid at pH 7 but greatly slowed at pH 3. At pH 3, the protein is highly positively charged and colloidal stability is dominant.[17] This effect is so pronounced that even in the presence of 0.9% benzyl alcohol, which causes partial unfolding of the protein, aggregation of GCSF is greatly inhibited at pH 3.[13] This is because charge–charge repulsion between the protein molecules makes the intermolecular interactions needed for aggregation greatly unfavorable.

Of course, if a given pH increases colloidal stability but greatly reduces conformational stability, the latter effect may dominate and aggregation could be accelerated. Thus, in practice, it is necessary to test protein stability across a relatively wide pH range to empirically determine the optimal pH for slowing aggregation rate.

Colloidal stability can be reduced by an increase in ionic strength. This is because charge shielding by ions can reduce the charge–charge repulsion between protein molecules. Therefore, in addition to screening for effects of pH, it is important to determine protein aggregation rate as a function of ionic strength.[17,28]

7.3 EFFECTS OF INTERFACES ON PROTEIN AGGREGATION

7.3.1 Effects of Air–Water Interface

Under solution conditions typical of those encountered in most commercial protein formulations, protein aggregates are thought to be more thermodynamically stable than properly folded, native protein molecules.[2,29–34] Yet, under carefully chosen formulation conditions, aggregation may proceed at a nearly negligible rate. Thus, it would be anticipated that there is a step in the reaction pathway for the formation of aggregates that provides a substantial activation energy barrier for aggregation. Frequently, it is presumed that this rate-limiting step is associated with unfolding or partial unfolding of the protein from its native conformation.

The presence of interfaces may open new reaction pathways for protein aggregation that reduce the energy barrier for unfolding or partial unfolding. Under typical solution conditions, the air–water interfacial region is a zone of about 2-nm thickness wherein the properties change from those of liquid water to air.[35] It is important to note that this dimension is of the same order of magnitude as the size of a protein molecule. Thus, when a protein adsorbs at an air–water interface, essentially the entire protein may experience forces due to surface tension that are dramatically different from those found in the bulk solution. For example, assuming an air–water interfacial tension of an order of 0.07 N/m, we can estimate that the transverse pressure within the 2-nm-wide interfacial region that acts on a plane perpendicular to the interface is about −700 atm. Thus, a protein that adsorbs to the air–water interface may experience a tension force perpendicular to the interface of the order of 140 pN, which may be sufficient to cause protein unfolding. For example, studies of individual protein molecules with an atomic force microscope have found that 150 pN is sufficient to mechanically unfold a protein.[36]

Due to the amphiphilic nature of proteins, their accumulation at air–liquid interfaces is expected; once at the interface, they may unfold. A single layer of unfolded protein at an interface would typically represent a negligible amount of protein when compared to the amount found in the bulk solution. However, if the solution is agitated, a new air–water interfacial area will be generated continuously, transferring unfolded molecules into the bulk solution. Agitated air–water interfaces may thus continuously produce unfolded or partially unfolded molecules that, in turn, may react to form aggregates at the interface and/or in the bulk solution.

7.3.2 Protein Aggregation at Solid–Liquid Interfaces

Solid–liquid interfaces may also damage proteins. Proteins are exposed to numerous solid–liquid interfaces during manufacturing processes and in the drug product container and delivery system. For example, these interfaces are found at the surface of membranes used in dialysis, diafiltration, and terminal sterilization; stainless steel–water interfaces present in pumps, piping, and storage vessels; the glass–water surface of glass vials and syringes; and the tubing–water interface during administration. Proteins can easily be damaged upon adsorption to such surfaces, as evidenced by the loss of activity of aminoacylase adsorbed on the surface of an ultrafiltration membrane,[37] by the aggregation of a monoclonal antibody (mAb) after exposure to stainless steel surfaces,[38] and by the loss of bioactivity of recombinant interleukin-2 exposed to silicone rubber–water interfaces in a peristaltic pump.[39]

In a simplified picture of protein adsorption to solid–liquid interfaces, we can think of a protein as a deformable droplet. When this droplet is placed on a surface, the contact angle θ between the protein and the solid surface is determined by a force balance such that[35]

$$\cos\theta = (\sigma_{ws} - \sigma_{ps})/\sigma_{pw}, \tag{7.1}$$

where σ_{ws} is the water–solid interfacial tension, σ_{ps} is the protein–solid interfacial tension, and σ_{pw} is the surface tension at the protein–water interface. When θ is near 180° (cos θ = −1), the protein structure is nearly unperturbed by its interaction with the surface, and when θ is near 0° ($\cos\theta = 1$), the protein spreads and is completely unfolded on the surface. Of course, for a real protein, such spreading is limited by intramolecular forces that hold the protein together; the protein may become distorted, partially unfolded rather than spreading completely. This unfolding may, in turn, create aggregation-prone intermediates that initiate aggregation. Furthermore, the adhesion energy per unit area of the protein–solid interface is

$$\Delta W_{swp} = \sigma_{pw}(1+\cos\theta). \tag{7.2}$$

As a first approximation, we can estimate the free energy of protein unfolding on the surface $\Delta G_{N_s \to U_s}$ as

$$\Delta G_{N_s \to U_s} = \Delta G_{N \to U} - \Delta A \Delta W_{swp}, \tag{7.3}$$

where $\Delta G_{N_s \to U_s}$ is the change in free energy of unfolding at the surface and ΔA is the difference in protein–solid interface upon unfolding at the surface per mol of protein.

7.3.3 Effects of Excipients on Structural Perturbation and Aggregation at Surfaces

Excipients such as sucrose and ammonium sulfate increase σ_{pw} because they are sterically excluded from the protein–water interface.[40–42] In the absence of specific surface–excipient interactions, the same phenomenon is expected to occur at the surface–water interface: σ_{ws} is expected to increase with increasing excipient concentration. In contrast, the interfacial tension at the protein–surface interface σ_{ps} is expected to be independent of the concentration of excipients in the bulk liquid (but of course dependent on the type of solid surface). Thus, we expect that $\Delta G_{N_s \to U_s}$ will decrease with increasing sucrose or ammonium sulfate, and the protein, although more stable in the bulk, will become less stable when adsorbed. Conversely, an additive that lowers surface tension at both the solid–water interface and the protein–water interface (decreasing both σ_{ws} and σ_{pw}) will increase the contact angle, resulting in less spreading and a more stable protein at the surface, albeit at the expense of stability in the bulk liquid. Such an effect is commonly seen with the addition of nonionic surfactants, which reduce surface-induced aggregation but can increase formation rates of protein aggregates in the bulk. Of particular interest are those excipients that are expected to increase surface tension at the water–surface interface (e.g., urea, arginine hydrochloride, lysine hydrochloride)[40,43,44] but, due to favorable interactions with the protein backbone, are thought to decrease surface tension at the protein–water interface.[45] In this case, the addition of such excipients may, at low excipient concentrations, increase θ, thus decreasing spreading, but at higher concentrations may show an opposite behavior.

For many excipients, the effect of their addition on surface tension is roughly linear with concentration[41] and can be characterized by their molar surface tension increment, $(\partial\sigma/\partial a_3)_T$:

$$\sigma(a_3) \approx \sigma_0 + (\partial\sigma/\partial a_3)_T\, a_3, \tag{7.4}$$

where σ_0 is the surface tension in the absence of the excipient, and α_3 represents the activity of the excipient. Molar surface tension increments for the protein–liquid interface are generally expected to be different from those at the solid–water interface. Substituting the expressions for surface tensions in terms of molar surface tension increments into Equations 7.1–7.3, assuming that the addition of a solute to the aqueous phase does not affect the solid–protein interfacial tension and taking the derivative with respect to excipient concentrations, yields the following simple result:

$$\frac{\partial \Delta W_{swp}}{\partial a_3} = \left(\frac{\partial \sigma_{sw}}{\partial a_3}\right) + \left(\frac{\partial \sigma_{pw}}{\partial a_3}\right). \tag{7.5}$$

Combining Equations 7.3–7.5 gives

$$\frac{\partial(\Delta G_{N_s \to U_s})}{\partial a_3} = \frac{\partial(\Delta G_{N \to U})}{\partial a_3} - \left[\left(\frac{\partial \sigma_{pw}}{\partial a_3}\right) + \left(\frac{\partial \sigma_{sw}}{\partial a_3}\right)\right]\Delta A. \tag{7.6}$$

Thus, as the activity of an excipient solute is increased, the free energy of unfolding of a protein adsorbed on a surface is decreased from that of the bulk by the sum of the molar surface tension increments at the water–solid and water–protein interfaces multiplied by the change in surface area that the protein covers as it adsorbs. For surfactants, these molar surface tension increments are expected to be negative. This means that as a surfactant is added, the free energy of unfolding on the surface becomes larger compared with the free energy of unfolding in the bulk; it becomes more difficult to unfold the protein at a surface relative to unfolding it in the bulk solution. Other mechanisms of action of surfactants may also lie in the competition of stabilizer with protein for the surface(s). Polysorbates actually bind more strongly to interfaces than protein.[46] Thus, this competition leads to a decreasing number of protein molecules at the interface.

Conversely, excipients such as sugars that have positive molar surface tension increments will make it relatively easier to unfold adsorbed proteins than proteins in the bulk. There is thus a compromise that must be struck: excluded excipients stabilize the protein in the bulk but destabilize it at interfaces, whereas surfactants will tend to stabilize against unfolding at interfaces but may destabilize the protein in the bulk.

7.3.4 Interactions of Proteins with Microparticle Surfaces

Microparticle–solution interfaces found in commercial protein formulations may induce protein aggregation and particle formation. Foreign microparticle contamination of protein formulations occurs for a number of reasons. For example, during the filling of vials, high-speed filling pumps can shed stainless steel microparticles directly into commercial formulations. In turn, these metallic microparticles may act as nucleation sites for heterogeneous protein aggregation and particle formation.[4] Shedding of microparticles of silica, an effect that is exacerbated by industrial depyrogenation procedures, occurs within glass vials.[47] Such microparticles may also induce heterogeneous protein aggregation in commercial protein formulations.[3] Formulations have been tested to determine the amount of protein lost from solution due to adsorption to the glass particles and have been incubated (usually at elevated temperature) to determine whether the presence of microparticle surfaces acts to accelerate protein aggregation.[38]

7.3.5 Protein Aggregation at Silicone Oil/Water Interfaces

In order to provide more convenience to patients, many protein-based therapeutics are being formulated in prefilled glass syringes. To allow the syringe plunger to move smoothly during injection, the barrel of the syringe must be lubricated.[48] Typically, silicone oil is used as the lubricant because of its good wetting properties, low surface tension, and low aqueous solubility.[49] However, the presence of silicone oil within formulations has been reported to be a risk for protein aggregation. In the late 1980s, there were a number of reports documenting that silicone oil released from disposable plastic syringes caused particle formation in insulin formulations.[50–52] These early reports have been augmented by more recent reports that emulsions of silicone oil caused aggregation of various model proteins[53] as well as a mAb.[54] In the latter case, agitation in the presence of silicone oil resulted in elevated levels of aggregated protein, and the secondary structure of the mAb in the precipitated aggregate was perturbed relative to that of the native protein.

The mechanism(s) for the synergy between agitation and silicone oil to cause protein aggregation is not well understood. Agitation may serve to increase the silicone oil surface area available for protein adsorption by emulsifying the oil and decreasing the size of oil droplets in suspension. In addition, agitation exposes protein to additional air–water interfaces, which may result in populations of partially unfolded protein molecules that, in turn, aggregate at silicone oil/water interfaces.

7.3.6 Shear Effects on Protein Aggregation

Therapeutic protein formulations can be exposed to shear stresses during their commercial production. Frequently, these shear stresses are assumed to be a culprit in protein aggregation. However, many studies have demonstrated that the magnitude and duration of shear exposure per se does not cause protein aggregation.[55–58] A recent study by Bee et al.[59] showed that even the shear rates expected in high-speed filling pumps (~20,000 s^{-1}) generate forces about an order of magnitude lower than those expected to be sufficient to unfold mAbs. Other processes expose proteins to much lower shear rates; for example, typical shear rates experienced during mixing operations or pipe flow are about 50 and 2000 s^{-1}, respectively.[60,61]

A probable reason for the frequent association of protein aggregation with processes that exert shear forces on fluids is the concomitant presence of interfaces in high-shear process equipment. For example, pumping of recombinant interleukin-2 in a peristaltic pump resulted in a dramatic loss of bioactivity, but the loss was associated with exposure to silicone rubber tubing surfaces and not to shear.[62] Sometimes, synergistic effects of shear and surface interactions have been observed. For example, inactivation of lysozymes at various surfaces was accelerated by stirring,[63] and aggregation of an IgG4 molecule due to contact with surfaces in a stainless steel chamber was accelerated when high shear was applied.[64]

High-shear processes may also result in generation of heterogeneous particulates, which may in turn result in protein aggregation. Stainless steel microparticles shed by a piston pump caused IgG aggregation.[65] High-shear processes such as filtration can also cause cavitation and/or bubble entrainment, both of which present air/water interfaces that may result in protein aggregation,[66,67] and aggregation may occur due to protein interactions with the filtration membrane itself.[37]

7.4 CRITICAL PROCESSING STEPS DURING DRUG PRODUCT MANUFACTURING OF BIOPHARMACEUTICALS

During drug product manufacturing, the protein encounters various critical processing steps that may have an impact on its stability. The following section describes relevant processing steps, and stresses arising during them, and how these steps may influence protein aggregation. The most important processing steps are summarized in Table 7.1.

7.4.1 Freezing and Thawing Operations

Drug substance bulk may be stored at temperatures allowing storage in a liquid state (e.g., at 2–8°C) or a frozen state. Temperature conditions of frozen storage need to be defined and may be, for example, "–20°C," "–40°C," "–60°C," "–70°C," or "–80°C" or similar with appropriate ranges. Alternatively, also temperature ranges such as "below –20°C" may be defined. The decision on storage conditions is governed by many factors, but the main factor should be maintainance of protein integrity. In this context, it needs to be mentioned

TABLE 7.1. Critical Processing Steps and Parameters during Drug Product Manufacturing of Biopharmaceuticals and Potential Implications for Protein Stability

Process Step (Parameters)	Potential Stresses to Proteins
Freeze/thaw of bulk drug solution	Cryoconcentration Solute crystallization Exposure to liquid–ice interface Low temperature pH shifts
Mixing of bulk drug solution after thawing and/or pooling	Air bubble entrainment, exposure to air–water interfaces Foaming Exposure to liquid–solid interfaces
Final formulation with excipients	Local exposure to pH extremes Exposure to liquid–solid interfaces
Filtration	Exposure to liquid–solid interfaces Regions of high protein concentration Air bubble intrainment, exposure to air–water interfaces
Filling into vials or syringe	Exposure to foreign microparticles Exposure to liquid–solid interfaces
Primary packaging	Exposure to liquid–solid interfaces Exposure to foreign microparticles Exposure to liquid–silicone oil interfaces Agitation Accidental freeze-thawing Exposure to leachates and/or extractables

that the material may not fully be frozen at –20°C and may rather be in a "slush-like" state, a condition that might contribute to protein instability. Adequate stability needs to be documented for drug substance as well as drug product at both real-time and accelerated temperature conditions.

The final storage temperature is one of many factors that affect protein stability during freezing, storage in the frozen state, and during thawing. It is well known also that the freezing conditions impact the protein stability. Critical parameters include (1) the temperature gradient and hence cooling rate used for freezing, (2) the contact surface by which the temperature (cooling gradient) is transferred to the solution to be frozen, (3) the volume of solution to be frozen as well as the volume of the vessel, and (4) the initiation point(s) of freezing and the impact on heterogeneity in composition throughout the vessel.

Initiation of ice formation, for example, at one end of the vessel, can result in an ice front moving away from the point of initiation. Consequently, cryoconcentration of solutes to differing degrees can occur at different points in the vessel. Protein aggregation may occur in a subfraction of the frozen bulk due to the higher concentration of protein and/or other solutes.

Furthermore, in the case of many buffers, and particularly for sodium phosphate buffers, cooling and freezing can result in selective crystallization of different buffer components, potentially resulting in pH shifts of the frozen solution.[68–70] For example, the pH of frozen solutions of sodium phosphate may be reduced to below pH 4 due to crystallization of the dibasic buffer salt. Citrate and succinate buffers may also show a pH change upon freezing. Such pH shifts can cause protein aggregation.

The degree of acidification of sodium phosphate buffered solution during freeze-thawing depends on the degree of dibasic salt crystallization, which, in turn, is dictated by processing parameters. For example, rapid cooling may allow the non-ice phase to reach a temperature at which a glass forms before extensive crystallization has occurred. But crystallization and acidification can occur during thawing if the warming rate is not sufficiently fast. Also, crystallization can occur during storage at high subzero temperatures. In general, sodium phosphate buffer should be avoided for bulk drug substance storage. In case this buffer is used, strict attention must be paid to processing conditions used during freeze-thawing to minimize acidification and resulting protein aggregation.

Even when buffers are employed that do not result in pH changes during freeze-thawing, alterations in processing parameters and systems can greatly impact protein stability. In this context, the influence of different cooling rates during freezing on protein aggregation levels of human growth hormone (hGH) has been reported[71] (Fig. 7.1).

Therefore, the size, material, volume, configuration, and freezing and thawing conditions on the bulk need to be carefully studied, selected, and controlled to minimize protein aggregation.

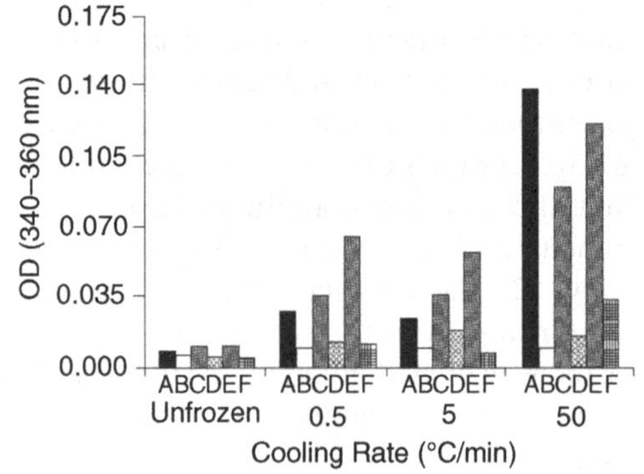

Figure 7.1. Effect of cooling rate on aggregate formation in different formulations (A–F) of human growth hormone (hGH) as assessed by the mean optical density in the 340- to 360-nm range (modified after Reference 71).

7.4.2 Mixing of Bulk Drug Solution after Thawing or Pooling

During drug product manufacturing, the bulk drug solution has to be mixed upon thawing or pooling of volumes from two or more containers. This procedure may be performed by using a magnetic stirrer (small scale), a paddle mixer (external or built-in), shaking (e.g., bags) by horizontal shakers, or by circulating pumping (e.g., bags). Importantly, all those procedures have in common agitational stress to the protein. This agitation is associated with the generation of air–water interfaces and enhanced contact with other wetted surfaces, which are stresses capable of inducing protein aggregation during the mixing procedure.[16] Any mixing procedure needs careful selection and control of parameters such as mixing time and speed.

7.4.3 Preparation of Final Formulation with Excipients

In the case where the bulk solution does not represent the final formulation in terms of composition and concentration of both excipients and protein, the bulk solution may be mixed with additional excipient stock solution(s) to yield the final formulation. This procedure offers some degrees of freedom as within certain limits: (1) the final protein concentration may be adjusted; (2) all final excipients may be added to reach their final concentration; and finally, (3) the final pH may be adjusted. However, there are certain limitations. In case of addition of solid excipients or concentrated acids or bases, the protein encounters a local microenvironment in the mixing zone (e.g., very low local pH), which may decrease protein stability and may even lead to protein precipitation. For example, Fig. 7.2 shows local precipitation of a therapeutic

Figure 7.2. Local precipitation of a monoclonal antibody during adjustment of pH by the use of a concentrated acid.

mAb in solution when using a concentrated acid for pH adjustment. This problem may be mitigated by more closely matching bulk drug substance formulation with final drug product formulation such that a concentrated acid or base may not be needed for final pH adjustment.

7.4.4 Filtration

Protein drug products are manufactured aseptically. Filtration is to reduce the solution's bioburden and to achieve sterility of the final drug product. Typically, the drug substance solution is filtered using filters with a pore size of 0.2 μm. Common filter materials used for drug product manufacturing encompass polyvinylidene difluoride (PVDF) (hydrophilized PVDF), polyethersulfone (PES), and cellulose acetate (CA). The choice of the filter is governed by various factors, including maximum flow rate and filtration speed, compatibility with the protein solution, and appropriate pore size.

One of the crucial challenges while dealing with a protein solution is protein adsorption to the filter material. This is especially relevant for low-concentration protein solutions (e.g., <0.1 mg/mL). At a low protein concentration, the adsorption to sterile filters (or other manufacturing equipment such as tubing)[39,72,73] may be especially problematic because of the difficulty in achieving the desired final protein concentration. Moreover, the filter material, like many other surfaces, will show a certain saturation threshold; as a consequence, protein adsorption will be highest in the beginning of the filtration process, which may be a source of inhomogeneity between containers with filtered drug products. Other components in the product may also adsorb to filters including excipients such as benzylalcohol[74] and polysorbates.[75,76]

In addition, adsorption of protein molecules to the filter surface may result in perturbation of the protein structure, with a concommitant risk of formation of soluble protein aggregates in the fitrate.

7.4.5 Filling

The filling of the bulk drug solution into the drug product primary packaging systems (e.g., glass vials) is usually performed in class A cleanrooms. The process parameters and equipment should be tested for compatibility with the therapeutic protein in its formulation. As an example, high filling speed may lead to foam in the drug product vial.

Various filling pumps are used in parenteral drug product manufacturing and each has its advantages and disadvantages. Positive displacement piston pumps are used by many facilities for their high precision during filling. They are accurate, easy to assemble, easy to clean and sterilize, and cause less dripping problems into the containers as they allow a suck-back of the solution. However, it is known that piston pumps might impact the protein stability, as reported, for example, by Cromwell and coworkers[77] (Fig. 7.3A). Also, recently, it has been documented by Tyagi and colleagues[65] that pumping of

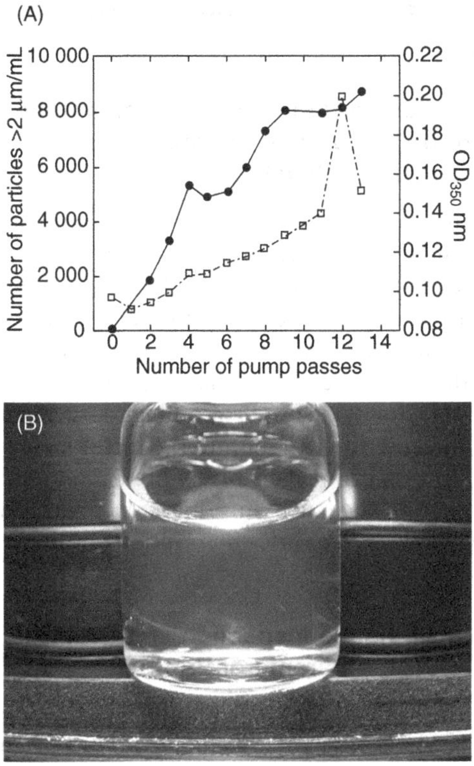

Figure 7.3. (A, modified after Reference 77): Increase in protein aggregation with the number of pump passes of a mAb solution through a piston pump. (B) Appearance of a mAb solution after filtration with a piston pump.

an IgG solution with a positive displacement piston pump caused formation of subvisible particles that were due to the adsorption of protein molecules to nanoparticles of stainless steel shed from the pump.

Peristaltic pumps are commonly used as an alternative for sensitive products. The pump tubing acts as a valve and for transporting the fluid, and it provides a sealed system from the protein solution supply vessel until the filling needle. Limitations may include filling accuracy and space requirements, just to name a few.

Other filling pump systems include time-pressure pumps and diaphragm pumps. Unpublished reports from our laboratories indicate that such pumps may be less likely to generate protein particles or aggregates than the piston pumps. But even with the alternative pumps, it is critical to assure that a formulation of a given therapeutic protein is compatible with the pump. This process requires measurement of protein aggregates and subvisible particles, as well as visible inspection for particles.

7.5 PARTICLES IN PARENTERAL PRODUCTS AND VISIBLE INSPECTION

7.5.1 Introduction: Understanding of Particles from the Parenteral Product Viewpoint is Important for Protein Biopharmaceutical Developers

Researchers working on developing therapeutic protein products often come from backgrounds in basic sciences such as biophysical chemistry, biochemistry, and protein chemistry. During their academic training, they usually have not been exposed to traditional areas of pharmaceutical sciences such as parenteral product development. As they gain experience in the industry, they may realize that their experiences of dealing with particles in therapeutic protein products are actually a subset of a much larger area of research and regulatory affairs. Because this insight into the history and practice of particle assessment for parenteral products is rarely presented in reviews or chapters on protein aggregation, and such information is valuable to the protein researcher, here we will provide brief descriptions of these topics.

7.5.2 The History of Visible Inspection and Particle Contamination Control in Parenteral Products

The sixth edition of the U.S. National Formulary (NF) (1936) provided for the first time guidance in terms of "cleanliness" of parenteral products for human use. It required "aqueous ampoule solutions to be clear, i.e. when observed over a bright light they shall be substantially free from precipitate, cloudiness or turbidity, specks or flecks, fibers or cotton hairs, or any undissolved material." Already at that time, the term "substantially free" in the NF VI statement caused a lot of confusion, leading to the following further statement in the next NF edition: "Substantially free shall be construed to mean a preparation which is free from foreign bodies that would be readily discernible by the unaided eye when viewed through a light reflected from a 100-W Mazda lamp using as a medium a ground glass and a background of black and white." The United States Pharmacopeia (USP) XIII, appearing at the same time around 1942, contained the following definition, which might appear comparable, however lacking the word "substantially" and therefore being more rigid: "Appearance of Solutions or Suspensions-Injections which are solutions of soluble medicaments must be clear, and free of any turbidity or undissolved material which can be detected readily without magnification when the solution is examined against black and white backgrounds with a bright light reflected from a 100-watt Mazda lamp or its equivalent." It is remarkable to say that at that time, soft drinks contained substantially fewer particles of suspended matter than parenteral products. During the time of the Second World War, when more and more parenteral products were required, this requirement led to a number of discussions[78] due to the friction between a desired, yet not manageable, state of absence of all impurities and the very subjective nature of visual inspection procedures.

7.5.3 Current Regulatory Requirements on Particulate Contamination and Control of Subvisible and Visible Particles in Parenteral Products in the EU, United States, and Japan

There are monographs and acceptance criteria/limits in the pharmacopoeias for visible particles and subvisible particles including USP <788>,[79] European Pharmacopoeia (Ph. Eur.) 2.9.19,[80] and Ph. Eur 2.9.20.[81] In the following sections, methodological details and acceptance criteria for subvisible and visible particles are further presented and discussed.

7.5.4 Subvisible Particle Measurements

The quantification of subvisible particles can be carried out via light obscuration, microscopic techniques, or the Coulter method.[82] Light obscuration or blockage particle counters are widely used in the pharmaceutical industry as the method of choice[83,84] and are described in the pharmacopoeias.[79,80] This method is able to detect and count individual particles by size in different size ranges from approximately 1 to 150 μm or larger with varying precision and reproducibility depending on the probe used. With the light obscuration method, the particle size is deduced from the quantity of light blocked as the particles pass in a single file fashion. Though this technique could be used for particles larger than about 1 μm,[85] however, the precision or reproducibility is low for these size ranges, as can be seen in Fig. 7.4.

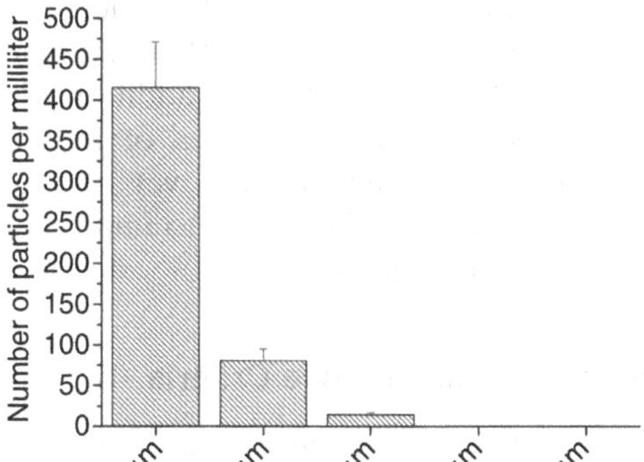

Figure 7.4. Variability of subvisible particle counting (light obscuration by HIAC) as assessed for a protein formulation at 10 mg/mL; subvisible particle counts over different size ranges are shown; for every sample, five measurements have been performed with five runs each, whereas the first run has been discarded.

Also, as the light obscuration method was developed for parenteral products in general, the typical method relies on a significantly high sample volume (i.e., 25 mL). However, recent work in some companies has led to the development of low-volume assays for light obscuration, which require as little as 0.4–1.0 mL of solution for analysis.[16,86]

A second method that Ph. Eur. allows for subvisible particle counting is the "microscopic method." The sample is filtered through a vacuum onto a grid lined filter and, once dried, is placed under the microscope for counting. The utility of this method for routine quantification of subvisible particles in therapeutic protein products is questionable because of the intensive time and effort required.

The methods for the detection of subvisible particles described above have also been used during formulation development of therapeutic proteins,[16,87] since protein aggregates may also be in the range of several micrometers. The rationale is to detect protein aggregates of this size as they are typically not picked up by methods such as size exclusion high pressure liquid chromatography (SE-HPLC). Moreover, it may provide additional information in terms of aggregation pathways. Interestingly, the formation of particles in the range of several micrometers has not correlated to aggregation observed in the size range of nanometers, that is, soluble aggregates.

A difficulty of the subvisible particle measurement in protein samples with light obscuration is that the instrument cannot differentiate between actual foreign particles, protein particles, and air bubbles, which can potentially lead to false positive results. For example, for the particle count testing of freshly reconstituted lyophilized protein samples, the reintroduction of water may result in many air bubbles, especially when the formulations contain a surfactant.

Recently, a technique such as microflow imaging has been used to count subvisible particles in therapeutic protein products. A study by Krishnamurthy et al.[88] demonstrated that this method was able to detect and quantify more protein particles, which were semitranslucent, than did light obscuration. The microflow imaging approach may also provide additional information on the nature of the particles because digital images of individual particles are obtained. However, these techniques are not yet fully evaluated on their potential use. Other, comparable technologies to improve subvisible particle measurement are also in development.

7.5.5 Subvisible Particle Acceptance Criteria

The acceptance criteria for subvisible particles are set in the pharmacopoeias (see also Table 7.2). These limits are probably derived from "historically relevant" contaminations of parenteral products, such as tiny glass particles or alike, potentially generated during sterile fill and finish processes. Probably, the limits were originally set based on risks such as emboli. However, a scientifically proven link between particles and specific safety implications

TABLE 7.2. Requirements of Subvisible Particulate Matter for Parenterals according to the Pharmacopoeias

Pharmacopoeia	Method	Volume	Limits
Ph. Eur. USP JP	Light obscuration	LVP	25 particles/mL ≥10 μm 3 particles/mL ≥25 μm
		SVP	6000 particles per container ≥10 μm 600 particles per container ≥25 μm
	Microscope	LVP	12 particles/mL ≥10 μm 2 particles/mL ≥25 μm
		SVP	3000 particles per container ≥10 μm 300 particles per container ≥10 μm

LVP, large-volume parenteral (>100 mL nominal fill volume); SVP, small-volume parenteral (≤100 mL nominal fill volume).

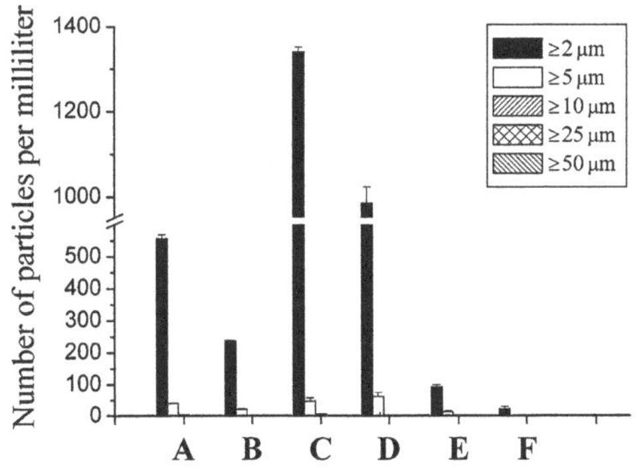

Figure 7.5. Subvisible particulate levels in marketed large-volume parenteral (LVP) products, namely, different 0.9% NaCl infusion bags (A–F). Subvisible particle levels were detected by means of light obscuration (HIAC); data represent mean ± SD ($n = 2$ measurements).

upon parenteral administration such as emboli after intravenous administration was, and still is, not available. More interestingly, the pharmacopoeial limits of small-volume parenteral products (<100 mL fill volume) also refer to the container and do not consider the administered dose or volume. Typical levels of subvisible particles in large-volume parenterals such as sodium chloride solution for infusion are given in Fig. 7.5. Levels of subvisible particles in different commercially available biotech drug products are given in Fig. 7.6.

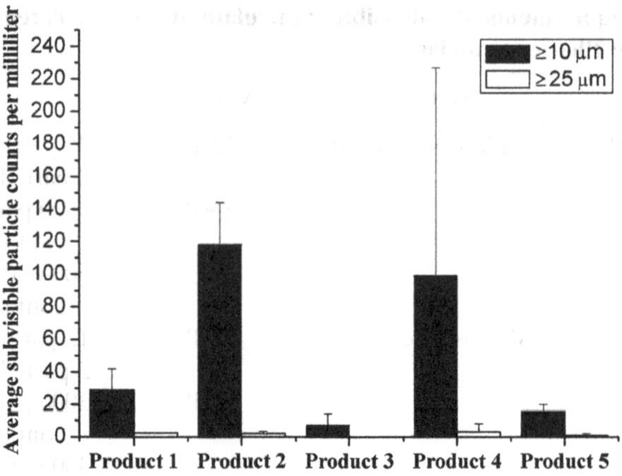

Figure 7.6. Average subvisible particle counts (≥10 μm and ≥25 μm/mL) for five different marketed biotech drug products (products 1–5) as assessed by light obscuration (HIAC), $n = 3$ measurements ± SD.

7.5.6 Visual Inspection Requirements

Visible inspection methods have been used now for decades to analyze for "extraneous particles" in parenteral drug products, with particles currently being defined as "contaminations consisting of extraneous, mobile, undissolved particulates other than gas bubbles, unintentionally present in the product" (e.g., glass particles). However, in case of sufficiently large protein aggregates, this method may also be suitable for the detection of protein particles.[16]

The Ph. Eur. monograph for parental preparations in conjunction with the visual particle monograph[80,81] requires "parenteral preparations" (which are not administered using a final filter) to be "practically free from (visible) particles." The USP requires parenteral products to be "essentially free of visible particles" and that for injectable products "every care should be exercised (…) to prevent contamination with microorganisms and foreign material." Neither USP nor Ph. Eur. provides a definition on what can be considered "essentially" or "practically." The authors rather consider that the requirement reflects the current capabilities of manufacturing and control.

Also the Japanese Pharmacopoeia defines that injections "must be clear and free from readily detectable foreign insoluble matters."[89] These requirements probably originated as mentioned in the introduction from safety concerns from foreign particulate matter after intravenous administration, as, for example, the risk of pulmonary embolism and creation of thrombi.[90–95] After intravenous injection, particles greater than ~8 μm in diameter are described to localize in the lung or in the pre-lung capillaries, while smaller particles may localize in the spleen, liver, or bone marrow.[96]

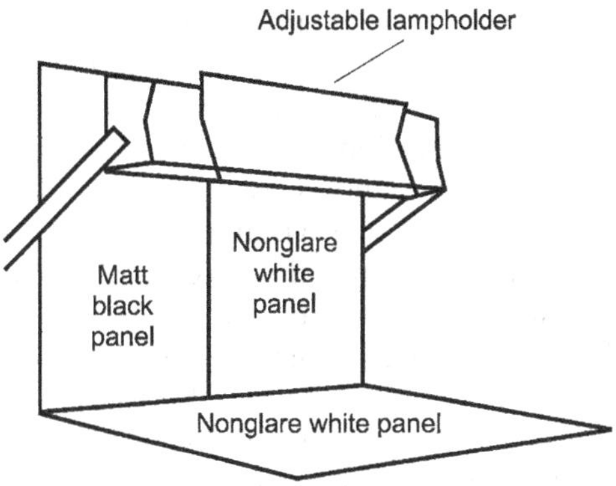

Figure 7.7. Apparatus for the visual inspections as per Ph. Eur. (current edition).

Interestingly, the general monograph of the Ph. Eur. on "Monoclonal antibodies for human use"[97] currently requires those final preparations to be "without visible particles." This criterion may be virtually impossible to meet. Moreover, there is not a published report that protein particles in this size range cause adverse events in patients after subcutaneous injection, which is the route of administration most often used for therapeutic proteins. Protein particles that happen to be large enough to be visible (e.g., >100 μm, but different definitions exist; see below) should be considered in the context of the total population of particles of all sizes. In this case, why should there be concern for one or a few particles greater than 100 μm when there may be several orders of magnitude greater numbers of smaller particles? And the total mass of protein in the subvisible particles may far exceed that in a few visible particles. Of course, such a case does not eliminate the need for visual inspection for foreign materials in 100% of manufactured vials and syringes.

Good manufacturing practice (GMP)[98–100] also requires that every container of a final drug product batch is to be subjected individually (100% of the batch) to a physical inspection, whenever the nature of the container permits, and that every container whose contents show evidence of contamination with visible foreign material is to be rejected.

7.5.7 Visual Inspection Methods

There are no specified viewing conditions or inspection time for visual inspection within the USP; however, the USP is currently drafting an informational chapter on visual inspection. The Ph. Eur. visual inspection method is carried out using an apparatus consisting of a box with a backboard comprising of two panels, the left one as a nonglare white panel and the other a matt black panel situated on the right with an observation time of about 5 s for each panel (Fig. 7.7).

An adjustable lamp provides the source of white light with an intensity of 2000–3750 lux. Differences exist in the visual inspection method between the pharmacopoeias. The Japanese Pharmacopeia[89] refers to white-light inspection at 1000 lux, with an observation time of 15 s against a black background. An automated rotation of the samples used with some inspection machines whirls up particulate matter in liquids within the container, and these particles are enhanced visually through a magnifying glass. A great advantage of the automated rotation systems is that the operator can see the full circumference of the container due the absence of clips or fingers, which could hinder the view. A concentrated light beam through the bottom of the container causes particles to reflect light due to the Tyndall effect[101–103] and therefore further facilitates inspection. From a GMP perspective, operators doing the inspection should pass regular eyesight checks and should be allowed frequent breaks from inspection. Other inspection methods would also be considered to be used; however, the process should be validated and the performance of the equipment is to be checked at intervals.[98–100] Furthermore, operators need to be carefully trained with training sets (containing "defects" but also "acceptable" vials), both initially and at specified intervals, in order to train them in the decision on whether a note on the final product container is a "fail" or a "pass." For commercial production or late-stage development, inspection may also use "automated" inspection machines, which, however, need to be carefully qualified and tested against the performance of trained (human) inspectors.

Another aspect of "appearance" testing is the determination of turbidity (or opalescence or clarity) of solutions for injection or infusion, for example, as defined in the Ph. Eur. chapter 2.2.1 determining the optical properties of a solution.[104] Considering turbidity in protein formulations, turbidity depends on protein concentration,[105] the presence of non-dissolved particles, the particle size, and particle number per volume unit. Aggregation has been reported to be indicated by a marked increase in turbidity over storage or stress time and has been used for the detection of aggregates.[16,87] However, turbidity has also been shown being a precursor to liquid–liquid phase separation.[106] Therefore, the relevance of turbidity data needs to be closely assessed on a case-by-case basis.

7.5.8 What Does "Visible" Mean?

It is described in the literature that the human eye has the ability to resolve objects slightly smaller than ~80 μm at a distance of ~25 cm.[107] There is, however, some doubt about the exact size limit of a suspended particle that is visible. It is noteworthy that illumination intensity, inspection time, inspection aids (magnifying lenses), automated handling (vial spinning), light polarization, inspection background, inspector training and experience, and number and type of particles may influence the probability of particle detection. Using different detection methods, it can be shown that the detection probability is significantly impacted by the above parameters and that extraneous particles as small as 25 µm, if present in sufficient number, could also be detected (P. Goldbach, pers. comm.). In general, particles in the range of 25–>100 µm may therefore be observed visually, depending on the setup.

With all the various systems available, the background, the light source, and the light intensity, as well the actual inspector all play a crucial role in the limit of detection of the particles with a major criterion being the subjective nature of visual inspection to each operator. Therefore, a "definition of visibility" is not a question only of particulate size, but rather a probabilistic term whether particulates are observed. Important parameters for the detection of visible particles by human eyes and therefore determining "visibility" of particulates include (1) the light conditions, (2) magnifying lenses used, (3) observation time, (4) type of background inspected against, (5) distance from samples, (6) automatic versus manual rotation, (7) the inspector's capability and training and eye conditions, and (8) the size, number, type, and refractive index of particles.

The inspection for visible defects as performed as a 100% control during drug product manufacturing therefore needs to be considered as probabilistic in nature. The requirement to produce material that is "essentially free of visible particles" therefore considers the limitations in manufacturing and control.

7.5.9 Potential Causes of Particulate Contamination in Parenteral Products

Sources for particulates in parenteral products are many. Generally, sources can be classified into "extrinsic" material, that is, particulates introduced accidentally from the outside, or "intrinsic" particulates, being a property of the product. Process steps where particulates may be introduced to the material administered to the patient may range from manufacturing, primary packaging, during storage, and administration (e.g., latent incompatibility between formulation and primary packaging). Protein aggregation and precipitation can be due to preparation of products in hospitals or can even occur during administration. Sources of contamination may be airborne particles, fractures of glass surfaces and edges, and so on. Particulates can take all shapes and sizes and can come from any material, organic or inorganic. Typically, particles are not due to one species only. Also, particles of foreign materials can be coated with adsorbed protein.

The description given in USP <788> already refers to potential causes and sources for particulate contamination in parenteral products, as it states that "particulate matter consists of mobile, <u>randomly-sourced</u>, extraneous substances, other than gas bubbles, that cannot be quantitated by chemical analysis due to the small amount. ..." Examples for types of some extraneous materials reported in parenteral solutions are described by Groves.[108] They encompass, for example, glass, metal, rubber, starch, zinc oxide, and cellulose

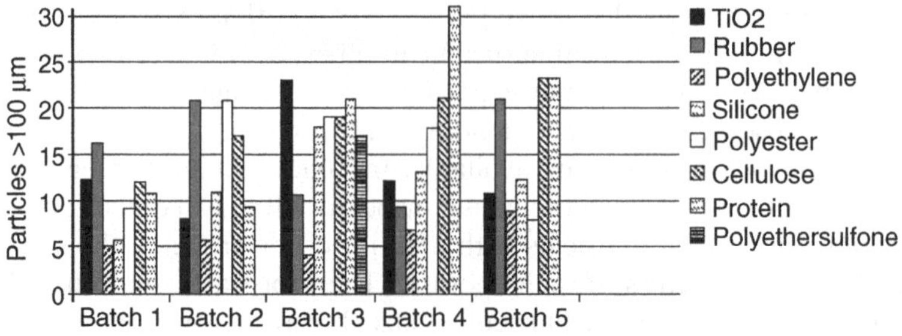

Figure 7.8. Contamination in parenteral products (batch-to-batch comparison) as assessed by Raman spectroscopy (modified from rapID with kind permission).

fibers. Though the data are from parenteral products in the mid-1970s, respective reports are still considered quite up to date, as recent data suggest in Fig. 7.8.

A particulate can be described by size and by all three dimensions (e.g., volume or cross-sectional area), but also by its transparency. Typical exogenous particles are usually not of spherical shape. Particulates also may or may not follow the Gaussian law with regard to their size distribution. An example for particulates with a narrow size distribution is bacteria. Theoretically, particulates in injectable products might be of sizes from very large to very small; however, chances are that larger particulates are statistically more easily detected during visual inspection. As already described above, various extraneous materials have been reported for parenteral solutions.[108] Such contaminants might be in the range of some micrometers and are thus invisible to the unaided eye or may be only detectable as a faint haze, if present in sufficient concentration.

In summary, parenteral products cannot be expected to be manufacturable excluding 100% of all particulates. It is noteworthy to mention that also during clinical and clinical hospital pharmacy practice, solutions and infusion sets that are prepared for parenteral administration for humans may differ in quality. The administration systems themselves are typically not essentially free of particulates. All the latter would counterbalance a requirement on administration fluids to be absolutely free of particulates. An unambiguous identification, differentiation, and assignment of whether particles are extrinsic or intrinsic in nature, despite best efforts using methods such as Raman spectroscopy or Fourier transform infrared (FTIR) microscopy, however, is not considered to be feasible.

7.6 SUMMARY AND OUTLOOK

A variety of definitions exist for protein aggregates. In this context, aggregates are typically defined as assemblies of protein molecules that increase in level with time of storage and/or in response to stress. Mechanistically, both conformational and colloidal stabilities play a role during aggregation pathways. Conditions such as temperature, air–liquid and solid–liquid interfaces, freezing/thawing, surface interactions, and extrinsic nuclei may facilitate or induce aggregation in biotech products. A number of potential stress conditions are encountered during bulk and drug product manufacturing, such as pH adjustment, filtration, and fill and finish, as well as storage and transport. Thus, it is essential both to understand conditions under which protein aggregation may occur and to counteract potential aggregation by adequate process and formulation development.

REFERENCES

1. Mahler H-C, Friess W, Grauschopf U, Kiese S. 2009. Protein aggregation: Pathways, induction factors and analysis. J Pharm Sci 98(9):2909–2934.
2. Roberts CJ. 2007. Non-native protein aggregation kinetics. Biotechnol Bioeng 98(5):927–938.
3. Chi EY, Weickmann J, Carpenter JF, Manning MC, Randolph TW. 2005. Heterogeneous nucleation-controlled particulate formation of recombinant human platelet-activating factor acetylhydrolase in pharmaceutical formulation. J Pharm Sci 94(2):256–274.
4. Tyagi A, Randolph T, Dong A, Maloney K Jr. CH, Carpenter J. 2009. IgG particle formation during filling pump operation: A case study of heterogeneous nucleation on stainless steel nanoparticles. J Pharm Sci 98(1):94–104.
5. Randolph TW, Carpenter JF. 2007. Engineering challenges of protein formulations. AIChE J 53(8):1902–1907.
6. Mahler HC, Muller R, Friess W, Delille A, Matheus S. 2005. Induction and analysis of aggregates in a liquid IgG1-antibody formulation. Eur J Pharm Biopharm 59(3):407–417.
7. Kendrick BS, Carpenter JF, Cleland JL, Randolph TW. 1998. A transient expansion of the native state precedes aggregation of recombinant human interferon-γ. Proc Natl Acad Sci U S A 95:14142–14146.
8. Webb JN, Webb SD, Cleland JL, Carpenter JF, Randolph TW. 2001. Partial molar volume, surface area, and hydration changes for equilibrium unfolding and formation of aggregation transition state: High-pressure and co-solute studies on recombinant human interferon-gamma. Proc Natl Acad Sci U S A 98(13): 7259–7264.
9. Corbo DC, Suddith RL, Sharma B, Naso RB. 1992. Stability, potency, and preservative effectiveness of epoetin alfa after addition of a bacteriostatic diluent. Am J Hosp Pharm 49(6):1455–1458.
10. Lam XM, Patapoff TW, Nguyen TH. 1997. The effect of benzyl alcohol on recombinant human interferon-γ. Pharm Res 14(6):725–729.
11. Roy S, Jung R, Kerwin BA, Randolph TW, Carpenter JF. 2005. Effects of benzyl alcohol on aggregation of recombinant human interleukin-1-receptor antagonist in reconstituted lyophilized formulations. J Pharm Sci 94(2):382–396.

12. Roy S, Katayama D, Dong AC, Kerwin BA, Randolph TW, Carpenter JF. 2006. Temperature dependence of benzyl alcohol- and 8-anilinonaphthalene-1-sulfonate-induced aggregation of recombinant human interleukin-1 receptor antagonist. Biochemistry 45(12):3898–3911.
13. Thirumangalathu R, Krishnan S, Brems DN, Randolph TW, Carpenter JF. 2006. Effects of pH, temperature, and sucrose on benzyl alcohol-induced aggregation of recombinant human granulocyte colony stimulating factor. J Pharm Sci 95(7):1480–1497.
14. Zhang Y, Roy S, Jones LS, Krishnan S, Kerwin BA, Chang BS, Manning MC, Randolph TW, Carpenter JF. 2004. Mechanism for benzyl alcohol-induced aggregation of recombinant human interleukin-1 receptor antagonist in aqueous solution. J Pharm Sci 93(12):3076–3089.
15. Tobler SA, Holmes BW, Cromwell MEM, Fernandez EJ. 2004. Benzyl alcohol-induced destabilization of interferon-gamma: A study by hydrogen-deuterium isotope exchange. J Pharm Sci 93(6):1605–1617.
16. Kiese S, Papppenberger A, Friess W, Mahler HC. 2008. Shaken, not stirred: Mechanical stress testing of an IgG1 antibody. J Pharm Sci 97(10):4347–4366.
17. Chi EY, Krishnan S, Kendrick BS, Chang BS, Carpenter JF, Randolph TW. 2003. Roles of conformational stability and colloidal stability in the aggregation of recombinant human granulocyte colony-stimulating factor. Protein Sci 12(5): 903–913.
18. Chi EY, Krishnan S, Randolph TW, Carpenter JF. 2003. Physical stability of proteins in aqueous solution: Mechanism and driving forces in nonnative protein aggregation. Pharm Res 20(9):1325–1336.
19. Pace CN. 1986. Determination and analysis of urea and guanidine hydrochloride denaturation curves. Methods Enzymol 131:266–280.
20. Wyman J. 1964. Linked functions and reciprocal effects in hemoglobin: A second look. Adv Protein Chem 19:223–286.
21. Wyman J Jr., Gill SJ. 1990. *Binding and Linkage. Functional Chemistry of Biological Macromolecules*. Mill Valley, CA: University Science Books.
22. Prestrelski SJ, Fox GM, Arakawa T. 1992. Binding of heparin to basic fibroblast growth factor induces a conformational change. Arch Biochem Biophys 293(2): 314–319.
23. Volkin DB, Verticelli AM, Bruner MW, Marfia KE, Tsai PK, Sardana MK, Middaugh CR. 1995. Deamidation of polyanion-stabilized acidic fibroblast growth-factor. J Pharm Sci 84(1):7–11.
24. Dabora JM, Sanyal G, Middaugh CR. 1991. Effect of polyanions on the refolding of human acidic fibroblast growth-factor. J Biol Chem 266(35): 23637–23640.
25. Volkin DB, Tsai PK, Dabora JM, Gress JO, Burke CJ, Linhardt RJ, Middaugh CR. 1993. Physical stabilization of acidic fibroblast growth-factor by polyanions. Arch Biochem Biophys 300(1):30–41.
26. Fan HH, Li HN, Zhang MY, Middaugh CR. 2007. Effects of solutes on empirical phase diagrams of human fibroblast growth factor 1. J Pharm Sci 96(6):1490–1503.

27. Ausar SF, Espina M, Brock J, Thyagarayapuran N, Repetto R, Khandke L, Middaugh CR. 2007. High-throughput screening of stabilizers for respiratory syncytial virus—Identification of stabilizers and their effects on the conformational thermostability of viral particles. Hum Vaccin 3(3):94–103.
28. Katayama DS, Nayar R, Chou DK, Valente JJ, Cooper J, Henry CS, Vander Velde DG, Villarete L, Liu CP, Manning MC. 2006. Effect of buffer species on the thermally induced aggregation of interferon-tau. J Pharm Sci 95(6):1212–1226.
29. Andrews JM, Roberts CJ. 2007. Non-native aggregation of alpha-chymotrypsinogen occurs through nucleation and growth with competing nucleus sizes and negative activation energies. Biochemistry 46(25):7558–7571.
30. Roberts CJ. 2006. Non native protein aggregation: Pathways, kinetics, and shelf life prediction. In: Murphy RM, Tsai AM, eds., *Misbehaving Proteins: Protein Misfolding, Aggregation, and Stability*. New York: Springer, p. 17.
31. Roberts CJ, Darrington RT, Whitley MB. 2003. Irreversible aggregation of recombinant bovine granulocyte-colony stimulating factor (bG-CSF) and implications for predicting protein shelf life. J Pharm Sci 92(5):1095–1111.
32. Roberts CJ. 2003. Kinetics of irreversible protein aggregation: Analysis of extended Lumry-Eyring models and implications for predicting protein shelf life. J Phys Chem B 107(5):1194–1207.
33. Gazit E. 2001. The "correctly folded" state of proteins: Is it a metastable state? Angew Chem Int Ed 41(2):257.
34. Onuchic JN, LutheySchulten Z, Wolynes PG. 1997. Theory of protein folding: The energy landscape perspective. Annu Rev Phys Chem 48:545–600.
35. Adamson AW, Gast AP. 1997. *Physical Chemistry of Surfaces*, 6th edition. New York: John Wiley & Sons.
36. Brockwell D. 2007. Probing the mechanical stability of proteins using the atomic force microscope. Biochem Soc Trans 35:1564–1568.
37. Bodalo A, Gomez JL, Gomez E, Maximo MF, Montiel MC. 2004. Study of L-aminoacylase deactivation in an ultrafiltration membrane reactor. Enzyme Microb Technol 35(2–3):261–266.
38. Bee J, Chiu D, Sawicki S, Stevenson J, Chatterjee K, Freund E, Carpenter J, Randolph T. 2009. Monoclonal antibody interactions with micro- and nanoparticles: Adsorption, aggregation and accelerated stability studies. J Pharm Sci 98(9):3218–3238.
39. Tzannis ST, Hrushesky WJM, Wood PA, Przybycien TM. 1996. Irreversible inactivation of interleukin 2 in a pump-based delivery environment. Proc Natl Acad Sci U S A 93(11):5460–5465.
40. Lin TY, Timasheff SN. 1996. On the role of surface tension in the stabilization of globular proteins. Protein Sci 5(2):372–381.
41. Timasheff SN. 1992. Water as ligand: Preferential binding and exclusion of denaturants in protein unfolding. Biochemistry 31(41):9857–9864.
42. Timasheff SN. 1993. The control of protein stability and association by weak interactions with water: How do solvents affect these processes? Annu Rev Biophys Biomol Struct 22:67–97.
43. Prakash V, Loucheux C, Scheufele S, Gorbunoff MJ, Timasheff SN. 1981. Interactions of proteins with solvent components in 8-M urea. Arch Biochem Biophys 210(2):455–464.

44. Arakawa T, Timasheff SN. 1984. Protein stabilization and destabilization by guanidinium salts. Biochemistry 23(25):5924–5929.
45. Stoner MR, Fischer N, Nixon L, Buckel S, Benke M, Austin F, Randolph TW, Kendrick BS. 2004. Protein-solute interactions affect the outcome of ultrafiltration/diafiltration operations. J Pharm Sci 93(9):2332–2342.
46. Katakam M, Bell, LN, Banga AK. 1995. Effect of surfactants on the physical stability of recombinant human growth hormone. J Pharm Sci 84:713–716.
47. Akers MJ. 2002. Excipient-drug interactions in parenteral formulations. J Pharm Sci 91(11):2283–2300.
48. Smith EJ. 1988. Siliconization of parenteral packaging components. J Parenter Sci Technol 42:S1–S13.
49. Rykke M, Rolla G. 1990. Effect of silicone oil on protein adsorption to hydroxyapatite in vitro and on pellicle formation in vivo. Scand J Dent Res 98(5):401–411.
50. Bernstein RK. 1987. Clouding and deactivation of clear (regular) human insulin—Association with silicone oil from disposable syringes. Diabetes Care 10(6):786–787.
51. Chantelau E, Berger M, Bohlken B. 1986. Silicone oil released from disposable insulin syringes. Diabetes Care 9(6):672–673.
52. Chantelau EA, Berger M. 1985. Pollution of insulin with silicone oil, a hazard of disposable plastic syringes. Lancet 1(8443):1459.
53. Jones LS, Kaufmann A, Middaugh CR. 2005. Silicone oil induced aggregation of proteins. J Pharm Sci 94(4):918–927.
54. Thirumangalathu R, Krishnan S, Speed Ricci M, Brems DM, Randolph TW, Carpenter JF. 2009. Silicone oil- and agitation-induced aggregation of a monoclonal antibody in aqueous solution. J Pharm Sci 98(9):3167–3181.
55. Jaspe J, Hagen SJ. 2006. Do protein molecules unfold in a simple shear flow? Biophys J 91(9):3415–3424.
56. Thomas CR, Dunnill P. 1979. Action of shear on enzymes—Studies with catalase and urease. Biotechnol Bioeng 21(12):2279–2302.
57. Thomas CR, Nienow AW, Dunnill P. 1979. Action of shear on enzymes—Studies with alcohol-dehydrogenase. Biotechnol Bioeng 21(12):2263–2278.
58. Harrison RG, Todd P, Rudge SR, Petrides DP. 2003. *Bioseparations Science and Engineering*. New York: Oxford University Press.
59. Bee J, Stevenson L, Mehta B, Svitel J, Pollastrini J, Platz R, Freund E, Carpenter J, Randolph T. 2010. Response of a concentrated monoclonal antibody formulation to high shear. Biotechnol Bioeng 105(1):121–129.
60. Williams PA, ed. 2007. *Handbook of Industrial Water Soluble Polymers*. Oxford: Blackwell Publishing.
61. Steffe JF. 1996. *Rheological Methods in Food Processing Engineering*. East Lansing, MI: Freeman Press.
62. Tzannis ST, Hrushesky WJM, Wood PA, Przybycien TM. 1997. Adsorption of a formulated protein on a drug delivery device surface. J Colloid Interface Sci 189(2):216–228.
63. Colombie S, Gaunand A, Lindet B. 2001. Lysozyme inactivation under mechanical stirring: Effect of physical and molecular interfaces. Enzyme Microb Technol 28(9–10):820–826.

64. Biddlecombe JG, Craig AV, Zhang H, Uddin S, Mulot S, Fish BC, Bracewell DG. 2007. Determining antibody stability: Creation of solid-liquid interfacial effects within a high shear environment. Biotechnol Prog 23(5):1218–1222.
65. Tyagi AK, Randolph TW, Dong A, Maloney KM, Hitscherich C, Jr., Carpenter JF. 2009. IgG particle formation during filling pump operation: A case study of heterogeneous nucleation on stainless steel nanoparticles. J Pharm Sci 98(1): 94–104.
66. van Reis R, Zydney A. 2007. Bioprocess membrane technology. J Memb Sci 297(1–2):16–50.
67. Narendranathan TJ, Dunnill P. 1982. The effect of shear on globular proteins during ultrafiltration: Studies of alcohol dehydrogenase. Biotechnol Bioeng 24:2103–2107.
68. van den Berg L. 1959. Effect of addition of sodium and potassium chloride to the reciprocal system: KH_2PO_4-Na_2HPO_4-H_2O on pH and composition during freezing. Arch Biochem Biophys 84:305–315.
69. van den Berg L. 1966. pH changes in buffers and foods during freezing and subsequent storage. Cryobiology 3(3):236–242.
70. van den Berg L, Rose D. 1959. Effect of freezing on the pH and composition of sodium and potassium phosphate solutions: The reciprocal system KH_2PO_4-Na_2HPO_4-H_2O. Arch Biochem Biophys 81:319–329.
71. Eckhardt BM, Oeswein JQ, Bewley TA. 1991. Effect of freezing on aggregation of human growth hormone. Pharm Res 8:1360–1364.
72. Wahlgren M, Arnebrant T. 1991. Protein adsorption to solid-surfaces. Trends Biotechnol 9(6):201–208.
73. Bonomini M, Pavone B, Sirolli V, Del Buono F, Di Cesare M, Del Boccio P, Amoroso L, Di Ilio C, Sacchetta P, Federici G, Urbani A. 2006. Proteomics characterization of protein adsorption onto hemodialysis membranes. J Proteome Res 5(10):2666–2674.
74. Van der Bruggen B, Kim JH, DiGgiano FA, Geens J, Vandecasteele C. 2004. Influence of MF pretreatment on NF performance for aqueous solutions containing particles and an organic foulant. Sep Purif Technol 36(3):203–213.
75. Mahler HC, Printz M, Kopf R, Schuller R, Muller R. 2008. Behaviour of polysorbate 20 during dialysis, concentration and filtration using membrane separation techniques. J Pharm Sci 97(2):764–774.
76. Mahler H-C, Huber F, Ravuri SKK, Reindl J, Rückert P, Müller R. 2009. Adsorption behaviour of a surfactant and a monoclonal antibody to sterilizing-grade filters. J Pharm Sci 99(6):2620–2627.
77. Cromwell M, Hilario E, Jacobson F. 2006. Protein aggregation and bioprocessing. AAPS J 8(3):E572–E579.
78. Godding EW. 1945. Foreign matter in solutions for injections. Pharm J 154:124.
79. USP. 2008. USP/NF General Chapter. United States Pharmacopeial Convention, Rockville, MD.
80. *Ph. Eur. 2.9.19*. 2008. European Directorate for the Quality of Medicine (EDQM).
81. *Ph. Eur. 2.9.20*. 2008. European Directorate for the Quality of Medicine (EDQM).

82. Henriquez RR, Ito T, Sun L, Crooks RM. 2004. The resurgence of Coulter counting for analyzing nanoscale objects. Analyst 129(6):478–482.

83. Pavanetto F, Conti B, Genta I, Ponci R, Montanari L, Grassi M. 1989. Particulate matter test in small volume parenterals—Critical aspects in sampling methodology. Farmaco 44(6):633–643.

84. Pavanetto F, Conti B, Genta I, Ponci R, Montanari L, Vianello S. 1989. Particulate matter contamination of small volume parenterals. Int J Pharm 51(1):55–61.

85. Etzler FM, Sanderson MS. 1995. Particle size analysis: A comparative study of various methods. Part Part Syst Charact 12(5):217–224.

86. Coa S, Pollastrini J, Jiao N, Jiang Y, Narhi L. 2007. Quantitative determination of subvisible particles in injections using a small volume light obscuration method. Poster presentation, AAPS National Biotechnology Conference, San Diego, CA.

87. Mahler H-C, Muller R, Friess W, Delille A, Matheus S. 2005. Induction and analysis of aggregates in a liquid IgG1-antibody formulation. Eur J Pharm Biopharm 59(3):407–417.

88. Huang CT, Sharma D, Oma P, Krishnamurthy R. 2009. Quantitation of protein particles in parenteral solutions using micro-flow imaging. J Pharm Sci 98(9): 3058–3071.

89. *Japanese Pharmacopoeia, JPXIV.* 2008. Tokyo: Yakuji Nippo.

90. Brewer JH, Dunning JHF. 1947. An in vitro and in vivo study of glass particles in ampules. J Am Pharm Assoc Sci Ed 36(10):289–293.

91. Vonglahn WC, Hall JW. 1949. The reaction produced in the pulmonary arteries by emboli of cotton fibers. Am J Pathol 25(4):575–595.

92. Bruening EJ. 1955. Origin and significance of intra-arterial foreign body emboli in lungs of children. Virchows Arch 327:460–479.

93. Stehbens WE, Florey HW. 1960. The behavior of intravenously injected particles observed in chambers in rabbits' ears. Q J Exp Physiol Cogn Med Sci 45:252–264.

94. Sarrut S, Nezelof C. 1960. A complication of intravenous therapy—The macrophagic pulmonary arteritis with giant cells. Presse Med 68(10):375–377.

95. Garvan JM, Gunner BW. 1964. The harmful effects of particles in intravenous fluids. Med J Aust 2:1.

96. DeLuca P. 2007. The science behind particulate matter from animal studies: consequences to specific organ systems. *American Society of Health-System Pharmacists.*

97. *Ph. Eur. General Monographs*. 2008. European Directorate for the Quality of Medicine (EDQM).

98. *EU Good Manufacturing Practices Guide.*

99. *FDA Code of Federal Regulations.*

100. *FDA Code of Federal Regulations.*

101. Kerker M. 1969. *The Scattering of Light and Other Electromagnetic Radiation.* New York: Academic Press.

102. Kerker M. 1987. Classics and classicists of colloid and interface science. J Colloid Interface Sci 119(2):602–604.

103. Gentry JW. 1997. The legacy of John Tyndall in aerosol science. J Aerosol Sci 28(8):1365–1372.

104. *Ph. Eur. 2.2.1.* 2008. European Directorate for the Quality of Medicine (EDQM).

105. Sukumar M, Doyle BL, Combs JL, Pekar AH. 2004. Opalescent appearance of an IgG1 antibody at high concentrations and its relationship to noncovalent association. Pharm Res 21(7):1087–1093.

106. Cromwell ME, Hilario E, Jacobson F. 2006. Protein aggregation and bioprocessing. AAPS J 8(3):E572–E579.

107. Blackwell HR. 1946. Contrast thresholds of the human eye. J Opt Soc Am 36(11):624–643.

108. Groves MJ. 1978. *Parenteral Products: The Preparation and Quality Control of Products for Injection*. London: William Heinemann Medical Books.

CHAPTER 8

Approaches to Managing Protein Aggregation in Product Development

WEI WANG and NICHOLAS W. WARNE

8.1 INTRODUCTION

Protein aggregation remains one of the key challenges in the development of many protein drug products. As discussed in previous chapters, protein aggregation can easily take place in various steps and stages during the drug manufacturing process as well as during storage. Various factors were found to contribute to protein aggregation, including environmental factors such as temperature, light, solution pH, buffer type and concentration, ionic strength, excipients and level, protein concentration, and metal ions, and various processing steps such as fermentation/expression, unfolding/refolding, purification, freeze-thaw, increased air–liquid interfaces due to processes such as mixing, formulation/filling, and drying.

Protein aggregates often exhibit either reduced or, in many cases, no biological activity.[1–4] More importantly, protein aggregates may have stronger immunogenicity and/or cellular toxicities,[5–7] as has been shown with proteins such as growth hormone,[8] insulin,[9] and interleukin-2 (IL-2).[10] Immunoglobulin aggregates have been shown to cause serious renal failure[11] and anaphylactoid reactions such as headache, fever, and chills.[12] Therefore, protein aggregation is a key parameter in screening protein formulations during the product development process. The scientific and regulatory challenges in overcoming protein aggregation issues remain one of the major hurdles in the commercialization of such products.[13]

Partly due to our lack of clear understanding of aggregation mechanisms, as well as a lack of the necessary tools to fully characterize protein aggregation at representative protein concentrations, the management of protein aggregation during the development of protein pharmaceuticals has been mostly

Aggregation of Therapeutic Proteins, Edited by Wei Wang and Christopher J. Roberts

experimental. Nonetheless, significant experience has been accumulated in the past two decades in managing protein aggregation during product development. This chapter discusses various approaches, which have been tested and proven to be practical and beneficial in managing protein aggregation during product development. These approaches can be divided into three main areas: (1) approaches in formulation development, (2) protection of proteins during various processing steps, and (3) aggregation control by structural modifications.

8.2 APPROACHES IN FORMULATION DEVELOPMENT

Generally, a viable protein product should have a shelf life of ≥18 months (preferably ≥24 months) under specified storage conditions. The protein product should contain a specified acceptable level of aggregates not only after initial product release but also during the product shelf life. The aggregation rate of the protein product needs to be estimated and controlled during product development so that such a requirement can be met during manufacture as well as during product storage conditions.

Controlling the rate of protein aggregation for long-term product storage is generally and often effectively accomplished through formulation development. Protein formulation development, however, has not been standardized for aggregation control partly due to the variety of aggregation behaviors (e.g., covalent vs. non-covalent, process induced vs. stability induced, etc.) and the mechanisms that are encountered. This section discusses three main formulation development methodologies that have been used in designing protein formulations for managing protein aggregation during long-term storage: (1) traditional formulation development (one factor at a time), (2) high-throughput formulation screening, and (3) computer-assisted design of formulations.

8.2.1 Traditional Formulation Development

There are two major product forms for commercial protein pharmaceuticals: liquid and freeze-dried solid. Because of the ease of processing, low cost of production, and convenience of administration, liquid protein products are usually preferred unless aggregation or some other instability (often chemical stability) is difficult to control adequately in a liquid state. Although a freeze-dried protein formulation generally offers more protection against protein modification, the freeze-drying process itself, however, could cause significant protein aggregation. Eventually, the protein's aggregation behavior in solution, process, and storage often dictates the selection of the final product form.

Development of a protein product with a minimal aggregation tendency usually goes through at least four stages: (1) preformulation, (2) formulation

TABLE 8.1. Formulation Development Stages and Tasks

Formulation Stages	Recommended Tasks
Preformulation	• Collect key physicochemical properties of the protein, such as size and purity of the protein, and pI • Predict aggregation behavior of the protein based on known structure, if such tools are available • Determine solubility and aggregation tendency at different pH levels and ionic strengths, and during multiple freeze/thaw processes and shaking
Formulation and stability evaluation	• Investigate the effect of formulation variables on protein aggregation, such as pH, buffer, salt, and excipients • Investigate the effect of various processing steps on protein aggregation • Define the product dosage form (liquid vs. solid)
Verification of long-term storage stability and administration protocol	• Conduct real-time stability studies • Verify compatibility of protein with the intended container/closure system and administration set • Verify the range or limits of formulation composition • Estimate/determine the shelf life of the product
Product definition	• Define the drug product composition, preparation steps, storage condition, and administration protocol

and stability evaluation, (3) verification of long-term storage stability and administration protocol, and (4) product definition. These stages and the recommended tasks (may not be exhaustive) are summarized in Table 8.1.

8.2.1.1 Preformulation In the initial stages of preformulation, formulation scientists need to collect the basic information on the protein candidate such as the source and expression system, size and purity of the protein, primary and higher-order structures, glycosylation state, and isoelectric point (pI). If computational tools are available, the hot aggregation spot(s) and sequence(s) can be identified, and the aggregation behavior of the protein may be tentatively predicted.

With the above information, the aggregation tendency of the protein candidate can then be quickly evaluated experimentally at different pH levels (e.g., 4–8) and ionic strengths (e.g., 0–500 mOsm/kg) under room and accelerated storage temperatures (generally at 25 and 40°C) as well as during multiple freeze/thaw processes (–80°C/5°C; two to five times). These protein solutions of different pH levels or ionic strengths can be prepared by ultrafiltration (UF) and diafiltration (DF) of the bulk protein solution and should be assessed at representative protein concentrations depending on the protein's solubility as well as the intended dose, route of administration, and storage conditions. The assessment of the appropriateness of various formulations can be determined by traditional techniques of size-exclusion high performance liquid chromatography (HPLC) and visual inspection. Recently, however, advanced technologies such as light scattering, the determination of second virial coefficients, and optical techniques have added tremendous insight into the aggregation behavior of protein-containing solutions. Some unstable proteins could form aggregates during the buffer exchange process under moderate conditions. Therefore, frequent and careful observations of the protein solution by visual inspection as well as spectroscopy during the UF/DF process are necessary.

In this preformulation stage, the aggregation tendency of the protein candidate under freeze-thaw and shaking conditions should be assessed. This can be achieved by subjecting the above protein solutions to multiple freeze/thaw cycles[14] as well as exposing the protein to significant air–liquid interfaces, which can induce denaturation. Often, shaker tables are used to induce these conditions (e.g., 200 rpm for 24 h using vialed protein with a substantial headspace).[15,16] The solubility of the protein candidate also needs to be estimated, if a high dose is projected for clinical use by subcutaneous injection. All of the above data are useful in designing future formulation studies.

8.2.1.2 Formulation and Stability Evaluation If the protein candidate has a tendency to aggregate in the above studies, formulation studies need to be conducted to identify a formulation with a minimal aggregation tendency in this second stage of the drug product development process. Four major formulation parameters can be varied sequentially at a time or in combinations—pH, buffering agent, salt, and other excipients. Since these factors can potentially interact with one another, it is advisable to verify the aggregation behavior not only by varying one parameter at a time but also in combinations. Formulation pH can have a significant effect on the aggregation of proteins. Table 8.2 lists several proteins that showed significant difference in the rate and/or extent of protein aggregation at different pH levels. Therefore, identification of optimum solution pH is critical in managing the aggregation behavior of protein drug candidates. A wide pH range needs to be examined as the protein-dependent relationship between protein aggregation and formulation pH is often complex and difficult to predict. Several types of relationships have been reported, including well shaped, bell shaped, sigmoidal, and roughly linear. The well-shaped relationships are widely reported—that is, presence of a stable pH, where minimum aggregation takes place. Examples include the aggregation of IL-1β at 0.1 mg/mL at pH 2–9 at 30°C,[17] and recombinant factor VIIa (rFVIIa) at 1 mg/mL in 10 mM bis-Tris, 10 mM $CaCl_2$, 0.05% sodium azide at 45°C within the pH range of 5–8.[18] In some cases, a pH of maximum aggregation exists with reduced protein aggregation at other pH levels in a wide pH range, that is, a bell-shaped relationship. Such examples include phenylmethanesulfonylfluoride (PMSF)-inhibited aggregation of α-chymotrypsin (CT) at 1.5 mg/mL at 65°C,[19] aggregation of insulin in the pH range of 3–9 at 0.1 mg/mL at room temperature.[20] Other relationships have also been reported. The heat-induced aggregation of hen egg white

TABLE 8.2. Comparison of Protein Aggregation at Different pH Levels

Proteins	Exp. Conditions	pH	Aggregation	Reference
Antibody (monoclonal)	37°C, 3 months	4.0	Aggregate readily	179
		7.0	Not detectable	
Apomyoglobin (W7FW14F)	20 μM at room temperature	4.0	Soluble	180
		7.0	Aggregates easily	
Botulinum neurotoxin (50 kD)	0.1 or 0.55 mg/mL in 20 mM sodium succinate or potassium phosphate buffer, 50°C, 4 h	5.0	Stable	29
		8.0	Aggregate rapidly	
rhGCSF	1.5 mg/mL in 150 mM NaCl and 10 mM phosphate buffer, 37°C, 5 days	3.5 (with HCl)	Not detectable	181
		7.0	40% aggregated	
Lysozyme (hen egg white)	2%, 57°C	2.0	Aggregates in 2 days	182
		4.0	No aggregates in 42 days	
Ovalbumin	37°C, 3 months	4.0	Aggregate readily	179
		7.0	Not detectable	

lysozyme (pI = 11) at 1 mg/mL at 98°C for 10 min showed a sigmoidal curve against pH.[21] In the study on aggregation of a model monoclonal antibody (IgG_2) at 50 mg/mL at an ionic strength of 4 mM at 37°C for 3 months, the increase in soluble aggregates was from 7% to 24% as pH was increased gradually from 4.0 to 9.0, and no clear relationship was seen.[22] In performing these experiments, however, one should take into account additional instabilities that may occur to the protein, which have their own independent pH dependence.[23] Examples could include direct deamidation at acidic pH or via the succinamide intermediate at more basic pH as well as acid-induced hydrolysis at Asp-Pro bonds. The formulator may need to optimize the appropriate pH on the basis of several instabilities (both chemical and physical) simultaneously, optimizing the pH at which the least significant level of instability is observed.

Stability studies are often conducted at several storage temperatures, for example, −40, 5, 25, and 40°C. While a high storage temperature is necessary for early evaluation of aggregation tendency under a tight development timeline, lower storage temperature would help to verify real-time aggregation

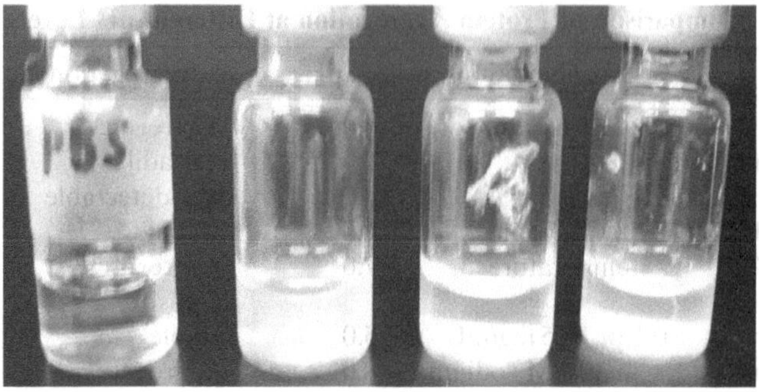

Figure 8.1. Multiple forms of protein aggregates (from left to right): PBS control, turbidity, skin formation (on the side wall), and particulate formation (on the side wall).

tendency and evaluate temperature dependency of protein aggregation. It is essential to demonstrate the predictive nature of accelerated stability studies to ensure that the relative rate of degradation follows Arrhenius kinetics.[24] If the accelerated conditions are not predictive, then lower temperatures should be utilized. The most aggressive storage temperature should be generally no greater than 10–15°C below the onset of the first thermal transition temperature. Selecting these conditions is critical if one is to utilize accelerated stability data as a representative means to predict behavior at lower temperatures. If samples are stored too closely to thermal denaturation conditions, then low levels of protein or domain unfolding can influence the representativeness of the data. A frozen condition is often included so that the protein candidate can be evaluated for feasibility of storing the protein as a frozen drug substance. Protein aggregates observed during stability studies should be characterized in terms of visibility, morphology (visible), size/structure, reversibility, activity, and quantity (see more details in Chapter 5). Protein aggregates often come in different forms, even for the same protein. Such an example is shown in Fig. 8.1, where multiple forms of aggregates are observed: turbidity, irregular particles, and surface skin (author's data). The skin layer can be thin and clear, and difficult to identify. Therefore, careful inspection of the sample vials is needed by gently swirling the vials under lighted conditions. It should be noted that foreign particles can be misinterpreted as protein particles and microbial contamination of a sample could make the sample vial appear cloudy. For these reasons, stability samples should be prepared in a clean environment, and special care should be taken to ensure aseptic technique in preparing these important stability samples.

The next step in the formulation development is to identify an optimum ionic strength to minimize protein aggregations, as the effect of ionic strength on protein aggregation is not only significant but also very much protein dependent.[25] Table 8.3 lists examples of protein aggregation under different

TABLE 8.3. Effect of Ionic Strength on Protein Aggregation

Proteins	Exp. Conditions	Ionic Strengths	Aggregation	Reference
α_1-Antitrypsin	In 20 mM phosphate, pH 7.5, 4°C	0.1 0.2 M NaCl 1.0 M NaCl	~1.5%/week ~1%/week	183
IL-1ra (recombinant human)	100 mg/mL in citrate, pH 6.5, 37°C for 4 weeks	I = 0.184 M I = 0.025 M	~60% ~90%	27
β-Lactoglobulin	Heating solution, pH 7	0 M NaCl 0.4 M NaCl	$C_g^* = 100$ g/L $C_g = 1$ g/L	184
β-Lactoglobulin (bovine)	Heating at pH 2.0 at 80°C	I = 13 mM I = 50 mM I = 100 mM	$C_a^{**} = 1.8\%$ $C_a = 1.3\%$ $C_a = 0.29\%$	185
rhG-CSF	Incubation at 1.5 mg/mL at pH 3.5 at 37°C for 5 days	0.32 nM HCl and 50–200 mM sodium acetate Above +150 mM NaCl	No aggregation Aggregated (5% loss of monomer)	186

*C_g=critical gelling concentration.
**C_a=critical association concentration.

ionic strengths, and illustrates that different proteins do have significantly different aggregation tendencies under varied ionic strengths. Additional examples include increased aggregation rate for holo-α-lactalbumin at 10 mg/mL at pH 7.0 with increasing NaCl concentrations from 0 to 60 mM[26] but reduced aggregation of recombinant human interleukin-1 receptor antagonist (rhIL-1ra) at 100 mg/mL at pH 6.5 at 37°C with increasing ionic strength from 0.025 to 0.184 M (4.5 mM citrate).[27] Sometimes, more complex relationship between aggregation rate and ionic strength can be seen, such as aggregation of insulin at pH 5.6 (pI = 5.5) at 0.1 mg/mL at room temperature in the ionic strength range of 0.5–100 mM.[20] It should be noted that the salt type can also make a significant difference in protein aggregation.[28]

The effect of ionic strength on protein aggregation may or may not be dependent on the solution pH, although pH always affects the charge state of a protein. Roy et al.[29] found that the aggregation of a botulinum neurotoxin (50 kD) at 0.1 and 0.55 mg/mL at pH 5 at 50°C could be induced by the addition of 150 mM NaCl, and this negative effect of NaCl was also seen at pH 8.0. Increasing the NaCl concentration from 10 to 200 mM facilitated thermal-induced aggregation of β-lactoglobulin-enriched whey protein isolate (WPI) and reduced the aggregation temperature both at pH 4.0 and 7.0, due to a charge screening effect.[30] On the other hand, Saluja et al.[22] found that the effect of ionic strength on the aggregation of a monoclonal antibody (IgG_2) is clearly pH dependent at 50 mg/mL, where the net change in the total amount of aggregates (soluble + insoluble) after incubation at 37°C for 3 months was positive at pH 4.0 when the ionic strength changed from 4 to 300 mM but negative at pH 7.4 under the same condition. In a different study, Soenderkaer et al.[18] demonstrated that inclusion of 150 mM NaCl increased the formation of soluble aggregates of rFVIIa at 1 mg/mL upon incubation at 45°C at pH levels 5, 6, and 8 but not at pH 7. These results suggest that the effect of ionic strength on protein aggregation needs to be evaluated in combination with solution pH.

The above formulation studies should reveal an optimum solution pH or pH range, and an optimum ionic strength or ionic strength range, where protein aggregation is minimal during storage. With this information, further experiments in formulation development can be performed to verify the optimum combination of solution pH and ionic strength conditions in the presence of different buffering systems. Although many buffering agents are available for pH adjustment, the aggregation behavior of proteins can be significantly different in different buffer systems[31–34] and/or at different buffer concentrations.[32–36] Therefore, further formulation development should identify the best buffering agent and its concentration.

Protein aggregation can be further minimized through excipient screening during formulation development. Many excipients can reduce protein aggregation, including sugars, polyols, amino acids, amines, salts, polymers, and surfactants.[37–40] These excipients are discussed in great detail in Chapters 4 and 6. Protein stabilization by excipients can be achieved through preferential interactions between excipients/additives and proteins, a widely accepted concept of protein stabilization.[41,42] Other mechanisms are also operable, such as increased rate of protein folding,[43,44] reduction of solvent accessibility and conformational mobility,[45] and increase in solvent viscosities.[46] Recent results indicate that protein stabilization can be achieved through binding between these additives and different aggregation intermediates,[47] although an earlier study did not find any detectable interaction between an additive (sorbitol) and a model protein.[48]

It should be noted that the degree of stabilization by different types of excipients can be significantly different for a particular protein, and excipients that stabilize one protein may actually destabilize another. For example, Bagger et al.[49] found that either sucrose (0.29 or 0.58 M) or sorbitol (0.31 or 0.62 M) reduced the initial aggregation rate of thermal aggregation of bovine serum albumin (BSA), but did not significantly reduce the extent of aggregation in the samples. While sucrose inhibited aggregation (formation of both soluble and insoluble aggregates) of rFVIIa at 1 mg/mL at pH 5 at 37°C in a concentration-dependent manner, methionine oxidation was actually increased at the same time.[18]

Non-ionic surfactants are a well-known class of excipients often used to prevent/inhibit protein aggregation.[40,50] Yet, the effect of non-ionic surfactants

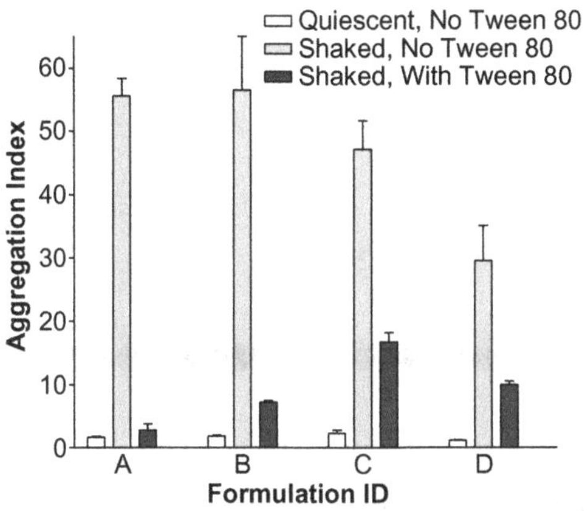

Figure 8.2. Effect of polysorbate 80 at 0.1% on shaking-induced aggregation of IL-2 mutein at 1 mg/mL in different formulations (see Wang et al.[53] for details). Samples were placed horizontally on an orbital shaker and shaken for 24 h at 350 circles/min at room temperature before UV measurement. Aggregation index (AI) was calculated based on the following equation: $AI = 100 \times OD_{350}/(OD_{280} - OD_{350})$.

on protein aggregation has not been consistent among different proteins, as their effects are multifaceted.[51] In fact, negative effects have been clearly demonstrated. For example, while polysorbate 20 was found to inhibit agitation-induced aggregation of polyethyleneglycol-derivatized granulocyte colony-stimulating factor (PEG-GCSF) in a concentration-dependent manner, it actually promoted the aggregation of PEG-GCSF during storage.[52] In a more recent study, polysorbate 80 was shown to inhibit effectively shaking-induced aggregation of IL-2 mutein in a formulation-dependent manner (Fig. 8.2) but promoted formation of both reducible and non-reducible aggregates of the mutein also in a formulation-dependent manner both at 5 and 40°C as shown in Fig. 8.3.[53] The negative effects of polysorbates on protein aggregation are likely due to their weak binding to proteins.[53–56] Another negative effect is the possible increase in oxidation of proteins catalyzed by the residual alkyl peroxides in the non-ionic surfactants. For example, recombinant human ciliary neurotrophic factor (CNTF) has been found to dimerize by alkyl peroxides present in polysorbate 80.[57] A correlation was found between the peroxide level in polysorbate 80 and the degree of oxidation in recombinant human granulocyte colony-stimulating factor (rhG-CSF), and the peroxide-induced oxidation appeared to be more serious than that induced by the atmospheric oxygen present in the vial headspace.[58] For these reasons, the quality and storage conditions for non-ionic surfactants need to be well controlled, and their quantity used in a protein formulation needs to be kept to a minimum, typically in a range between 0.005% and 0.02%.

It should be mentioned that excipients may contain residual amounts of metals. Some metals can directly facilitate protein aggregation and/or

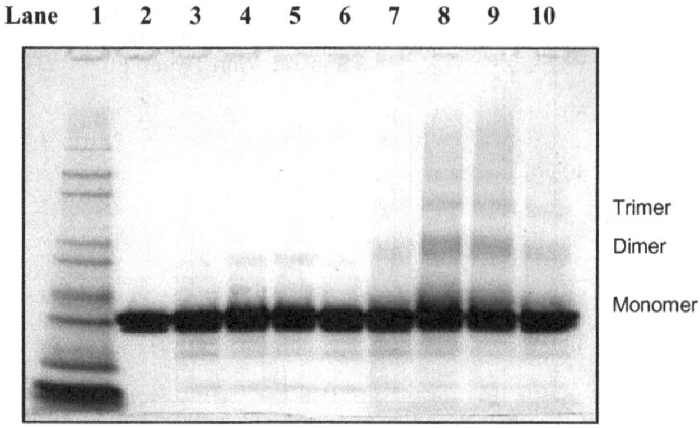

Figure 8.3. SDS-PAGE analysis of the aged IL-2 mutein stability samples (2 months at 40°C) under *reduced* conditions. Lanes 1 and 2 are the molecular weight marker (Mark 12) and IL-2 mutein reference. Lanes 3–6 are formulations A, B, C, and D, and lanes 7–10 are the same four formulations with 0.1% Tween 80.

accelerate protein oxidation, leading to increased aggregation. Aluminum(III) at concentrations above 0.1 mM was found to induce aggregation of BSA at 2.5 μM at pH 7.4 under room conditions.[59] Zinc at 1 mM promotes the formation of aggregates of β-lactoglobulin in solution at 50°C based on light scattering detection.[60] $CaCl_2$ at 10 mM was found to promote heat-induced unfolding and aggregation of β-lactoglobulin at pH 7.0 at 78°C[61] and, at a 1:1 molar ratio, to promote heat-induced aggregation of holo-α-lactalbumin at pH 7.0 at 95°C.[26] Therefore, a chelating agent is sometimes included in a protein formulation to minimize metal-induced protein aggregation.

If proteins have a high tendency to aggregate in a liquid state and no effective formulation approaches can be found to control it for successful commercialization, preparation of a dried product form would be a potential alternative. The commonly used method for preparing a solid product form is lyophilization. Description of the lyophilization process and selection of excipients useful in the inhibition of lyophilization-induced and storage-induced protein aggregation can be found in Chapter 4 and elsewhere.[62] Briefly, the following key formulation- and process-related factors need to be considered. First, the aggregation behavior of the protein candidate upon lyophilization needs to be evaluated (see more details next). An amorphous stabilizer is usually needed to protect the protein from significant aggregation during lyophilization and long-term storage. Disaccharides, which are the most widely used stabilizers in solid-state protein products, have effects that vary depending upon both the protein and quantity of stabilizer used.[63–67] Other excipients can also be effective in inhibiting protein aggregation in a solid state such as hydroxylethyl starch for a botulinum neurotoxin[29] or IL-11,[68] carboxymethyl cellulose, dextran, or polyethylene glycol (PEG) for lyophilized BSA,[69] or human albumin (HA),[70] and HP-β-CD for several solid proteins during storage, including a mouse monoclonal antibody,[71] insulin,[72] IL-2,[73] and tumor necrosis factor (TNF),[74] and NaCl for dried whey protein.[75] The effect of solid-state "pH" on storage-induced protein aggregation needs to be evaluated, and an optimum solid-state "pH" needs to be identified for minimization of protein aggregation both during lyophilization and long-term storage. It should be noted that the solid-state "pH" can be different from that of the liquid state in terms of aggregation tendency, and a balanced pH might have been chosen to minimize protein aggregation both in the liquid state and solid state, as well as during the freezing step of the lyophilization process.[29] Lastly, the effect of moisture content on the aggregation of the protein needs to be assessed, and an optimum content of moisture should be determined based on storage stability studies, as the relationship between moisture content of the formulation and protein aggregation rate is often complex.[66,69,72,75–77]

8.2.1.3 Verification of Long-Term Storage Stability During formulation development, stability studies are often conducted under accelerated storage conditions, for example, high temperatures. The key issue is the low-temperature storage stability predictability of the aggregation data obtained under such conditions. Arrhenius kinetics suggests that protein stability, within a defined set of conditions (formulation, protein concentration), should follow the aggregate kinetics of the individual molecular instabilities. For example, the rate of protein aggregation may reflect the sum of the rates of protein denaturation that may expose hydrophobic residues, chemical modifications that may result in active side chains that are able to cross-link different molecules, as well as the critical forces that keep a protein in its native state. The poor predictability is therefore complicated by the presence of multiple and changing protein degradation pathways at different temperatures.[78] Experimental data should be collected to begin to decipher these complex relationships. Information on aggregation behavior derived from accelerated stability studies should be approached with caution.

In reality, protein aggregation may follow the Arrhenius relationship in a narrow temperature range. Rezaei-Ghaleh et al.[79] found that trifluoroethanol (TFE)-induced aggregation of α-chymotrypsin at 0.25 mg/mL at pH 7.0 followed Arrhenius relationship in a narrow temperature range of 5–25°C. In a recent study, the aggregation behavior of four model proteins at 25°C in lyophilized, sucrose-based formulations could be accurately predicted based on the Arrhenius plot of protein aggregation obtained at 40 and 50°C.[80] The temperature-induced protein aggregation in a solid state could be better predicted than in a liquid state due to limited hydrophobic interactions.

Nonetheless, once the formulation is defined based on accelerated stability studies, long-term storage stability of the protein, in terms of aggregation tendency, needs to be verified. Only real-time stability studies are acceptable to define the expiration date for the final protein product. Such stability studies should be conducted with the final selected container/closure system after its compatibility has been verified.

8.2.2 High-Throughput Formulation Development (HTFD)

The concept of high-throughput screening was formed many years ago. Its application was later expanded to pharmaceutical formulation development, thanks to the availability of a variety of high-throughput analytical methodologies. An early example of such application was the use of a high-throughput UV measurement to screen the aggregation tendency of transforming growth factor-β in 1999.[81] The concept of HTFD was later described in detail for the development of biotherapeutics.[82,83] HTFD is highly desired in protein product development in today's environment because of the ever-increasing pressure for pharmaceutical companies to bring novel biotherapeutics to the market early and the fierce competition among pharmaceutical companies.

A common practical HTFD format is the use of commercially available multiwell (up to 1536 wells) microplates.[83] Such microplates can be placed in automated liquid-processing systems for sample preparation, dispensing, and handling. The low sample volume needed for analysis can potentially save a large amount of protein drug substance, which is an obvious advantage particularly for drug substance of limited quantity. Sample microplates can be analyzed immediately after preparation. Alternatively, they can be sealed and subjected to a variety of experimental stresses, such as high-temperature incubation and shaking, and analyzed later.

The HTFD microplate format is especially useful for rapid screening of protein aggregation through turbidity measurement, as this can be conveniently and non-invasively analyzed by a spectrophotometer. A variety of plates are available to meet the assay requirements of different spectrophotometers. Recently, a successful example was reported on the rapid screening of salmon calcitonin aggregation.[84] One hundred aqueous formulations of salmon calcitonin having different buffer compositions and a wide range of pH values (2.5 and 10.5) were prepared in microplates and incubated for 1 week. Samples were analyzed by several assays: turbidity at 350 nm, intrinsic fluorescence, and 1-anilinonaphthalene 8-sulfonate (ANS) fluorescence. Although the aggregation data were collected easily, it was noticed that growth in protein aggregation can be seen on the side wall of the sample wells, and measurement needs to be taken at different positions of the sample wells to minimize variations in the results. In addition, the small sample volume in microplates may not be sufficient to allow detailed analysis of other degradation pathways by traditional analytical methods, and protein surface adsorption/evaporation may contribute significantly to the variations in the results.

8.2.3 Computer-Assisted Design of Formulations

Traditionally, examination of protein aggregation behavior in formulation development is performed by changing one factor at a time (OFAAT). Such experimental methodology can certainly generate clean results and easily examine the effects of a factor at many different levels. Such a method, however, is generally not very efficient, and it is difficult to detect interactions among different factors. Statistical design of experiments (DOEs) can potentially overcome these disadvantages, allowing easy detection of interactions and characterization of multiple responses to a particular factor in the form of a contour and 3D plots. Most modern computer DOE programs are user-friendly and can be applied conveniently in the design of protein formulations. The experimental results can be processed, interpreted, and plotted in a presentable format with these programs.

DOE has been used successfully in the design and study of protein formulations, such as development of a formulation for recombinant PEGylated staphylokinase SY161[85] and a multidose formulation for a humanized monoclonal antibody.[86] DOE is especially useful in the evaluation of the effect of various formulation process variables on the final protein products such as preparation of inhalable insulin[87] and BSA[88] by spray-drying process and insulin-loaded polylactide microspheres.[89]

DOE also has its own limitations. When there are many formulation variables (>4), it becomes unwieldy to evaluate all these variables using a multilevel factorial design. This type of design would generate a huge and unreasonable number of experiments, and the results would be difficult to interpret. Therefore, a more simplified design is normally adopted to evaluate many factors initially, assuming a linear response for all the factors. A second limitation of DOE is that the experimental results are limited to a range between the upper and lower boundaries of a formulation factor chosen by a formulation scientist. The selection of these limits may vary based on the preference and experience of the formulation scientist. If these limits are not chosen properly, the experimental results may reveal a local optimum condition, while the global optimum condition could reside outside those limits.

Due to these limitations, we do not recommend the blind use of DOE in formulation development. We suggest using OFAAT during initial formulation development, especially for unpredictable molecules and applying DOE at a later stage, when one understands the basic behavior of the protein molecule and is seeking to optimize the formulation. The key factors (e.g., solution pH) should cover a wide range not to miss the potential optimum level or region. It should be mentioned that both methods (OFAAT and DOE) can eventually lead to definition of the final protein product as well as provide clarity regarding the design space. Selection of one of these two approaches to formulation development depends on many factors, including the availability of materials, the behavior of the protein drug candidate, the development timeline, and the preference of formulation scientists.

8.3 PROTECTION OF PROTEINS IN VARIOUS PROCESSING STEPS

Product development involves many processing steps, including shaking/shipping, freeze-thaw, drug product manufacturing operations, and drying. Drug products also need to be administered properly to patients. All these steps could potentially cause protein aggregation. Detailed discussion of these aggregation-influencing factors can be found in Chapter 4. This section discusses possible formulation/process approaches and the pitfalls in minimizing process-induced protein aggregation during product development.

8.3.1 Shaking

When protein solutions are moved or vibrated during shipping or mixing, proteins are subject to a certain degree of shaking stress. Shaking could potentially create air/water interfaces, and the hydrophobic property of air relative to water can induce exposure of the hydrophobic residues and promote aggregation (see also Chapter 7).[90] Many proteins easily aggregate during shaking/shipping/mixing, such as Albutropin™ (human growth hormone genetically fused to human albumin),[56] rFXIII,[91] human growth hormone (hGH),[92–94] insulin,[9,95] IgGs,[15,96] and IL-2 mutein.[53]

Non-ionic surfactants are often used and can be very effective to protect proteins from shaking/shipping/mixing-induced aggregation. The common surfactants used in protein formulations include polysorbates and Pluronics.[52,94,97] Such examples include polysorbates for IL-2 mutein,[53] factor XIII[91] or rhGH,[93] and Pluoronic F68 or Brij 35 for rhGH.[95] Other successful examples include inhibition of shear/shaking-induced rhGH aggregation by Pluronic F88, polysorbates, and Brij 35,[94,97] PEG-GCSF aggregation by polysorbate 20,[52] aggregation of factor XIII by polysorbate 20,[91] and insulin aggregation by glyceryl monooleate.[98]

The working mechanisms of these surfactants include competitive (with proteins) accumulation at hydrophobic surfaces/interfaces, direct binding to proteins, or alteration of the physical properties of the protein solution.[55,94,99,100] Certain surfactants such as poloxamers may also increase the viscosity of a protein solution, putatively restraining the motion of protein backbone to inhibit protein aggregation.[100] While non-ionic surfactants can be very effective in protecting proteins against shaking-induced aggregation, they might increase protein aggregation during storage (see previous Section 8.2.1.2 for details). Therefore, both shaking and long-term storage stability studies need to be evaluated when such surfactants are used.

8.3.2 Freezing/Thawing

Drug substance often needs to be stored in a frozen state to protect the protein from rapid aggregation and/or degradation until the time of drug product manufacturing. Freeze-thawing of a protein solution may potentially cause protein aggregation due to several possible freezing-induced stresses: formation of ice–water interfaces, solute concentration, low-temperature destabilization, and potential pH changes or phase separation.[62] Protein examples of freeze-thaw-induced aggregation include albumin or IgG (–20°C for 12h),[101] antibodies,[36,102–104] factor IX,[105] and factor XIII.[91]

Various formulation/process approaches are available to prevent/inhibit freeze-thaw-induced protein aggregation depending on the aggregation mechanism(s). For freezing-sensitive proteins, understanding the cause(s) of protein aggregation is the first step for possible aggregation prevention/inhibition during freeze-thaw. For example, a slower freezing rate usually generates larger ice crystals, and such a process could be implemented to inhibit protein aggregation if the ice–water interface is mainly responsible for protein aggregation during freeze-thaw.[106] On the other hand, a lower freezing rate could facilitate crystallization of other solution components, promoting crystallization-induced protein aggregation.[107]

Many formulation excipients can be used to inhibit freeze-thaw-induced protein aggregation. These excipients may inhibit protein aggregation through minimization of protein–ice interactions, suppression of pH changes,[108] preferential interaction, and/or other mechanisms such as increased solution viscosity and steric hindrance of protein–protein interactions.[109] Among these excipients, surfactants seem to be the most commonly used, as they are effective in minimizing protein–ice interactions, a cause for aggregation of many proteins.[91,105] Examples include use of polysorbate 80 to inhibit the aggregation of factor IX during freeze-thaw,[105] and lactate dehydrogenase (LDH) and glutamate dehydrogenase (GDH) during quench freeze-thaw,[106] and polysorbate 20 in reducing the formation of insoluble aggregates during freeze-thawing of rFXIII at 1, 5, or 10 mg/mL.[91] A variety of sugars, such as sucrose, trehalose, and glucose, have been shown to reduce the freeze/thaw-induced aggregation of a number of antibodies.[36,102–104]

It should be mentioned that concentrated protein solutions are often more resistant against freezing-induced protein aggregation such as labile LDH.[110,111] Increasing the concentration of a monoclonal antibody from 20 to 80 mg/mL reduced the level of precipitates induced upon freezing in an isotonic NaCl solution at pH 5.5.[112] Concentration-dependent inhibition of freezing-induced protein aggregation was also observed in mixtures of different proteins. Quench freezing-induced aggregation of citrate synthase (CS) and LDH can be inhibited by addition of 0.5 mg/mL late embryogenesis abundant (LEA) proteins or BSA.[113] The increased stability at high protein concentrations during freezing is likely due to one or more of the following factors—limited area of ice–water interface for interfacial denaturation, protein's self-stabilization due to its amorphous nature,[114] and conversion of monomers to active and more stable dimers or multimers.[115,116]

8.3.3 Manufacturing Processes for Liquid Drug Product

The challenges in protein fill-finish operations have been recently reviewed.[117] The manufacturing process for a liquid drug product consists of several independent operations, including filtration, filling, capping, and inspection. Each of these processes can generate some degree of stress, such as increased air–liquid interfacial denaturation, exposure of proteins to solid interfaces, shedding of particles of construction materials, and shaking. All these stresses can potentially cause protein aggregation. For example, filling of a mAb solution through a radial piston pump (shearing effect) facilitated formation of aggregates and particulates.[118] More detailed discussion on protein aggregation and particulate formation during drug product manufacturing can be found in Chapter 7. Such operation-associated protein aggregation should be investigated and addressed at each independent operation step.

8.3.4 Drying

Proteins are frequently made into a dried form to inhibit storage-induced protein aggregation. Drying removes at least part of protein's hydration layer, disrupts the native state of a protein, and potentially causes protein aggregation.[119,120] Several drying processes have been reported in the literature, including vacuum-drying, freeze-drying, spray-drying, and spray freeze-drying. Detailed discussion of these methods and their effect on protein aggregation can be found in Chapter 4. Their relative effects on protein aggregation can be different due to the differences in processing conditions, as shown in Table 8.4.

Drying-induced aggregation can be inhibited to different degrees through the use of formulation excipients. Their relative effect on protein aggregation is strongly dependent on the excipient type and concentration, as shown in Table 8.4. Another example of the concentration effect is the increased recovery of β-galactosidase after freeze-drying at 3.3 mg/mL with an increasing weight fraction of trehalose:protein.[121] Excipients can serve as water substitutes, capable of forming hydrogen bonds with the protein, which could effectively prevent or inhibit protein aggregation during drying.[111,122–124] The formation of hydrogen bonds has been demonstrated by IR spectroscopy between carbohydrates and many freeze-dried proteins.[122–124] These excipients have to be amorphous to form intimate interactions.[125–127] The most commonly used excipients against drying-induced aggregation by forming hydrogen bonds are saccharides, especially disaccharides.[125,128] Even the most commonly used disaccharides, sucrose, and trehalose may show different degrees of aggregation protection, depending on the protein as well as the protein:disaccharide ratio. This is critical not only in the solution state but in particular in the freeze-dried state. Lyophilization of hGH at 1 mg/mL in 2 mg/mL sucrose and 5 mM potassium phosphate buffer at pH 7.4 led to formation of 0.35% insoluble aggregates, but can increase to 5% if sucrose is replaced by trehalose.[129] The same level of protection could be achieved by altering the ratio of excipient:protein.

Another important class of stabilizing excipients during drying is polymers.[128,130] Polymers may stabilize proteins against aggregation by increasing the formulation Tg[131,132] and/or by inhibiting the crystallization of other formu-

TABLE 8.4. Factors Affecting Drying-Induced Protein Aggregation

Factors	Proteins	Exp. Conditions	Aggregation Effect	Reference
Drying methods	Met-hGH	Freeze-drying or spray-drying at 2 mg/mL in 1.1 mM sodium phosphate buffer at pH 7.8	More aggregation with spray-drying	187
Protein concentration	rhIL-1ra	Freeze-drying at 15, 35, and 70 mg/mL in 10 mM potassium phosphate buffer at pH 7.0	11%, 5%, and 3% insoluble aggregates, respectively	135
Sugar type	hGH	Freeze-drying at 1 mg/mL in 5 mM potassium phosphate buffer at pH 7.4 with 2 mg/mL sucrose or trahalose	0.35% or 5.0% insoluble aggregates	129
Salt	rhIL-1ra	Freeze-drying at 15 mg/mL in 10 mM potassium phosphate buffer at pH 7.0 Above +140 mM NaCl	11% and 15% insoluble aggregates	135
Tweens	rhIL-1ra	Freeze-drying at 15 mg/mL in 10 mM potassium phosphate buffer at pH 7.0 Above + 140 mM NaCl	11% and 5% insoluble aggregates	135
Excipient level	IgG1 (human)	Vacuum drying at 32°C at 24 h at 0.1 mbar at IgG1 : mannitol weight ratios of 2:8; 3:7; and 4:6	0%, 2.8%, and 3.3%	188

lation excipients in a solid formulation.[133] Many proteins, which we could consider amorphous biopolymers in the freeze-dried state, protect themselves from aggregation during freezing as well as freeze-drying.[134,135] Therefore, protein aggregation can often be inhibited simply by increasing the protein concentration. For example, increasing the concentration of bovine IgG from 0.5 to 8 mg/mL gradually decreased the formation of insoluble aggregates during freeze-drying of the protein in 0.22 M NaCl and 10 mM phosphate at pH 7.1.[136] Freeze-drying of rhIL-1ra at 15, 35, and 70 mg/mL in 10 mM potassium phosphate buffer at pH 7.0 resulted in the formation of insoluble aggregates of 11%, 5%, and 3%, respectively.[135] Exceptions do exist. Many antibodies have been shown to be less stable both during lyophilization and storage at high concentrations. For example, lyophilization of a human monoclonal antibody caused significant aggregation in phosphate buffered saline (PBS) or physiological saline, and increasing the protein concentration from 1 to 2.5 or 5 mg/mL increased the level of lyophilization-induced aggregation as measured by OD600.[137]

8.3.5 Reconstitution

Dried protein products generally need a reconstitution step before drug administration. Such a step could lead to different levels of protein aggregation, depending on the reconstitution procedure, reconstitution medium, and presence of other excipients. For example, reconstitution of freeze-dried rhIL-1ra with 0.9% (w/v) benzyl alcohol caused a greater degree of protein aggregation than reconstitution with water.[135] Reconstitution of lyophilized β-galactosidase with water alone recovered 97% monomer relative to the prelyophilization level and addition of 0.5% 2-hydroxylpropyl-β-cyclodextrin in the reconstitution medium reduced recovery to 95%.[121] Similarly, reconstitution of lyophilized interferon (IFN)-γ at 0.4 mg/mL in PBS pH 7.0 with water or water +0.03% polysorbate 20 led to a different extent of protein aggregation with the polysorbate-containing formulation being noticeably less opalescent.[138] It is generally recommended that reconstitution be conducted carefully not to create any air bubbles. In this regard, high vacuum in the product container should be avoided to allow adequate control of the reconstitution medium stream in contact with the inside wall of the product container. Inclusion of a protein stabilizer(s) in the reconstitution medium often minimizes protein aggregation.[138]

8.3.6 Preparation of Controlled Protein Delivery Systems

Most protein drug products are formulated as immediate release systems and administered by injection. Such products generally require multiple injections for chronic or life-threatening indications. Controlled release systems for protein drugs are therefore being developed to minimize the frequency of injections, to increase patient acceptance or compliance, or to expand the life cycle of the protein product. A common and relatively mature method of preparing such a delivery system is emulsification in the presence of a non-aqueous solvent.

Unfortunately, mixing any non-aqueous solvent with an aqueous protein solution can easily cause protein aggregation at the interface of the immiscible solvents.[95,139] Selection and proper use of a protein stabilizer can be effective in inhibiting protein aggregation during the emulsification process. Hydroxypropyl-β-cyclodextrin was shown to protect both ovalbumin and lysozyme against methylene chloride/water interface-induced denaturation and aggregation.[140] Pluronic F127 is effective in protecting rhGH against emulsification (with methylene chloride)-induced aggregation in a concentration-dependent manner.[141] In a recent report, the emulsification-induced erythropoietin aggregation was significantly minimized when the emulsification process was modified with only two aqueous phases involved.[142]

8.3.7 Miscellaneous Processes

Many other processes may induce protein aggregation, such as jet nebulization,[143] and irradiation.[144] Proper use of a stabilizing agent could inhibit or prevent protein aggregation under these processing conditions. Small liposomes were shown to inhibit jet nebulization-induced IFN-γ aggregation.[143] Polysorbate 20 significantly inhibited nebulization-induced aggregation of recombinant consensus α-interferon (rConIFN).[145] L-tyrosine was effective in preventing aggregation of ceruloplasmin and hemoglobin during irradiation.[144]

8.4 AGGREGATION CONTROL BY STRUCTURAL MODIFICATIONS

The previous sections have dealt with various formulation or process approaches for inhibiting protein aggregation. Of course, these measures often achieve a limited degree of aggregation inhibition. If a protein drug candidate has a very high tendency to aggregate, changing the formulation or process-related factors may not be sufficient to minimize protein aggregation to an acceptable level. In this case, a more drastic approach can be taken: structural modification of the protein either genetically or chemically. There are several methods that can be used to achieve this goal: site-specific mutagenesis (even with one amino acid substitution), PEGylation, glycosylation, sialylation, and protein fusion.[146] A major limitation of these approaches is the possible reduction or even complete loss in protein activity.[147,148] Therefore, a key issue is the preservation of protein activity during and after the modification.

8.4.1 Mutagenesis

Protein aggregation is often controlled by one or a few amino acids or an aggregation-prone sequence(s) in proteins.[149] Reduction of protein aggregation could in principle be achieved through modification or mutation of regions of the protein that have been implicated as prone to aggregation, if they can be accurately predicted through sequence analysis.[150] If these aggregation prone regions are located outside the active center(s), they could be mutated into proteins with the same activity but different aggregation tendency.[151–153] Understanding these models is of keen interest to those developing proteins, especially at elevated protein concentrations (100+ mg/mL); however, this area of research is still in need of much investment. Such point mutations are made often effectively on the surface hydrophobic residues[154] and sometimes hydrophilic residues, as well.[155,156] The key advantage of site-specific mutation is the effectiveness in changing the aggregation behavior of a protein, even with one amino acid substitution.[157] In addition, the mutein generally does not need extra processing steps for purification or characterization. A major limitation is the possible increase in immunogenicity in addition to a possible reduction or loss in protein activity and conformational stability.

8.4.2 PEGylation

There are several PEGylated commercial protein drug products on the market.[158,159] The purpose of PEGylation for these products is mostly to increase blood stability and reduce renal elimination. The resultant increase in blood half-life allows less dosing frequency and more patient acceptance. For example, attachment of a single 20, 30, or 40 kDa PEG moiety to recombinant human granulocyte-macrophage colony-stimulating factor (GM-CSF) increased the blood elimination half-lives in rats, respectively, by 32-, 27-, or 40-fold.[160]

Additional benefits of PEGylation are the possible reduction in aggregation tendency, and general improvement in protein solubility and immunogenicity.[161,162] Therefore, a protein with a short blood half-life and/or high tendency of aggregation, PEGylation may be a reasonable option.[163] Conjugation of methoxy polyethylene glycol (mPEG) to insulin substantially increased its resistance to shaking-induced fibrillation.[164] N-terminal attachment of a 5-kD or 20-kDa PEG moiety to GCSF had the ability to (1) prevent protein precipitation by rendering the aggregates soluble and (2) slow down the rate of aggregation relative to GCSF under physiological pH and temperature.[165] The favorable solvation of water molecules around the PEG group is likely responsible for the enhanced PEG-GCSF solubility, and steric hindrance of PEG influences the aggregation tendency.

On the other hand, PEGylation can potentially reduce the protein activity and is associated with a variety of potential issues, including the added manufacturing cost, non-specificity of PEGylation reactions, separation, purification, and characterization burdens due to the inherent polydispersity of PEG reagents and presence of multiple PEGylated products.[166]

8.4.3 Glycosylation

Glycosylation of proteins generally increases protein solubility due to the hydrophilic nature of saccharides, leading to reduced tendency of protein aggregation.[3,167–169] Potential reduced aggregation can be attributed to an increase in the overall molecular solvent-accessible surface area (SASA)[169] and/or steric hindrance of the glycans during the aggregation process.[170,171]

Therefore, proteins can be properly glycosylated either genetically or chemically to inhibit at least solubility-limited protein aggregation. For example, the glycosylated α_1-antitrypsin has been found to be more resistant to thermally induced aggregation than non-glycosylated form.[172] Glycosylation at N65 in protein O-fucosyltransferase 1 is critical in preventing rapid aggregation.[173] Addition of O-linked N-acetylglucosamine (O-GlcNAc) on polyhistidine tagged-Sp1 transcription factor reduces its tendency for thermal aggregation.[171] The tendency of insulin fibrillation/aggregation in solution can be inhibited by glucosylation[174] or formation of monosubstituted, disubstituted, and trisubstituted p-succinamidophenylglucopyranoside derivatives (SAPG-insulin).[174] Coupling of dextran to a model protein α-chymotrypsin increased protein unfolding temperature and inhibited the formation of insoluble aggregates at 20 mg/mL in 10 mM potassium phosphate buffer at pH 7.1 at 65°C.[175] In a recent study, glycosylated *Peniophora lycii* phytase was shown to aggregate approximately 200 times slower than the deglycosylated protein.[170] In addition, glycosylation may reduce the immunogenicity of a protein.[176]

Similarly, the negative effect of glycosylation on protein aggregation was also reported. Glycosylation of IL-2 mutein accelerated its aggregation during storage at 40°C at pH 4.0 but inhibited its aggregation at pH 7.5.[177] Glucosylation of bovine β-lactoglobulin accelerated its thermal aggregation but led to inhibition of urea-induced aggregation at 10 mg/mL.[178]

8.5 SUMMARY

Various approaches have been discussed in inhibition/prevention of protein aggregation during product development. The fundamental approach is a systematic evaluation of the protein aggregation tendency, investigation of the aggregation mechanism(s), and identification of an optimum formulation composition during the product development process. Product development can follow the traditional formulation development process (one factor at a time), high-throughput formulation screening, computer-assisted design of formulations, or a combination of these processes. Aggregation-prone proteins need to be protected carefully throughout various processing steps, such as shaking/shipping, freeze-thaw, drug product manufacturing operations, drying, reconstitution, and preparation of controlled protein delivery systems. Several methods of structural modification of proteins either genetically or chemically for aggregation inhibition/prevention are relatively mature,

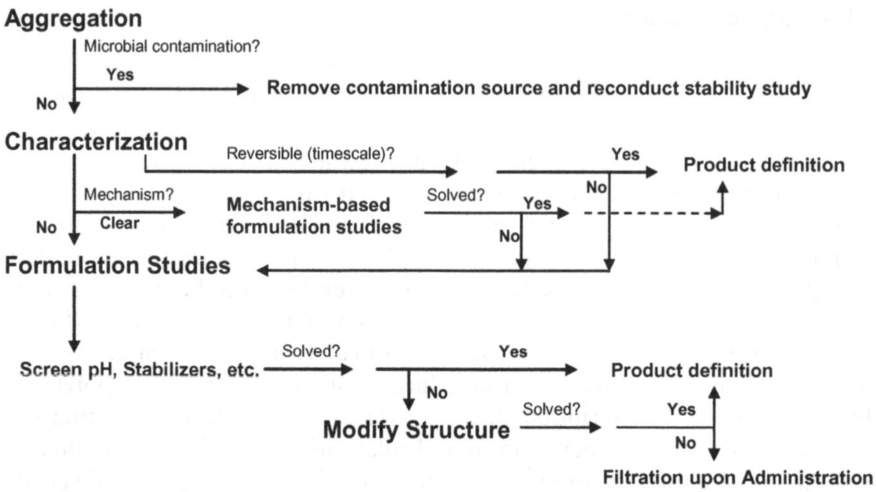

Figure 8.4. Recommended steps in addressing protein aggregation during product development.

including site-specific mutagenesis, PEGylation, glycosylation, and protein fusion, but they are generally not recommended unless it is difficult to overcome the aggregation tendency of the protein candidate through the formulation development process. If everything fails in efforts to control protein aggregation, the administration process offers the last chance to remove (insoluble) aggregates present in the drug products. In this regard, a syringe filter or in-line filter can be used to filter the aggregates during drug administration. Figure 8.4 shows the recommended steps in addressing protein aggregation in developing protein drug products.

REFERENCES

1. Townsend MW, Byron PR, DeLuca PP. 1990. The effects of formulation additives on the degradation of freeze-dried ribonuclease A. Pharm Res 7(10):1086–1091.
2. Cleland JL, Wang DI. 1990. Refolding and aggregation of bovine carbonic anhydrase B: Quasi-elastic light scattering analysis. Biochemistry 29(50):11072–11078.
3. Runkel L, Meier W, Pepinsky RB, Karpusas M, Whitty A, Kimball K, Brickelmaier M, Muldowney C, Jones W, Goelz SE. 1998. Structural and functional differences between glycosylated and non-glycosylated forms of human interferon-beta (IFN-beta). Pharm Res 15(4):641–649.
4. Nesta DP, Elliott JS, Warr JP. 2000. Supercritical fluid precipitation of recombinant human immunoglobulin from aqueous solutions. Biotechnol Bioeng 67(4):457–464.
5. Ellis RJ, Minton AP. 2006. Protein aggregation in crowded environments. Biol Chem 387(5):485–497.

6. Gsponer J, Vendruscolo M. 2006. Theoretical approaches to protein aggregation. Protein Pept Lett 13(3):287–293.
7. Hull E, Spoja C, Cordova M, Cohlberg JA. 2008. Neurofilament protein aggregation in a cell line model system. Biochem Biophys Res Commun 366(1):73–79.
8. Moore WV, Leppert P. 1980. Role of aggregated human growth hormone (hGH) in development of antibodies to hGH. J Clin Endocrinol Metab 51(4):691–697.
9. Brange J, Andersen L, Laursen ED, Meyn G, Rasmussen E. 1997. Toward understanding insulin fibrillation. J Pharm Sci 86(5):517–525.
10. Curatolo L, Valsasina B, Caccia C, Raimondi GL, Orsini G, Bianchetti A. 1997. Recombinant human IL-2 is cytotoxic to oligodendrocytes after in vitro self-aggregation. Cytokine 9(10):734–739.
11. Demeule B, Gurny R, Arvinte T. 2006. Where disease pathogenesis meets protein formulation: Renal deposition of immunoglobulin aggregates. Eur J Pharm Biopharm 62(2):121–130.
12. Ryan ME, Webster ML, Statler JD. 1996. Adverse effects of intravenous immunoglobulin therapy. Clin Pediatr 35(1):23–31.
13. Weinberg WC, Frazier-Jessen MR, Wu WJ, Weir A, Hartsough M, Keegan P, Fuchs C. 2005. Development and regulation of monoclonal antibody products: Challenges and opportunities. Cancer Metastasis Rev 24(4):569–584.
14. Warne N, Koval R. 1998. Freeze-Thaw of rhIL-11. Conference on Srotein Stability, Breckenridge, CO.
15. Kiese S, Papppenberger A, Friess W, Mahler HC. 2008. Shaken, not stirred: Mechanical stress testing of an IgG1 antibody. J Pharm Sci. doi: 10.1002/jps.21328.
16. Tyagi AK, Randolph TW, Dong A, Maloney KM, Hitscherich C Jr., Carpenter JF. 2009. IgG particle formation during filling pump operation: A case study of heterogeneous nucleation on stainless steel nanoparticles. J Pharm Sci 98(1):94–104.
17. Gu LC, Erdos EA, Chiang HS, Calderwood T, Tsai K, Visor GC, Duffy J, Hsu WC, Foster LC. 1991. Stability of interleukin 1 beta (IL-1 beta) in aqueous solution: Analytical methods, kinetics, products, and solution formulation implications. Pharm Res 8(4):485–490.
18. Soenderkaer S, Carpenter JF, Van De Weert M, Hansen LL, Flink J, Frokjaer S. 2004. Effects of sucrose on rFVIIa aggregation and methionine oxidation. Eur J Pharm Sci 21(5):597–606.
19. Rezaei-Ghaleh N, Ramshini H, Ebrahim HA, Moosavi MAA, Nemat GM. 2008. Thermal aggregation of alpha-chymotrypsin: Role of hydrophobic and electrostatic interactions. Biophys Chem 132(1):23–32.
20. Giger K, Vanam RP, Seyrek E, Dubin PL. 2008. Suppression of insulin aggregation by heparin. Biomacromolecules 9(9):2338–2344.
21. Hamada H, Takahashi R, Noguchi T, Shiraki K. 2008. Differences in the effects of solution additives on heat- and refolding-induced aggregation. Biotechnol Prog 24(2):436–443.
22. Saluja A, Badkar AV, Zeng DL, Kalonia DS. 2007. Ultrasonic rheology of a monoclonal antibody (IgG2) solution: Implications for physical stability of proteins in high concentration formulations. J Pharm Sci 96(12):3181–3195.
23. Kenley RA, Warne NW. 1994. Acid-catalyzed peptide bond hydrolysis of recombinant human interleukin 11. Pharm Res 11(1):72–76.

24. Webb C, Horhota M. 2004. Using Arrhenius Kitetics to Predict Shelf-Life of Proteins. IBC Conference on Protein Formulations, Boston.
25. Wang W. 1999. Instability, stabilization, and formulation of liquid protein pharmaceuticals. Int J Pharm 185(2):129–188.
26. McGuffey MK, Epting KL, Kelly RM, Foegeding EA. 2005. Denaturation and aggregation of three alpha-lactalbumin preparations at neutral pH. J Agric Food Chem 53(8):3182–3190.
27. Alford JR, Kendrick BS, Carpenter JF, Randolph TW. 2008. High concentration formulations of recombinant human interleukin-1 receptor antagonist: II. Aggregation kinetics. J Pharm Sci 97(8):3005–3021.
28. Kita Y, Arakawa T. 2002. Salts and glycine increase reversibility and decrease aggregation during thermal unfolding of ribonuclease-A. Biosci Biotechnol Biochem 66(4):880–882.
29. Roy S, Henderson I, Nayar R, Randolph TW, Carpenter JF. 2008. Effect of pH on stability of recombinant botulinum serotype A vaccine in aqueous solution and during storage of freeze-dried formulations. J Pharm Sci. doi: 10.1002/jps.21409.
30. Unterhaslberger G, Schmitt C, Sanchez C, Appolonia NC, Raemy A. 2006. Heat denaturation and aggregation of beta-lactoglobulin enriched WPI in the presence of arginine HCl, NaCl and guanidinium HCl at pH 4.0 and 7.0. Food Hydrocolloids 20(7):1006–1019.
31. Narhi LO, Philo JS, Sun B, Chang BS, Arakawa T. 1999. Reversibility of heat-induced denaturation of the recombinant human megakaryocyte growth and development factor. Pharm Res 16(6):799–807.
32. Wang YJ, Shahrokh Z, Vemuri S, Eberlein G, Beylin I, Busch M. 1996. Characterization, stability, and formulations of basic fibroblast growth factor. Pharm Biotechnol 9:141–180.
33. Pikal MJ, Dellerman KM, Roy ML, Riggin RM. 1991. The effects of formulation variables on the stability of freeze-dried human growth hormone. Pharm Res 8(4):427–436.
34. Eberlein GA, Stratton PR, Wang YJ. 1994. Stability of rhbFGF as determined by UV spectroscopic measurements of turbidity. PDA J Pharm Sci Technol 48(5):224–230.
35. Won CM, Molnar TE, McKean RE, Spenlehauer GA. 1998. Stabilizers against heat-induced aggregation of RPR 114849, an acidic fibroblast growth factor (aFGF). Int J Pharm 167:25–36.
36. Paborji M, Pochopin NL, Coppola WP, Bogardus JB. 1994. Chemical and physical stability of chimeric L6, a mouse-human monoclonal antibody. Pharm Res 11(May):764–771.
37. Tsai PK, Volkin DB, Dabora JM, Thompson KC, Bruner MW, Gress JO, Matuszewska B, Keogan M, Bondi JV, Middaugh CR. 1993. Formulation design of acidic fibroblast growth factor. Pharm Res 10(5):649–659.
38. Gonzalez M, Murature DA, Fidelio GD. 1995. Thermal stability of human immunoglobulins with sorbitol. A critical evaluation. Vox Sang 68(1):1–4.
39. Zhang MZ, Wen J, Arakawa T, Prestrelski SJ. 1995. A new strategy for enhancing the stability of lyophilized protein: The effect of the reconstitution medium on keratinocyte growth factor. Pharm Res 12(10):1447–1452.

40. Wang W. 2005. Protein aggregation and its inhibition in biopharmaceutics. Int J Pharm 289(1–2):1–30.
41. Arakawa T, Kita Y, Carpenter JF. 1991. Protein–solvent interactions in pharmaceutical formulations. Pharm Res 8(3):285–291.
42. Timasheff SN. 1998. Control of protein stability and reactions by weakly interacting cosolvents: The simplicity of the complicated. Adv Protein Chem 51:355–432.
43. Wang A, Robertson AD, Bolen DW. 1995. Effects of a naturally occurring compatible osmolyte on the internal dynamics of ribonuclease A. Biochemistry 34(46):15096–15104.
44. Frye KJ, Royer CA. 1997. The kinetic basis for the stabilization of staphylococcal nuclease by xylose. Protein Sci 6(4):789–793.
45. Kendrick BS, Chang BS, Arakawa T, Peterson B, Randolph TW, Manning MC, Carpenter JF. 1997. Preferential exclusion of sucrose from recombinant interleukin-1 receptor antagonist: Role in restricted conformational mobility and compaction of native state. Proc Natl Acad Sci U S A 94(22):11917–11922.
46. Jacob M, Schmid FX. 1999. Protein folding as a diffusional process. Biochemistry 38(42):13773–13779.
47. Necula M, Kayed R, Milton S, Glabe CG. 2007. Small molecule inhibitors of aggregation indicate that amyloid beta oligomerization and fibrillization pathways are independent and distinct. J Biol Chem 282(14):10311–10324.
48. Petersen SB, Jonson V, Fojan P, Wimmer R, Pedersen S. 2004. Sorbitol prevents the self-aggregation of unfolded lysozyme leading to an up to 13 degrees C stabilisation of the folded form. J Biotechnol 114(3):269–278.
49. Bagger HL, Ogendal LH, Westh P. 2007. Solute effects on the irreversible aggregation of serum albumin. Biophys Chem 130(1–2):17–25.
50. Kerwin BA. 2008. Polysorbates 20 and 80 used in the formulation of protein biotherapeutics: Structure and degradation pathways. J Pharm Sci 97(8):2924–2935.
51. Chi EY, Krishnan S, Randolph TW, Carpenter JF. 2003. Physical stability of proteins in aqueous solution: Mechanism and driving forces in nonnative protein aggregation. Pharm Res 20(9):1325–1336.
52. Treuheit MJ, Kosky AA, Brems DN. 2002. Inverse relationship of protein concentration and aggregation. Pharm Res 19(4):511–516.
53. Wang W, Wang YJ, Wang DQ. 2008. Dual effects of Tween 80 on protein stability. Int J Pharm 347(1–2):31–38.
54. Randolph TW, Jones LS. 2002. Surfactant-protein interactions. Pharm Biotechnol 13:159–175.
55. Bam NB, Randolph TW, Cleland JL. 1995. Stability of protein formulations: Investigation of surfactant effects by a novel EPR spectroscopic technique. Pharm Res 12(1):2–11.
56. Chou DK, Krishnamurthy R, Randolph TW, Carpenter JF, Manning MC. 2005. Effects of Tween 20 and Tween 80 on the stability of Albutropin during agitation. J Pharm Sci 94(6):1368–1381.
57. Knepp VM, Whatley JL, Muchnik A, Calderwood TS. 1996. Identification of antioxidants for prevention of peroxide-mediated oxidation of recombinant

human ciliary neurotrophic factor and recombinant human nerve growth factor. PDA J Pharm Sci Technol 50(3):163–171.

58. Herman AC, Boone TC, Lu HS. 1996. Characterization, formulation, and stability of Neupogen (Filgrastim), a recombinant human granulocyte-colony stimulating factor. Pharm Biotechnol 9:303–328.
59. Long X, Zhang C, Cheng J, Bi S. 2008. A novel method for study of the aggregation of protein induced by metal ion aluminum(III) using resonance Rayleigh scattering technique. Spectrochim Acta A Mol Biomol Spectrosc 69(1):71–77.
60. Navarra G, Leone M, Militello V. 2007. Thermal aggregation of beta-lactoglobulin in presence of metal ions. Biophys Chem 131(13):52–61.
61. Mounsey JS, O'Kennedy BT. 2007. Conditions limiting the influence of thiol-disulphide interchange reactions on the heat-induced aggregation kinetics of beta-lactoglobulin. Int Dairy J 17(9):1034–1042.
62. Wang W. 2000. Lyophilization and development of solid protein pharmaceuticals. Int J Pharm 203(1–2):1–60.
63. Chang LL, Shepherd D, Sun J, Ouellette D, Grant KL, Tang XC, Pikal MJ. 2005. Mechanism of protein stabilization by sugars during freeze-drying and storage: Native structure preservation, specific interaction, and/or immobilization in a glassy matrix? J Pharm Sci 94(7):1427–1444.
64. Andya JD, Hsu CC, Shire SJ. 2003. Mechanisms of aggregate formation and carbohydrate excipient stabilization of lyophilized humanized monoclonal antibody formulations. AAPS PharmSci 5(2):E10.
65. Pikal MJ, Rigsbee D, Roy ML, Galreath D, Kovach KJ, Wang BS, Carpenter JF, Cicerone MT. 2008. Solid state chemistry of proteins: II. The correlation of storage stability of freeze-dried human growth hormone (hGH) with structure and dynamics in the glassy solid. J Pharm Sci 97(12):5106–5121.
66. Breen ED, Curley JG, Overcashier DE, Hsu CC, Shire SJ. 2001. Effect of moisture on the stability of a lyophilized humanized monoclonal antibody formulation. Pharm Res 18(9):1345–1353.
67. Andya J, Cleland JL, Hsu CC, Lam XM, Overcashier DE, Shire SJ, Yang JY-F, Wu SS-Y. 2004. *Protein Formulation*. Genentech, Inc. US6685940 B2.
68. Garzon-Rodriguez W, Koval RL, Chongprasert S, Krishnan S, Randolph TW, Warne NW, Carpenter JF. 2004. Optimizing storage stability of lyophilized recombinant human interleukin-11 with disaccharide/hydroxyethyl starch mixtures. J Pharm Sci 93(3):684–696.
69. Liu WR, Langer R, Klibanov AM. 1990. Moisture-induced aggregation of lyophilized proteins in the solid state. Biotechnol Bioeng 37:177–184.
70. Costantino HR, Griebenow K, Mishra P, Langer R, Klibanov AM. 1995. Fourier-transform infrared spectroscopic investigation of protein stability in the lyophilized form. Biochim Biophys Acta 1253(1):69–74.
71. Ressing ME, Jiskoot W, Talsma H, Van Ingen CW, Crommelin DJ. 1992. Influence of sucrose, dextran, and hydroxypropyl-beta-cyclodextrin as lyoprotectants for a freeze-dried mouse IgG2a monoclonal antibody (MN12). Pharm Res 9(February):266–270.
72. Katakam M, Banga AK. 1995. Aggregation of insulin and its prevention by carbohydrate excipients. PDA J Pharm Sci Technol 49(4):160–165.

73. Hora MS, Rana RK, Wilcox CL, Katre NV, Hirtzer P, Wolfe SN, Thomson JW. 1992. Development of a lyophilized formulation of interleukin-2. Dev Biol Stand 74:295–303.
74. Hora MS, Rana RK, Smith FW. 1992. Lyophilized formulations of recombinant tumor necrosis factor. Pharm Res 9(1):33–36.
75. Zhou P, Liu X, Labuza TP. 2008. Moisture-induced aggregation of whey proteins in a protein/buffer model system. J Agric Food Chem 56(6):2048–2054.
76. Jain NK, Roy I. 2008. Role of trehalose in moisture-induced aggregation of bovine serum albumin. Eur J Pharm Biopharm 69(3):824–834.
77. Separovic F, Lam YH, Ke X, Chan HK. 1998. A solid-state NMR study of protein hydration and stability. Pharm Res 15(12):1816–1821.
78. Sek D, Luisi D, Warne NW. 2008. When Correlations Fail—Understanding the Differences between Differential Scanning Calorimetry and Real Time Stability. IIR Conference on Formulation and Forced Degradation, San Diego, CA.
79. Rezaei-Ghaleh N, Ebrahim-Habibi A, Moosavi-Movahedi AA, Nemat-Gorgani M. 2007. Role of electrostatic interactions in 2,2,2-trifluoroethanol-induced structural changes and aggregation of alpha-chymotrypsin. Arch Biochem Biophys 457(2):160–169.
80. Wang B, Tchessalov S, Cicerone MT, Warne NW, Pikal MJ. 2008. Impact of sucrose level on storage stability of proteins in freeze-dried solids: II. Correlation of aggregation rate with protein structure and molecular mobility. J Pharm Sci. doi: 10.1002/jps.21622.
81. Pellaud J, Schote U, Arvinte T, Seelig J. 1999. Conformation and self-association of human recombinant transforming growth factor-beta3 in aqueous solutions. J Biol Chem 274(12):7699–7704.
82. Nayar R, Manning MC. 2002. High throughput formulation: Strategies for rapid development of stable protein products. Pharm Biotechnol 13:177–198.
83. Capelle MA, Gurny R, Arvinte T. 2007. High throughput screening of protein formulation stability: Practical considerations. Eur J Pharm Biopharm 65(2):131–148.
84. Capelle MA, Gurny R, Arvinte T. 2009. A high throughput protein formulation platform: case study of salmon calcitonin. Pharm Res 26(1):118–128.
85. Bedu-Addo F, Moreadith R, Advant SJ. 2002. Preformulation development of recombinant pegylated staphylokinase SY161 using statistical design. AAPS PharmSci 4(4):E19.
86. Gupta S, Kaisheva E. 2003. Development of a multidose formulation for a humanized monoclonal antibody using experimental design techniques. AAPS PharmSci 5(2):E8.
87. Maltesen MJ, Bjerregaard S, Hovgaard L, Havelund S, van de Weert M. 2008. Quality by design—Spray drying of insulin intended for inhalation. Eur J Pharm Biopharm 70(3):828–838.
88. Prinn KB, Costantino HR, Tracy M. 2002. Statistical modeling of protein spray drying at the lab scale. AAPS PharmSciTech 3(1):E4.
89. Hu YQ, Guo JX, Wang LJ, Tan R, Zhen LY. 2000. Preparation and evaluation of insulin-loaded polylactide microspheres using factorial design. Drug Dev Ind Pharm 26(12):1309–1313.

90. Volkin DB, Klibanov AM. 1989. Minimizing protein inactivation. In Creighton TE, ed., *Protein Function: A Practical Approach*. Oxford, England: Information Press, pp. 1–24.
91. Kreilgaard L, Jones LS, Randolph TW, Frokjaer S, Flink JM, Manning MC, Carpenter JF. 1998. Effect of Tween 20 on freeze-thawing-and agitation-induced aggregation of recombinant human factor XIII. J Pharm Sci 87(12):1597–1603.
92. Katakam M, Banga AK. 1997. Use of poloxamer polymers to stabilize recombinant human growth hormone against various processing stresses. Pharm Dev Technol 2(2):143–149.
93. Katakam M, Bell LN, Banga AK. 1995. Effect of surfactants on the physical stability of recombinant human growth hormone. J Pharm Sci 84(6):713–716.
94. Bam NB, Cleland JL, Yang J, Manning MC, Carpenter JF, Kelley RF, Randolph TW. 1998. Tween protects recombinant human growth hormone against agitation-induced damage via hydrophobic interactions. J Pharm Sci 87(12):1554–1559.
95. Kwon YM, Baudys M, Knutson K, Kim SW. 2001. In situ study of insulin aggregation induced by water-organic solvent interface. Pharm Res 18(12):1754–1759.
96. Kueltzo LA. 2004. *Formulation Strategies for Protein Therapeutics*. Boston: IBC Life Sciences.
97. Maa Y-F, Hsu CC. 1997. Protein denaturation by combined effect of shear and air-liquid interface. Biotechnol Bioeng 54:503–512.
98. Sadhale Y, Shah JC. 1999. Stabilization of insulin against agitation-induced aggregation by the GMO cubic phase gel. Int J Pharm 191(1):51–64.
99. Webb SD, Golledge SL, Cleland JL, Carpenter JF, Randolph TW. 2002. Surface adsorption of recombinant human interferon-gamma in lyophilized and spray-lyophilized formulations. J Pharm Sci 91(6):1474–1487.
100. Wang PL, Johnston TP. 1993. Enhanced stability of two model proteins in an agitated solution environment using poloxamer 407. J Parenter Sci Technol 47(4):183–189.
101. Holmberg M, Stibius KB, Ndoni S, Larsen NB, Kingshott P, Hou XL. 2007. Protein aggregation and degradation during iodine labelling and its consequences for protein adsorption to biomaterials. Anal Biochem 361(1):120–125.
102. Lam XM, Oeswein JQ, Ongpipattanakul B, Shahrokh Z, Wang SX, Weissburg RP, Wong RL. 2000. *Stabilized Antibody Formulation*. Genentech.
103. Chen B, Bautista R, Yu K, Zapata GA, Chamow SM. 2003. Influence of histidine on the stability and physical properties of a fully human antibody in aqueous and solid forms. Pharm Res 20(12):1952–1960.
104. Lee LS. 1997. *Stabilized Monomeric Protein Compositions*. US5656730.
105. Bush L, Webb C, Bartlett L, Burnett B. 1998. The formulation of recombinant factor IX: Stability, robustness, and convenience. Semin Hematol 35(2 Suppl. 2):18–21.
106. Chang BS, Kendrick BS, Carpenter JF. 1996. Surface-induced denaturation of proteins during freezing and its inhibition by surfactants. J Pharm Sci 85(12):1325–1330.
107. Eckhardt BM, Oeswein JQ, Bewley TA. 1991. Effect of freezing on aggregation of human growth hormone. Pharm Res 8(11):1360–1364.

108. Anchordoquy TJ, Carpenter JF. 1996. Polymers protect lactate dehydrogenase during freeze-drying by inhibiting dissociation in the frozen state. Arch Biochem Biophys 332(2):231–238.
109. Farruggia B, Garcia G, D'Angelo C, Pico G. 1997. Destabilization of human serum albumin by polyethylene glycols studied by thermodynamical equilibrium and kinetic approaches. Int J Biol Macromol 20(1):43–51.
110. Carpenter JF, Pikal MJ, Chang BS, Randolph TW. 1997. Rational design of stable lyophilized protein formulations: Some practical advice. Pharm Res 14(8):969–975.
111. Carpenter JF, Crowe JH, Arakawa T. 1990. Comparison of solute-induced protein stabilization in aqueous solution and in frozen and dried state. J Dairy Sci 73:3527–3636.
112. Lashmar UT, Vanderburgh M, Little SJ. 2007. Bulk freeze-thawing of macromolecules. Bioprocess Int 5:44–54.
113. Goyal K, Walton LJ, Tunnacliffe A. 2005. LEA proteins prevent protein aggregation due to water stress. Biochem J 388(1):151–157.
114. Kantor A, Ho K, Warne N. 2007. Monoclonal Antibody Frozen Drug Substance: Reduction of Freeze-Thaw and Storage Instability. IBC Conference on Protein Formulations, Boston.
115. Allison SD, Dong A, Carpenter JF. 1996. Counteracting effects of thiocyanate and sucrose on chymotrypsinogen secondary structure and aggregation during freezing, drying, and rehydration. Biophys J 71(4):2022–2032.
116. Mozhaev VV, Martinek K. 1984. Structure-stability relationships in proteins: New approaches to stabilizing enzymes. Enzyme Microb Technol 6:50–59.
117. Patro SY, Freund E, Chang BS. 2002. Protein formulation and fill-finish operations. Biotechnol Annu Rev 8:55–84.
118. Cromwell ME, Hilario E, Jacobson F. 2006. Protein aggregation and bioprocessing. AAPS J 8(3):E572–E579.
119. Kuhlman B, Yang HY, Boice JA, Fairman R, Raleigh DP. 1997. An exceptionally stable helix from the ribosomal protein L9: Implications for protein folding and stability. J Mol Biol 270(5):640–647.
120. Rupley JA, Careri G. 1991. Protein hydration and function. Adv Protein Chem 41:37–172.
121. Yoshioka S, Miyazaki T, Aso Y, Kawanishi T. 2007. Significance of local mobility in aggregation of beta-galactosidase lyophilized with trehalose, sucrose or stachyose. Pharm Res 24(9):1660–1667.
122. Crowe JH, Crowe LM, Carpenter JF. 1993. Preserving dry biomaterials: The water replacement hypothesis, part 2. BioPharm 6(May):40–43.
123. Carpenter JF, Crowe JH. 1989. An infrared spectroscopic study of the interactions of carbohydrates with dried proteins. Biochemistry 28(9):3916–3922.
124. Prestrelski SJ, Tedeschi N, Arakawa T, Carpenter JF. 1993. Dehydration-induced conformational transitions in proteins and their inhibition by stabilizers. Biophys J 65(2):661–671.
125. Lueckel B, Helk B, Bodmer D, Leuenberger H. 1998. Effects of formulation and process variables on the aggregation of freeze-dried interleukin-6 (IL-6) after lyophilization and on storage. Pharm Dev Technol 3(3):337–346.

126. Franks F. 1994. Long-term stabilization of biologicals. Biotechnology (N Y) 12(3):253–256.

127. Fox KC. 1995. Biopreservation. Putting proteins under glass. Science 267(5206): 1922–1923.

128. Kreilgaard L, Frokjaer S, Flink JM, Randolph TW, Carpenter JF. 1998. Effects of additives on the stability of recombinant human factor XIII during freeze-drying and storage in the dried solid. Arch Biochem Biophys 360(1):121–134.

129. Salnikova MS, Middaugh CR, Rytting JH. 2008. Stability of lyophilized human growth hormone. Int J Pharm 358(1–2):108–113.

130. Volkin DB, Middaugh CR. 1996. The characterization, stabilization, and formulation of acidic fibroblast growth factor. Pharm Biotechnol 9:181–217.

131. te Booy MP, de Ruiter RA, de Meere AL. 1992. Evaluation of the physical stability of freeze-dried sucrose-containing formulations by differential scanning calorimetry. Pharm Res 9(1):109–114.

132. Prestrelski SJ, Pikal KA, Arakawa T. 1995. Optimization of lyophilization conditions for recombinant human interleukin-2 by dried-state conformational analysis using Fourier-transform infrared spectroscopy. Pharm Res 12(9):1250–1259.

133. Izutsu K, Yoshioka S, Kojima S. 1995. Increased stabilizing effects of amphiphilic excipients on freeze-drying of lactate dehydrogenase (LDH) by dispersion into sugar matrices. Pharm Res 12(6):838–843.

134. Sarciaux J-M, Mansour S, Hageman MJ, Nail SL. 1998. Effect of species, processing conditions and phosphate buffer composition on IgG aggregation during lyophilization. PharmSci 1(Suppl.):S-545.

135. Roy S, Jung R, Kerwin BA, Randolph TW, Carpenter JF. 2005. Effects of benzyl alcohol on aggregation of recombinant human interleukin-1-receptor antagonist in reconstituted lyophilized formulations. J Pharm Sci 94(2):382–396.

136. Sarciaux JM, Mansour S, Hageman MJ, Nail SL. 1999. Effects of buffer composition and processing conditions on aggregation of bovine IgG during freeze-drying. J Pharm Sci 88(12):1354–1361.

137. Hagiwara H, Yuasa H, Yamamoto Y. 2000. Stabilized human monoclonal antibody preparation. US6165467.

138. Webb SD, Cleland JL, Carpenter JF, Randolph TW. 2002. A new mechanism for decreasing aggregation of recombinant human interferon-gamma by a surfactant: Slowed dissolution of lyophilized formulations in a solution containing 0.03% polysorbate 20. J Pharm Sci 91(2):543–558.

139. Castellanos IJ, Crespo R, Griebenow K. 2003. Poly(ethylene glycol) as stabilizer and emulsifying agent: A novel stabilization approach preventing aggregation and inactivation of proteins upon encapsulation in bioerodible polyester microspheres. J Control Release 88(1):135–145.

140. Sah H. 1999. Stabilization of proteins against methylene chloride/water interface-induced denaturation and aggregation. J Control Release 58(2):143–151.

141. Wei G, Lu LF, Lu WY. 2007. Stabilization of recombinant human growth hormone against emulsification-induced aggregation by Pluronic surfactants during microencapsulation. Int J Pharm 338(1–2):125–132.

142. Geng Y, Yuan W, Wu F, Chen J, He M, Jin T. 2008. Formulating erythropoietin-loaded sustained-release PLGA microspheres without protein aggregation. J Control Release 130(3):259–265.

143. Kanaoka E, Nagata S, Hirano K. 1999. Stabilization of aerosolized IFN-gamma by liposomes. Int J Pharm 188(2):165–172.

144. Assemand E, Lacroix M, Mateescu MA. 2003. L-tyrosine prevents aggregation of therapeutic proteins by gamma-irradiation. Biotechnol Appl Biochem 38(Pt 2):151–156.

145. Ip AY, Arakawa T, Silvers H, Ransone CM, Niven RW. 1995. Stability of recombinant consensus interferon to air-jet and ultrasonic nebulization. J Pharm Sci 84(10):1210–1214.

146. Sachdev D, Chirgwin JM. 1999. Properties of soluble fusions between mammalian aspartic proteinases and bacterial maltose-binding protein. J Protein Chem 18(1):127–136.

147. Shirley BA, Stanssens P, Steyaert J, Pace CN. 1989. Conformational stability and activity of ribonuclease T1 and mutants. Gln25—Lys, Glu58—Ala, and the double mutant. J Biol Chem 264(20):11621–11625.

148. Shoichet BK, Baase WA, Kuroki R, Matthews BW. 1995. A relationship between protein stability and protein function. Proc Natl Acad Sci U S A 92(2):452–456.

149. Tanaka M, Machida Y, Nishikawa Y, Akagi T, Morishima I, Hashikawa T, Fujisawa T, Nukina N. 2002. The effects of aggregation-inducing motifs on amyloid formation of model proteins related to neurodegenerative diseases. Biochemistry (Mosc) 41(32):10277–10286.

150. Rousseau F, Schymkowitz J, Serrano L. 2006. Protein aggregation and amyloidosis: Confusion of the kinds? Curr Opin Struct Biol 16(1):118–126.

151. Soto C, Castano EM, Frangione B, Inestrosa NC. 1995. The alpha-helical to beta-strand transition in the amino-terminal fragment of the amyloid beta-peptide modulates amyloid formation. J Biol Chem 270:3063–3067.

152. Teschke CM. 1999. Aggregation and assembly of phage P22 temperature-sensitive coat protein mutants in vitro mimic the in vivo phenotype. Biochemistry 38(10):2873–2881.

153. Fields G, Alonso D, Stiger D, Dill KA. 1992. Theory for the aggregation of proteins and copolymers. J Phys Chem 96:3974–3981.

154. Patro SY, Przybycien TM. 1994. Simulations of kinetically irreversible protein aggregate structure. Biophys J 66:1274–1289.

155. Fraser PE, McLachlan DR, Surewicz WK, Mizzen CA, Snow AD, Nguyen JT, Kirschner DA. 1994. Conformation and fibrillogenesis of Alzheimer A beta peptides with selected substitution of charged residues. J Mol Biol 244(1):64–73.

156. Nielsen L, Frokjaer S, Brange J, Uversky VN, Fink AL. 2001. Probing the mechanism of insulin fibril formation with insulin mutants. Biochemistry (Mosc) 40(28):8397–8409.

157. Wang H, He L, Lensch M, Gabius HJ, Fee CJ, Middelberg AP. 2008. Single-site Cys-substituting mutation of human lectin galectin-2: Modulating solubility in recombinant production, reducing long-term aggregation, and enabling site-specific monoPEGylation. Biomacromolecules 9(11):3223–3230.

158. Veronese FM, Mero A. 2008. The impact of PEGylation on biological therapies. BioDrugs 22(5):315–329.

159. Harris JM, Martin NE, Modi M. 2001. Pegylation: A novel process for modifying pharmacokinetics. Clin Pharmacokinet 40(7):539–551.

160. Lee DL, Sharif I, Kodihalli S, Stewart DI, Tsvetnitsky V. 2008. Preparation and characterization of monopegylated human granulocyte-macrophage colony-stimulating factor. J Interferon Cytokine Res 28(2):101–112.

161. Katre NV. 1990. Immunogenicity of recombinant IL-2 modified by covalent attachment of polyethylene glycol. J Immunol 144(1):209–213.

162. Hinds KD, Kim SW. 2002. Effects of PEG conjugation on insulin properties. Adv Drug Deliv Rev 54(4):505–530.

163. Pepinsky RB, Shapiro RI, Wang S, Chakraborty A, Gill A, Lepage DJ, Wen D, Rayhorn P, Horan GS, Taylor FR, Garber EA, Galdes A, Engber TM. 2002. Long-acting forms of Sonic hedgehog with improved pharmacokinetic and pharmacodynamic properties are efficacious in a nerve injury model. J Pharm Sci 91(2):371–387.

164. Hinds K, Koh JJ, Joss L, Liu F, Baudys M, Kim SW. 2000. Synthesis and characterization of poly(ethylene glycol)-insulin conjugates. Bioconjug Chem 11(2):195–201.

165. Rajan RS, Li T, Aras M, Sloey C, Sutherland W, Arai H, Briddell R, Kinstler O, Lueras AM, Zhang Y, Yeghnazar H, Treuheit M, Brems DN. 2006. Modulation of protein aggregation by polyethylene glycol conjugation: GCSF as a case study. Protein Sci 15(5):1063–1075.

166. Gaberc-Porekar V, Zore I, Podobnik B, Menart V. 2008. Obstacles and pitfalls in the PEGylation of therapeutic proteins. Curr Opin Drug Discov Dev 11(2): 242–250.

167. Liu DT. 1992. Glycoprotein pharmaceuticals: Scientific and regulatory considerations, and the US Orphan Drug Act. Trends Biotechnol 10(4):114–120.

168. Baek WO, Vijayalakshmi MA. 1997. Effect of chemical glycosylation of RNase A on the protein stability and surface histidines accessibility in immobilized metal ion affinity electrophoresis (IMAGE) system. Biochim Biophys Acta 1336(3):394–402.

169. Sola RJ, Griebenow K. 2008. Effects of glycosylation on the stability of protein pharmaceuticals. J Pharm Sci. doi: 10.1002/jps.21504.

170. Hoiberg-Nielsen R, Fuglsang CC, Arleth L, Westh P. 2006. Interrelation ships of glycosylation and aggregation kinetics for *Peniophora lycii* phytase. Biochemistry (Mosc) 45(15):5057–5066.

171. Lim KH, Chang HI. 2006. O-linked N-acetylglucosamine suppresses thermal aggregation of Sp1. FEBS Lett 580(19):4645–4652.

172. Kwon KS, Yu MH. 1997. Effect of glycosylation on the stability of alpha1-antitrypsin toward urea denaturation and thermal deactivation. Biochim Biophys Acta 1335(3):265–272.

173. Loriol C, Audfray A, Dupuy F, Germot A, Maftah A. 2007. The two N-glycans present on bovine Pofut1 are differently involved in its solubility and activity. FEBS J 274(5):1202–1211.

174. Baudys M, Uchio T, Mix D, Wilson D, Kim SW. 1995. Physical stabilization of insulin by glycosylation. J Pharm Sci 84(1):28–33.

175. Sola RJ, Al-Azzam W, Griebenow K. 2006. Engineering of protein thermodynamic, kinetic, and colloidal stability: Chemical glycosylation with monofunctionally activated glycans. Biotechnol Bioeng 94(6):1072–1079.

176. Hattori M, Numamoto K, Kobayashi K, Takahashi K. 2000. Functional changes in beta-lactoglobulin by conjugation with cationic saccharides. J Agric Food Chem 48(6):2050–2056.

177. Wang W, Antonsen K, Wang YJ, Wang DQ. 2008. pH dependent effect of glycosylation on protein stability. Eur J Pharm Sci 33(2):120–127.

178. Broersen K, Elshof M, de Groot J, Voragen AGJ, Hamer RJ, de Jongh HHJ. 2007. Aggregation of beta-lactoglobulin regulated by glucosylation. J Agric Food Chem 55(6):2431–2437.

179. Bajaj H, Sharma VK, Badkar A, Zeng D, Nema S, Kalonia DS. 2006. Protein structural conformation and not second virial coefficient relates to long-term irreversible aggregation of a monoclonal antibody and ovalbumin in solution. Pharm Res 23(6):1382–1394.

180. Vilasi S, Iannuzzi C, Portaccio M, Irace G, Sirangelo I. 2008. Effect of trehalose on W7FW14F apomyoglobin and insulin fibrillization: New insight into inhibition activity. Biochemistry (Mosc) 47(6):1789–1796.

181. Thirumangalathu R, Krishnan S, Brems DN, Randolph TW, Carpenter JF. 2006. Effects of pH, temperature, and sucrose on benzyl alcohol-induced aggregation of recombinant human granulocyte colony stimulating factor. J Pharm Sci 95(7):1480–1497.

182. Arnaudov LN, de Vries R. 2005. Thermally induced fibrillar aggregation of hen egg white lysozyme. Biophys J 88(1):515–526.

183. Vemuri S, Yu CT, Roosdorp N. 1993. Formulation and stability of recombinant alpha 1-antitrypsin. Pharm Biotechnol 5:263–286.

184. Baussay K, Bon CL, Nicolai T, Durand D, Busnel JP. 2004. Influence of the ionic strength on the heat-induced aggregation of the globular protein beta-lactoglobulin at pH 7. Int J Biol Macromol 34(1–2):21–28.

185. Arnaudov LN, de Vries R. 2006. Strong impact of ionic strength on the kinetics of fibrilar aggregation of bovine beta-lactoglobulin. Biomacromolecules 7(12): 3490–3498.

186. Chi EY, Krishnan S, Kendrick BS, Chang BS, Carpenter JF, Randolph TW. 2003. Roles of conformational stability and colloidal stability in the aggregation of recombinant human granulocyte colony-stimulating factor. Protein Sci 12(5): 903–913.

187. Abdul-Fattah AM, Lechuga-Ballesteros D, Kalonia DS, Pikal MJ. 2008. The impact of drying method and formulation on the physical properties and stability of methionyl human growth hormone in the amorphous solid state. J Pharm Sci 97(1):163–184.

188. Schule S, Schulz-Fademrecht T, Garidel P, Bechtold-Peters K, Friess W. 2008. Stabilization of IgG1 in spray-dried powders for inhalation. Eur J Pharm Biopharm 69(3):793–807.

CHAPTER 9

Case Studies Involving Protein Aggregation

RAHUL S. RAJAN, TIANSHENG LI, and TSUTOMU ARAKAWA

9.1 INTRODUCTION

Protein aggregation is a complex degradation pathway[1] that has elicited deep interest in both academia as well as the biotechnology industry.[2–7] Protein aggregation appears to be a universal phenomenon occurring both *in vivo* and *in vitro*. When it occurs *in vivo*, its relationship to various neurodegenerative diseases has been an intense area of research,[4–6] with the ultimate goal of developing a therapeutic drug. When it occurs *in vitro*, it affects the quality of the product and raises potential safety concerns for pharmaceutical proteins.[7] *In vitro*, protein aggregation can occur under a variety of conditions related to drug development. These include process development, transportation, product storage, and handling, both for the drug substance and the drug product, as well as delivery into the patient.[2,3] The dosage form of the product can also dictate the nature and extent of aggregation. For instance, a protein may be prone to aggregation in the liquid state but be resilient to aggregation in the lyophilized state. Certain processing conditions, such as freeze/thaw and drug product filling, may pose unique challenges for certain sensitive proteins.[3]

Due to the prevalence of protein aggregation during biotechnology product development, there is widespread interest in understanding and mitigating it. Six case studies are presented in this chapter that cover diverse areas of protein aggregation. Five of the six case studies revolve around monoclonal antibodies. This is because at present, antibodies and antibody-based therapeutic agents represent a significant area of development in the biotechnology industry.[2] In general, the action of antibodies is specific, which is generally thought to lead to fewer side effects. The class of antibodies that are of

Aggregation of Therapeutic Proteins, Edited by Wei Wang and Christopher J. Roberts

greatest interest are the IgGs, which contain two identical heavy chains of ~50 kDa and two identical light chains of ~25 kDa, making the total molecular weight around 150 kDa.[8] There are disulfide bonds linking the two heavy chains, linking the heavy and light chains, as well as residing inside the chains. IgGs are divided into four different subclasses. In order of relative abundance in the human plasma, these are IgG1, IgG2, IgG3, and IgG4. The structural differences among these subtypes reside in the number and location of interchain disulfide bonds, as well as the length of the hinge region. The sequence of the first 100 amino acids or so in each chain differs from antibody to antibody. These are called the variable (V) regions, and they create antigen specificity for the molecule. The remaining sequences are constant (C) for a given IgG class. Many IgGs can be cleaved at the hinge region, resulting in three fragments. The one with the heavy chain regions C-terminal of the hinge region is called Fc (for fragment crystallizable), which is responsible for effector recognition and binding. The other two fragments, with the light chain linked with the region of the heavy chain N-terminal of the hinge, are called Fab (for antigen-binding fragment).[8]

The first case involves studying the aggregation behavior of a much smaller protein, keratocyte growth factor (KGF), which delves into the principles of protein unfolding that would precede aggregation and the preferential exclusion mechanism to explain the stabilizing effect of osmolytes. The remaining five cases are focused on antibody aggregation, under different conditions. The second case study by Perico et al. describes antibody aggregation in the liquid state, under accelerated temperature storage conditions, that has a complex dependence on fragmentation and inverse temperature dependence. In the third case study, protein aggregation in an Fc fusion protein is caused by crystallization of a well-known stabilizer, namely sorbitol, in the frozen state. The fourth and fifth case studies exemplify aggregation challenges that were faced during formulation development for lyophilized antibodies. The sixth and last case study shows insoluble particle formation due to metal shedding from filling pumps, as well as structural characterization of insoluble antibody aggregates. In each study, the protein aggregation phenomenon observed, along with the mechanisms involved and the implications of that work, is discussed in depth.

9.2 CASE STUDY 1: AGGREGATION IN THE LIQUID STATE: THE ROLE OF OSMOLYTES IN STABILIZING KGF TOWARD AGGREGATION

9.2.1 Background

Aggregation of proteins can be suppressed by either increasing the stability of the proteins or decreasing the aggregation rate. A simple model of protein aggregation may be expressed as

$$N \rightleftarrows U \rightarrow Agg$$

according to Lumry and Eyring.[9] Here the native protein (N) is in equilibrium with the unfolded protein (U), which aggregates into "Agg." Formation of aggregates can then be suppressed by stabilizing the native structure or reducing the rate constant of aggregation. Protein structure can be stabilized by a variety of compounds, the so-called cosolvents.[10–12] Sugars, polyols and certain amino acids stabilize the structure of native proteins at moderate (0.1 M) to high concentrations (1 M). These cosolvents are also referred to as osmolytes or compatible solutes since they are utilized in nature to raise the osmotic pressure of the cellular environment, yet are compatible with the macromolecular function and cell viability.[13,14] Here we present a case of KGF, which has a strong tendency to aggregate, and explore the role of osmolytes in modulating the stability of this protein.

9.2.2 Model Protein and Conditions

KGF is a member of the fibroblast growth factor (FGF) family and is an approved protein therapeutic used as a wound healing agent.[15,16] Recombinant KGF was expressed in *Escherichia coli* and contains 163 amino acid residues. Previous studies employing deletion, and mutational and structural analysis have identified homology between KGF and other members of the FGF family, especially the presence of the β-trefoil motif in the crystal structure.[17] Irreversible aggregation of KGF was induced by thermal denaturation, as reported earlier.[18] In the past, Chen et al.[18,19] used circular dichroism (CD) to monitor structural changes during thermal unfolding of KGF and to measure the melting temperature of KGF. To monitor the effect of various excipients, changes in UV absorbance and CD signals as a function of temperature have been used to monitor thermal unfolding of KGF in this study. The various osmolytes used were sucrose, dimethyl sulfoxide (DMSO), proline, betaine, and trehalose.

9.2.3 Results and Interpretation

Figure 9.1 illustrates a typical absorbance change, monitored at 280 nm, due to protein unfolding. When protein thermally unfolds, but without leading to aggregation, it normally results in a fairly small decrease (~10%) in absorbance due to exposure of aromatic groups (shown by dotted curve). KGF instead exhibited a sharp increase in absorbance due to insoluble aggregation of thermally unfolded protein (dark line), generally much larger than the absorbance change observed for thermal unfolding. The CD signal also showed a change at the temperature at which the absorbance increase occurred, indicating that aggregation was due to thermal unfolding. In this case, only the onset temperature was meaningful, because melting temperatures cannot be easily deduced from either the CD signal or absorbance changes, due to the fact that these signals could be impacted by competing aggregation processes.

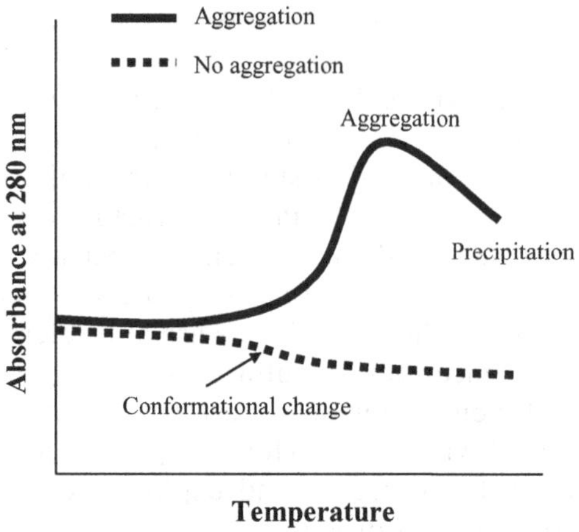

Figure 9.1. Model presentation of thermal unfolding of protein. Onset of insoluble protein aggregation results in increase in absorption with increase in temperature (solid line) due to increase in light scatter. Once quantitative protein precipitation occurs, absorbance at 280 nm decreases due to decrease in the amount of soluble absorbing species. The dotted line depicts conformational change that occurs due to melting as a function of temperature, without aggregation or precipitation.

Figure 9.1 illustrates that absorbance change is a complex function of unfolding rate, aggregation rate, and precipitation. The onset temperature of KGF was thus determined to be 41°C, by both CD and absorbance changes, in 10 mM phosphate, pH 7.0, in the absence of stabilizing cosolvents. As this onset temperature was independent of the protein concentration, it could be inferred that a conformational change was essential for the observed aggregation. Figure 9.2 shows the effects of sugars, trehalose (open square), and sucrose (open triangle) on the onset temperature. These two sugars increased the onset temperature, as has been observed for many other proteins,[12,20] indicative of stabilization toward unfolding. Incubation of KGF at 37°C resulted in aggregation and loss of protein with a "shelf life" (defined by time taken for loss of 50% monomer) of 0.35 days. When the shelf life of KGF was determined at 37°C in the presence of sugars, both trehalose and sucrose increased the shelf life equally effectively (Fig. 9.2). Thus, it was evident that these sugars increase both melting temperature and shelf life at 37°C. Interestingly, while the increase in onset temperature for unfolding was linear with increasing sugar concentration (left panel), there seemed to be a threshold sugar concentration (~0.5 M) needed to appreciably increase shelf life at 37°C (right panel).

Such a correlation was tested for sugars, polyols, DMSO, and proline. Figure 9.3 shows a plot of shelf life as a function of melting temperature (onset temperature). Different concentrations of these excipients were used to change the melting temperature and the shelf life. There are several points to

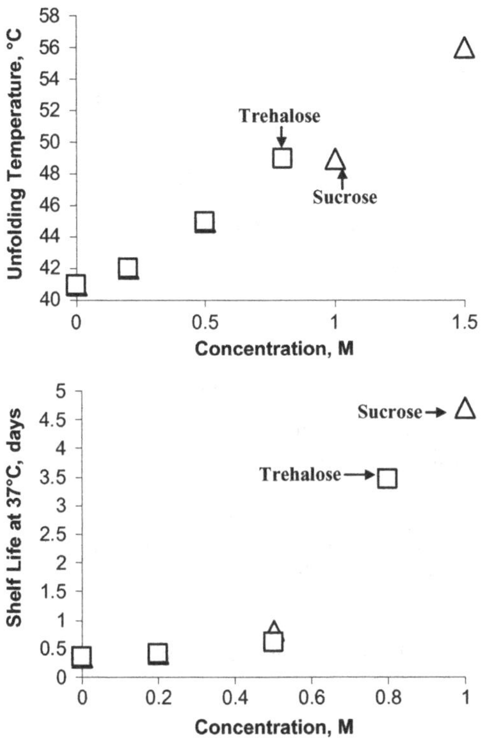

Figure 9.2. Unfolding temperature (top panel) and shelf life at 37°C (bottom panel) of KGF as a function of sucrose (triangle) and trehalose (square) concentration. Shelf life in this case refers to the length of time needed to drop the monomer concentration by 50%. The stabilizing effect of both sugars can be seen as an increase in both the unfolding temperature as well as the 37°C half-life with increasing sugar concentration.

be deduced from this plot. Sugars increased melting temperature of KGF, which in turn increased the shelf life. As the shelf life was determined from the remaining soluble protein, it was clear that sugars increased the conformational stability of KGF, thereby reducing aggregation. Protein-stabilizing polyols had the same property, namely a correlation between shelf life and conformational stability. On the other hand, proline was unique. It increased the conformational stability of KGF with increasing concentration (in this case, melting of the protein), consistent with its protein-stabilizing property.[21] However, the shelf life of the protein at 37°C remained the same, demonstrating no correlation between conformational stability and shelf life. Thus, at elevated temperature, it enhanced the stability of KGF, while it had no effect on the storage stability at a lower temperature of 37°C. This suggested that proline may interact with KGF differently at 37°C and elevated temperature,

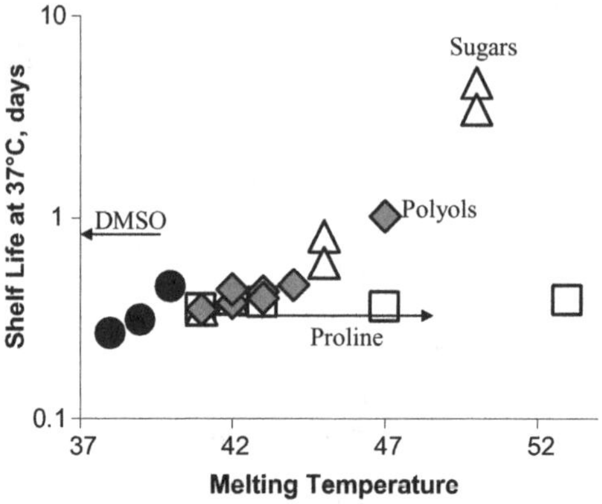

Figure 9.3. Correlation between shelf life and melting temperature for the effects of sugars (triangle), polyols (gray-filled diamonds), DMSO (filled circle), and proline (square) on KGF. For sugars and polyols, each symbol indicates a different sugar. For DMSO and proline, each symbol indicates a different concentration. For DMSO, increasing concentration decreased the melting temperature (leftward arrow), while for proline, increasing concentration increased melting temperature (rightward arrow).

as illustrated in Fig. 9.3. DMSO decreased both the melting temperature and shelf life (see filled circles in Fig. 9.3).

9.2.4 Discussion and Lessons Learned

As shown for KGF, sugars and polyols are often used to stabilize many other proteins against aggregation.[22–24] Among sugars, sucrose and trehalose are commonly used. Extensive studies of protein–solvent interaction analysis, pioneered by Timasheff and his coworkers, made a significant contribution to understanding the mechanism of protein stabilization by these solvent additives.[25–28] The protein-stabilizing effect of osmolytes now finds applications in pharmaceutical formulations for suppressing aggregation.[29] Numerous examples exist for their use in enhancing storage stability and reducing aggregation. The effects of cosolvents on protein stability can be readily explained from the thermodynamic analysis of their interactions with the proteins. At high concentrations of cosolvents, many different effects determine their mode of interaction with the proteins, including hydration of cosolvents and protein, excluded volume effects and affinity of cosolvents for certain amino acid side chains and backbones. Excluded volume effect arises from the steric exclusion of solvent components from the surface of a protein molecule with a radius of R_P, and increases with the size of the cosolvent (Fig. 9.4). There is a layer of excess water surrounding the protein surface due to this effect (shown by the dotted line in Fig. 9.4). A large flexible polymer, for example, polyethylene glycol (PEG), is more strongly excluded than osmolyte cosolvents, as the radius of PEG (R_{PEG}) is much larger than the radius of the osmo-

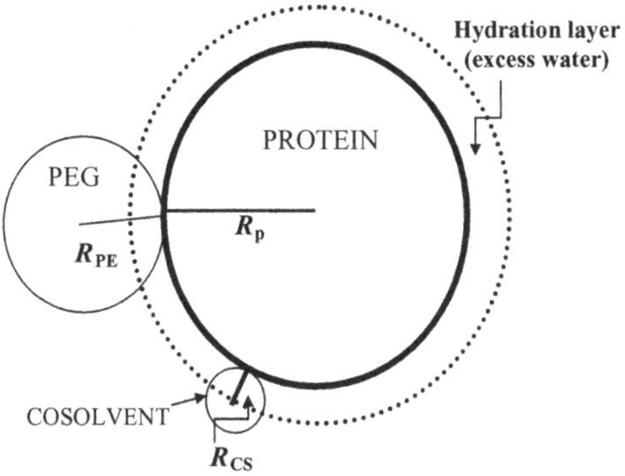

Figure 9.4. Schematic presentation of excluded volume effect. The large circle in the middle represents a protein molecule with R_P, and two kinds of excipients are depicted: a large polymer like PEG, with radius R_{PEG}, and an osmolyte cosolvent with radius R_{CS}. The hydration layer surrounding the protein is also depicted. Due to size differences, large flexible polymers like PEG are more strongly excluded than small cosolvent molecules.

lyte cosolvents (R_{CS}). Various weak and strong interactions contribute to the overall interaction of cosolvents with the protein. Such interactions can be determined by equilibrium dialysis experiments and may be formally grouped into two different modes. First, there is excess of cosolvent in the vicinity of the protein surface compared with its concentration in the bulk phase, as illustrated in Fig. 9.5 (left panel). This is called "preferential interaction," indicating that the cosolvent concentration is higher at the protein surface than in the bulk phase. The opposite case is also illustrated in Fig. 9.5 (right panel), in which there is excess water at the protein surface. This is called "preferential hydration" or "preferential exclusion" of cosolvent, indicating a deficiency of cosolvent in the vicinity of the protein molecule. Osmolytes show preferential hydration of proteins; in other words, osmolytes are preferentially excluded from the protein surface.[21,30] Many sugars and polyols and certain salts, which are known to stabilize proteins, are all preferentially excluded from the proteins.[11,12,21,30]

Figure 9.6 plots hypothetical temperature-dependent effects of excipients on protein degradation. The black line shows the temperature dependence of degradation in the absence of excipients. In some cases, excipients may have no effect on degradation at high temperature (dotted lines), but may have either a stabilizing (SE-LT) or a destabilizing (DE-LT) impact on the rate of degradation under recommended or lower storage temperatures, with respect to the rate of degradation in the absence of excipients. Case DE-LT enhances degradation at low temperature, while case SE-LT reduces degradation. The opposite class of excipients is depicted by the dashed lines in Fig. 9.6: at high temperature, it either stabilizes (SE-HT) or destabilizes (DE-HT) the

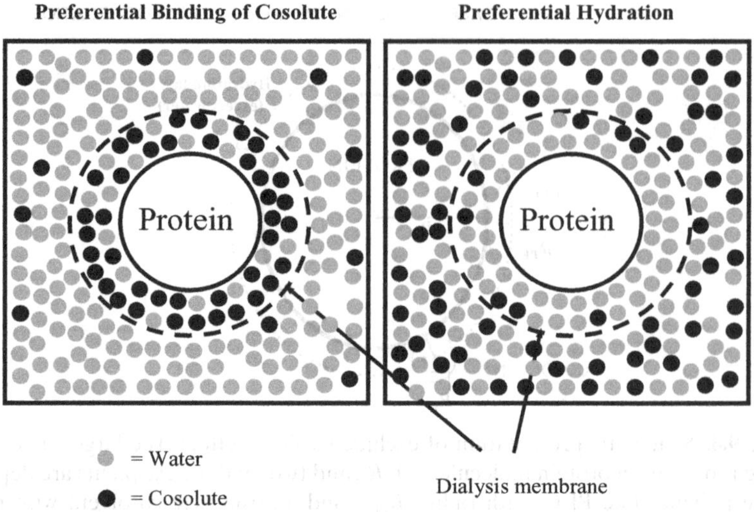

Figure 9.5. Typical model cases of preferential interaction. Left, preferential interaction. Right, preferential hydration. The protein molecule is represented as a large sphere in the center of each panel. The hydration layer immediately surrounding the protein is indicated as the area under the black dashed circle. If there is an excess of cosolute (small, filled black circle) around the protein compared with water (small, filled gray circle), this situation is called preferential interaction. Excess water in the hydration layer compared with the cosolute is called preferential hydration or preferential exclusion of the cosolute.

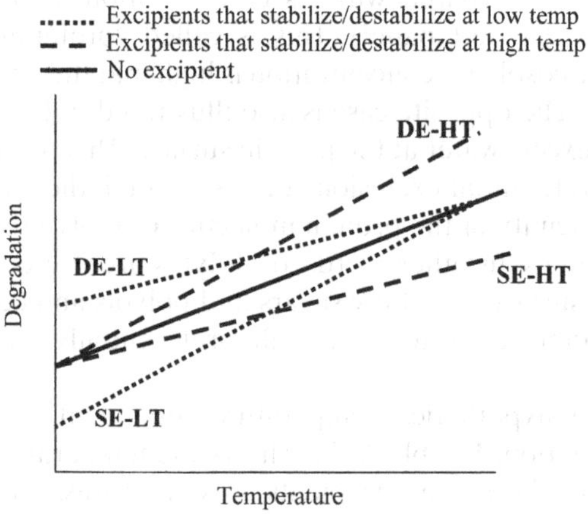

Figure 9.6. Model to explain four classes of excipients. The solid line is the degradation of the protein in the absence of any excipients. The dotted line shows those excipients that at low temperatures have stabilizing (SE-LT) or destabilizing (DE-LT) effects but are indistinguishable from the excipient-free formulation degradation rate at higher temperature. The dashed line shows excipients that at high temperatures only manifest either stabilizing (SE-HT) or destabilizing (DE-HT) effects relative to the excipient-free formulation, but not at lower temperature.

molecule with respect to the rate of degradation in the absence of an excipient, but has no difference at low temperature. It is to be noted that qualitative results illustrated in Fig. 9.6 are expected to hold for a degradation route in which unfolding to some degree is required: either a precursor to the rate-limiting step, or is itself the rate-limiting step. Cases where the excipient at hand either stabilizes or destabilizes at high or low temperatures have not been included here for simplicity.

Temperature dependence of excipient effects occurs due to temperature dependence of the interactions between excipient and protein. As protein molecules fluctuate under equilibrium between the native and unfolded states, the excipient interaction with protein depends on the state of the protein as well as the temperature. An extreme case of such temperature-dependent effects of excipients can be seen for certain cryoprotectants. Organic solvents, such as DMSO, show unfavorable interactions with the native protein structure; that is, they are preferentially excluded from the protein surface,[31] similar to structure-stabilizing cosolvents. Nevertheless, they destabilize proteins, as shown in the KGF case (Fig. 9.3). At high temperature, where protein structure is destabilized, they further destabilize the structure by binding more strongly to the unfolded state, leading to decreased melting temperature (see Fig. 9.7). This is presumably due to hydrophobic binding of organic solvents, which is generally thought to increase with temperature.[32] In the frozen state, such hydrophobic solvent interaction diminishes, and only solvent exclusion effects occur with the proteins and stabilize the native structure against freezing-induced stresses,[33] as depicted in Fig. 9.7. It should be noted that in certain cases, the native structure of enzymatic proteins can be altered by organic solvents in such a way that it tremendously increases the catalytic activity of the protein.[34] This is a complex and evolving area with work in progress in several laboratories.

What is the mechanism by which these osmolytes increase the stability of proteins? The osmolytes are preferentially excluded from the protein surface, meaning that the interaction between osmolytes and protein is thermodynamically unfavorable (cf. Chapter 2). This increases the free energy of the native state of the protein as schematically depicted in Fig. 9.8. Although not experimentally determined, a greater exclusion of osmolytes is expected from the unfolded state, because it has a greater surface area compared with the folded native state. Thus, the unfavorable interaction and hence free energy would increase even more so for the unfolded state in the presence of the cosolvent. This leads to a greater energy difference between the native and unfolded states in the presence of osmolytes, that is, more energy is required to unfold proteins in the presence of preferentially excluded cosolvents. As preferential exclusion and unfavorable interaction increase with osmolyte concentration, the native structure is effectively more stabilized at higher osmolyte concentrations (Fig. 9.8). These concepts can be extended to the situation where there is self-association. When proteins aggregate, the surface area per protein molecule decreases, which in turn results in less unfavorable interaction between

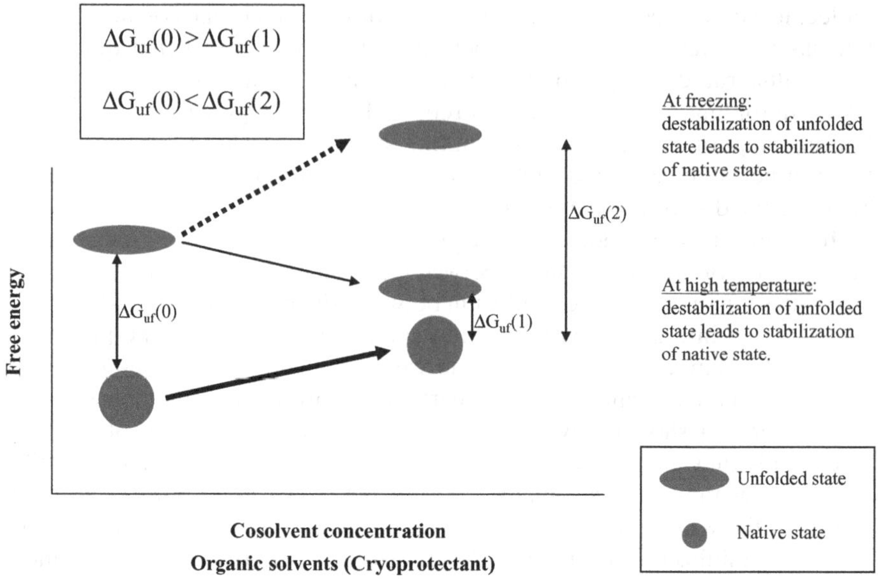

Figure 9.7. This figure depicts the effect of organic solvents and cryoprotectants as a function of temperature. Under both temperature conditions, the free energy of the native state increases (thick arrow). At high temperature, the free energy of the unfolded state decreases (thin arrow), decreasing the free energy barrier between the native and unfolded state (ΔGuf(1)) compared with the value in the absence of an excipient (ΔGuf(0)), thus favoring unfolding. At lower temperatures, the free energy of the unfolded increases (ΔGuf(2), thick dotted arrow), thus making it less favorable for unfolding.

the osmolyte and protein and hence stabilization of the associated state in the presence of osmolytes. Thus, osmolytes do stabilize the final self-associated state over the native state. However, this final self-associated or aggregated state in proteins usually requires a first step of protein unfolding. Since the unfolded intermediate is much more destabilized than the native state, as explained earlier, it is much harder to reach the aggregated state in the presence of osmolytes. In this manner, preferentially excluded cosolvents are thought to be able to stabilize the native structure and reduce aggregation.

9.3 CASE STUDY 2: AGGREGATION IN THE LIQUID STATE: HETEROGENEITY AND NON-LINEARITY IN IgG2 AGGREGATION DURING LONG-TERM STORAGE[35]

9.3.1 Background

The second case study describes non-Arrhenius kinetics observed for antibody aggregation in the liquid state. During formulation development, protein sta-

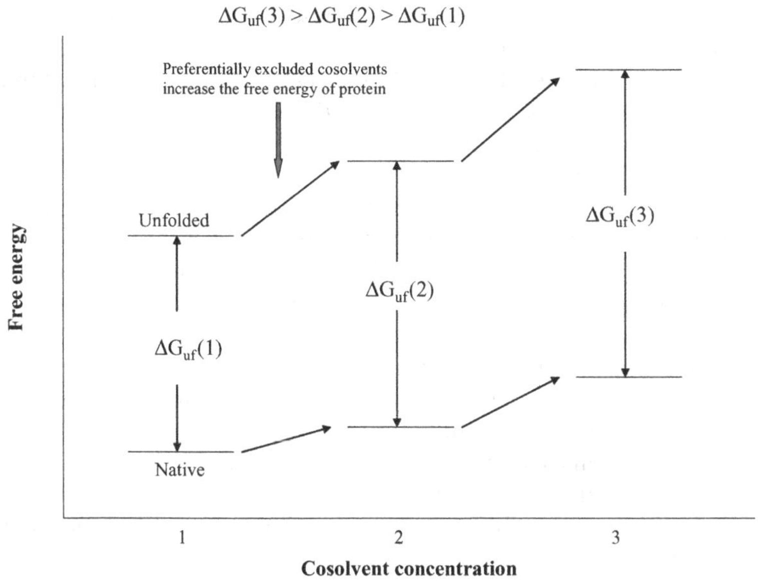

Figure 9.8. Change in free energy of native and unfolded protein by the addition of preferentially excluded cosolvents. In the x-axis, 1, 2, and 3 represent three increasing levels of cosolvent concentration (3 > 2 > 1) in an arbitrary scale. In this model, the free energy of the native state increases with increasing cosolvent concentration, but the free energy of the unfolded state increases much more. Thus, the free energy of unfolding (ΔG_{uf}), defined by the difference in the free energies of the unfolded and folded states, increases with increasing cosolvent concentration. Thus, there is greater stabilization at higher cosolvent concentration.

bility is tested in a variety of formulation conditions, such as pH, buffer, and excipient.[36] Proteins are commonly stored at temperatures above the recommended storage temperature to accelerate aggregation and other degradation phenomena. Formulation selection based on accelerated conditions is often carried out because the difference in stability afforded by the various formulations under consideration may not manifest themselves at the recommended storage conditions of 2–8°C for many months or years. Thus, protein formulations are typically exposed to higher temperature conditions, such as 29 and 37°C, well below the onset of melting, and the trends are extrapolated to the degradation that could be expected at the end of the shelf life at 2–8°C using linear Arrhenius kinetics. Storing at higher temperatures is also carried out as part of forced degradation, which is a regulatory requirement to establish analytical comparability and method suitability.[37] Appearance of small amounts of degradation products may depend at least in part on the sensitivity of detection technique. An extremely sensitive technique, such as dynamic light scattering (DLS), may be able to detect aggregate and particulate formation in a much earlier timescale[38] (http://www.ncbi.nlm.nih.gov/pubmed/17025274?ordinalpos=2&itool=EntrezSystem2.PEntrez.Pubmed.Pubmed_ResultsPanel.Pubmed_DefaultReportPanel.Pubmed_RVDocSum). It should be noted that as DLS can detect minute amount of particulates, possible precursors of precipitation, quantitation in this technique is highly unreliable.

As mentioned above, formulation development strategies often involve assumption of linear Arrhenius kinetics to extrapolate short-term higher temperature data to long-term trends at the lower, recommended storage temperature.[36] Even if a reaction shows linear Arrhenius behavior in the absence of excipients, this could change when these compounds are added and therefore must be assessed in the relevant formulation. If there are competing pathways, then the trends obtained from higher temperature data may not be representative of long-term stability at the lower temperatures, unless more complex kinetic schemes are used, such as a second-order Arrhenius reaction. The case study of Perico et al.[35] exemplifies one such behavior from a monoclonal antibody toward soluble aggregation as monitored by size-exclusion chromatography (SEC).

9.3.2 Model Protein and Conditions

These authors sought to select the optimum pH for stability toward soluble aggregation based on accelerated temperature for a fully human monoclonal IgG2 antibody stored in 10 mM acetate and 5% sorbitol. This antibody was produced in Chinese hamster ovary cells and purified at Amgen, Inc. As in other antibodies, it was a homodimer with 6 interchain and 12 intrachain disulfide bonds. The pH range under consideration was 5.0–5.5, which has been documented in several articles as an optimum storage pH for antibodies.[38,39] They stored the IgG2 antibody formulated at either 30 or 100 mg/mL at pH 5.0, 5.2, and 5.5 in the acetate/sorbitol formulation mentioned above at 4, 29, and 37°C for a time period of 1 year. At each time point, samples were monitored for aggregation by SEC. After quantifying aggregation by SEC, the rate of aggregation at each temperature for each pH was calculated. An Arrhenius plot was then generated by plotting the log of the rate of aggregation against the inverse of the temperature. Other analytical techniques used were denatured SEC to quantify covalent aggregates, reversed-phase chromatography to elucidate IgG2 disulfide isoforms and to estimate clips, differential scanning calorimetry (DSC) to probe thermal transitions of antibody domains, and limited proteolysis to generate the (Fab)2 domain.

9.3.3 Results and Interpretation

The authors observed an unexpected non-linearity in the Arrhenius plots that increased with decreasing pH. Importantly, the plots showed that the trend in the rates at 4°C was opposite of that observed at 37°C. Under the 4°C storage condition, the rate of aggregation was lowest at pH 5.0. However, at 37°C, the rate of aggregation was highest at pH 5.0. Interestingly, at the 30 mg/mL concentration, this inversion in trend was not observed. To explore whether the 37°C condition induced unfolding in the protein, the authors examined DSC thermograms of this antibody and observed that the onset of thermal melting did not begin until beyond 50°C. This indicated that the amount of unfolded protein was infinitesimally small at 37°C, although its contribution to protein aggregation could be significant if it was prone to aggregate and the rate of unfolding was not slow compared with the timescale of the experiment.

To explain this phenomenon, the authors explored the role of the recently discovered IgG2 disulfide-bonded isoforms.[40,41] In this work, the authors discovered that IgG2s existed as a mixture of isoforms ("IgG2-A," "IgG2-A/B," and "IgG2-B") that differed in the disulfide connectivity within the hinge region. Further, using a redox refolding approach, they were able to identify two sets of conditions: one of the conditions resulted in a solution enriched in the "IgG2-A" form, while the other condition resulted in enrichment of the "B" form. Under these conditions, the model IgG2 antibody was dialyzed from its formulation buffer to one with a high pH (~8.0). Following this, the antibody underwent a redox treatment using a combination of cysteine and cystamine. The authors discovered that the addition of a small level of guanidine hydrochloride (0.9 M) during the redox treatment caused enrichment of IgG2-A, while absence of this chaotrope resulted in enrichment of IgG2-B. Following incubation with the redox agents with or without guanidine hydrochloride, the antibody was dialyzed back to its original formulation buffer, and it resulted in isoform enrichment depending on which procedure was adopted. Using these enriched pools, the authors again conducted a stability study, this time incubating the enriched "IgG2-A" and "IgG2-B" isoforms solutions. After 1 year of incubation time, they found that at 4°C, there was more aggregate in the "IgG2-B" form over the "IgG2-A" form as determined by SEC. However, in the 37°C incubation condition, there was more aggregate in the "IgG2-A" form over the "IgG2-B" form. A close examination of the SEC profile revealed that the stability profile of the two isoforms was different in the small oligomer peak of the SEC chromatogram compared with the void volume aggregate peak. Characterization data showed that the small oligomer peak was predominantly dimer, while the early eluting peak was composed of higher-order aggregate structures. The difference between the two isoforms at 37°C manifested itself only in the higher-order aggregate peak, while the dimer peaks were equivalent. However, at 4°C, the "B" form had higher levels of both the dimer peak as well as the higher aggregate peak.

Another aspect to this work was the difference in fragmentation profiles of the "IgG2-A" and "IgG2-B" forms. At 4°C, there were minimal and comparable levels of clips for both isoforms. However, there was a distinct increase in the clip peaks for the "IgG2-A" form, which also correlated with the increase in the higher-order aggregate peak for this isoform. Moreover, when Arrhenius plots of aggregation and clipping tendencies were generated,

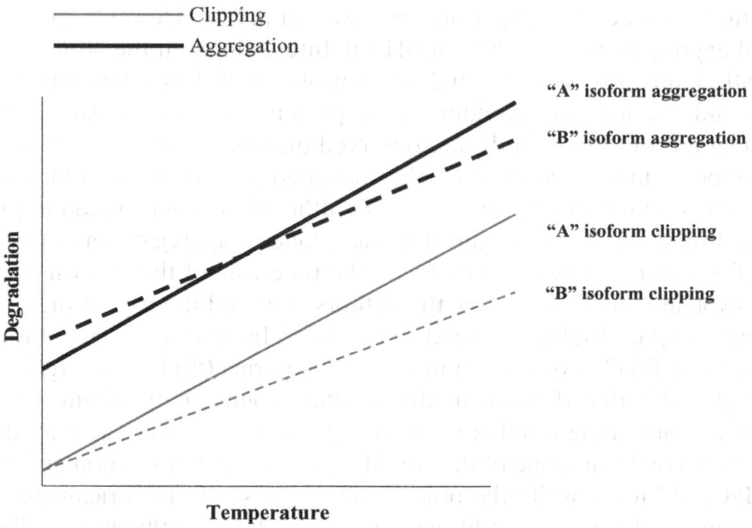

Figure 9.9. Model to explain temperature-dependent effects on aggregation of the IgG2-A and IgG2-B isoforms. At lower temperatures, the IgG2-A isoform has lower aggregation than IgG2-B (thick lines). However, the clipping rate for IgG2-A is higher than IgG2-B as a function of temperature (thin lines). Due to participation of clips in aggregation, this increases the rate of aggregation for the IgG2-A form (thick solid line) compared with that for IgG2-B (thick dashed line). The net result is that IgG2-A aggregates to a greater extent than IgG2-B at higher temperatures.

there was a distinct deviation from linearity for the aggregation peaks. However, the Arrhenius plots for the clips were linear, with much higher rates at 37°C compared with 4°C. This showed that clipping followed Arrhenius kinetics, and further suggested that possibly that the deviation from linearity in the Arrhenius plots for the aggregate may occur due to a clipping reaction in addition to aggregation. This was also supported by the linear pH dependence for aggregation at 30 mg/mL concentration at both 4 and 37°C, presumably because the clips did not associate to elute in the high-molecular-weight region, but rather appeared as post peaks in SEC as expected. Figure 9.9 is a schematic representation of this phenomenon: at low temperatures, the IgG2-B isoform has more aggregate than the IgG2-A isoform. With increase in temperature, the A isoform undergoes greater fragmentation that in turn participates in aggregate formation. As a result, at higher temperatures, the IgG2-A isoform has greater aggregate composition than the IgG2-B isoform.

9.3.4 Discussion and Lessons Learned

What were the lessons behind this study? One conclusion was to caution against assuming that the trends in aggregation under accelerated conditions are always representative of those at lower temperature. The main contributor to the non-linearity in the Arrhenius plots was that the IgG2 antibody was not homogenous, but rather a composite of different disulfide-bonded isoforms with different stability properties. In addition, a fragmentation phenomenon was observed, that potentially caused the aggregation phenomenon to not follow linear Arrhenius kinetics. The differences in clipping between the two isoforms could be due to the more flexible nature of the "IgG2-A" isoform as assessed by the disulfide structure around the hinge region. Clearly, more work would need to be carried out to understand why the aggregation reaction had this unique temperature dependence, as well as to understand the difference in the stability behavior between the two isoforms. Nonetheless, this study illustrates the complexity of antibody aggregation, which would be useful to remember for future formulation studies.

Even when degradation follows a linear relation with temperature, the effects of excipients may be temperature dependent. This means that even with a linear Arrhenius relationship in the absence of excipient, the temperature dependence must also be evaluated in the presence of excipient. For example, excipients that reduce hydrophobic protein–protein interaction may have stronger effects at higher temperature, wherein such excipients would interact more strongly with the proteins and prevent aggregation. The effects of osmolytes may have an opposite consequence of enhancing hydrophobic association at high temperature. Their effects on protein aggregation may diminish at normal low-temperature storage conditions. This study demonstrates that linear Arrhenius kinetics may not always be the case for aggregation. It should also be kept in mind that while multiple contributing factors may exist for a given degradation route, one factor may dominate, and thus make the process appear to follow linear Arrhenius kinetics. It is also possible that the relative contribution of different underlying elements may change with temperature.

9.4 CASE STUDY 3: AGGREGATION IN THE FROZEN STATE: THE ROLE OF EXCIPIENT CRYSTALLIZATION[42]

9.4.1 Background

The third case study focuses on protein stability in the frozen state. Proteins are commonly stored frozen to minimize degradation pathways, such as aggregation and covalent modifications, that may be otherwise active in the liquid state. However, it is well documented that the process of freezing itself can be damaging to proteins. Stresses that arise during the process of freeze/thaw include unfolding at the ice–water interface, pH changes due to selective crystallization of buffer components, and an increase in effective protein concentration due to ice formation. To minimize protein denaturation, cryoprotectants such as sucrose and sorbitol have historically been included in the formulation.[12,13,32,42–44]

For proteins in the liquid state, evidence has been gathered over the years to show that these cryoprotectants stabilize proteins via preferential exclusion, thus providing a relatively inert local environment to maintain the native structure of the protein.[10–12,21,26,28] The mechanism of their effects on protein solutions has been described in case 1 (the KGF study). Briefly, the structure-stabilizing cosolvents are preferentially excluded from the protein surface, which leads to a greater free energy difference between the native and unfolded state, as shown in Fig. 9.8. Conversely, organic solvents, such as DMSO and alcohol, are also excluded from the native protein, leading to destabilization of the native protein (thick arrow in Fig. 9.7). These solvents bind more favorably to the unfolded state, leading to the stabilization of the unfolded state[31] (thin arrow). However, protein destabilizing organic solvents, which are preferentially excluded from the native structure, often cause cryostabilization.[32] This can be readily explained from the temperature-dependent effects of excipients, leading to destabilization of the unfolded state (see thick dotted arrow in Fig. 9.7). This case study discusses the work of Piedmonte et al.,[42] titled "Sorbitol crystallization can lead to protein aggregation in the frozen state." This study revolves around the puzzling observation that the model protein studied by the authors aggregated in frozen formulations in spite of using sorbitol, which is a well-known cryoprotectant.

9.4.2 Model Protein and Conditions

The authors chose to employ an N-terminal Fc fusion protein that had a molecular weight of about 60 kDa. The protein was expressed in *E. coli* and therefore did not undergo posttranslational modifications such as glycosylation. It was maintained in 10 mM sodium acetate, pH 4.0 with tonifying levels of the excipient sorbitol, at a concentration of 2 mg/mL, in 3-cc glass vials. Aggregation was analyzed by SEC, while sub-ambient DSC was used to monitor thermal transitions.

9.4.3 Results and Interpretation

The authors observed that aggregation of this Fc fusion protein did not occur during freeze/thaw cycles or in the liquid state, but rather manifested during long-term storage at –30°C (in this case, 32 weeks and beyond). To make matters more complicated, there was very little aggregation at –70 and –20°C compared with the –30°C storage temperature. This phenomenon appeared to be restricted to the sorbitol excipient, as this effect did not manifest itself in sucrose, though several levels of excipient were tested. The aggregation of this protein also increased when there was a lower level of sorbitol (100 mM) as opposed to an isotonic (300 mM) level of sorbitol. One simple explanation could have been that this was related to the glass transition temperature (Tg′) of sorbitol versus sucrose. However, the measured Tg′ of sucrose was –32°C and that for sorbitol was –45°C. Thus, both –20°C as well as –30°C were above the Tg′ for sorbitol, yet protein aggregation only occurred at the –30°C storage temperature.

To investigate the underlying phenomenon, the authors incubated the protein in the sorbitol formulation at –20°C as well as –30°C, and as a function of time, recorded sub-ambient DSC thermograms. To facilitate sample handling and avoid unnecessary thawing and transfer, the authors cleverly incubated the protein in the DSC pans as opposed to glass vials, and recorded DSC melting transitions (the samples were first cooled from –20 or –30°C to –60°C, then warmed to +15°C, and the –60 to +15°C region was reported). For the samples stored at –30°C, a melting endotherm was observed at –20°C between the 5-day and 4-week time points, but not beyond. Past 4 weeks, the –20°C endotherm was replaced by another endotherm at –8°C, which persisted throughout the remaining time course. In contrast, none of these endotherms were observed in the samples stored at –20°C. In additional experiments, this –8°C endotherm increased with increasing sorbitol concentration. Importantly, none of the –30°C pans showed the Tg′ transition for sorbitol that was present when liquid samples were used as starting material.

The interpretation that most consistently fits these data, which is also the one suggested by the authors, is that the –30°C incubation results in partitioning of sorbitol from an amorphous to a crystalline phase. Not only was this supported by the endothermic transitions at –20 and –8°C, but also the lack of a Tg′ transition, which is typical of amorphous solutes and the sorbitol concentration-dependent increase in these endothermic events. The two endothermic events were likely two polymorphic forms of sorbitol, with the –8°C form being the more stable one, and being converted from the –20°C form. Presumably, for the samples stored at –20°C, the first polymorph was "melted" and thus prevented from forming, and consequently the second polymorph, dependent on the first one, never formed—thus resulting in sorbitol always being in the amorphous phase.

9.4.4 Discussion and Lessons Learned

The consequences for protein aggregation can then be explained by lack of cryoprotection at –30°C as sorbitol converts into a crystalline form, thus separating from the amorphous phase where the protein resides. Once stabilization provided by sorbitol was lowered, the protein may have been subjected to the various denaturing forces as mentioned earlier, such as denaturation at the ice/water interface, which would lead to the observed aggregation phenomenon. In addition to providing mechanistic evidence into why this aggregation may have occurred, the authors also offer solutions to this problem. They found that in the presence of protein, compared with buffer, the extent of sorbitol crystallization was suppressed. At higher concentrations of protein, the crystallization of sorbitol may be negligible, thus rendering the protein protected from aggregation. Of course, another option might be to use another excipient instead of sorbitol, such as sucrose.

The above sorbitol case is apparently opposite to the behavior of DMSO described in the first case study. DMSO typically destabilizes proteins at high temperature (as in Fig. 9.3) but stabilizes proteins during freezing.[31] In this case, sorbitol stabilizes proteins at higher temperature and in the liquid state, but destabilizes this model Fc fusion protein during freezing. However, the mechanism of this temperature dependence for sorbitol is not the same as DMSO. It appears that sorbitol should have stabilized the protein during freezing, if its concentration would have been maintained in the frozen state, according to its stabilizing property as observed at high temperature. Instead, its crystallization resulted in decreased effective concentration of sorbitol to confer protein stabilization against freezing stress, and thus the stabilizing effect of sorbitol decreased with lower availability in the freeze concentrate, which is the phase where the protein resides in the frozen state.

While this chapter offers evidence to explain why protein aggregation occurred—which was due to the lowered level of sorbitol in the protein rich amorphous phase—a deeper question would be why sorbitol stabilized the protein against aggregation in the first place. As pointed out earlier, excipients such as sorbitol stabilize proteins in the liquid state by the preferential exclusion mechanism (see Figs. 9.4 and 9.5). Based on their observations, Carpenter and Crowe proposed that this was the same mechanism of stabilization present in the frozen state.[45] One aspect to consider is the effective levels of tonifying excipient, protein, and water in the freeze/concentrate in the frozen state. During freezing, the majority of water separates into the ice phase, which is pure, that is, does not contain protein or excipient. The "amorphous" or non-frozen phase is the one that contains residual water, excipient, and protein. Thus, it is reasonable to assume that the effective concentration of protein in this phase would be orders of magnitude greater than that in the liquid phase. This, coupled with the greater opportunities for the protein to denature due to contacts with the ice surface, renders the freeze/thaw process a suitable environment for aggregation. In fact, recent observations with lactate dehydrogenase as a model protein suggest that ice formation, rather than the composition of the freeze concentrate, played a crucial role in protein denaturation in the frozen state.[46] If stabilization in the freeze/concentrate were due to preferential exclusion, it would stand to reason that since the interaction parameter is directly dependent on excipient concentration, it therefore becomes extremely important that the stabilizing excipient stays in the same amorphous phase as that of the protein to confer cryoprotection by this mechanism. Even small levels of crystallization may have a negative impact on protein stability.

However, it should be noted that the idea of preferential exclusion requires the presence of water, which becomes increasingly scarce as we go to the amorphous freeze concentrate with most of the water crystallized as ice. Further, the observation in this case study that increasing protein concentration suppressed sorbitol crystallization suggests that the chemical potential for sorbitol in the amorphous phase was lowered, thus making crystallization less favorable. Higher protein concentration would thus also raise the Tg′ of the protein–sorbitol mixture. One possibility is that sorbitol binds to native proteins in the freeze concentrate, which stabilizes proteins and makes it more difficult to crystallize sorbitol as the protein concentration is raised. However, since the freeze concentrate still contains residual water, it is not clear how sorbitol would bind to proteins in the presence of this water. The hypothesis that preferential exclusion is the mechanism of stabilization even in the frozen state[32,45] is based on the observation that widely different compounds, which are preferentially excluded, stabilize proteins in frozen state. Another possibility is that increasing protein concentration "dilutes" effective sorbitol concentration and diminishes the intermolecular contacts that are needed to pack into crystals. Since it is difficult to analyze the freeze concentrate, some of these questions still await greater clarity.

9.5 CASE STUDIES 4 AND 5: AGGREGATION IN THE LYOPHILIZED STATE: ROLE OF RESIDUAL MOISTURE AND MECHANISMS OF EXCIPIENT STABILIZATION[47,48]

9.5.1 Background

Protein therapeutics are often lyophilized to achieve an acceptable shelf life. The process of lyophilization is geared toward removing most of the water in the drug product, with the final product typically containing less than 1% moisture. This residual moisture is water that is presumably bound to the protein and is therefore needed for its native conformation. There are numerous observations that proteins are hydrated at different degrees and such hydration is essential for the protein's function. For example, enzymes regain activity when hydrated at 0.2–0.3 g water/g protein.[49] However, it is not clear what the essential level of hydration should be for the most stable structure. The presence of bulk water causes many degradation phenomena, such as residue fragmentation and isomerization events.[50] Protein aggregation in solution, which relies on diffusive encounters between smaller units, can in principle also be minimized by removal of bulk water.[51] Thus, the idea behind lyophilization is to overcome these various potential instabilities by keeping the protein in a dry state.

There are four key steps in lyophilization, namely freezing, annealing, primary drying, and secondary drying.[51] Freezing, as the name suggests, converts bulk water into ice, while all the protein and solutes typically remain in a single amorphous phase. After freezing, an additional annealing step can be included in the process, where the system is held for a period of time at a temperature between the ice melt and the glass transition temperature of the freeze concentrate.[52] This step helps to crystallize out bulking agents such as glycine and mannitol, and allows larger ice crystals to grow via a process called Ostwald ripening.[51–53] The increase in the size of ice crystals can help create more pores by which water vapor can diffuse out, thereby potentially decreasing the primary drying time. During the primary drying step, the temperature is held 2–5°C below the glass transition temperature (Tg′) of the freeze/concentrate, and there is a vacuum applied, which shifts the ice/water/vapor equilibrium in the pressure/temperature phase diagram in such a way that ice is removed by conversion into water vapor via sublimation.[51,53] After the primary drying process, the shelf temperature is ramped up, and most of the remaining water is removed by desorption, in a process called secondary drying.[51,53]

However, the process of lyophilization can destabilize proteins, causing them to aggregate.[54] As mentioned in the earlier case study, the process of freezing can induce damage to proteins. As elaborated in the first case study, polyols, such as sucrose and other carbohydrate excipients, have long been known to stabilize proteins during freeze/thaw stress by a mechanism of preferential exclusion, which is conducive to proper folding and keeping the protein away from the destabilizing effects of an ice–water interface.[32,43,51,54] These excipients typically remain in the amorphous phase and create a high-viscosity environment to minimize protein–protein contacts that would lead to aggregation. Further, during the process of secondary drying, removal of bound water could potentially disrupt hydrogen bonds that stabilize the native structure, causing unfolding due to exposure of the hydrophobic core. To substantiate this hypothesis, it has been proposed that carbohydrate and polyol excipients may stabilize proteins via a potential hydrogen bond replacement mechanism where the –OH group from water is substituted by the –OH group from the polyol.[55,56]

In the 1990s, Andya et al. showed that aggregation was indeed a concern for monoclonal antibodies in the lyophilized state, and that sucrose and trehalose were effective at reducing it.[57] The work of Cleland et al. revealed that the ratio of stabilizer to protein was a critical parameter to consider for the stabilizing effects of these excipients.[58] In this section, we consider two case studies from the work of Shire and colleagues.[47,48] The first study is from Breen et al. where the effect of moisture on protein aggregation in the solid state was investigated.[47] The level of moisture in the sample is a critical stability parameter as the glass transition temperature (Tg) generally decreases as the moisture content increases.[59] In other words, as the moisture content increases, it becomes increasingly facile to go from a viscous state, which would prevent molecular mobility and intermolecular collisions, to a more "rubbery" state, which would allow molecular motion to occur. While the effect of moisture on other model protein systems was studied,[59] this was one of the first studies to determine the effect of moisture on the stability of a recombinant human monoclonal antibody.

9.5.2 Model Protein and Conditions (Case Study 4)

These authors prepared lyophilized antibody samples containing moisture levels that varied from 1% to 8%, by either subjecting them to high relative humidity (64% for 3h or 52% for 24h), opening stoppered vials during the cold lyophilization cycle step and then restoppering them, or by a shortened lyophilization cycle. All vials contained a fixed formulation of 40 mg/mL antibody, 85 mM sucrose, 5 mM histidine, pH 6.0, and 0.01% polysorbate 20. These vials were then stored under accelerated temperature conditions (15, 30, 40, and 50°C), as well as 5°C. They were monitored after 6 and 12 months by a variety of stability-indicating techniques: SEC, sodium dodecyl sulfate–polyacrylamide gel electrophoresis (SDS-PAGE) and turbidity to quantify aggregate content, hydrophobic interaction chromatography to assess aspartate isomerization, and Fourier transform infrared (FTIR) spectroscopy to monitor secondary structural changes. As alluded to earlier, there is a large change in molecular mobility at the glass transition temperature. An aspect that is considered is the temperature dependence of the molecular mobility. Substances that show relatively small changes in molecular mobility with temperature, especially around the Tg, are said to be "strong" glass formers. It turns out that proteins are strong glass-forming substances, and consequently, Tg values in protein–sugar systems are difficult to measure using conventional DSC. In this chapter, the authors pointed this out and measured protein Tg values using modulated DSC.

9.5.3 Results and Interpretation (Case Study 4)

Using DSC, the authors confirmed that the Tg of the sample decreased with increasing moisture content. All the samples exhibited a robust cake structure in spite of prolonged storage above the Tg, especially for the high-moisture samples. This could be due to the small change in heat capacity at the glass transition temperature, indicating a strong glass. One of the first signs that moisture impacted protein aggregation was observed in the turbidity measurements at temperatures ≥15°C. All samples that had a moisture level greater than 1% were turbid to varying extents upon reconstitution, indicating formation of insoluble protein aggregates, with the 8% moisture sample showing the highest aggregation at temperatures ≥30°C, which was consistent with the glass transition of this sample being 25°C. These observations were in agreement with the SEC observations: there was greater soluble aggregate formation in the different moisture-containing samples only when stored above the Tg for that sample. For instance, the 8% moisture sample with a Tg of 25°C had a higher soluble aggregate level than the others at ≥30°C, while the 5% moisture sample with a Tg of 40°C had a higher level of aggregate than the 2% moisture sample only at the 50°C temperature, that is, when the temperature was above its Tg. However, the same samples did not differ as appreciably when stored at temperatures below their respective Tg's.

Interestingly, the 2% and 3% moisture samples had lower aggregate level than their 1% moisture counterparts when these samples were stored below the Tg. The authors explain this by theoretically estimating the total number of "water-binding sites" on the protein. Their calculations yielded a number of 500 water-binding sites on the protein, and the ratio of protein:sugar was less than 1:500 in the formulation that they used. They therefore hypothesized that the residual water in the 1% water samples was perhaps not enough to satisfy these unsatisfied binding sites, which was possible in the 2% and 3% samples. To further support this, they also found that increasing the protein:sugar ratio to 1:500 increased the stability of the protein toward aggregation, but increasing the ratio any further did not.

The authors also examined the secondary structure of this antibody using FTIR. FTIR is a reliable technique to examine the secondary structure of proteins in the dried state. For FTIR spectroscopy, one of its major advantages is that the technique is not limited by the physical states of the analyzed samples.[59–63] Proteins in essentially all physical states such as liquid, gel, colloidal, and solid state can be analyzed by using FTIR spectroscopy. Amide I/I′ bands in FTIR spectra of proteins are particularly sensitive to different types of β-strand structures. Interestingly, they did not observe any significant changes in the amide I pattern in the prelyophilized sample compared with the samples that in the study were either the most stable (2% at 5°C for 9 months) or least stable (8% at 50°C for 9 months). At first glance, these data seem to suggest that protein aggregation in the 8% moisture samples at 50°C may not involve structural changes. However, it should be kept in mind that the total aggregation level was less than 10%—thus a small structural change in 10% of the sample undergoing aggregation would be masked by 90% of the sample that did not undergo aggregation and therefore did not have any structural difference. Further, it may be possible that although there were marginal changes in secondary structure, there could be substantial changes in the tertiary structure. Such changes in tertiary structure are often sufficient to cause extensive aggregation and precipitation of proteins.[64] Therefore, when the structure changes of antibodies are analyzed by FTIR alone, the absence of changes does not necessarily mean the absence of tertiary structure changes. Tertiary structural changes in the solid state may be picked up by Raman spectroscopy[65] or solid-state fluorescence.[66]

9.5.4 Model Protein and Conditions (Case Study 5)

Similar to the previous case, the authors utilized a recombinant human monoclonal antibody manufactured by Genentech for their studies. The antibody was formulated by dialysis either into water (without excipient) or a pH 6 buffer containing sucrose or trehalose at different excipient to protein ratios. To lyophilize the vials, primary drying was carried out at –30°C at a pressure of 100 mTorr, followed by two hold steps at –20°C and at –10°C while ramping up to the secondary drying phase, which was performed at 5°C for 10 h, also at 100 mTorr. Residual moisture was estimated using the Karl Fischer titration method. Soluble aggregates were quantitated by analytical SEC, while preparative SEC was employed to fraction collect the aggregates. An online light scattering system (SEC-LS) was used to determine average solution molecular weights of the various SEC peaks. Secondary structural changes in both liquid and lyophilized samples were monitored mainly by FTIR, though CD was also employed for liquid samples. To study intermolecular disulfide linkages, SDS-PAGE was carried out, while free thiol was estimated using Ellman's reagent. Conventional DSC was utilized to estimate Tg in the various formulations. Samples were incubated at 30°C for 1 year to generate aggregates.

9.5.5 Results and Interpretation (Case Study 5)

This study from Andya et al.[48] further explores the mechanism of stabilization of aggregation in the lyophilized state by the carbohydrate excipient. In this case, all formulations contained a fixed 3% moisture level, which from the previous study was shown to be optimal for storage temperature less than the Tg. To compare the effect of excipient, these authors lyophilized the antibody under consideration in an excipient-free formulation. As expected, the absence of excipient led to sizeable increases in aggregate formation as assessed by SEC as well as SDS-PAGE. Using online light scattering measurements, the two high-molecular-weight peaks observed on SEC were characterized as dimer and trimer. In the presence of 500:1 molar ratio of sucrose : protein, the trimer peak was absent in the reconstituted sample by SEC as well as by SDS-PAGE, thus showing the stabilization afforded by sucrose. The authors also monitored the kinetics of aggregate formation at various excipient to protein ratios, compared with the excipient-free formulation. A rapid decline in the rate of aggregation was seen when the excipient : protein molar ratio was increased from 0 to 500. Beyond the 500:1 excipient : protein ratio, the rate of aggregation did decrease further, but much slower. Thus, the 500:1 excipient : protein molar ratio appeared to be an important parameter for aggregate stabilization.

To determine the role of disulfides in the oligomerization of the monoclonal antibody, samples were analyzed by SDS-PAGE under reducing conditions. It was found that both excipient-free as well as the sucrose-containing samples reduced to heavy and light chain, leaving no detectable high-molecular-weight species. In contrast, under non-reducing conditions, the excipient-free sample had a significant proportion of high-molecular-weight species, which was absent in the sucrose-containing sample. This suggested that the aggregates formed under these lyophilization conditions were predominantly covalently linked disulfides. To further explore the role of cysteines in the aggregation process, the monomer, dimer, and trimer peaks were fractionated and estimated for free thiol content using Ellman's reagent and monitoring absorbance at 412 nm. Interestingly, while the monomer sample had a free thiol : protein ratio of 1.0, in both dimer and trimer samples, this ratio was less than 0.1%, indicating that the free thiols in the monomer played a role in the oligomerization process, presumably by their involvement in disulfide bond formation.

To determine if lyophilization affected the structure of their antibody, Andya and coworkers compared the second derivative FTIR spectrum of their antibody to that of the same protein prior to lyophilization.[48] FTIR spectra in this case revealed that in the absence of stabilizing excipients such as sucrose or trehalose, the lyophilized human monoclonal antibody contained significantly different secondary structures from the prelyophilized form. The changes in secondary structures involved mostly the unfolding of the native β-barrel to unordered structures. However, the same antibody maintained mostly native-like structures when sucrose or trehalose was present during lyophilization. The authors thus concluded that excessive dehydration through lyophilization led to unfolding of antibody structures, and that the carbohydrates such as sucrose and trehalose conferred a stabilizing effect on the secondary structures of the antibody, presumably by offering complete coverage of surface hydrogen bonding sites.

Andya and coworkers also employed FTIR spectroscopy to evaluate the long-term stability of lyophilized human monoclonal antibody after 1 year of storage.[48] The presence of sucrose also conferred a stabilizing effect on the structure of the antibody during long-term storage. Furthermore, both FTIR and CD spectra showed that in the presence of sucrose, the antibody maintained its structure even after reconstitution. These studies thus showed the usefulness of FTIR spectroscopy in the evaluation of secondary structures of monoclonal antibodies in the solid state and the effect of the lyophilization process on the structures of antibodies.

9.5.6 Discussion and Lessons Learned

These case studies highlighted two ways in which protein stability in the lyophilized state could be modulated: by changing the moisture content and by varying the excipient composition. The authors inferred that these data provided further evidence for stabilization of carbohydrate excipients by a "water substitution" hypothesis, an extension of their suggestion from the previous work. In these two pieces of work, the authors linked several lines of evidence to come up with this model: the rapid decrease in aggregate rate up to the 500:1 excipient:protein ratio, the calculated water substitution sites in the protein being around 500–550, and the increased stability seen in 2% and 3% moisture samples compared with the 1% sample (the latter being postulated to not satisfy all the water-binding sites). The effect of storage temperature relative to the glass transition temperature had been studied prior to this work, but for non-antibody proteins.[67–69] In some cases, storage above the Tg resulted in greater protein degradation,[67,68] while in the case of Mazzobre et al., there was no significant change in enzyme activity, a measure of protein stability, with storage above Tg. Thus, this chapter was the first to show that for the antibody studied, storage above Tg correlated with enhanced degradation. Without a doubt, the specific degradation pathway and extent of degradation will vary for the specific antibody at hand, along with whether the storage temperature was slightly above the Tg versus significantly above the Tg. More recent work from Michael Pikal's lab (Tang and Pikal[70]) confirmed that the samples with the lowest moisture level may not be the most optimum in terms of protein stability to aggregation, even though the measured relaxation times suggest much higher molecular mobility at higher moisture levels. However, the observation that the addition of a small amount of sorbitol to either sucrose or trehalose formulations increased stability to aggregation in spite of changes in structure in the sorbitol/sucrose formulation but no significant structural changes in the sorbitol/trehalose formulation suggests that the complexities of glass dynamics are yet to be fully explored.

9.6 CASE STUDY 6: PROTEIN PARTICULATION DUE TO NUCLEATION BY FOREIGN MATERIAL IN FILL/FINISH MANUFACTURING OPERATIONS[72]

9.6.1 Background

In addition to forming soluble aggregates, as seen in the previous case studies, protein solutions are also prone to forming insoluble aggregates and particulates. Regulatory authorities have voiced concerns over the presence of protein particles due to a possible potential to elicit an immunogenic response (Carpenter et al.[72]). If the level of particulation is severe, they may also have the potential to clog pores during IV administration and decrease the bioavailability of the therapeutic. Measuring, understanding, and mitigating protein particles have now become an area of intense activity in the pharmaceutical sciences.

9.6.2 Model Protein and Conditions

The model protein chosen here was a monoclonal IgG antibody manufactured by Biogen Idec, with a pI of 8.0. This product was formulated at 20 mg/mL in a pH 6 isotonic buffer with 0.02% polysorbate 80 and filled up to 5.3 mL using a National Instrument Filmatic pump. This pump had its piston rod and cylinder made of stainless steel, while the piston head was composed of polyvinylidene fluoride and the O-ring was made of silicone. Protein particles were obtained by pumping protein through this pump (FUS-10) at a line speed of 39 cycles per minute. The solution was incubated for 10 h at room temperature prior to analysis. Particles were detected using a Beckman Coulter Counter. Soluble aggregates and total soluble protein were quantitated using SEC. Elemental analysis was performed using an inductively coupled plasma optical emission spectrometer. For model particle testing and heterogenous nucleation experiments, nanoparticles were obtained from Argonide Corporation with an average particle size of 100 nm. Secondary structural measurements were made using FTIR.

9.6.3 Results and Interpretation

This case study investigates the formation of particles that were observed in a clinical manufacturing facility during filling of an IgG monoclonal antibody.[71] This problem was traced to the use of a positive displacement pump that was used in the filling operation. The importance of this problem was underscored by the fact that the last filtration step occurred when the protein solution entered the holding unit prior to entering the filling apparatus, and there were no mitigation steps for particles formed during the filling operation. The questions raised in this study revolved around the reason for the formation of particles during this filling operation, coupled with characterizing the size distribution and structural features of the observed particles.

The authors formulated the hypothesis that nano- and microparticles shed from the contact surfaces of the filling pump were responsible for nucleating the association of protein molecules to form detectable particles. The basis for this hypothesis was that in a given filling operation, every vial contained a few visible particles. Further, changing the primary container type did not change the observation of particles, suggesting that it was not due to the components of the glass. To test this hypothesis, the authors isolated these particles and analyzed them by FTIR. The presence of a strong amide I band at 1639 cm^{-1} indicated that the particles were composed at least partially of protein. However, the FTIR spectrum of the particles was slightly altered compared with that of the native protein, with an extra band at 1662 cm^{-1} indicative of a β-turn and changes in the intensity and profile of the main 1639 cm^{-1} band. Yet, the particle spectrum resembled the native spectrum much more than that obtained for a fully heat-denatured aggregate of the same protein. Therefore, FTIR data suggested that the formation of particles could result from perturbation to the structure of the studied antibody.

The second observation that pointed to the direct participation of protein molecules in particle formation was that there was a 20-fold lower level of particles during pumping in the same buffer without protein. This suggested that protein was definitely involved in particle formation, but also pointed out a low, basal level of particle formation in the absence of protein. To determine why this was the case, the authors performed an elemental analysis of the isolated particles. The results showed that in addition to protein, these particles contained detectable quantities of silicon, iron, nickel, and chromium, which were not detected in the buffer prior to pumping. To then directly test for the ability of stainless steel particles to nucleate protein particulation, the authors performed an experiment where they mixed a protein solution that did not contain particles with a solution containing stainless steel particles sized at 0.1 μm. In a matter of a few hours, there was a large increase in particles in the 1.5- to 3-μm range. Concomitantly, there was also a ~4% drop in the level of soluble IgG in solution. Interestingly, the number of particles found in this larger range (1.5–3 μm, compared with the 0.1-μm size of the stainless steel bead) was similar in number (30,000) to the starting level of stainless steel beads, suggesting that the growth in particle size was due to several layers of IgG adsorbing on to these particles. Taken together, these observations suggested a shedding of stainless steel from the pump, which acted as a nucleus to drive protein association to form particles.

9.6.4 Discussion and Lessons Learned

There were two other interesting angles to this work. One was the size distribution of these particles. The authors assessed particle size using two instruments: a Beckman Z1 coulter counter and a Beckman LS230 particle sizer. In the case of the Beckman Z1 counter, these particles were maximally present in the lower size range of the instrument, namely 1.5–3 μm, and there were minimal particles observed beyond 10 μm. The Beckman LS 230, which has a lower limit of 0.04 μm, showed that 99% of the particles were in the 0.25- to 0.95-μm range. Thus, these protein particles were very small in size, compared with foreign particles that are typically measured in the ≥10- and ≥25-μm ranges. However, given this size distribution, an outstanding question from this case study that remains to be answered is how particles that were primarily under 3 μm could be visually detected. One possibility is that these particles coalesced together to appear visually detectable while dissociating during the quantitative measurement. It is also possible that light scattering effects might have rendered them barely visible.

The second feature of interest was that in the SEC profile, the authors did not observe any changes in the high-molecular-weight region for solutions that contained particles compared with solutions that did not. In this respect, the behavior of the IgG was similar to that observed for stirring-induced particle formation by Kiese et al.[73]

The behavior seen in this case study can perhaps be explained by protein aggregation due to adsorption and denaturation of the molecule on the surface of the stainless steel nanoparticle. Presumably, protein molecules would undergo a conformational change due to surface denaturation on the metal particle, rendering them aggregation prone. There are two possible scenarios from this point on. One possibility could be that proteins desorbed from the stainless steel surface in this altered conformation and promoted protein aggregation with other similar conformers, thus "recycling" the stainless steel particle in a heterogeneous nucleation mechanism. The other possibility is that once the proteins adsorbed, they stayed on the surface and induced other protein molecules to adsorb, thereby increasing the size of the resulting particle. Further work would need to be performed, such as forensic analysis to detect for the presence of stainless steel in the particulates, to potentially discriminate these two different mechanisms. In sum, this case study illustrated how the manufacturing process can play a role in generating protein particles, and the steps taken to understand and mitigate particles.

9.7 OVERALL DISCUSSION

In summary, these case studies highlighted that protein aggregation could proceed through multiple possible routes. Each case had a different stress-causing aggregation. The first case study characterized the role of osmolytes, such as polyols in stabilizing the aggregation of KGF upon thermal denaturation, and elucidated the "preferential exclusion" mechanism. The second case study showed the differential temperature dependencies toward aggregation that could be adopted by IgG2 isoforms and clip-mediated aggregation that could cause non-linearity in Arrhenius plots. In the third case study, excipient crystallization in the frozen state was responsible for aggregation while in the fourth and fifth case studies, the stress of lyophilization and the excipients involved caused aggregation. Finally, the sixth case study showed that stainless steel leaching during manufacturing fill and finish processes could form protein-containing particles.

9.7.1 Are Aggregates Created Equal?

The previous case studies showed that aggregation can occur under a variety of conditions. One question that arises is if the nature of the aggregates is the same or different as a result of different stresses on the protein. For example, are particles generated by agitation stress on the same pathway as accelerated temperature-induced soluble aggregates? To understand this, we present a case study with a monoclonal IgG1 antibody that was subject to two different types of stress: heat and agitation. For heat stress, the antibody was incubated at pH 5 and 45°C for a period of 3 months. The same antibody in a pH 8 solution was also placed inside a room temperature incubator that moved in a rolling fashion to produce agitation.

Figure 9.10 compares the second derivative FTIR spectrum of human IgG1 monoclonal antibody in native condition, that is, pH 5.0 buffer, with those of the antibody after undergoing heat and agitation stress treatment. Spectrum A in Fig. 9.10 is the second derivative FTIR spectrum of the IgG1 antibody in buffer at pH 5, panel B shows the FTIR spectrum of the IgG1 antibody in the supernatant after agitation stress in buffer at pH 8, spectrum C is the FTIR spectrum of the IgG1 soluble aggregate induced by heat treatment and purified by SEC, spectrum D shows the FTIR spectrum of the corresponding insoluble aggregate form of the IgG1 induced by agitation stress, and spectrum E shows the FTIR spectrum of insoluble aggregates that were also formed after the heat treatment. Insoluble antibody aggregates were collected by centrifugation and the corresponding FTIR spectra were collected by using a single-bounce attenuated total reflectance (ATR) device mounted on a Bomem FTIR spectrometer. Spectrum A in Fig. 9.10 exhibits a major amide I band at $1637\,cm^{-1}$, and weak amide I bands at 1660, 1675, and $1689\,cm^{-1}$, which are characteristic of the native β-barrel fold in the IgG1 antibody at pH 5. The FTIR spectrum (spectrum B) of soluble antibody remaining in the

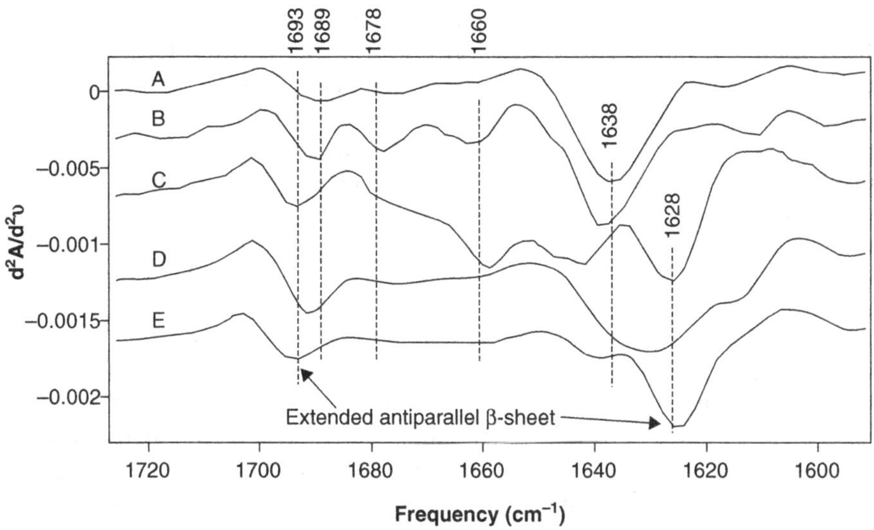

Figure 9.10. Second derivative FTIR spectra of human IgG1 monoclonal antibody. Spectrum A: IgG1 in solution at pH 5; spectrum B: supernatant of IgG1 in solution at pH 8 after shear stress; spectrum C: heat-induced soluble aggregate of IgG1; spectrum D: shear stress-induced insoluble aggregate of IgG1; spectrum E: heat-induced insoluble aggregate of IgG1.

supernatant appears similar to that of the native antibody, but with more discrete amide I bands associated with β-turns at 1660 and 1678 cm^{-1}. This indicates that the structure of the soluble form of IgG1 still maintains most of the native features, despite the fact that a significant amount of the IgG1 forms insoluble aggregates when agitation stress is applied. In contrast to the spectrum of the soluble form of IgG1 after agitation stress, the spectrum (spectrum D) of the insoluble IgG1 agitation-induced aggregates exhibits drastically different spectral features from those of the native IgG1 and the soluble IgG1 aggregates induced by agitation force. The amide I band centered at 1630 cm^{-1} is very broad, in comparison to other amide I bands. This indicates that the secondary structures of the insoluble IgG1 aggregates induced by agitation contain many different types of β-sheets structures, or that the insoluble aggregates contain many heterogeneous populations of aggregates with each population containing different types of β-sheet.

The FTIR spectrum (spectrum C) of the isolated soluble aggregates induced by heat exhibits very different spectral features from those in the spectrum (spectrum E) of the insoluble IgG1 aggregates isolated after heat stress. In the spectrum (spectrum E) of the insoluble aggregates induced by heat, the major amide I band is centered at 1628 cm^{-1}, together with a minor band at 1693 cm^{-1}, which are characteristic of extended antiparallel β-sheet associated with denatured proteins. The amide I band at 1638 cm^{-1} in spectrum E is greatly diminished, in comparison to that of the soluble IgG1 (spectrum B) or native form (spectrum A). This indicates that the structure of the insoluble IgG1 aggregates induced by heat lacks most of the native β-barrel fold and contains mostly extended antiparallel β-sheet. However, the spectrum (spectrum C) of the soluble IgG1 aggregates shows significant intensities in the region 1640–1670 cm^{-1}, with discrete amide I bands at 1645 and 1660 cm^{-1}. The distinct spectral features suggest that the soluble IgG1 aggregates induced by heat contain significant amount of irregular and loop structures, in addition to the antiparallel β-sheet. It is likely that the irregular and loop structures are eventually converted to extended antiparallel β-sheet when the soluble aggregates form insoluble aggregates during heat exposure.

The above case study not only showed, for the IgG1 at hand, the secondary structural transitions that were afforded during aggregate formation, but also revealed that the aggregate structures formed by heat treatment differed significantly from those formed by agitation. Thus, a given antibody in principle may have different pathways of aggregation depending on the stress applied, and this in turn may have important consequences for drug development.

9.7.2 Future Directions

Each case study was found to be rigorous in its investigational approach, and the insights offered by these studies may aid researchers if there was a similar phenomenon encountered. In the future, one may expect greater advances in the technology to detect aggregates and protein particles. In particular, there is a gap in our ability to quantify protein particles in the 0.1- to 1-μm range, and it is hoped that future technologies may help us measure aggregates in this range. Technology advancements are also needed for accurate determination of aggregates in high-concentration protein formulations, for which none of the current technologies are adequate. Another aspect that would be important to clarify, along with evolving data, is the immunogenic potential of protein aggregates and particles. Finally, greater advances could be expected in our ability to predict aggregation propensities, and this would greatly help researchers develop robust processes and formulations.

ACKNOWLEDGMENTS

The authors would like to acknowledge Drew N. Kelner, Linda Narhi, Margaret Ricci, David Brems, Wei Wang, and Chris Roberts for critical reading of this manuscript and helpful discussions.

REFERENCES

1. Dobson CM. 2004. Principles of protein folding, misfolding and aggregation. Semin Cell Dev Biol 15:3–16.

2. Wang W, Singh S, Zeng DL, King K, Nema S. 2007. Antibody structure, instability and formulation. J Pharm Sci 96:1–26.
3. Rathore N, Rajan RS. 2008. Current perspectives on stability of protein drug products during formulation, fill and finish operations. Biotechnol Prog 24:504–514.
4. Dobson CM. 2006. Protein aggregation and its consequences for human disease. Protein Pept Lett 13:219–227.
5. Chen S, Ferrone FA, Wetzel R. 2002. Huntington's disease age-of-onset linked to polyglutamine aggregation nucleation. Proc Natl Acad Sci U S A 99:11884–11889.
6. Illing ME, Rajan RS, Bence NF, Kopito RR. 2002. A rhodopsin mutant linked to autosomal dominant retinitis pigmentosa is prone to aggregate and interacts with the ubiquitin proteasome system. J Biol Chem 277:34150–34160.
7. Rosenberg AS. 2006. Effects of protein aggregates: An immunologic perspective. AAPS J 8:E501–E507.
8. Janeway CA, Travers PA, Walport M, Shlomchik MJ. 2001. *Immunobiology*. New York: Garland Publishing.
9. Lumry R, Eyring HJ. 1954. Conformational changes of proteins. J Phys Chem 58:110–120.
10. Arakawa T, Kita Y, Carpenter JF. 1991. Protein–solvent interactions in pharmaceutical formulations. Pharm Res 8(3):285–291.
11. Gekko K, Timasheff SN. 1981. Mechanism of protein stabilization by glycerol: Preferential hydration in glycerol-water mixtures. Biochemistry 20(16):4667–4676.
12. Lee JC, Timasheff SN. 1981. The stabilization of proteins by sucrose. J Biol Chem 256(14):7193–7201.
13. Yancey PH, Clark ME, Hand SC, Bowlus RD, Somero GN. 1982. Living with water stress: Evolution of osmolyte systems. Science 217(4566):1214–1222.
14. Milner JL, McClellan DJ, Wood JM. 1987. Factors reducing and promoting the effectiveness of proline as an osmoprotectant in *Escherichia coli* K12. J Gen Microbiol 133(7):1851–1860.
15. Soler PM, Wright TE, Smith PD, Maggi SP, Hill DP, Ko F, Jimenez PA, Robson MC. 1999. In vivo characterization of keratinocyte growth factor-2 as a potential wound healing agent. Wound Repair Regen 7(3):172–178.
16. Gomperts BN, Belperio JA, Fishbein MC, Keane MP, Burdick MD, Strieter RM. 2007. Keratinocyte growth factor improves repair in the injured tracheal epithelium. Am J Respir Cell Mol Biol 37(1):48–56.
17. Osslund TD, Syed R, Singer E, Hsu EW, Nybo R, Chen BL, Harvey T, Arakawa T, Narhi LO, Chirino A, Morris CF. 1998. Correlation between the 1.6 A crystal structure and mutational analysis of keratinocyte growth factor. Protein Sci 7(8):1681–1690.
18. Chen BL, Arakawa T, Hsu E, Narhi LO, Tressel TJ, Chien SL. 1994. Strategies to suppress aggregation of recombinant keratinocyte growth factor during liquid formulation development. J Pharm Sci 83(12):1657–1661.
19. Chen BL, Arakawa T. 1996. Stabilization of recombinant human keratinocyte growth factor by osmolytes and salts. J Pharm Sci. 85(4):419–426.

20. Back JF, Oakenfull D, Smith MB. 1979. Increased thermal stability of proteins in the presence of sugars and polyols. Biochemistry 18(23):5191–5196.
21. Arakawa T, Timasheff SN. 1985. The stabilization of protein by osmolytes. Biophys J 47:411–414.
22. Kendrick BS, Carpenter JF, Cleland JL, Randolph TW. 1998. A transient expansion of the native state precedes aggregation of recombinant human interferon-gamma. Proc Natl Acad Sci U S A 95(24):14142–14146.
23. Tiwari A, Bhat R. 2006. Stabilization of yeast hexokinase A by polyol osmolytes: Correlation with the physicochemical properties of aqueous solutions. Biophys Chem 124(2):90–99.
24. Krishnan S, Chi EY, Webb JN, Chang BS, Shan D, Goldenberg M, Manning MC, Randolph TW, Carpenter JF. 2002. Aggregation of granulocyte colony stimulating factor under physiological conditions: Characterization and thermodynamic inhibition. Biochemistry 41(20):6422–6431.
25. Timasheff SN. 1998. Control of protein stability and reactions by weakly interacting cosolvents: The simplicity of the complicated. Adv Protein Chem 51:355–432.
26. Timasheff SN. 2002. Protein-solvent preferential interactions, protein hydration, and the modulation of biochemical reactions by solvent components. Proc Natl Acad Sci U S A 99(15):9721–9726.
27. Timasheff SN. 2002. Thermodynamic binding and site occupancy in the light of the Schellman exchange concept. Biophys Chem 101–102:99–111.
28. Timasheff SN. 2002. Protein hydration, thermodynamic binding, and preferential hydration. Biochemistry 41(46):13473–13482.
29. Arakawa T, Tsumoto K, Kita Y, Chang B, Ejima D. 2007. Biotechnology applications of amino acids in protein purification and formulations. Amino Acids 33(4):587–605.
30. Arakawa T, Timasheff SN. 1983. Preferential interactions of proteins with solvent components in aqueous amino acid solutions. Arch Biochem Biophys 224(1):169–177.
31. Arakawa T, Kita Y, Timasheff SN. 2007. Protein precipitation and denaturation by dimethyl sulfoxide. Biophys Chem 131:62–70.
32. Fujita Y, Noda Y. 1983. The effect of organic solvents on the thermal denaturation of lysozyme as measured by differential scanning calorimetry. Bull Chem Soc Jpn 56(1):233–237.
33. Carpenter JF, Arakawa T, Crowe JH. 1992. Interactions of stabilizing additives with proteins during freeze-thawing and freeze-drying. Dev Biol Stand 74:225–238.
34. Klibanov AM. 2001. Improving enzymes by using them in organic solvents. Nature 409(6817):241–246.
35. Perico N, Purtell J, Dillon TM, Ricci MS. 2008. Conformational implications of an inversed pH-dependent antibody aggregation. J Pharm Sci. September 19. Epub ahead of print.
36. Chang BS, Hershenson S. 2002. Practical approaches to protein formulation development. Pharm Biotechnol 13:1–25.
37. Geigel MA. 2003. The stability testing program. In Medina C, ed., *Compliance Handbook for Pharmaceuticals, Medical Devices and Biologics*, Part of Drug and Pharmaceutical Sciences, Vol. 136. New York: Informa Healthcare.

38. Philo JS. 2006. Is any measurement method optimal for all aggregate sizes and types? AAPS J 8:E564–E571.

39. Alexander AJ, Hughes DE. 1995. Monitoring of IgG antibody thermal stability by micellar electrokinetic capillary chromatography and matrix-assisted laser desorption/ionization mass spectrometry. Anal Chem 67:3626–3632.

40. Wypych J, Li M, Guo A, Zhang Z, Martinez T, Allen MJ, Fodor S, Kelner DN, Flynn GC, Liu YD, Bondarenko PV, Ricci MS, Dillon TM, Balland A. 2008. Human IgG2 antibodies display disulfide mediated structural isoforms. J Biol Chem 283:16194–16205.

41. Dillon TM, Ricci MS, Vezina C, Flynn GC, Liu YD, Rehder DS, Plant M, Henkle B, Li Y, Deechongkit S, Varnum B, Wypych J, Balland A, Bondarenko PV. 2008. Structural and functional characterization of disulfide isoforms of the human IgG2 subclass. J Biol Chem 283:16206–16215.

42. Piedmonte DM, Summers C, McAuley A, Karamujic L, Ratnaswamy G. 2007. Sorbitol crystallization can lead to protein aggregation in frozen protein formulations. Pharm Res 24:136–146.

43. Yancey PH. 2005. Organic osmolytes as compatible, metabolic and counteracting cytoprotectants in high osmolarity and other stresses. J Exp Biol 208:2819–2830.

44. Arakawa T, Timasheff SN.1982. Stabilization of protein by sugars. Biochemistry 21:6536–6544.

45. Carpenter JF, Crowe JH. 1988. The mechanism of cryoprotection of proteins by solutes. Cryobiology 25(3):244–255.

46. Bhatnagar BS, Pikal MJ, Bogner RH. 2008. Study of the individual contributions of ice formation and freeze-concentration on isothermal stability of lactate dehydrogenase during freezing. J Pharm Sci 97(2):798–814.

47. Breen ED, Curley JG, Overcashier DE, Hsu CC, Shire SJ. 2001. Effect of moisture on the stability of a lyophilized humanized monoclonal antibody formulation. Pharm Res 18:1345–1353.

48. Andya JD, Hsu CC, Shire SJ. 2003. Mechanisms of aggregate formation and carbohydrate excipient stabilization of lyophilized humanized monoclonal antibody formulations. AAPS PharmSci 5:E10.

49. Careri G, Gratton E, Yang PH, Rupley JA. 1980. Correlation of IR spectroscopic, heat capacity, diamagnetic susceptibility and enzymatic measurements on lysozyme powder. Nature 284:572–573.

50. Wang W. 1999. Instability, stabilization and formulation of liquid protein pharmaceuticals. Int J Pharm 185:125–188.

51. Carpenter JF, Chang BS, Garzon-Rodriguez W, Randolph TW. 2002. Rational design of stable lyophilized protein formulations: Theory and practice. Pharm Biotechnol 13:109–133.

52. Lu X, Pikal MJ. 2004. Freeze-drying of mannitol-trehalose-sodium chloride-based formulations: The impact of annealing on dry layer resistance to mass transfer and cake structure. Pharm Dev Technol 9:85–95.

53. Carpenter JF, Pikal MJ, Chang BS, Randolph TW. 1997. Rational design of stable lyophilized protein formulations: some practical advice. Pharm Res 14:969–975.

54. Arakawa T, Prestrelski SJ, Kenney WC, Carpenter JF. 2001. Factors affecting short-term and long-term stabilities of proteins. Adv Drug Deliv Rev 46:307–326.
55. Carpenter JF, Izutsu K-I, Randolph TW. 2004. Freezing- and drying-induced perturbations of protein structure and mechanisms of protein protection by stabilizing additives. (Freeze-Drying/Lyophilization of Pharmaceutical and Biological Products). Drugs Pharm Sci 137:147–186.
56. Pikal MJ. 1994. Freeze-drying of proteins: Process, formulation, and stability. *ACS Symp Ser* (Formulation and Delivery of Proteins and Peptides) 567:20–133.
57. Andya JD, Maa YF, Costantino HR, Nguyen PA, Dasovich N, Sweeney TD, Hsu CC, Shire SJ. 1999. The effect of formulation excipients on protein stability and aerosol performance of spray-dried powders of a recombinant humanized anti-IgE monoclonal antibody. Pharm Res 16:350–358.
58. Cleland JL, Lam X, Kendrick B, Yang J, Yang TH, Overcashier D, Brooks D, Hsu C, Carpenter JF. 2001. A specific molar ratio of stabilizer to protein is required for storage stability of a lyophilized monoclonal antibody. J Pharm Sci 90:310–321.
59. Pikal MJ, Dellerman K, Roy ML. 1992. Formulation and stability of freeze-dried proteins: Effects of moisture and oxygen on the stability of freeze-dried formulations of human growth hormone. Dev Biol Stand 74:21–37.
60. Huang K, Dong J, Phillips NB, Carey PR, Weiss MA. 2003 Pro-insulin is refractory to protein fibrillation. J Biol Chem 280:42345–42355.
61. Hua QX, Weiss MA. 2004. Mechanism of insulin fibrillation: The structure of insulin under amyloidogenic conditions resembles a protein-folding. J Biol Chem 279:21449–21460.
62. Przybycien TM, Dunn JP, Valax P, Georgiou G. 1994. Secondary structure characterization of β-lactamase inclusion bodies. Protein Eng 7:131–136.
63. Li T, Talvenheimo L, Zeni L, Rosenfeld R, Stearns G, Arakawa T. 2002. Changes in protein conformation and dynamics upon formation of brain-derived neurotrophic factor and its receptor. Biopolymers 67:10–19.
64. Ejima D, Tsumoto K, Fukada H, Yumioka R, Nagase K, Arakawa T, Philo J. 2007. Effects of acid exposure on the conformation, stability, and aggregation of monoclonal antibodies. Proteins 66:954–962.
65. Kim Y, Rose CA, Liu Y, Ozaki, Y, Datta G, Tu AT. 1994. FT-IR and near-infared FT-Raman studies of the secondary structure of insulinotropin in the solid state: α-Helix to β-sheet conversion induced by phenol and/or by high shear force. J Pharm Sci 83:1175–1180.
66. Ramachander R, Jiang Y, Li C, Eris T, Young M, Dimitrova M, Narhi L. 2008. Solid state fluorescence of lyophilized proteins. Anal Biochem 376:173–182.
67. Chang BS, Beauvais RM, Dong AC, Carpenter JF. 1996. Physical factors affecting the storage stability of freeze-dried interleukin-1 receptor antagonist—Glass transition and protein conformation. Arch Biochem Biophys 331:249–258.
68. Strickley G, Anderson BD. 1997. Solid-state stability of human insulin. 2. Effect of water on reactive intermediate partitioning in lyophiles from pH 2–5 solutions—Stabilization against covalent dimer formation. J Pharm Sci 86:645–653.
69. Mazzobre MF, Buera MD, Chirife J. 1997. Glass transition and thermal stability of lactase in low-moisture amorphous polymeric matrices. Biotechnol Prog 13:195–199.

70. Tang X, Pikal MJ. 2004. Design of freeze-drying processes for pharmaceuticals: Practical advice. Pharm Res 21:191–200.
71. Tyagi AK, Randolph TW, Dong A, Maloney KM, Hitscherich C Jr., Carpenter JF. 2009. IgG particle formation during filling pump operation: A case study of heterogeneous nucleation on stainless steel nanoparticles. J Pharm Sci 98:94–104.
72. Carpenter JF, Randolph TW, Jiskoot W, Crommelin DJ, Middaugh CR, Winter G, Fan YX, Kirshner S, Verthelyi D, Kozlowski S, Clouse KA, Swann PG, Rosenberg A, Cherney B. 2008. Overlooking subvisible particles in therapeutic protein products: Gaps that may compromise product quality. J Pharm Sci. August 14. Epub ahead of print.
73. Kiese S, Papppenberger A, Friess W, Mahler HC. 2008. Shaken, not stirred: Mechanical stress testing of an IgG1 antibody. J Pharm Sci 97:4347–4366.

CHAPTER 10

Aggregation and Immunogenicity of Therapeutic Proteins

VASCO FILIPE, ANDREA HAWE, HUUB SCHELLEKENS, and WIM JISKOOT

10.1 INTRODUCTION

The aim of this chapter is to summarize the state of the art regarding knowledge on the role of protein aggregates in unwanted immunogenicity of therapeutic proteins. A general overview on protein immunogenicity and the relevant immunological background are provided in order to better understand aggregate-related immunogenicity. Different aspects relating protein aggregates to immunogenicity are then discussed, and our current understanding about this subject is summarized.

Therapeutic proteins have proven efficacy in treating a wide range of diseases and are now established as a major drug class. However, a major concern associated with the use of these drugs is that repeated administration of such proteins to patients often leads to the induction of undesirable antidrug antibodies (ADAs). These can be classified as binding or neutralizing antibodies, each of which can lead to a wide range of consequences, varying from none to life-threatening situations.

Unwanted immunogenicity has been a major issue ever since proteins were first used as therapeutics. Although some advancement has been made in understanding the immunogenicity of therapeutic proteins, opinions are often conflicting and, to a large extent, are based on assumptions rather than on facts. Among the several factors playing a role in immunogenicity, the presence of impurities, in particular aggregates, has been put forward as a major concern. Several studies have suggested that formulations with a high amount of aggregates tend to be more immunogenic.[1] However, little is known about the nature of the aggregate species responsible for such effects.

Aggregation of Therapeutic Proteins, Edited by Wei Wang and Christopher J. Roberts

10.2 IMMUNOGENICITY OF THERAPEUTIC PROTEINS

10.2.1 Incidence of Immunogenicity during Therapy

The first reports of unwanted immunogenicity go alongside with the first use of therapeutic proteins, more than a century ago, when animal antiserum was introduced in humans for the treatment and prevention of infections. A few decades later, more precisely in 1922, insulin of bovine and porcine origin started to be used in the treatment of diabetes mellitus. All these products triggered immune responses involving ADAs, even after a single or few injections, and sometimes led to fatal anaphylactic reactions.[2] These effects were at that time explained by the protein's foreign nature and impurities present in the formulations. In fact, the most severe adverse reactions to porcine and bovine insulin decreased after purification of the products.

The identification of new therapeutic targets, associated with the need to reduce unwanted immunogenicity, led to the next generation of therapeutic proteins, this time of human origin. Examples of these proteins include the plasma-derived clotting factors and growth hormone isolated from the pituitary glands of cadavers. However, these proteins also induced a strong antibody response in many patients.[3,4] This was ascribed to the high level of impurities present in these products. Moreover, given that many of the patients were treated with these products because they had an innate deficiency of an endogenous protein, they also lacked the immune tolerance to the therapeutic protein.

Profiting from the advancement of recombinant DNA techniques and sequencing of the human genome, most therapeutic proteins introduced in the market nowadays are at least human homologues, with some even produced in human cell lines. However, in contrast to initial expectations, these proteins still induce ADAs, and in some cases in the majority of patients. Therapeutic proteins, for which cases of immunogenicity after repeated administration have not been reported, are extremely rare. It is now clear that a highly pure and stable formulation of a fully human therapeutic protein is not enough to avoid ADA formation.

Clinical ADA formation outcome is very difficult to predict and it often contradicts well-accepted hypotheses. For example, infliximab, a chimeric monoclonal antibody (mAb) against the tumor necrosis factor-alpha (TNF-α), shows good results in treating diseases like rheumatoid arthritis and Crohn's disease. However, after a few months of treatment, this drug has an antibody formation rate that ranges from 20% to 60%, depending on the studies.[5,6] One may think that the mouse component can be the main reason for these high values, but similar results have been observed with recently approved fully human antibodies like adalimumab. This mAb has the same therapeutic target as infliximab and has shown levels of ADA formation of about 20%.[7]

10.2.2 Consequences of Protein Immunogenicity for Patients

The immunogenicity of therapeutic proteins is mainly characterized by the formation of binding or neutralizing antibodies. The biological and clinical consequences of these antibodies vary according to their levels, specificity, and the role of the therapeutic protein. Binding antibodies bind to the drug in sites that may or may not interfere with its pharmacokinetic/pharmacodynamic (PK/PD) behavior. Neutralizing antibodies interfere directly or indirectly with the active site of the drug, neutralizing its therapeutic effect. In some cases, these antibodies can cross-react or even neutralize endogenous proteins, and this can have very serious consequences.[8]

In many cases, the formation of ADAs has very little biological and clinical consequences. The most severe general immune reactions, such as acute and delayed infusion-like reactions, have become less common, owing in part to the high purity of new products. Acute infusion reactions normally occur within the first 48h after an infusion and include type I hypersensitivity, nausea, flushing, and dyspnea. Late infusion reactions typically occur after a longer period of treatment and they are more frequent, particularly for high-dose mAbs. They include serum sickness-like reactions and other type III and delayed-type hypersensitivity reactions.[9]

The most common biological effect of ADAs is loss of efficacy. In some cases, this might be overcome by increasing the dose, for instance, with factor VIII in the treatment of hemophilia A patients.[10] Normally, the interference of binding antibodies leads to a reduced half-life of the drug, resulting in lower activity. However, some cases have been reported where binding antibodies actually increase the half-life of therapeutic proteins and thereby increase their activity, as reported for human growth hormone (hGH).[11] Persistent levels of neutralizing antibodies lead to complete loss of activity. In such cases, the treatment has to be stopped and alternative treatments, if available, should be considered. The most dramatic consequences of immunogenicity occur when neutralizing antibodies neutralize an endogenous protein with an essential biological function.

One of the most severe cases of unwanted immunogenicity registered in the last decade occurred with erythropoietin (EPO), a hormone that regulates erythropoiesis and is normally used to treat anemia. When neutralizing antibodies start to cross-react with native EPO, patients cease the production of red blood cells and may develop pure red cell aplasia (PRCA).[12] Although the percentage of patients who develop neutralizing antibodies against EPO is low, the consequences of this incident can be disastrous.

Concerns about unwanted immunogenicity have to consider both probability and consequences, with higher relevance on the second consideration. In the case of therapeutic insulin, which has relatively high rates of ADA incidence, it is not fully understood why they seem to have no major effect on the drug's performance.[13] However, despite these high incidence rates, the low consequences attenuate the concern regarding the immunogenicity of this protein. In contrast, despite its low rates of antibody formation, EPO has been the center of many immunogenicity discussions, especially after an upsurge in the incidence of antibody-mediated PRCA between 2000 and 2002.[14] Therefore, the levels and rates of antibody formation against a certain therapeutic protein have to be carefully balanced with the impact that they may have on the patient and on the progression of the disease.

10.2.3 Factors Playing a Role in Protein Immunogenicity

There have been several product-, treatment-, and analysis-related factors identified and discussed to play a role in unwanted immunogenicity (Fig. 10.1). The particular relevance and the relative contribution of each of these factors to immunogenicity, and the underlying mechanisms, have not been fully elucidated. Moreover, there may also be unknown factors playing a role in this area.[15]

10.2.3.1 Product Related It is well-known that the human immune system reacts against nonhuman proteins and that sequence variation can explain the high immunogenicity of therapeutic proteins like streptokinase,[16] bovine

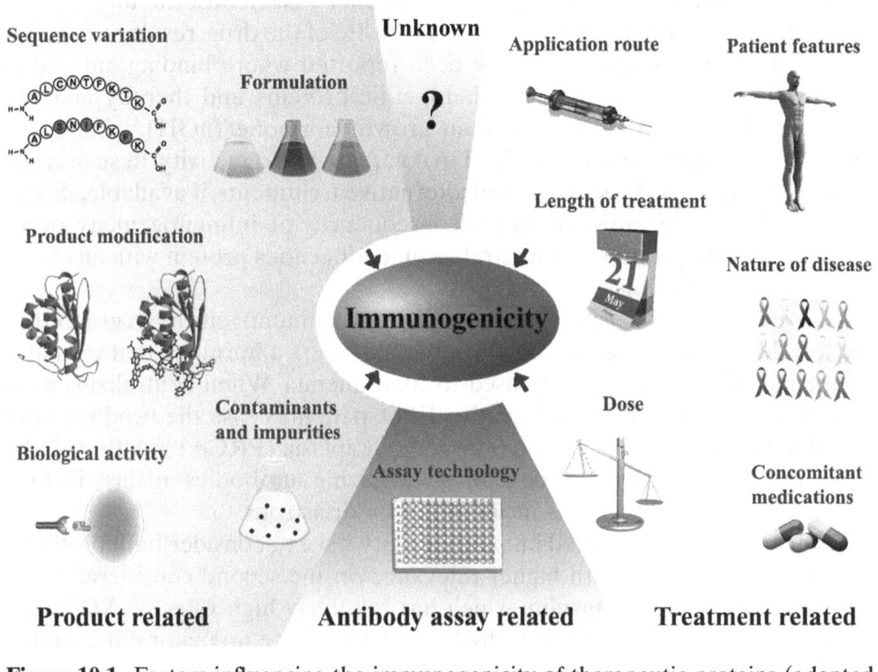

Figure 10.1. Factors influencing the immunogenicity of therapeutic proteins (adapted from Reference 15).

adenosine deaminase,[17] and salmon calcitonin.[18] However, as mentioned earlier in this chapter, similarity to the human sequence does not necessarily imply a lack of immunogenicity. Conversely, there are several examples of recombinant therapeutic proteins with sequences differing from the naturally occurring human sequence that have not led to an increased immune response.[19,20]

Product modification intended to increase half-life can have a positive or negative impact on immunogenicity. Deglycosylation of glycoproteins may expose new epitopes that are normally hidden by the glycan or may expose new hydrophobic regions of the protein and alter its solubility, thereby affecting immunogenicity.[21,22] Moreover, posttranslational modifications by the host cell, such as the addition of functional groups or structural changes, may also affect immunogenicity. Both natural, glycosylated interferon-α2 (IFN-α2) and interleukin-2 (IL-2) were reported to be less immunogenic than their nonglycosylated counterparts made in *Escherichia coli*.[23,24]

PEGylation is claimed to reduce the immunogenicity of therapeutic proteins by shielding epitopes.[25] In fact, this was verified in some nonhuman PEGylated proteins like bovine adenosine deamidase and asparginase.[26] However, there are reports of high immunogenicity of PEGylated human proteins, such as megakaryocyte growth and differentiation factor (MGDF), but a comparison with unPEGylated MGDF is lacking.[27]

The biological activity of the therapeutic protein can also influence immunogenicity. An immune-stimulating protein is more likely to induce ADAs than an immune-suppressive protein. mAbs targeted to cell-bound epitopes are expected to induce a higher immune response than those with a target in solution, allegedly due to their higher probability of gathering the necessary immunological components to trigger an immune response.[26]

The product formulation is another important factor for immunogenicity. For instance, human serum albumin (HSA) has been commonly added as a stabilizing agent, but it has been related with immunogenicity increase.[28] On the other hand, the replacement of HSA by glycine and polysorbate 80 in the formulation of EPO coincided with the upsurge of antibody-mediated PRCA outside the United States.

The integrity of therapeutic proteins profoundly depends on the formulation, and some degradation can occur during handling and storage. Degraded products have proven to be more immunogenic, as will be discussed later in this chapter. In particular, aggregates have shown to enhance immunogenicity and they are now one of the main concerns in the development of new therapeutic products. Ironically, downstream processing of a product can contribute to the introduction of impurities. For example, antibodies to factor VIII were associated with the introduction of a new pasteurization stage in the manufacturing process.[29] Although the purity of biopharmaceuticals has increased significantly over the past few decades, impurities are still present in low amounts and they are thought to play an important role in immunogenicity.

Evidence for the importance of product variables can be found in significant variations in immunogenicity reported for different IFN-β products. For example, Avonex and Betaferon, which differ in formulation composition, level of impurities, sequence of amino acids, glycosylation, and producing cell type, have reported a percentage of patients with neutralizing antibodies of 13% for Avonex and 43% for Betaferon.[30]

10.2.3.2 Treatment Related There are various factors related to the treatment itself that can strongly influence unwanted immunogenicity. Several clinical and preclinical studies registered significant differences in ADAs according to the route of administration. Evidence suggests that subcutaneous (s.c.) administration is most likely to elicit an immune response, followed by intramuscular (i.m.) administration and intravenous (i.v.) administration, and the least immunogenic seems to be local (topical) administration.[31] For instance, it has been clinically shown that i.m. administration of IFN-β is less immunogenic than s.c. administration.[32] However, there have been no published cases whereby a change in the administration route completely eliminated the immunogenicity of a therapeutic protein.[29]

The dose and duration of the treatment also play a role since they determine the level of exposure to the immune system. Higher doses or prolonged duration of treatment increases exposure, thereby increasing the risk of immunogenicity. However, these two factors appear to be independent of each other. IFN-β formulations produced in *E. coli* and Chinese hamster ovary (CHO) cells both induced antibodies after 6–12 months of treatment, despite the 10-fold difference in protein amount between the two formulations.[31]

Patient features like genetic background, gender, and age have all been reported to influence the incidence of ADAs,[26] although for most of them, the correlation is not consistent. Individuals who are deficient for functional genes may lack the immune tolerance for certain therapeutic proteins, making them more prone to develop an immune response. This may be the reason for the high immunogenicity levels reported for proteins used to treat genetic diseases, such as factor VIII and growth hormone.[26]

Concomitant medications and the nature of the disease can also influence immunogenicity. For instance, the chimeric mAb rituximab is practically nonimmunogenic in patients with non-Hodgkin's lymphoma, whereas the incidence of ADA formation against the same drug has been reported to be 27% and 65% in patients with primary Sjogren's syndrome and systemic lupus erythematosus, respectively.[33] Cancer patients are less likely to produce antibodies against therapeutic proteins. This is likely due to their compromised immune system, caused both by the disease and by immune-suppressive concomitant medications.[26] It has also been suggested that the average survival of cancer patients treated with therapeutic proteins may be too short to develop antibody responses.

10.2.3.3 Antibody Assay Related Adopting an appropriate strategy for the assessment of unwanted immunogenicity of therapeutic proteins is essential. A major factor leading to the highly variable and sometimes conflicting results concerning immunogenicity is most likely the assay technology used. For most cases, patient sera are first screened for the presence of binding antibodies. Positive samples are then submitted to confirmation assays and, if positive, antibody specificity and the presence of neutralizing antibodies are assayed. However, the lack of international standardization of assays and reference preparations makes it virtually impossible to compare test results from different laboratories.[26]

For instance, screening of ADAs is extremely difficult in the presence of a drug in the sample because it often prevents the ADAs to bind to the detection antibodies. This is particularly problematic for drugs with a long half-life, such as mAbs. The most recent assays for ADA screening include an acid pretreatment of the collected serum before the detection step in order to separate the binding or neutralizing antibodies from the drug.[34] Many immunogenicity studies from the past are now being questioned, and there is an urgent need for optimization and standardization of all immunogenicity assays.

10.3 IMMUNE MECHANISMS RELATED TO PROTEIN IMMUNOGENICITY

The immune system constitutes a natural defense that has evolved to respond to all types of invasive threats, like bacteria and viruses. This defense is a complex system that involves the coordinated interaction of numerous elements. Many basic aspects of immunological functions have already been clarified to date, but there are still numerous questions to be answered. In order to better address the role of aggregates in the immunogenicity of therapeutic proteins, it is important to understand the relevant immune responses possibly involved in this process.

From a classical point of view, the immune system can be divided into two arms: the innate and the adaptive immune systems. The innate immune system comprises the cells and mechanisms that constitute the first line of defense against threats, in a fast and nonspecific manner. Phagocytic cells, like macrophages that ingest and kill pathogens, together with the complement system, are primarily responsible for the innate immune response. The innate immune system strongly depends on pattern recognition receptors (PRRs) that have evolved to recognize highly conserved pathogen-associated molecular patterns (PAMPs). The most important and well-characterized class of these receptors is the toll-like receptors (TLRs).[35]

The adaptive immune system is capable of mounting highly specific responses against pathogenic agents. It consists of a complex and diverse surveillance cell network capable of developing long-term memory cells against specific molecular determinants. The fundamental mechanisms of this type of response are mediated primarily by B- and helper T lymphocytes. They circulate between the bloodstream, the lymph nodes, and peripheral lymphoid tissues in a state referred to as naive cells. Upon contact with antigens, they can get activated and transform into effector cells. Both lymphocytes carry unique antigen-specific receptors on their surface. Through complex signaling mechanisms, B cells are able to generate specific antibodies against any conceivable antigen. Given that unwanted immunogenicity of therapeutic proteins is primarily based on the detection of antibodies, this chapter will further focus on the adaptive arm of the immune system.

10.3.1 Immune Responses to Therapeutic Proteins

Therapeutic proteins can activate the immune system through two different mechanisms: classical immune response and breaking of immune tolerance.

10.3.1.1 Classical Immune Response The classical activation of the immune system is triggered by the presence of non-self epitopes, for example, those present in nonhuman therapeutic proteins like bacterial-derived asparginase and streptokinase.[36] The mechanism typically involves T-cell-dependent activation of B cells with subsequent production of different high-affinity antibody isotypes and induction of memory cells. Resembling a vaccination-type immune reaction, the classical immune response may be enhanced by adjuvant equivalent "danger" signals, such as impurities or aggregates.[37]

This type of immune response is also found in patients who lack immune tolerance to the therapeutic protein, for example, due to a genetic defect, and is sometimes observed in patients treated with modified human proteins. In general, it is hard to predict if modified human proteins will trigger a classical immune response. Immunogenicity consequences of sequence deviation from the human amino acid sequence are less dependent on the level of divergence and depend more on the type of amino acid changed and its location in the sequence.[38]

The classical response usually originates a fast and robust immune reaction, often after the first injection. The ADAs formed are normally neutralizing antibodies, and memory effect is very often observed. In most cases, there is considerable loss of efficacy of the therapeutic, and sometimes the treatment has to be stopped due to anaphylactic-type reactions. Similar to vaccines, retreatment with the same product often triggers very fast and efficient immune responses.[39]

10.3.1.2 Breaking of Immune Tolerance The immune system has developed several mechanisms in order to tolerate self-proteins. Given that most recombinant human proteins are homologous to their endogenous counterparts, the immune response to such therapeutic proteins involves breaking of immune tolerance, the mechanism of which is not yet fully understood.

Antibody formation via this process is slow and it often takes several months of chronic administration to break immune tolerance. ADAs normally disappear after stopping the treatment and sometimes disappear even during the treatment. ADA formation through this process often has no major consequences but sometimes results in a reduction of efficacy, neutralization of activity, and neutralization of endogenous proteins.[36]

10.3.2 Immune Tolerance

T- and B lymphocytes have the ability of randomly generating an infinite variety of receptors on their surface, commonly termed T-cell receptors (TCRs) and B-cell receptors (BCRs). This is achieved by random gene rearrangements in the receptor loci. It ensures a vast receptor repertoire providing recognition of any conceivable antigen. Very often, a subset of these lymphocytes expresses self-reactive receptors, and these cells must be downregulated in order to avoid autoimmune responses. A complex regulating mechanism, known as central tolerance, is responsible for verifying and suppressing self-reactive cells through several checkpoints in the pathway of lymphocyte differentiation, occurring in the thymus and in the bone marrow.[40]

Despite the action of this mechanism, many autoreactive lymphocytes manage to escape into the peripheral pool of circulating lymphocytes. Here, an additional regulation mechanism, not yet fully understood, is responsible for keeping them inactive, known as peripheral tolerance.[40] Autoreactive circulating lymphocytes are thought to be either eliminated by apoptosis upon contact with self-antigens or to enter several possible states of anergy.[41] In any case, an immune reaction against a human protein requires breaking of this peripheral tolerance.

With concern to protein immunogenicity of human homologues, breaking of B-cell tolerance seems to be the most relevant mechanism.[26] Given that only B cells are capable of producing antibodies, B-cell tolerance has to be broken for the production of self-antibodies to occur. This mechanism is not fully elucidated and many pieces of the puzzle seem to be missing. For example, the role of regulatory T cells (Treg), which are known to have immune-suppressive regulatory functions over other cells, in breaking immune tolerance is not clear. Moreover, the impact of many signaling proteins still remains uncertain, and several immunological pathways are still to be characterized. In any case, breaking of B-cell tolerance ultimately involves the activation of autoreactive B cells.[26]

10.3.3 B-Cell Activation Mechanisms

In order to understand the proposed models for immunogenicity of human homologues in therapy, it is important to realize that there are different ways of activating B cells. These activation mechanisms can be classified as T cell dependent (Td) or T cell independent (Ti), depending on the contribution of helper CD4(+) T cell.

10.3.3.1 Td Td activation of B cells is normally triggered by antigens with a small number of copies of several different epitopes. Classical examples of these antigens include microbial proteins, several other non-self proteins, and some altered self-proteins. This mechanism requires a sequence of coordinated events involving different cell types and several stimulatory signals (Fig. 10.2).[42]

Td activation of B cells requires antigens with epitopes capable of interacting with the protein-processing machinery of professional antigen-presenting cells (APCs), such as dendritic cells (DCs), and both TCRs and BCRs. Any failure in delivering the stimulatory signals mentioned in Fig. 10.2 may lead to B- or T-cell anergy or apoptosis.[42] It is important to note that B cells are also considered professional APCs, and this fact adds some variants to the mechanism already described. The presentation of processed peptides by B cells to naive T cells can activate both cells, resulting in an accelerated version of Td activation.[43]

In general, the dependence on other cells and the requirement of several signaling steps make Td activation a slower process compared with Ti activation.[1] However, Td activation of B cells normally results in a more robust immune response, exclusively leading to the formation of memory B- and T cells.[42] It also promotes antibody affinity maturation and isotype switching, from the classes IgM and IgD into IgG, IgE, and IgA.[44]

10.3.3.2 Ti Antigens that are expressed on the surface of pathogens in an organized and highly repetitive form can directly stimulate B cells to produce antibodies without the help of T cells. Examples of these antigens include both relatively simple polymers of peptides and polysaccharides, as well as higher-order structures such as viral capsids, composed of repetitive arrays of multiple protein components.[1] This mechanism is not fully understood, as there are still several unknown signaling pathways. Currently, Ti activation of B cells is divided in two different categories, commonly termed type 1 and type 2 activations (Fig. 10.3). Type 1 activation is also known as polyclonal B-cell activation, because these antigens are capable of directly inducing the proliferation and differentiation of several different B cells into antibody-secreting cells.[45] This mechanism is thought to involve the recognition of antigens by both BCRs and TLRs. The antibodies produced through this type of Ti activation are normally nonspecific and preferentially recognize heterologous antigens.[45] Despite being commonly classified as a Ti mechanism, polyclonal activation may also have some variants that require T-cell help.[46] Conversely, type 2 activation takes place when several antigen clusters are presented to B cells in a manner that causes cross-linking of BCRs in a multivalent fashion.[47] This occurrence triggers a complex signaling pathway, such as the activation of Bruton's tyrosine kinases,[48] eventually leading to B-cell activation (Fig. 10.3).

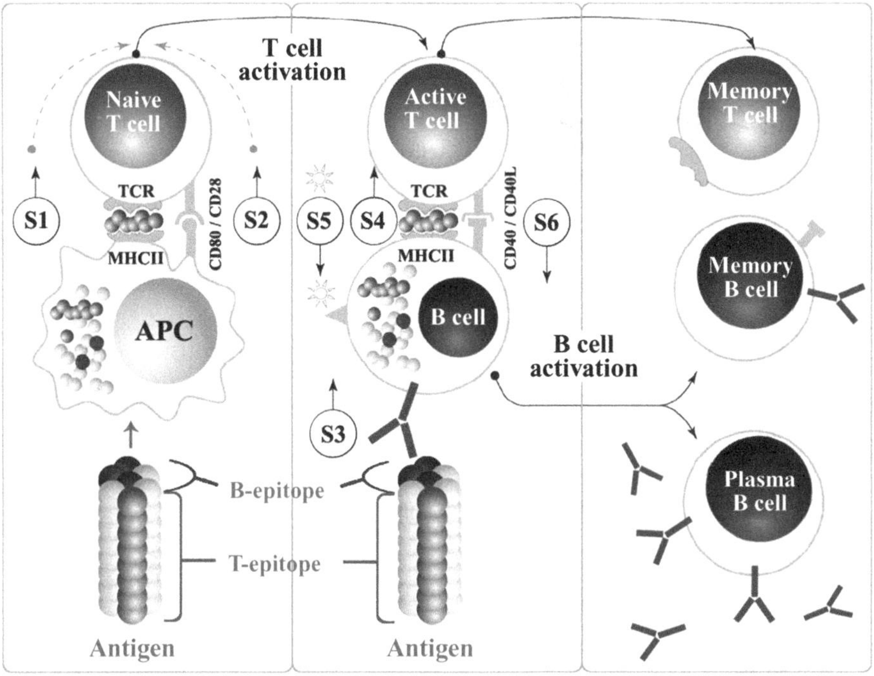

Figure 10.2. Common T-cell-dependent B-cell activation mechanism (adapted from Reference 42). The first step in this process involves the nonspecific internalization of an antigen by a professional APC. It processes the antigen into peptides, which are then presented to a naive T cell, in major histocompatibility complex class II (MHC II) molecules on the surface of the APC. Two signals must be given by the APC for the T cell to become fully activated: the interaction of the MHC II–peptide complex with the TCR (S1) and the interaction of costimulatory molecules from the APC, such as CD80 and CD86, with costimulatory molecules from the T cell, such as CD28 (S2). Once fully activated, the T cell divides and starts producing an array of cytokines with various activities. Meanwhile, activation of naive B cells starts with the recognition of the cognate antigen by BCRs (S3), more specifically IgM and IgD receptors. The B cell internalizes and processes the antigen–Ig complex and presents cognate T-cell epitopes in MHC II molecules on its surface. Upon recognition of this MHC II–epitope complex by the effector T cell (S4), two other signals must occur for the complete activation of B cells: the delivery of several cytokines from T cells to B cells (S5) and the interaction of the CD40 costimulatory molecule from the B cell with the recently expressed CD40L costimulatory molecule from the T cell (S6). Once fully activated, the B cell proliferates and differentiates into memory and antibody-secreting plasma cells, with the ability to perform antibody affinity maturation and isotype switching.

It has been reported that Ti activation of B cells can be modulated by accessory cells, like macrophages and DCs, which can be stimulated via TLRs to produce several stimulatory signals.[45] Independence of T cells and the direct activation of B cells by the antigen make this response much faster than Td

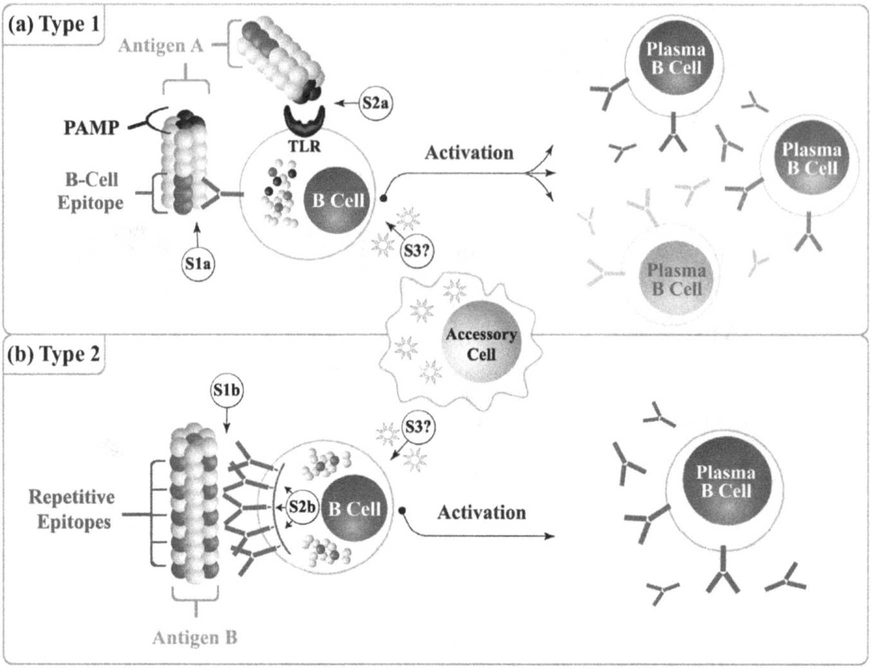

Figure 10.3. T-cell-independent B-cell activation mechanisms. (a) Type 1 activation requires the recognition of a polyclonal activator antigen (antigen A) by both BCRs (S1a) and TLRs (S2a), which leads to the production of several activation signals. B cells are activated in a polyclonal fashion. (b) Type 2 activation requires the recognition of a type 2 antigen (antigen B), which normally displays repetitive epitopes, by BCRs (S1b). Clustering of bound BCRs leads to multivalent cross-linking (S2b), which triggers the production of several activation signals. In either case, B cells are activated, proliferate, and become antibody-producing plasma B cells. The contribution of external stimulatory factors by accessory cells, such as professional APCs, is not clear (S3?).

responses.[1] However, Ti activation does not lead to affinity maturation or to the generation of memory B cells, and it has often been mentioned that B cells activated via this mechanism do not undergo isotype switching.[42] This last feature of B cells is often associated with the engagement of CD40 on B cells by CD40L on T cells, a step exclusive for Td activation. However, some studies have suggested that isotype switching resulting from Ti activation is possible. HIV-1 envelope glycoproteins have been reported to directly activate B cells, triggering isotype class switching from IgM to IgA or IgG, in a CD40-independent manner.[49] It has been described that several interleukins and activation factors such as interleukin-10 (IL10) and B-cell activating factor belonging to the TNF family (BAFF), delivered by accessory cells, play crucial roles in B-cell development and activation.[49,50] However, knowledge regarding the exact targets and function of these proteins remains very limited. Still, these signals may hold the key to answer many questions about breaking immune tolerance by human protein homologues in therapy.

10.3.4 Breaking of B-Cell Immune Tolerance

Over the last century, as the comprehension of the immune system increased, different models have been proposed to explain the immune response against therapeutic proteins. While the mechanisms of classical immune responses seem to be consensual, divergences start emerging with the mechanisms of breaking immune tolerance. It has already been shown that tolerance may be broken both by Td and Ti activation of autoreactive B cells.[51] The main task is now to clarify which of these two mechanisms is the most relevant for breaking immune tolerance during therapy and to identify the main trigger of this process.

Among the proposed silencing mechanisms of B-cell peripheral tolerance, the most feasible seems to be the induction of anergy.[40] Autoreactive anergic B cells may start to produce antibodies if they are given a certain danger signal as trigger. For example, bacterial endotoxins, capable of being recognized by TLRs, have proven to trigger the production of antibodies against self-antigens. However, this particular Td activation of B cells has shown to result in a feeble immune response. When self-antigens are coupled to foreign T-cell epitopes, only a weak IgM response is induced unless multiple high doses of antigen are given together with immune adjuvants.[26]

Alternatively, the danger signal may be given to the B cell in a Ti way. It has been shown that the most potent way to induce high levels of IgG through Ti activation against a self-antigen is to present it arrayed on virus or virus-like particles (VLPs).[52–54] An ordered presentation of repetitive epitopes is characteristic of microbials and the immune system has apparently learned to react vigorously to this type of antigen presentation. Protein aggregates may also present epitopes in an array form, and they are often pointed to as the most likely risk factor involved in breaking immune tolerance,[26] as will be later described.

In any case, there is not enough evidence to fully support a higher emphasis of either the Ti or Td mechanism in breaking immune tolerance. Our current lack of knowledge regarding several pathways involved in breaking immune tolerance makes it very difficult to substantiate exact mechanisms and main triggers of this process. Nevertheless, several studies are now being carried out in this field supporting the relevance of protein aggregates in breaking immune tolerance.

10.4 AGGREGATES AND IMMUNOGENICITY

The ability of protein aggregates to trigger an antibody response against the monomeric form of the protein has been known for more than half a century. However, despite this obvious connection, little is known about the particularities that make protein aggregates potent inducers of the immune system.

It has been shown that not all aggregates are immunogenic[55] and, in fact, there are several intrinsic and extrinsic aspects to consider. Aggregate features such as size, molecular weight (MW), amount, solubility, and resemblance to microbial structures are each thought to play a role in their immunogenicity. Moreover, product-, treatment-, and host-related factors may also influence the immune response to protein aggregates. Examples of these factors include product origin, route of administration, host immune status, and, in the case of therapeutic versions of endogenous proteins, the abundance and manner of presentation of the endogenous protein.[1] Unfortunately, for most cases, the way these factors may affect an aggregate-related immune response is uncertain and their relevance remains unclear.

Protein aggregates can be formed by various mechanisms and there are several different ways to classify them. From an immunological point of view, and according to their propensity to trigger immune responses against the constituent monomer, aggregates can be roughly classified into three different types: (1) an assembly of native proteins in a polymeric-like structure; (2) an assembly of denatured or partially denatured protein irreversibly associated (within the given environment); and (3) covalently linked proteins, which could be either in a native or denatured state.[1]

Given that the activity of microbial pathogenic proteins depends critically on the conformation of their active site, the immune system has evolved to preferentially make conformation-dependent antibodies for this site. This preference for making antibodies to conformational rather than to linear determinants, that is, independent of spatial arrangement, appears to be a general property of the immune system.[56–58] Regarding immunogenicity of therapeutic proteins, it has been documented that neutralizing antibodies to EPO tend to be conformation dependent since most of them fail to bind to completely denatured EPO containing mainly linear determinants.[12] Aggregate-induced antibodies that also bind to the monomeric protein are a major concern for the immunogenicity of therapeutic proteins. For these reasons, native-like aggregates are thought to be more dangerous than conformationally changed aggregates.[1]

In the field of protein vaccination, where a robust immune response to the administered protein is the desired effect, the presence of aggregates in the formulations can have diverse effects. Since most prophylactic and therapeutic vaccines aim to create antibodies against native-state proteins, which can be presented in a monomeric state or organized in polymeric structures, it is important to ensure that the antibodies formed against aggregates can cross-react with the corresponding target protein. Vaccine formulations with high percentages of aggregates may elicit strong immune responses, but not necessarily with the desired outcome. Therefore, vaccine aggregation should only be allowed in a controlled and well-characterized way.

The correlation between protein aggregates and both wanted and unwanted immunogenicity is a crucial topic for the pharmaceutical industry and it needs to be promptly clarified. Given that each protein is structurally unique and that aggregates can take countless forms, general conclusions can probably only be drawn based on a larger number of case-by-case studies. Understanding the effect of aggregates in immunogenicity requires far more research and it can only be achieved by extensive collaboration between academia, industry, and the regulatory authorities.

10.4.1 Evidence of Aggregates as a Risk Factor for Immunogenicity

Throughout the last century, several studies have drawn attention to relationships between protein aggregates and immunogenicity. In order to better understand these concerns, some relevant clinical and nonclinical studies will be addressed.

10.4.1.1 Nonclinical The first documented nonclinical studies directly relating therapeutic protein aggregates to immunogenicity date back to the 1960s, in which immunoglobulin (Ig) from different sources was injected in rodents. In the research performed by Dresser in 1962, while studying induced immunological paralysis in mice through small intraperitoneal injections of bovine γ globulin (BGG), he realized that immune responses could be eliminated by removing the high MW material from the preparations.[59] Accordingly, in 1963, Claman noticed that in order to properly induce immune tolerance to BGG in mice, the preparations had to be ultracentrifuged and the pellet removed.[60] Administration of the resuspended pellet resulted in high production of antibodies and fast clearance of the antigen. In 1966, Gamble decided to further study the correlation between the amount of soluble aggregates and the corresponding immune response.[61] Therefore, preparations of human γ globulin (HGG) with increasing amounts of heat-generated aggregates were injected in mice and the immune response was monitored. The results showed a clear dose-dependent correlation between the amount of aggregates administrated and the immune response intensity (Fig. 10.4).[61] Once again, most mice treated with centrifuged HGG failed to induce a detectable immune response.

The choice of Ig as a model antigen for studying the effects of aggregates on the immune response was an interesting one since these proteins differ from most other proteins in terms of complex immune system interactions. Ig can modify the immune response to itself through binding and colligation of multiple receptors on the B-cell surface.[62] Whether this may have influenced the immunogenicity of the aggregated species in these studies is not certain, as the affinity and binding of human and bovine Ig with murine BCRs is not clear. Nevertheless, these studies have drawn attention to the fact that high MW species are potent inducers of the immune system, and they have emphasized the need to obtain more pure and aggregate-free therapeutic products.

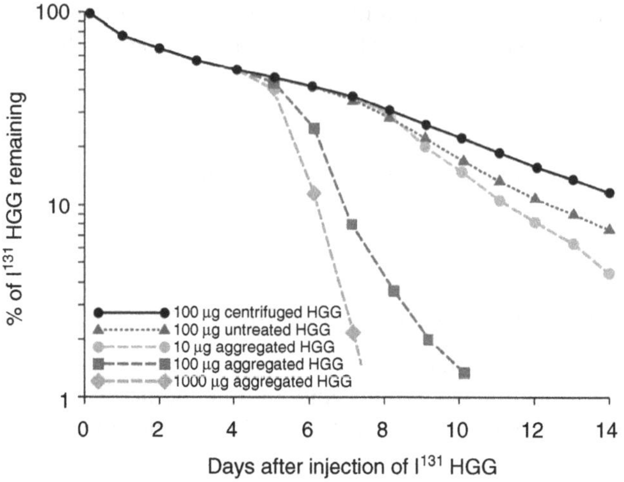

Figure 10.4. Immune elimination of iodine I-131 tagged HGG (I^{131} HGG) following administration of aggregated HGG in mice.[61]

Most nonclinical studies involving therapeutic proteins have involved the administration of foreign proteins. Based on the principles reviewed earlier in this chapter, this probably influences the immune response to aggregates. The ability of aggregates to elicit antibody responses to self-proteins was evaluated by Braun et al. in 1997.[63] They used human interferon-α (hIFN-α) transgenic mice to study the capacity of aggregated hIFN-α products to break immune tolerance. Various administration routes were tested with two types of aggregates: hIFN-α cross-linked with either human or mouse serum albumin (IFN-α-albumin) and homogeneous aggregates of hIFN-α (IFN-α-IFN-α). IFN-α-albumin aggregates were prepared by artificial cross-linking with glutaraldehyde at a 1:1 molar ratio. IFN-α-IFN-α aggregates were obtained from a fraction of an old IFN-α bulk solution. The results clearly showed that both these aggregates were able to induce antibodies to hIFN-α in the transgenic mice, whereas monomeric hIFN-α failed to do so (within the time frame of the study). It became clear that immune tolerance to self-proteins can be broken by aggregates. Unfortunately, the aggregate species responsible for this occurrence were not comprehensively characterized in the context of the study. However, the study provided valuable information regarding the role of aggregates, dosing frequency, and route of administration on antibody induction to hIFN-α.

The propensity of different types of aggregates to elicit immune responses to self-proteins was further investigated by Hermeling et al.[64] They evaluated the ability of different well-characterized degradation products of hIFN-α containing different types and amounts of aggregates to break the immune tolerance of hIFN-α transgenic mice. The different types of aggregates were obtained by metal-catalyzed oxidation, hydrogen peroxide oxidation, glutaraldehyde cross-linking, and extreme-heat treatment. Only metal-catalyzed oxidized aggregates were found to break immune tolerance in transgenic mice. Oxidation through hydrogen peroxide treatment failed to generate aggregates and to induce antibodies against native hIFN-α in transgenic mice. Thus, the hIFN-α aggregates present in the metal-catalyzed oxidation product probably play an important role in breaking immune tolerance. It now remains to be evaluated if these findings for hIFN-α can be generalized for other proteins.

10.4.1.2 Clinical Clinical evidence that aggregates in therapeutic protein products play a crucial role in immunogenicity was apparent from very early studies. In the 1950s and 1960s, HGG preparations containing substantially aggregated material were reported to trigger severe anaphylactic reactions in patients.[65] In 1969, Ellis and Henney decided to more carefully study the correlation between the aggregates present in different available HGG preparations and antibody induction in patients.[66] They verified that antibody responses to these preparations were directed against the high MW fraction (i.e., the pellet after 2-h centrifugation at 30,000 rpm). The antibodies were thought to be specific for novel or cryptic determinants present only on aggregated species. Also of relevance, immune responses to therapeutic HGG were far more potent and prevalent in antibody-deficient patients, which drew attention to the importance of immune tolerance in immunogenicity.[65]

Anaphylactic reactions were also observed in patients treated with early commercial preparations of HSA and pasteurized plasma (PP) solutions. In 1979, Ring et al. studied the immune responses of patients suffering from anaphylaxis upon administration of either of these two therapeutic products.[67] All the administered preparations contained between 5% and 15% aggregates. ADAs detected in most patients appeared to be specific for these aggregates and failed to bind to monomeric HSA and deaggregated PP.

One of the most intriguing clinical cases relating protein aggregates to immunogenicity occurred with hGH. Originally purified from formalin-fixed pituitary glands in the early 1960s, early therapeutic hGH preparations contained substantial amounts of aggregates (40–70%) and induced antibody responses in up to 50% of the patients.[11,68] After revising the manufacturing process and substantially reducing the level of aggregates down to 5–10%, the incidence of antibody formation decreased yet prevailed. Interestingly, the level of aggregates appeared not to determine whether there would be an immune response, but rather the nature of the antibody response. In 1980, studies carried out by Moore and Leppert showed that patients treated with the heavily aggregated hGH product revealed persistent ADAs, whereas those treated with the less aggregated products developed transient ADAs.[11] Furthermore, the appearance of binding antibodies to hGH was far more frequent than that of neutralizing antibodies.[69] The latter type of antibodies was much more common in "knock-out" patients for hGH who lack immune tolerance to this protein.

Recombinant human IL-2 (rhIL-2) has been used for therapy since the 1980s and can be viewed as a good clinical model to test the general principles correlating aggregates and immunogenicity. rhIL-2 products have been reported to contain a high amount of small aggregates, with an average of more than 20 constituent monomers/aggregate, and to have an incidence of binding antibody responses of about 60%.[1,24] However, despite the high immunogenicity rate verified for rhIL-2, induction of neutralizing antibodies seems to be uncommon. Interestingly, neutralizing antibodies largely arose in a group of patients treated with rhIL-2 in an immunologically provocative fashion, that is, s.c. administration or concomitant treatment with IFN-α, and extensive treatment over a prolonged time period.[24] Nevertheless, neutralization *in vitro* did not seem to predict neutralization *in vivo*, as patients with antibody-mediated loss of response to rhIL-2 were extremely rare. Of interest, neutralizing antibody responses generally arise several months after the appearance of binding antibodies, which may involve epitope spreading in a hypothetical Td activation scenario.[1]

10.4.2 Proposed Mechanisms of Antibody Induction by Protein Aggregates

Protein aggregates can activate the immune system through different mechanisms. Recognition and processing of aggregates has been reported to be accomplished through nonspecific uptake by APCs[70,71] and specific uptake by B cells.[43] Protein aggregates are thought to play a key role in breaking immune tolerance, but they may also influence classical immune responses. However, the exact mechanisms by which aggregates may trigger unwanted immunogenicity of therapeutic proteins are still unknown.

The increased MW of aggregates compared to the monomeric protein has been pointed as one of the main causes for their elevated immunogenicity since it fosters their exposition to the immune system and enhances their engulfment by APCs.[72] Interestingly, it has been described that some blood-borne DCs capture particulate (insoluble) antigens in the periphery, migrate to the spleen, and can induce splenic marginal zone B cells and B_1 B cells to proliferate and produce antibodies in a Ti fashion.[73,74] However, mass and size per se are not enough to justify the propensity of aggregates to influence immune responses to therapeutic proteins. Protein aggregates are believed to trigger immunogenicity because they are thought to originate a set of new characteristics that enhance their recognition by the immune system (Fig. 10.5). Aggregates can originate neo-epitopes, which may arise from their quaternary structure.[75] They are also thought to expose epitopes, usually hidden in the native protein (cryptic epitopes), as a result of partial or complete unfolding of their constituents.[1] However, the most concerning feature of aggregates seems to be the fact that they may present epitopes in a repetitive arrayed form.[26] Repetitive epitope presentation has been pointed as the most likely cause of aggregate-related immunogenicity, since it might mimic

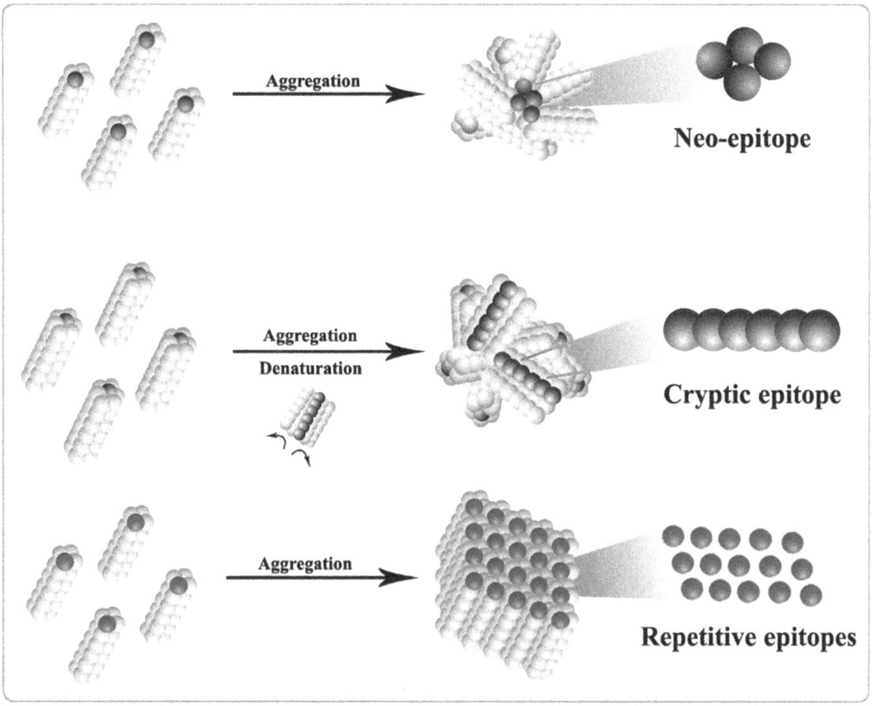

Figure 10.5. Main concerns regarding epitope presentation in protein aggregates.

well-conserved microbial patterns, to which the immune system has learned to vigorously react.[54]

Intrinsic characteristics of aggregates may determine the mechanism by which they trigger the immune system. Protein aggregates containing neo-epitopes or exposed cryptic epitopes may enhance immunogenicity via presentation of these epitopes to T cells by professional APCs or via direct recognition by B cells. Nevertheless, ADAs formed against this type of aggregates are less likely to cross-react with the monomeric form since native protein monomers will lack the referred neo-epitopes and exposed cryptic determinants. However, immune responses against these aggregates may be responsible for some cases of acute infusion reactions[67] more commonly associated with classical immune responses.

On the other hand, protein aggregates that present epitopes in a repetitive arrayed form, mimicking well-conserved microbial patterns, may lead to multivalent cross-linking of BCRs.[1,26] Once it occurs, the immune response can then follow two different pathways that may or may not require T-cell help. The first pathway results in acceleration of B-cell antigen processing and its presentation to T cells, as a consequence of rapid targeting of BCR–aggregate complexes to the MHC II peptide loading compartment.[43] The second mechanism is Ti and involves a complex signaling pathway, such as the activation of

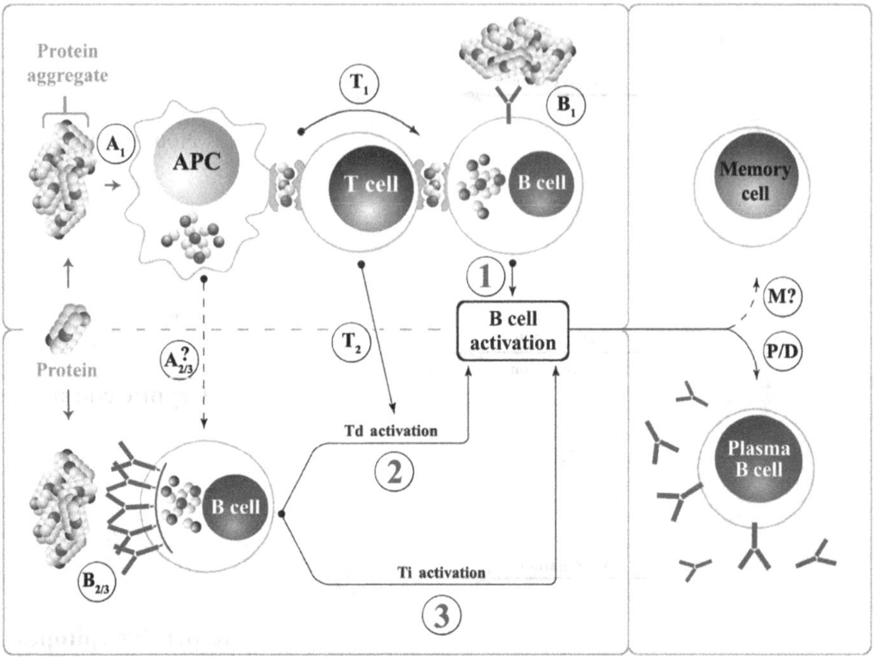

Figure 10.6. Proposed mechanisms of antibody induction by protein aggregates. This mechanism follows a typical Td activation (see Fig. 9.2 for details). It involves the uptake and processing of protein aggregates by professional APCs (A_1), followed by T-cell activation (T_1), accompanied by aggregate uptake and processing by B cells (B_1), which eventually leads to B-cell proliferation and differentiation (P/D) and to the induction of memory cells (M?). (1) This mechanism occurs when aggregates trigger multivalent cross-linking of B-cell receptors ($B_{2/3}$). B cells may then receive activating signals from accessory cells ($A_{2/3}$). T cells are then recruited to activate B cells (T_2), which leads to their proliferation and differentiation (P/D) and possibly to the induction of memory cells (M?). (2) This mechanism also occurs when aggregates trigger multivalent cross-linking of B-cell receptors ($B_{2/3}$), which triggers the production of several activation signals. B cells may then receive additional signals from accessory cells ($A_{2/3}$). T-cell help is not required for this type of B-cell activation. B cells are then activated, which leads to their proliferation and differentiation (P/D). The induction of memory cells via this mechanism remains unclear (M?). See color insert.

Bruton's tyrosine kinases, which mediate B-cell proliferation (Fig. 10.6).[48] Protein aggregates exhibiting repetitive epitope presentation are more prone to trigger the formation of ADAs, which can cross-react with the monomeric protein.

Therapeutic protein aggregates have been pointed to as the most likely risk factor necessary to break immune tolerance during therapy. Anergic self-reactive B cells may become active upon cross-linking of BCRs by repetitive epitopes presented on self-protein aggregates. This theory would explain the late emergence of ADAs to human homologues during therapy and the frequent lack of memory cell induction. The probability of a protein aggregate that presents immunologically relevant repetitive epitopes to be recognized by anergic self-reactive B cells is so low that it may take several months before breaking of immune tolerance occurs. On the other hand, in the eventual scenario of Ti type 2 B-cell activation, induction of a memory effect would be less likely.[8] The involvement of T cells in this process still remains unclear, yet their eventual participation would also require breaking of T-cell tolerance. It has been reported that T-cell tolerance can be broken by autoreactive B cells whose tolerance had previously been broken by self-protein aggregates.[76] However, there is not enough evidence to either support or refute T-cell involvement in breaking immune tolerance by therapeutic protein aggregates.

10.4.3 Possible Factors Responsible for the Immunogenicity of Protein Aggregates

Several intrinsic and extrinsic aspects are thought to play a role in the tendency of protein aggregates to trigger immune responses. Extrinsic factors that may affect the immune response to protein aggregates are mostly the same as the ones affecting the immunogenicity of proteins in general (see Section 9.2.3). The presence of impurities, frequency and route of administration, host immune tolerance to the monomeric protein, activity of concomitant immunomodulators, and immunomodulatory activity of the protein itself are some of the extrinsic factors suggested to directly affect immune responses to protein aggregates.[1] Furthermore, intrinsic aggregate features such as size, amount, and other inherent characteristics are also expected to influence their immunogenicity, and they will be thoroughly discussed below. However, very few studies have directly addressed this subject and the current knowledge originates mainly from vaccinology studies. For most cases, the way these intrinsic and extrinsic factors influence an aggregate-related immune response is poorly understood. Moreover, it is imperative to remember that each protein is unique to some degree and so are the aggregates that they may form. Thus, the search for general guidelines of aggregate-related immunogenicity is extremely challenging.

Both qualitative and quantitative factors are discussed to influence the immune responses to protein aggregates. However, it is very difficult to determine a maximum allowable "immunologically safe" dose limit of protein aggregates present in therapeutic products. Some products may have very low immunological consequences despite certain levels of aggregates, while for other products, very small changes in aggregate levels may profoundly affect these consequences. As for qualitative factors, there are several aspects to consider.

One of the most important factors thought to influence aggregate-related immunogenicity is the conformation of the aggregate protein constituents.

Aggregates formed by native-like proteins are expected to pose a higher threat for immunogenicity than the ones formed by denatured or partially denatured proteins.[1] Native-like conformation of aggregate constituents is considered to be crucial for cross-reactivity of aggregate-induced ADAs with the monomeric protein. Furthermore, as previously mentioned, the immune system has evolved to preferentially make antibodies to conformational rather than to linear determinants. These assumptions were confirmed by Hermeling et al. in 2006, who challenged tolerant mice with progressively denatured aggregates of hIFN-α and observed that the more native-like aggregates exhibited a higher immunogenicity.[77] Conformational factors may also explain why aggregate-induced ADAs, triggered by challenging animal models with denatured protein aggregates obtained by extreme-heat stress, normally fail to cross-react with the native protein.[55,78]

The amino acid sequence of protein aggregate constituents is another factor thought to play a role in their immunogenicity as it determines the epitopes that will be exposed to the immune system. A single amino acid change has been reported to induce a completely new epitope,[38] which may hypothetically have increased immunogenicity once presented in a repetitive manner on aggregated forms. Aggregates formed by already immunogenic proteins are also expected to be more efficient in triggering immune responses. Thus, aggregates of a foreign origin protein are expected to be more immunogenic than self-protein aggregates.[1]

Protein aggregate size and solubility are also among the main factors believed to influence immune responses. However, little is known about the immunogenicity of small oligomers compared to large multimers. Nevertheless, multimerization seems to be crucial to immunogenicity, as larger-sized monomers are not necessarily more immunogenic than small monomeric proteins. Moreover, it has been long known that particulate antigens are rapidly endocytosed by APCs, which initiate immune responses.[72]

The number of constituent monomers in an aggregate that is necessary to efficiently trigger immune responses remains a controversial issue. In 1990, clinical studies performed by Ratner et al. showed that pure crystalline dimers of human insulin were able to trigger immunogenicity in a dose-dependent manner.[79] Conversely, the year before, while investigating the molecular requirements for Ti type 2 antibody responses to polymers carrying polysaccharides, Dintzis et al. verified that the presentation of at least 20 repetitive epitopes is required to successfully trigger an immune response.[80] They hypothesized that the B-cell stimulatory signal is "quantized" in that a minimum number of BCRs must be connected together in a spatially contiguous cluster, which they defined as an "immunon." In that study, the polymers had to have an MW higher than 100 kDa and to possess a minimum of 20 epitope repeats in order to induce antibody formation via this mechanism. Thus, according to this theory, aggregates that present more than 20 repetitive epitopes on their surface and have an MW bigger than 100 kDa may be more immunogenic, assuming that Ti type 2 B-cell activation is the preferred mechanism.

Successive work by many in the field has indicated that the generation of a signaling complex leading to Ti type 2 B-cell activation and antibody production depends also on epitope affinity, binding kinetics, and polymer rigidity.[47] Rigid polymers have proven to immobilize BCRs more efficiently than flexible polymers,[52] which is a vital step for this type of B-cell activation. Accordingly, it is possible that rigid protein aggregates (i.e., with strong internal interactions) pose a higher immunological risk than flexible aggregates. However, one could also argue that flexible aggregates may adapt more easily to bind to multiple BCRs, making them the most immunogenic type of aggregates.

Several studies have addressed the effect of self-antigen organization on antibody responses. In 1997, Bachmann and Zinkernagel extensively examined the effects of challenging viral coat protein (VCP) tolerant mice with differently organized VCP products. They verified that soluble monomeric VCP was toleragenic; poorly organized VCPs were able to trigger mainly Td responses; and highly organized VCPs induced strong Ti responses.[81] They have proposed that repetitive determinants at a spacing of 5–10 nm are the most immunogenic presentation, since it is unique to microbial structures, to which the immune system has evolved to strongly respond.[81–83] More recently, Chackerian et al. observed that self-protein aggregates arrayed on VLPs in a dense and highly organized manner managed to trigger long-lasting high IgG titers, whereas less dense aggregate arrays failed.[53] Thus, according to these results, dense and organized aggregates with an average epitope spacing of 5–10 nm would be expected to be more immunogenic. However, in these studies, the self-protein aggregate arrays were artificially created and involved viral components. Hence, the generalization of these findings to natural occurring aggregates should be addressed with caution.

Hydrophobicity of protein aggregates has also been suggested to play a role in their immunogenicity. It has been proposed that hydrophobic portions of biological molecules act, when exposed, as universal damage-associated molecular patterns to initiate repair, remodeling, and immunity.[84] Thus, it is possible that more hydrophobic aggregates of therapeutic proteins represent a bigger threat for unwanted immunogenicity.

Another intrinsic factor that may affect aggregate-related immunogenicity is the aggregate nature, that is, if they are formed only by the protein itself or by a mixture of protein and impurities, excipients, or degraded protein that may be present in the formulation. For example, protein adsorption to nano- or microparticles, often released from pumps, pipes, vessels, filters, or primary packaging material, may create aggregates with different epitope presentation than self-aggregates, and these may pose different immunogenicity risks. Moreover, glycosylation and PEGylation may exert influence not only on the formation of aggregates but also on aggregate-related immunogenicity.[1] However, identification of the immunological hazards posed by different aggregate types has not yet been explored.

10.4.4 Risk Analysis of Protein Aggregates

The most relevant clinical risk posed by immune reactions to protein aggregates in therapeutic products may be narrowed down to the generation of cross-reacting antibodies that may affect the efficacy of the product or may even neutralize an endogenous protein counterpart. For the formation of cross-reacting antibodies between aggregates and monomeric protein to occur, it is important that protein aggregate constituents maintain a native-like conformation. This requirement is more evident for neutralizing antibodies, which are directed against the active site of the protein. While aggregates of denatured protein may potently induce antibodies, they would not be expected to induce neutralizing antibodies because protein conformation was lost on denaturation. Antibodies directed to linear epitopes within the binding site of the protein could effectively neutralize it if they bind to a linear determinant critical in ligand–receptor binding, but this occurrence seems to be rare.[1]

Epitope spreading has been suggested as a possible mechanism involved in the appearance of neutralizing antibody responses. Epitope spreading is normally defined as the diversification of epitope specificity from the initial dominant epitope-specific immune response to subdominant epitopes on that protein, normally derived by repetitive antigen presentation.[85] Thus, through epitope spreading, the presence of binding antibodies may facilitate the generation of neutralizing antibodies. In fact, for many therapeutic protein products, development of neutralizing antibodies is preceded by high titer and prolonged binding antibody responses. In such cases, according to this theory, preventing a neutralizing antibody response may be possible by limiting the titer and duration of the primary response.[1]

The amount, size, and type of aggregates necessary to trigger immune responses are a major concern for pharmaceutical companies and regulatory agencies. In 2008, the hypothetical immunological significance of normally overlooked "subvisible particles" present in protein formulations was discussed by experienced researchers in the field.[86] Current United States Pharmacopeia (USP) particulate requirements state that particles >10 μm in size should be controlled bellow 6000 particles per container, but no regulations have been established for the smaller sizes. Therefore, it is possible that immunogenically relevant protein aggregates have been routinely ignored by these regulations. For every aggregate >10 μm present in a formulation, there can be considerable amounts of slightly smaller aggregates, and each may contain hundreds or thousands of protein units. Thus, given the lack of knowledge regarding the most immunogenic aggregate sizes, it is important to develop instruments, protocols, and regulations that properly address a broader range of particle sizes, especially in the subvisible range.

10.4.5 Predictive Models to Access the Immunogenicity of Aggregates

Reliable models for prediction of aggregate-related immunogenicity are not available at the moment. The proper model would require a thorough understanding of the underlying mechanisms, and this is far from the current situation. There are some predictive models for unwanted immunogenicity of therapeutic proteins, but none of them has proven to be entirely effective and all of them have considerable downsides. Furthermore, the application of available techniques to predict the immunogenicity caused by protein aggregates is still very limited. The available models to predict immunogenicity may be divided into three classes: *in vitro*, *in silico*, and *in vivo*. However, given that *in vitro* and *in silico* techniques are mostly based on the aptitude of proteins to actively interact with immune or innate cells, basically identifying T- or B-cell epitopes,[87] these techniques are not suitable for predicting aggregate-related immunogenicity. Given the number of other factors that may equally influence immunogenicity, it seems very unlikely that these two methods may one day fully predict an immune response to either monomeric or aggregated therapeutic proteins.[26]

The use of animal models to predict immunogenicity *in vivo* seems to be the most promising method to predict aggregate-related immune responses.[77] *In vivo* prediction has the advantage of including the entire immune machinery and environment necessary to better simulate the extremely complex scenario that results in complete immune responses, especially when protein aggregates are involved. Furthermore, with these models, it is possible to include other immunogenicity-influencing factors other than hazardous epitopes, such as route and frequency of administration, concomitant medication, and possibly disease state. However, the vast majority of proteins used for therapeutic purposes in humans are foreign to animals, and this feature greatly affects their immune response. Therefore, in order to circumvent this problem, transgenic animal models have been developed, and they have a promising predictive potential.[26] These animals may be tolerant to certain human homologues or may possess more humanlike immunological features, such as the case of transgenic mice expressing the human MHC II complex (human leukocyte antigen [HLA] system). Thus, *in vivo* testing with transgenic animal models may have the potential to predict aggregate-related immunogenicity in the near future. However, much work still needs to be done before any type of model can properly predict immune responses to protein aggregates.

10.5 CONCLUSIONS

Protein aggregates are potent inducers of immune responses to therapeutic protein products. Thoroughly understanding the connection between aggregates and immunogenicity is extremely relevant to optimize the efficacy and safety of biopharmaceutical products. However, current knowledge regarding this subject is very limited and is mostly based on assumptions.

There are several aspects to consider when analyzing the impact of aggregates in the immunogenicity of therapeutic proteins. According to current knowledge, high-order multimeric protein aggregates containing native-like molecules with repetitive epitope presentation seem to be the most immunogenic type of aggregates. However, given that therapeutic proteins may have different aggregation profiles and may pose unique immunological risks, a case-by-case approach is needed in order to better understand the immunogenicity of aggregates.

The risk caused by the presence of aggregates should be evaluated not only by the tendency of aggregates to trigger ADAs but also by the clinical consequences that these ADAs may have. Antibodies against aggregates may not bind to the monomeric active form of the drug and, in case they do, it may not have relevant clinical consequences.

According to the type of immunogenic threat posed by protein aggregates, several immunological mechanisms have been proposed, which may or may not require T-cell help. The development of reliable predictive models to assess the immunogenicity of protein aggregates is highly necessary but is dependent on the elucidation of these mechanisms.

REFERENCES

1. Rosenberg AS. 2006. Effects of protein aggregates: An immunologic perspective. AAPS Journal 8(3):E501–E507.
2. Schernthaner G. 1993. Immunogenicity and allergenic potential of animal and human insulins. Diabetes Care 16(Suppl. 3):155–165.
3. Milner RD. 1985. Growth hormone 1985. British Medical Journal (Clinical Research Ed.) 291(6509):1593–1594.
4. Jacquemin MG, Saint-Remy JM. 1998. Factor VIII immunogenicity. Haemophilia 4(4):552–557.
5. Gottlieb AB, Evans R, Li S, Dooley LT, Guzzo CA, Baker D, Bala M, Marano CW, Menter A. 2004. Infliximab induction therapy for patients with severe plaque-type psoriasis: A randomized, double-blind, placebo-controlled trial. Journal of the American Academy of Dermatology 51(4):534–542.
6. Baert F, Noman M, Vermeire S, Van Assche G, D'Haens G, Carbonez A, Rutgeerts P. 2003. Influence of immunogenicity on the long-term efficacy of infliximab in Crohn's disease. New England Journal of Medicine 348(7):601–608.
7. West RL, Zelinkova Z, Wolbink GJ, Kuipers EJ, Stokkers PC, van der Woude CJ. 2008. Immunogenicity negatively influences the outcome of adalimumab treatment in Crohn's disease. Alimentary Pharmacology & Therapeutics 28(9):1122–1126.
8. De Groot AS, Moise L. 2007. Prediction of immunogenicity for therapeutic proteins: State of the art. Current Opinion in Drug Discovery & Development 10(3):332–340.
9. Cheifetz A, Smedley M, Martin S, Reiter M, Leone G, Mayer L, Plevy S. 2003. The incidence and management of infusion reactions to infliximab: A large center experience. American Journal of Gastroenterology 98(6):1315–1324.
10. Patten PA, Schellekens H. 2003. The immunogenicity of biopharmaceuticals. Lessons learned and consequences for protein drug development. Developments in Biologicals 112:81–97.

11. Moore WV, Leppert P. 1980. Role of aggregated human growth hormone (hGH) in development of antibodies to hGH. Journal of Clinical Endocrinology and Metabolism 51(4):691–697.
12. Casadevall N, Nataf J, Viron B, Kolta A, Kiladjian JJ, Martin-Dupont P, Michaud P, Papo T, Ugo V, Teyssandier I, Varet B, Mayeux P. 2002. Pure red-cell aplasia and antierythropoietin antibodies in patients treated with recombinant erythropoietin. New England Journal of Medicine 346(7):469–475.
13. Meager A. 1994. Human antibodies to insulin in diabetes. Journal of Interferon Research 14(4):181–182.
14. Schellekens H. 2005. Immunologic mechanisms of EPO-associated pure red cell aplasia. Best Practice & Research Clinical Haematology 18(3):473–480.
15. Schellekens H. 2002. Bioequivalence and the immunogenicity of biopharmaceuticals. Nature Reviews. Drug Discovery 1(6):457–462.
16. Rosenschein U, Lenz R, Radnay J, Ben Tovim T, Rozenszajn LA. 1991. Streptokinase immunogenicity in thrombolytic therapy for acute myocardial infarction. Israel Journal of Medical Sciences 27(10):541–545.
17. Chaffee S, Mary A, Stiehm ER, Girault D, Fischer A, Hershfield MS. 1992. IgG antibody response to polyethylene glycol-modified adenosine deaminase in patients with adenosine deaminase deficiency. Journal of Clinical Investigation 89(5):1643–1651.
18. Grauer A, Frank-Raue K, Schroth J, Raue F, Ziegler R. 1994. Neutralizing antibodies against salmon calcitonin. The cause of a treatment failure in Paget's disease. Deutsche Medizinische Wochenschrift (1946) 119(14):507–510.
19. Kontsek P, Liptakova H, Kontsekova E. 1999. Immunogenicity of interferon-alpha 2 in therapy: Structural and physiological aspects. Acta Virologica 43(1):63–70.
20. Girard F, Gourmelen M. 1986. Clinical experience with Somatonorm. Acta Paediatrica Scandinavica 325:29–32.
21. Gribben JG, Devereux S, Thomas NS, Keim M, Jones HM, Goldstone AH, Linch DC. 1990. Development of antibodies to unprotected glycosylation sites on recombinant human GM-CSF. Lancet 335(8687):434–437.
22. Karpusas M, Whitty A, Runkel L, Hochman P. 1998. The structure of human interferon-beta: Implications for activity. Cellular and Molecular Life Sciences 54(11):1203–1216.
23. Antonelli G, Simeoni E, Currenti M, De Pisa F, Colizzi V, Pistello M, Dianzani F. 1997. Interferon antibodies in patients with infectious diseases. Anti-interferon antibodies. Biotherapy (Dordrecht, Netherlands) 10(1):7–14.
24. Prummer O. 1997. Treatment-induced antibodies to interleukin-2. Biotherapy (Dordrecht, Netherlands) 10(1):15–24.
25. Veronese FM, Pasut G. 2005. PEGylation, successful approach to drug delivery. Drug Discovery Today 10(21):1451–1458.
26. Schellekens H. 2008. How to predict and prevent the immunogenicity of therapeutic proteins. Biotechnology Annual Review 14:191–202.
27. Vadhan-Raj S. 2000. Clinical experience with recombinant human thrombopoietin in chemotherapy-induced thrombocytopenia. Seminars in Hematology 37(2 Suppl. 4):28–34.

28. Palleroni AV, Aglione A, Labow M, Brunda MJ, Pestka S, Sinigaglia F, Garotta G, Alsenz J, Braun A. 1997. Interferon immunogenicity: Preclinical evaluation of interferon-alpha 2a. J Interferon Cytokine Res 17(Suppl. 1):S23–S27.
29. Schellekens H. 2005. Factors influencing the immunogenicity of therapeutic proteins. Nephrology, Dialysis, Transplantation 20(Suppl. 6):vi3–vi9.
30. Sominanda A, Rot U, Suoniemi M, Deisenhammer F, Hillert J, Fogdell-Hahn A. 2007. Interferon beta preparations for the treatment of multiple sclerosis patients differ in neutralizing antibody seroprevalence and immunogenicity. Multiple Sclerosis (Houndmills, Basingstoke, England) 13(2):208–214.
31. Ross C, Clemmesen KM, Svenson M, Sorensen PS, Koch-Henriksen N, Skovgaard GL, Bendtzen K. 2000. Immunogenicity of interferon-beta in multiple sclerosis patients: Influence of preparation, dosage, dose frequency, and route of administration. Danish Multiple Sclerosis Study Group. Annals of neurology 48(5):706–712.
32. Perini P, Facchinetti A, Bulian P, Massaro AR, Pascalis DD, Bertolotto A, Biasi G, Gallo P. 2001. Interferon-beta (INF-beta) antibodies in interferon-beta1a- and interferon-beta1b-treated multiple sclerosis patients. Prevalence, kinetics, cross-reactivity, and factors enhancing interferon-beta immunogenicity in vivo. European Cytokine Network 12(1):56–61.
33. Van Walle I, Gansemans Y, Parren PW, Stas P, Lasters I. 2007. Immunogenicity screening in protein drug development. Expert Opinion on Biological Therapy 7(3):405–418.
34. Sickert D, Kroeger K, Zickler C, Chokote E, Winkler B, Grenet JM, Legay F, Zaar A. 2008. Improvement of drug tolerance in immunogenicity testing by acid treatment on Biacore. Journal of Immunological Methods 334(1–2):29–36.
35. Akira S, Takeda K. 2004. Functions of toll-like receptors: Lessons from KO mice. Comptes Rendus Biologies 327(6):581–589.
36. Schellekens H. 2003. Immunogenicity of therapeutic proteins. Nephrology, Dialysis, Transplantation 18(7):1257–1259.
37. Matzinger P. 1994. Tolerance, danger, and the extended family. Annual Review of Immunology 12:991–1045.
38. Ottesen JL, Nilsson P, Jami J, Weilguny D, Duhrkop M, Bucchini D, Havelund S, Fogh JM. 1994. The potential immunogenicity of human insulin and insulin analogues evaluated in a transgenic mouse model. Diabetologia 37(12):1178–1185.
39. Hermeling S, Crommelin DJ, Schellekens H, Jiskoot W. 2004. Structure-immunogenicity relationships of therapeutic proteins. Pharmaceutical Research 21(6):897–903.
40. Tarner IH, Fathman CG. 2006. Does our current understanding of the molecular basis of immune tolerance predict new therapies for autoimmune disease? Nature Clinical Practice 2(9):491–499.
41. Singh NJ, Schwartz RH. 2006. Primer: Mechanisms of immunologic tolerance. Nature Clinical Practice 2(1):44–52.
42. De Groot AS, Scott DW. 2007. Immunogenicity of protein therapeutics. Trends in Immunology 28(11):482–490.
43. Cheng PC, Steele CR, Gu L, Song W, Pierce SK. 1999. MHC class II antigen processing in B cells: Accelerated intracellular targeting of antigens. Journal of Immunology 162(12):7171–7180.

44. Silver K, Cornall RJ. 2003. Isotype control of B cell signaling. Science's STKE: Signal Transduction Knowledge Environment 2003(184):pe21.
45. Montes CL, Acosta-Rodriguez EV, Merino MC, Bermejo DA, Gruppi A. 2007. Polyclonal B cell activation in infections: Infectious agents' devilry or defense mechanism of the host? Journal of Leukocyte Biology 82(5):1027–1032.
46. Stevenson PG, Doherty PC. 1999. Non-antigen-specific B-cell activation following murine gammaherpesvirus infection is CD4 independent in vitro but CD4 dependent in vivo. Journal of Virology 73(2):1075–1079.
47. Vos Q, Lees A, Wu ZQ, Snapper CM, Mond JJ. 2000. B-cell activation by T-cell-independent type 2 antigens as an integral part of the humoral immune response to pathogenic microorganisms. Immunological Reviews 176:154–170.
48. Fluckiger AC, Li Z, Kato RM, Wahl MI, Ochs HD, Longnecker R, Kinet JP, Witte ON, Scharenberg AM, Rawlings DJ. 1998. Btk/Tec kinases regulate sustained increases in intracellular Ca^{2+} following B-cell receptor activation. EMBO Journal 17(7):1973–1985.
49. He B, Qiao X, Klasse PJ, Chiu A, Chadburn A, Knowles DM, Moore JP, Cerutti A. 2006. HIV-1 envelope triggers polyclonal Ig class switch recombination through a CD40-independent mechanism involving BAFF and C-type lectin receptors. Journal of Immunology 176(7):3931–3941.
50. Mackay F, Schneider P, Rennert P, Browning J. 2003. BAFF and APRIL: A tutorial on B cell survival. Annual Review of Immunology 21:231–264.
51. Melchers F, Rolink AR. 2006. B cell tolerance–How to make it and how to break it. Current Topics in Microbiology and Immunology 305:1–23.
52. Bachmann MF, Rohrer UH, Kundig TM, Burki K, Hengartner H, Zinkernagel RM. 1993. The influence of antigen organization on B cell responsiveness. Science (New York, N.Y.) 262(5138):1448–1451.
53. Chackerian B, Lenz P, Lowy DR, Schiller JT. 2002. Determinants of autoantibody induction by conjugated papillomavirus virus-like particles. Journal of Immunology 169(11):6120–6126.
54. Fehr T, Bachmann MF, Bucher E, Kalinke U, Di Padova FE, Lang AB, Hengartner H, Zinkernagel RM. 1997. Role of repetitive antigen patterns for induction of antibodies against antibodies. Journal of Experimental Medicine 185(10):1785–1792.
55. Hermeling S, Aranha L, Damen JM, Slijper M, Schellekens H, Crommelin DJ, Jiskoot W. 2005. Structural characterization and immunogenicity in wild-type and immune tolerant mice of degraded recombinant human interferon alpha2b. Pharmaceutical Research 22(12):1997–2006.
56. Ito HO, Nakashima T, So T, Hirata M, Inoue M. 2003. Immunodominance of conformation-dependent B-cell epitopes of protein antigens. Biochemical and Biophysical Research Communications 308(4):770–776.
57. Nath A, Hall E, Tuzova M, Dobbs M, Jons M, Anderson C, Woodward J, Guo Z, Fu W, Kryscio R, Wekstein D, Smith C, Markesbery WR, Mattson MP. 2003. Autoantibodies to amyloid beta-peptide (Abeta) are increased in Alzheimer's disease patients and Abeta antibodies can enhance Abeta neurotoxicity: Implications for disease pathogenesis and vaccine development. Neuromolecular Medicine 3(1):29–39.

58. O'Nuallain B, Wetzel R. 2002. Conformational Abs recognizing a generic amyloid fibril epitope. Proceedings of the National Academy of Sciences of the United States of America 99(3):1485–1490.
59. Dresser DW. 1962. Specific inhibition of antibody production. II. Paralysis induced in adult mice by small quantities of protein antigen. Immunology 5:378–388.
60. Claman HN. 1963. Tolerance to a protein antigen in adult mice and the effect of nonspecific factors. Journal of Immunology 91:833–839.
61. Gamble CN. 1966. The role of soluble aggregates in the primary immune response of mice to human gamma globulin. International Archives of Allergy and Applied Immunology 30(5):446–455.
62. Getahun A, Heyman B. 2006. How antibodies act as natural adjuvants. Immunology Letters 104(1–2):38–45.
63. Braun A, Kwee L, Labow MA, Alsenz J. 1997. Protein aggregates seem to play a key role among the parameters influencing the antigenicity of interferon alpha (IFN-alpha) in normal and transgenic mice. Pharmaceutical Research 14(10):1472–1478.
64. Hermeling S, Jiskoot W, Crommelin D, Bornaes C, Schellekens H. 2005. Development of a transgenic mouse model immune tolerant for human interferon beta. Pharmaceutical Research 22(6):847–851.
65. Barandun S, Kistler P, Jeunet F, Isliker H. 1962. Intravenous administration of human gamma-globulin. Vox Sanguinis 7:157–174.
66. Ellis EF, Henney CS. 1969. Adverse reactions following administration of human gamma globulin. Journal of Allergy 43(1):45–54.
67. Ring J, Stephan W, Brendel W. 1979. Anaphylactoid reactions to infusions of plasma protein and human serum albumin. Role of aggregated proteins and of stabilizers added during production. Clinical Allergy 9(1):89–97.
68. Underwood LE, Voina SJ, Van Wyk JJ. 1974. Restoration of growth by human growth hormone (Roos) in hypopituitary dwarfs immunized by other human growth hormone preparations: Clinical and immunological studies. Journal of Clinical Endocrinology and Metabolism 38(2):288–297.
69. Frasier SD. 1983. Human pituitary growth hormone (hGH) therapy in growth hormone deficiency. Endocrine Reviews 4(2):155–170.
70. Daha MR, van EL, Hazevoet HM, Kijlstra A. 1982. Degradation of soluble immunoglobulin aggregates in vitro by monocytes from normal subjects and from patients with systemic lupus erythematosus. Scandinavian Journal of Immunology 16(2):117–122.
71. Weltzien RB, Pachter JS. 2000. Visualization of beta-amyloid peptide (Abeta) phagocytosis by human mononuclear phagocytes: Dependency on Abeta aggregate size. Journal of Neuroscience Research 59(4):522–527.
72. Frei PC, Benacerraf B, Thorbecke GJ. 1965. Phagocytosis of the antigen, a crucial step in the induction of the primary response. Proceedings of the National Academy of Sciences of the United States of America 53:20–23.
73. Balazs M, Martin F, Zhou T, Kearney J. 2002. Blood dendritic cells interact with splenic marginal zone B cells to initiate T-independent immune responses. Immunity 17(3):341–352.

74. Martin F, Oliver AM, Kearney JF. 2001. Marginal zone and B1 B cells unite in the early response against T-independent blood-borne particulate antigens. Immunity 14(5):617–629.
75. Robbins DC, Cooper SM, Fineberg SE, Mead PM. 1987. Antibodies to covalent aggregates of insulin in blood of insulin-using diabetic patients. Diabetes 36(7):838–841.
76. Mamula MJ, Lin RH, Janeway CA Jr., Hardin JA. 1992. Breaking T cell tolerance with foreign and self co-immunogens. A study of autoimmune B and T cell epitopes of cytochrome c. Journal of Immunology 149(3):789–795.
77. Hermeling S, Schellekens H, Maas C, Gebbink MF, Crommelin DJ, Jiskoot W. 2006. Antibody response to aggregated human interferon alpha2b in wild-type and transgenic immune tolerant mice depends on type and level of aggregation. Journal of Pharmaceutical Sciences 95(5):1084–1096.
78. Purohit VS, Middaugh CR, Balasubramanian SV. 2006. Influence of aggregation on immunogenicity of recombinant human factor VIII in hemophilia A mice. Journal of Pharmaceutical Sciences 95(2):358–371.
79. Ratner RE, Phillips TM, Steiner M. 1990. Persistent cutaneous insulin allergy resulting from high-molecular-weight insulin aggregates. Diabetes 39(6):728–733.
80. Dintzis RZ, Okajima M, Middleton MH, Greene G, Dintzis HM. 1989. The immunogenicity of soluble haptenated polymers is determined by molecular mass and hapten valence. Journal of Immunology 143(4):1239–1244.
81. Bachmann MF, Zinkernagel RM. 1997. Neutralizing antiviral B cell responses. Annual Review of Immunology 15:235–270.
82. Bachmann MF, Kundig TM, Speiser DE, McKall-Faienza K, Kishara K, Mak TW, Ohashi PS. 1997. T-cell-independent antiviral B cell responses in CD45-deficient mice. Cellular Immunology 175(1):12–15.
83. Bachmann MF, Zinkernagel RM. 1996. The influence of virus structure on antibody responses and virus serotype formation. Immunology Today 17(12): 553–558.
84. Seong SY, Matzinger P. 2004. Hydrophobicity: An ancient damage-associated molecular pattern that initiates innate immune responses. Nature Reviews 4(6): 469–478.
85. Lehmann PV, Sercarz EE, Forsthuber T, Dayan CM, Gammon G. 1993. Determinant spreading and the dynamics of the autoimmune T-cell repertoire. Immunology Today 14(5):203–208.
86. Carpenter JF, Randolph TW, Jiskoot W, Crommelin DJ, Middaugh CR, Winter G, Fan YX, Kirshner S, Verthelyi D, Kozlowski S, Clouse KA, Swann PG, Rosenberg A, Cherney B. 2008. Overlooking subvisible particles in therapeutic protein products: Gaps that may compromise product quality. Journal of Pharmaceutical Sciences 98(4):1201–1205.
87. Roggen EL, Soni NK, Verheyen GR. 2006. Respiratory immunotoxicity: An in vitro assessment. Toxicology In Vitro 20(8):1249–1264.

CHAPTER 11

Regulatory Perspective on Aggregates as a Product Quality Attribute

WENDY C. WEINBERG, LINAN HA, SUSAN L. KIRSHNER, and DANIELA I. VERTHELYI

11.1 INTRODUCTION

Regulatory oversight of protein products must render a balance between the desire to rapidly make novel therapeutics available and the need to carefully evaluate clinical efficacy and potential safety risks. As with all drugs, the regulatory review of biological therapeutics involves a multidisciplinary approach encompassing the characterization of product quality, preclinical safety, and clinical trial design. However, the risk assessment of protein products addresses specific scientific and regulatory concerns. These relate to the inherent heterogeneity of protein products, with the associated issues of maintaining lot-to-lot consistency, the instability of protein products relative to small molecular entities, and the potential impact of manufacturing changes on product quality. These concerns are exacerbated by the failure of current biochemical and biophysical analytical methods to fully characterize larger complex proteins. As such, it is necessary to define, monitor, and control for product-specific critical quality attributes. This often necessitates a case-by-case approach to ensure the availability of a consistent, safe, and effective product.

As discussed elsewhere in this book, proteins may aggregate under certain conditions, potentially impacting product yield, bioavailability, potency, and immunogenicity. Therefore, protein aggregation is considered a critical quality attribute for protein products. Moreover, changes in protein aggregation can be viewed as a hallmark of product quality, and therefore monitoring product aggregation is important in assessing process control, lot-to-lot consistency, and product stability. The correct assessment of aggregates is challenging as they can vary in size from small dimers to multimeric visible particulates, be

Aggregation of Therapeutic Proteins, Edited by Wei Wang and Christopher J. Roberts

linked by covalent or non-covalent interactions, be soluble or insoluble, form amorphous or fibrillar complexes, and have native or non-native conformations depending on the protein and its environment.

From a risk perspective, several elements come into play regarding aggregation of protein products: (1) the product's inherent tendency to form aggregates during manufacturing and storage, (2) the ability to adequately monitor and characterize aggregate species, and (3) the potential clinical consequences brought about by the presence of aggregates. This chapter will provide an overview of the regulatory review process and will address the principal regulatory challenges and technical limitations encountered by reviewers when assessing aggregates in protein products. Furthermore, case studies will be provided to highlight the relevance and potential impact of aggregate formation in product development.

11.2 AN OVERVIEW OF THE REGULATORY PROCESS

11.2.1 Drug Development to Licensure

Most biotechnology and many complex natural source products are regulated under The Public Health Service Act (1944), Section 351, as biologics. As such, the marketing application for a biologic is the Biologics License Application (BLA), which for approval must provide evidence that the product meets applicable requirements to ensure its continued safety, purity, and potency, as codified in the Code of Federal Regulations (CFR) (21 CFR).[1] However, some biotechnology and complex natural source products are regulated under the Federal Food, Drug, and Cosmetic (FD&C) Act (and as amended as part of the Food and Drug Administration Modernization Act of 1997 [FDAMA or Modernization Act]). Despite this difference in regulatory pathways, the scientific and risk considerations are similar for protein products regulated under both acts, so, for the purposes of this discussion, we will not distinguish between products regulated under New Drug Application (NDA) or Biologics License Application (BLA).

The clinical drug development process is commonly divided into three phases, reviewed as part of the investigational new drug (IND) application process, and if successful, culminates in a BLA or NDA. Although in each phase of the investigation sufficient information should be submitted to assure the proper identification, quality, purity, specificity, and potency of the investigational drug, the amount of information needed to make that assurance varies with the phase of the investigation. The processes used in manufacturing biologic products may evolve as the studies progress from first-in-human phase I studies through pivotal phase III trials. The control of product quality is reassessed at each stage of the IND process, with controls appropriate for the stage of development. This evaluation continues post-marketing, and comparability studies are frequently necessary to confirm the consistency of the product derived from new processes or following changes in raw materials with the product used in the clinical trials that supported licensure.

The emphasis in the Chemistry, Manufacturing, and Controls (CMC) section of an initial phase I submission should be placed on providing information that will allow evaluation of the safety of subjects recruited to the proposed study. As described in 21 CFR 312.21, phase I studies are designed to determine the metabolism and pharmacological actions of the drug in humans and the side effects associated with increasing doses, and, if possible, to gain early evidence on effectiveness. The second stage of testing, the phase II trial, includes controlled clinical studies to determine the common short-term side effects and risks associated with the drug and preliminary information on the efficacy of the product in the target patient population. Finally, the pivotal phase III trial must be designed with sufficient power to better define the efficacy and safety of the product. A phase IV clinical trial refers to studies conducted after a product receives market approval. The need for such studies will be determined during the review of the marketing application, depending on the available data as well as on the clinical indication. Typically, phase IV studies include the determination of safety and efficacy in a larger patient population or in a different patient population (such as a pediatric population).

With regard to product quality assessments, the sponsor must establish the ability to manufacture safe, potent, and consistent products in compliance with the general statutory current good manufacturing practice (CGMP) and IND authority.[2] As per 21 CFR 210.2(c), products are exempt from complying with the requirements of the regulations (21 CFR 211) (CGMPs) in phase I.[2] However, important quality parameters should be identified and controlled prior to the initiation of phase I. This includes a safety assessment of raw materials, as well as a thorough testing of the cell substrates for sterility, the presence of adventitious agents, and the ability of the manufacturing process to clear viral and process-related contaminants. Specifications should be set based on available knowledge, with the expectation that the testing parameters and acceptance criteria may be adjusted as additional understanding of product quality attributes is attained and as additional manufacturing and clinical experience is gained. From the outset, lot release testing for protein characterization should include measures of biochemical purity, to capture critical safety-related product and process impurities, and of potency. Potency (or bioactivity), typically assessed in the format of a cell-based bioassay, should capture to the extent possible the mechanism of action of the product. In addition, the stability of the product should be assessed as part of an ongoing stability protocol to ensure safety and potency following a designated storage time.[3,4] Details should be provided in the IND to support that the study design and the test methods used will assure the continued integrity of the product. Inadequate demonstration of product stability may raise safety concerns. In addition, preclinical animal studies are vital for establishing a safe starting dose and a toxicology profile. Important additional safety concerns must be considered in the case of therapeutic proteins of foreign origin or with endogenous counterparts (refer to Section 11.3.1).

11.2.2 Pre- and Post-Marketing Comparability

It is recognized that modifications to the manufacture of the drug substance or drug product are likely as product development progresses. When planning manufacturing changes or scale-up programs, sponsors should consider that comparability will have to be demonstrated.[5,6] Comparability establishes that the physicochemical properties, biological activity, and immunochemical properties of the product derived from the modified process are highly similar to those of the predecessor product in the IND phase or to the approved product in the case of a licensed or approved product. Should differences be detected, evidence will be needed to demonstrate that such differences will not adversely impact the safety or efficacy of the product. Additional characterization and/or bridging studies may be required to support the comparability of the material. If such studies fail to provide convincing evidence of comparability, a clinical study will be required. In the case of new marketing indications, the product will be reassessed to confirm safety for the proposed route of administration and patient population.

11.3 PRODUCT AGGREGATES AND SAFETY CONCERNS

The presence of aggregates can impact the product's biodistribution, pharmacokinetics, and activity, but a primary concern is that aggregates may enhance the product's immunogenicity[7–11](see also Chapter 10). In the case of therapeutic antibodies or multimeric receptors, aggregation can also directly modify the affinity and, consequently, the potency and safety profile of the product.[12–14]

11.3.1 Aggregates and Immunogenicity

Immune responses to biologics are observed frequently.[15] Numerous host and product-related factors determine whether a product will be immunogenic in a group of patients,[12] and many may still be unknown. However, the presence of product aggregates can facilitate the induction of an immune response to the product[10,12,16–18] (see also Chapter 10). The increased immunogenicity of proteins, when available to the immune system as aggregates, is well-known—albeit not perfectly understood—and has been exploited in vaccine development.[19–23] In fact, substantial understanding of the immunogenicity of protein aggregates and relative tolerogenicity of soluble monomeric proteins come from the field of vaccine development.[19,24] The increased immunogenicity of aggregates may stem from changes in pharmacokinetics and tissue concentrations (product depositions and renal clearance), altered cellular uptake (endocytosis vs. phagocytosis, uptake by antigen-presenting cells, and autophagy), or shifts in intracellular trafficking and processing.[25–27] Alternatively, aggregates can lead to cross-linking of surface receptors, elicit inflammatory responses, or mimic an invading pathogen resulting in direct stimulation of receptors of the immune system.[28,29] Lastly, the immune system is armed with gene-encoded pattern recognition receptors poised to respond to invasive microbial threats that have inherent "signatures" consisting of repetitive displays of proteins, polysaccharides, nucleic acids, or lipids on their external surfaces.[30–32] Highly organized large aggregates (>100 kD) of native rather than denatured proteins may be more effective both at triggering innate immune responses and at establishing effective adaptive immune responses.[33–35]

The risk associated with the presence of aggregates can be modeled in animals. Hermeling et al.[36] showed that recombinant human IFN-α preparations containing large-molecular-weight (MW) aggregates in native-like configurations were able to break the tolerance of the transgenic mice expressing recombinant human interferon α-2b. Consistent with observations regarding the potency of large MW aggregates in animal models, patients who received treatment with heavily aggregated preparations of human growth hormone protein developed long-lasting antibody responses, while those treated with a less aggregated preparation developed a more transient response.[16] Similar findings have been observed in patients treated with the cytokine interleukin-2 (IL-2), where the product, formulated as a large MW aggregate, elicited antibody responses in over 80% of patients.[37]

Clinically, an immune response to a therapeutic product can be characterized by the presence of circulating antibodies to the product, inflammation, and/or complement activation. While often no clinical impact is observed, immune responses have been associated with clinical manifestations that can be severe, such as anaphylaxis, loss of efficacy for lifesaving products, or development of factor deficiency syndromes when the responses cross-reacted with non-redundant endogenous factors.[38–40] Historically, severe hypersensitivity reactions were observed following clinical administration of protein products, such as intravenous immunoglobulins (IVIGs) or human serum albumin (HSA), and have been linked to higher levels of protein aggregates.[41,42] Infusion reactions are commonly seen following administration of therapeutic antibodies, some of which may be attributed to complement fixation and complement mediated cytotoxicity.[12,43] Further, an immune response raised against the therapeutic protein may target an endogenous protein counterpart of the therapeutic, leading to serious adverse events, particularly if the endogenous protein has a "unique" or non-redundant function.[15,38] In this potentially devastating scenario, prior to initiating phase I studies, the agency expects an immune response assay to be in place before clinical use and may request that testing of patient samples be performed in real time during phase I and II trials. Additional emphasis is placed on seeking the most relevant and quantitative bioassay possible for assessing neutralizing activity. In these cases, the allowable levels of aggregates, which could influence immunogenicity, are often more stringent. It should be noted that antibody responses can also lead to changes in pharmacokinetics (PK) and pharmacodynamics (PD) and require dose adjustment to avoid loss in efficacy or undue toxicities.[38,44,45]

11.4 THE ASSESSMENT OF AGGREGATES: REGULATORY APPROACHES TO CONTROLLING PRODUCT AGGREGATION

11.4.1 Factors Contributing to the Formation of Aggregates

At each stage of the manufacturing process, therapeutic proteins are exposed to a variety of environments with the potential to cause aggregate formation. The causes of protein aggregation have been extensively addressed in other chapters of this book. Briefly, protein aggregates often result from the interaction of partially unfolded or misfolded molecules at high concentrations as occurs during pharmaceutically relevant production scales, but can also result from exposure to interfaces (e.g., air–liquid and solid–liquid) or metal ions[46] (see Case Study 1). In addition, aggregates can be generated from protein alone or can result from heterogeneous nucleation by foreign micro- and nanoparticles shed, for example, from filling pumps and product containers/closures.[47] Examples of contributing factors that may be encountered, from fermentation, harvest, purification, formulation, and storage, are presented in Table 11.1.

Case Study 1

An investigation revealed out-of-trend and out-of-specification results for aggregates in syringes. Inductively coupled plasma–mass spectrometry (ICP-MS) studies of the syringes revealed highly variable amounts of tungsten oxide residue from the manufacturing process in the syringes, indicating container heterogeneity. Incubation of the active pharmaceutical ingredient (API) with tungsten wires that had been used in the syringe manufacturing process led to an increase in oxidized API, which accelerated aggregate formation. Incubation of the API with unused (elemental) tungsten wires did not result in product oxidation, indicating that the increased aggregation likely resulted from API oxidation due to variable exposure to the tungsten oxides formed during the syringe manufacturing.

11.4.2 Selection and Validation of Detection Methods

11.4.2.1 Common Methods Used to Assess Aggregates Because protein aggregation is a critical quality attribute, protein therapeutic products have aggregate assessment as part of their drug substance and drug product release specifications, as well as for assessment of product stability. Although a desirable strategy involves the use of one test to detect all potential aggregates formed by the product with acceptable sensitivity and precision, often more than one method is likely to be required, as methods that accurately

TABLE 11.1. Factors Contributing to Aggregation during the Manufacturing Process

Primary Sequence
Protein modifications + tertiary structure
• Sequence
• Free thiol groups
• Glycosylation
• PEGylation
• Hydrophobic moieties
Manufacture
• Misfolded protein
• Inorganic impurities
• Temperature
Oxidation
Folding
Formulation
• Concentration
• Solvents
• pH
• Liposomes
Filling and container closure
• Headspace
• Mechanical stress
• Impurities
Silicones
Tungsten
Glass
Storage
• Photostability
• Agitation
• Temperature
Degradation
Deamidation
Oxidation
Freezing/thawing
• Crystallization
Mechanical stress (stirring, pumping)
Adventitious agent removal steps
Shearing, local temperature shifts, cavitation

detect submicron-sized aggregates (e.g., size-exclusion high-performance liquid chromatography [SE-HPLC]) do not detect micron-sized aggregates and vice versa[48] (see also Chapter 5). When selecting the tests to be utilized, one should consider (1) the nature and characteristics of the aggregates formed and include the necessary controls to ensure that the assessment is accurate and complete; and (2) the use of methods that do not require sample alteration for aggregate assessment, such as field flow fractionation, analytical ultracentrifugation (AUC), and use of light scattering to enhance sensitivity of some of these separation techniques in detecting aggregates. Early in the IND process, wider specifications for aggregates can be set, with the expectation that they will be tightened as clinical and manufacturing experience is gained. For products where there is a risk of severe adverse reactions due to product aggregation, the agency may require more stringent specifications earlier in development. It is important to note that it is sometimes difficult to accurately assess aggregates of the final drug product due to formulation issues, for example, the presence of chaperone proteins such as HSA in the formulation mixture, formulation in creams, or extremely low quantities of the API in the drug product.

In addition to the control of aggregates as part of lot release, the formation of product aggregates during shipping and storage has been observed and is a concern that should be addressed in stability studies and incorporated into an established stability protocol. An understanding of the breakdown pathways of a product is important in identifying optimal test methods for assessing the formation of aggregates during stability. To address this, accelerated stability studies in which products are monitored under stress conditions using a variety of methods[3,49] are usually informative and recommended. As in the case of lot release, assessment of aggregates during stability often requires the use of multiple methods that separate and detect aggregates based on different physical principles to ensure detection of the entire spectrum of aggregates present in a sample.

11.4.2.2 Validation of Test Methods for Detecting and Measuring Aggregates All analytical methods used to assess aggregates for release and/or stability evaluations must be demonstrated to be suitable for their intended purpose. Critical aspects of assay validation for the assessment of aggregates include specificity, range, linearity, accuracy, precision, limits of detection and quantitation, robustness, and system suitability testing. The validation of analytical methods is covered in detail in the "International Conference on Harmonization (ICH) Q2: Validation of Analytical Procedures: Text and Methodology."[50] It is recommended that sponsors consider assessing the impact of forced degradation (oxidation, UV exposure, high/low pH), as well as stressed (elevated temperature) conditions on the formation of aggregates, and confirm that the assays selected are suitably sensitive to detect aggregates formed under such conditions. In addition, the agency recommends the use of assays that utilize orthogonal methods for product characterization and validation of release tests that monitor aggregation. Orthogonal methods use a different operating principle to obtain data regarding a particular attribute such as aggregate content. For example, AUC measuring sedimentation rate is an orthogonal method from size exclusion chromatography, both utilized to assess and characterize aggregate content. These characterization tests should be scientifically sound and should provide results that are meaningful but may not necessarily entail full method validation. We highlight below several validation issues that have been particularly problematic for companies:

- *Combining Samples.* Sponsors will routinely pool two drug product containers for aggregate testing. This is unlikely to accurately reflect container closure heterogeneity, as well as possible lack of homogeneity of the drug product. The latter concern may be adequately addressed by validating product homogeneity. The former issue may be addressed by appropriate raw materials qualification programs, including robust extractable/leachable studies designed to assess heterogeneity in the container closure system as noted in Case Study 1. Alternatively, analysis of individual samples from a larger, statistically based sample size can be performed. Vendor notification programs to ensure sponsors will be aware of changes to the manufacture of primary container closure that could impact product aggregation should also be established, rather than relying on combining samples to ensure consistent quality of the raw material.
- *Recovery.* The agency has encountered instances in which companies have implemented aggregate detection assays with low recoveries, for example, 80% of the total protein. Understanding the assay system and its strengths and limitations is important. While no percentage of absolute recovery is mandated, it is useful to characterize assays for recovery and recovery bias (i.e., aggregates, particularly high MW aggregates, disproportionately lost in the process) to understand the information provided by the assay. In addition, system suitability controls need to be in place to ensure that assay sensitivity and bias are appropriately managed.
- *Assay Matrix.* As noted in Case Study 2 (directly below), it is critical to utilize an assay matrix that allows for accurate quantitation of all aggregates present in the drug product when stored under its recommended storage conditions. Thus, the use of orthogonal methods for the detection of protein aggregates is highly recommended.
- *Formulation.* Excipients used in drug product formulation (e.g., HSA) can interfere with the assessment of aggregates; therefore, it is important that any impact of these excipients on the methods used be understood. To minimize this concern, different methods may be required for the assessment of aggregates in the drug substance and in the drug product.
- *Container Closure.* In addition to testing container closure leachables and extractables for their inherent toxicity, they should be assessed for their impact on product quality, including aggregate formation.

Case Study 2

For many years, a company routinely performed an SE-HPLC-based method for assessing aggregates. The agency requested that the company verify the ability of the test to detect aggregates using an orthogonal method; in this case, AUC was chosen. AUC studies showed a large increase (>10%) in aggregate content after a single freeze/thaw. An investigation determined that the running buffer for the SE-HPLC method solubilized the aggregates. The SE-HPLC method was modified to allow detection of the aggregates and storage conditions were altered.

11.4.3 Setting Specifications for Aggregates as Part of Lot Release and Stability Testing

A specification is defined as a list of tests, references to analytical procedures, and appropriate acceptance criteria, which are numerical limits, ranges, or other criteria for the tests described.[51] Methods should be selected that have been demonstrated to be sensitive for detecting aggregates at the time of product release and during stability. Setting appropriate limits for release and stability ought to take into consideration data derived from manufacture, clinical experience, and stability studies (including real-time data under the proposed storage conditions and stressed and forced degradation studies). Because aggregated proteins are an immunogenicity risk, it is critical to link the acceptance limits to clinical experience. The presence of aggregates in the final product or the emergence of aggregates resulting from manufacturing changes or during storage could raise safety concerns. Similarly, inadequate tests or specifications can result in a clinical hold, particularly in those products where the development of an immune response can have a greater adverse impact on patient safety. The acceptance criteria are generally tightened as product development progresses and manufacturing experience is gained.

Case Study 3

Antidrug antibodies were found to develop in a proportion of patients receiving repeated, long-term treatment of a protein therapeutic. The antibodies were capable of neutralizing product activity in an *in vitro* assay, raising the concern that clinical efficacy would be reduced over time of treatment. The sponsor determined that by reformulating the product with a new stabilizing agent, aggregate formation could be reduced. Clinical trials with the new formulation were undertaken to establish whether reduction of aggregates would reduce the incidence of a neutralizing immune response and would favor prolonged clinical efficacy of this product.

TABLE 11.2. Relevant Regulatory Documents

FDA Guidance Documents
Points to Consider in the Manufacture and Testing of Monoclonal Antibody Products for Human Use (February 1997)
Monoclonal Antibodies Used as Reagents in Drug Manufacturing (March 2001)
Bioanalytical Method Validation (May 2001)
Good Review Management Principles and Practices for PDUFA (Prescription Drug User Fee Act) Products (April 2005)
CGMP for Phase 1 Investigational Drugs (July 2008)
PAT—A Framework for Innovative Pharmaceutical Development, Manufacturing, and Quality Assurance (September 2004)
Guidance for Industry for the Submission of Chemistry, Manufacturing, and Controls Information for a Therapeutic Recombinant DNA-Derived Product or a Monoclonal Antibody Product for *In Vivo* Use (August 1996)
ICH Guidance Documents
Q1A(R2) Stability Testing of New Drug Substance and Products (Second Revision)
Q1B Stability Testing: Photostability Testing of New Drug Substances and Products
Q1E Evaluation of Stability Data
Q2A Text on Validation of Analytical Procedures
Q2B Validation of Analytical Procedures: Methodology
Q3A Impurities in New Drug Substances (Revised)
Q3B Impurities in New Drug Products (Revised)
Q5C Quality of Biotechnological Products: Stability Testing of Biotechnological/Biological Products
Q5E Comparability of Biotechnological/Biological Products Subject to Changes in their Manufacturing Process
Q6B Specifications: Test Procedures and Acceptance Criteria for Biotechnological/Biological Products
Q7 Good Manufacturing Practice Guide for Active Pharmaceutical Ingredients
Q8(R2) Pharmaceutical Development
Q9 Quality Risk Management
Q10 Pharmaceutical Quality System

11.4.4 Commonly Used Guidance Documents Relating to Control of Aggregates

IND submissions must conform to federal regulations as per title 21 of the Code of Federal Regulations (21 CFR). All manufacturing steps should comply with CGMPs, as appropriate for the stage of product development.[2] For guidance, the Food and Drug Administration (FDA) has published a series of documents that reflect the agency's opinion regarding specific topics. The ICH also publishes documents on a variety of quality, safety, efficacy, and cross-cutting topics to facilitate global drug development. A select list of FDA and ICH guidance documents that apply to control of protein aggregates are listed in Table 11.2.

11.5 FUTURE CHALLENGES

11.5.1 Control of Subvisible Particulates: A Gap in Regulatory Science

The limits set by the United States Pharmacopeia (USP) for particulates ≥10 μm or ≥25 μm in size, as outlined in <USP 788>, were originally established to contend with the possibility of capillary occlusion and did not contemplate the biological consequences of the administration of a broad size range of aggregates containing protein. Thus, acceptable levels of smaller protein aggregates ranging from oligomers to subvisible particles are not well defined. The limits for soluble aggregates are set on a case-by-case basis, depending on clinical experience and manufacturing history.

For most protein therapeutics, companies routinely specify submicron-sized aggregates both to address the risk of immunogenicity and to control process consistency. Aggregates smaller than 0.1 μm are routinely monitored by a variety of GMP-suited methods, primarily by size exclusion at present, although other methods are also employed. However, aggregates that range in size from 0.2 to 10.0 μm have not been routinely specified or monitored, although they may increase risk of immunogenicity. Indeed, the activity of the most widely used vaccine adjuvant, alum, is thought to be partly due to its ability to present the antigens as particulates in the subvisible range. Aggregates of this size range may also be indicative of problems with product consistency and/or stability. Currently, the range of methods that can accurately measure aggregates between 2 and 10 μm in a GMP environment is limited (primarily to light obscuration and Coulter principle), and they do not provide detail into the chemical composition of the aggregates. However, companies are rapidly developing newer methods or modifications to existing methods to address these gaps in aggregate quantification and characterization.[48,52,53] Capturing this information could be very useful for assessing risk relevant to the formation of aggregates in important lifesaving products such as coagulation factor VIII, IL-2, or replacement enzymes.

11.5.2 Assessment of Biosimilars

Much emphasis has recently been placed on the development of "follow-on" proteins and related products, with the expectation that their similarity to currently approved products might reduce the need for extensive clinical trials for marketing applications. As noted in Table 11.1, a variety of changes in the manufacturing of recombinant protein products can result in alterations in critical quality attributes not always foreseen. This makes a determination of the structural similarity of biologic therapeutics all the more challenging and distinct from small molecule drugs. Technological advances in manufacturing science and analytical methodologies are expected to enhance the control and evaluation of these differences. The ultimate evaluation of these protein products by the FDA will require consideration of numerous factors in a science-driven approach, to include the manufacturing processes used, the complexity of the product and available physicochemical characterization studies, the degree of mechanistic understanding of the product's activity, pharmacodynamic and pharmacokinetic data, and clinical experience.

11.6 SUMMARY

Control of aggregate formation is an important component of ensuring the safety and efficacy of protein products. Aggregates can form at numerous steps throughout the manufacturing process and must be controlled to attain consistency in product quality, reduced lot-to-lot variability, and stability during distribution and storage. The presence of product aggregates must be closely controlled and monitored during manufacture, at release, and during storage using suitable methods to ensure the safety and potency of the product using validated assays. Acceptable levels of aggregates in protein products should be consistently achieved by tightening process control and by establishing appropriate acceptance criteria for lot release and stability studies. The impact of manufacturing changes on product aggregation should always be assessed. Several orthogonal techniques should be used to assess aggregate content in this scenario. These techniques must be able to detect and quantify high MW aggregates, as well as subvisible particles, which may be lost in sample preparation for or during performance of analytical techniques such as size exclusion chromatography (SEC). Importantly, the presence of aggregates is a known factor in the development of antidrug antibodies, which can lead to life-threatening hypersensitivity reactions, modify pharmacokinetics, or even neutralize a product, thereby reducing its therapeutic efficacy. In worst-case scenarios, the antibodies to therapeutic proteins may cross-neutralize the endogenous counterpart and can result in serious adverse events.

ACKNOWLEDGMENTS

We wish to acknowledge Subramanian Muthukkumar for helpful discussions and Drs. Ruth Cordoba-Rodriguez, Kathleen Clouse, Barry Cherney, Amy Rosenberg, Patrick Swann, and Steven Kozlowski for critically reading the manuscript.

DISCLAIMER

The information presented here reflects the current thinking and experiences of the authors. However, this is not an official policy document and should not be used in lieu of regulations, published FDA guidance, or direct discussions with the agency.

REFERENCES

1. U.S. Government. 2009. *Code of Federal Regulations.* Title 21, Food and Drugs.
2. CGMP. 2008. *Guidance for Industry, CGMP for Phase 1 Investigational Drugs.* Center for Drug Evaluation and Research, Center for Biologics Evaluation and Research, and Office of Regulatory Affairs.
3. International Conference on Harmonization. 1996. ICH Topic Q1B: Stability testing: Photostability testing of new drug substances and products.
4. International Conference on Harmonization. 1996. ICH Topic Q5C: Quality of biotechnological products: Stability testing of biotechnological/biological products.
5. Center for Biologics Evaluation and Research. 1997. *Points to Consider in the Manufacture and Testing of Monoclonal Antibody Products for Human Use.*
6. International Conference on Harmonization. 2005. ICH Topic Q5E: Comparability of biotechnological/biological products subject to changes in their manufacturing process.
7. Verbaan F, van Dam I, Takakura Y, Hashida M, Hennink W, Storm G, Oussoren C. 2003. Intravenous fate of poly(2-(dimethylamino)ethyl methacrylate)-based polyplexes. Eur J Pharm Sci 20:419–427.
8. Basu A, Yang K, Wang M, Liu S, Chintala R, Palm T, Zhao H, Peng P, Wu D, Zhang Z, Hua J, Hsieh MC, Zhou J, Petti G, Li X, Janjua A, Mendez M, Liu J, Longley C, Zhang Z, Mehlig M, Borowski V, Viswanathan M, Filpula D. 2006. Structure-function engineering of interferon-beta-1b for improving stability, solubility, potency, immunogenicity, and pharmacokinetic properties by site-selective mono-PEGylation. Bioconjug Chem 17:618–630.
9. Fradkin AH, Carpenter JF, Randolph TW. 2009. Immunogenicity of aggregates of recombinant human growth hormone in mouse models. J Pharm Sci 98:3247–3264.
10. Braun A, Kwee L, Labow MA, Alsenz J. 1997. Protein aggregates seem to play a key role among the parameters influencing the antigenicity of interferon alpha (IFN-alpha) in normal and transgenic mice. Pharm Res 14:1472–1478.
11. Arakawa T, Boone T, Davis JM, Kenney WC. 1986. Structure of unfolded and refolded recombinant derived [Ala125]interleukin 2. Biochemistry 25:8274–8277.
12. Rosenberg AS. 2006. Effects of protein aggregates: An immunologic perspective. AAPS J 8:E501–E507.
13. Wang W, Kelner DN. 2003. Correlation of rFVIII inactivation with aggregation in solution. Pharm Res 20:693–700.
14. Sisu C, Baron AJ, Branderhorst HM, Connell SD, Weijers CA, de Vries R, Hayes ED, Pukin AV, Gilbert M, Pieters RJ, Zuilhof H, Visser GM, Turnbull WB. 2009. The influence of ligand valency on aggregation mechanisms for inhibiting bacterial toxins. Chembiochem 10:329–337.
15. Koren E, Smith HW, Shores E, Shankar G, Finco-Kent D, Rup B, Barrett YC, Devanarayan V, Gorovits B, Gupta S, Parish T, Quarmby V, Moxness M, Swanson SJ, Taniguchi G, Zuckerman LA, Stebbins CC, Mire-Sluis A. 2008. Recommendations on risk-based strategies for detection and characterization of antibodies against biotechnology products. J Immunol Methods 333:1–9.

16. Moore WV, Leppert P. 1980. Role of aggregated human growth hormone (hGH) in development of antibodies to hGH. J Clin Endocrinol Metab 51:691–697.
17. Hochuli E. 1997. Interferon immunogenicity: Technical evaluation of interferon-alpha 2a. J Interferon Cytokine Res 17(Suppl. 1):S15–S21.
18. Schellekens H. 2008. How to predict and prevent the immunogenicity of therapeutic proteins. Biotechnol Annu Rev 14:191–202.
19. Alexander J, Oseroff C, Dahlberg C, Qin M, Ishioka G, Beebe M, Fikes J, Newman M, Chesnut RW, Morton PA, Fok K, Appella E, Sette A. 2002. A decaepitope polypeptide primes for multiple CD8+ IFN-gamma and Th lymphocyte responses: Evaluation of multiepitope polypeptides as a mode for vaccine delivery. J Immunol 168:6189–6198.
20. O'Hagan DT, Singh M, Ulmer JB. 2006. Microparticle-based technologies for vaccines. Methods 40:10–19.
21. O'Hagan DT, Singh M. 2003. Microparticles as vaccine adjuvants and delivery systems. Expert Rev Vaccines 2:269–283.
22. Zaborsky N, Brunner M, Wallner M, Himly M, Karl T, Schwarzenbacher R, Ferreira F, Achatz G. 2010. Antigen aggregation decides the fate of the allergic immune response. J Immunol 184:725–735.
23. Eyles JE, Carpenter ZC, Alpar HO, Williamson ED. 2003. Immunological aspects of polymer microsphere vaccine delivery systems. J Drug Target 11:509–514.
24. Valdes I, Bernardo L, Gil L, Pavon A, Lazo L, Lopez C, Romero Y, Menendez I, Falcon V, Betancourt L, Martin J, Chinea G, Silva R, Guzman MG, Guillen G, Hermida L. 2009. A novel fusion protein domain III-capsid from dengue-2, in a highly aggregated form, induces a functional immune response and protection in mice. Virology 394:249–258.
25. Demeule B, Gurny R, Arvinte T. 2006. Where disease pathogenesis meets protein formulation: Renal deposition of immunoglobulin aggregates. Eur J Pharm Biopharm 62:121–130.
26. Kubota H. 2009. Quality control against misfolded proteins in the cytosol: A network for cell survival. J Biochem 146:609–616.
27. Ishida Y, Yamamoto A, Kitamura A, Lamande SR, Yoshimori T, Bateman JF, Kubota H, Nagata K. 2009. Autophagic elimination of misfolded procollagen aggregates in the endoplasmic reticulum as a means of cell protection. Mol Biol Cell 20:2744–2754.
28. Andrews NL, Pfeiffer JR, Martinez AM, Haaland DM, Davis RW, Kawakami T, Oliver JM, Wilson BS, Lidke DS. 2009. Small, mobile FcepsilonRI receptor aggregates are signaling competent. Immunity 31:469–479.
29. Bachmann MF, Zinkernagel RM. 1997. Neutralizing antiviral B cell responses. Ann Rev Immunol 15:235–270.
30. Kumar H, Kawai T, Akira S. 2009. Pathogen recognition in the innate immune response. Biochem J 420:1–16.
31. Klinman DM, Currie D, Gursel I, Verthelyi D. 2004. Use of CpG oligodeoxynucleotides as immune adjuvants. Immunol Rev 199:201–216.
32. Ishii KJ, Akira S. 2007. Toll or toll-free adjuvant path toward the optimal vaccine development. J Clin Immunol 27:363–371.

33. Dintzis RZ, Middleton MH, Dintzis HM. 1983. Studies on the immunogenicity and tolerogenicity of T-independent antigens. J Immunol 131:2196–2203.

34. Dintzis RZ, Okajima M, Middleton MH, Greene G, Dintzis HM. 1989. The immunogenicity of soluble haptenated polymers is determined by molecular mass and hapten valence. J Immunol 143:1239–1244.

35. Bachmann MF, Rohrer UH, Kundig TM, Burki K, Hengartner H, Zinkernagel RM. 1993. The influence of antigen organization on B cell responsiveness. Science 262:1448–1451.

36. Hermeling S, Schellekens H, Maas C, Gebbink MF, Crommelin DJ, Jiskoot W. 2006. Antibody response to aggregated human interferon alpha2b in wild-type and transgenic immune tolerant mice depends on type and level of aggregation. J Pharm Sci 95:1084–1096.

37. Prummer O. 1997. Treatment-induced antibodies to interleukin-2. Biotherapy 10:15–24.

38. Worobec A, Rosenberg AS. 2004. A risk-based approach to immunogenicity concerns of therapeutic protein products. Part 1: Considering consequences of the immune response to a protein. BioPharm Int 17:22–26.

39. Casadevall N. 2002. Antibodies against rHuEPO: Native and recombinant. Nephrol Dial Transplant 17(Suppl. 5):42–47.

40. Li J, Yang C, Xia Y, Bertino A, Glaspy J, Roberts M, Kuter DJ. 2001. Thrombocytopenia caused by the development of antibodies to thrombopoietin. Blood 98:3241–3248.

41. Gamble CN. 1966. The role of soluble aggregates in the primary immune response of mice to human gamma globulin. Int Arch Allergy Appl Immunol 30:446–455.

42. Ring J, Stephan W, Brendel W. 1979. Anaphylactoid reactions to infusions of plasma protein and human serum albumin. Role of aggregated proteins and of stabilizers added during production. Clin Allergy 9:89–97.

43. Tawara T, Hasegawa K, Sugiura Y, Harada K, Miura T, Hayashi S, Tahara T, Ishikawa M, Yoshida H, Kubo K, Ishida I, Kataoka S. 2008. Complement activation plays a key role in antibody-induced infusion toxicity in monkeys and rats. J Immunol 180:2294–2298.

44. Rosenberg AS. 2003. Immunogenicity of biological therapeutics: A hierarchy of concerns. Dev Biol (Basel) 112:15–21.

45. Shankar G, Shores E, Wagner C, Mire-Sluis A. 2006. Scientific and regulatory considerations on the immunogenicity of biologics. Trends Biotechnol 24:274–280.

46. Ricchelli F, Buggio R, Drago D, Salmona M, Forloni G, Negro A, Tognon G, Zatta P. 2006. Aggregation/fibrillogenesis of recombinant human prion protein and Gerstmann-Straussler-Scheinker disease peptides in the presence of metal ions. Biochemistry 45:6724–6732.

47. Boven K, Stryker S, Knight J, Thomas A, van Regenmortel M, Kemeny DM, Power D, Rossert J, Casadevall N. 2005. The increased incidence of pure red cell aplasia with an Eprex formulation in uncoated rubber stopper syringes. Kidney Int 67:2346–2353.

48. Narhi LO, Jiang Y, Cao S, Benedek K, Shnek D. 2009. A critical review of analytical methods for subvisible and visible particles. Curr Pharm Biotechnol 10:373–381.

49. International Conference on Harmonization. 2003. ICH Topic Q1A(R2): Stability testing of new drug substances and products (second revision).
50. International Conference on Harmonization. 1995. ICH Topic Q2(R1): Validation of analytical procedures: Text and methodology.
51. International Conference on Harmonization. 1999. ICH Topic Q6B: Specifications: Test procedures and acceptance criteria for biotechnological/biological Products.
52. International Conference on Harmonization. 2009. ICH Topic Q4B Annex 3: Evaluation and recommendation of pharmacopoeial texts for use in the ICH regions on test for particulate contamination: Sub-visible particles general chapter.
53. Carpenter JF, Randolph TW, Jiskoot W, Crommelin DJ, Middaugh CR, Winter G, Fan YX, Kirshner S, Verthelyi D, Kozlowski S, Clouse KA, Swann PG, Rosenberg A, Cherney B. 2009. Overlooking subvisible particles in therapeutic protein products: Gaps that may compromise product quality. J Pharm Sci 98:1201–1205.

INDEX

Page numbers in *italics* refer to Figures; those in **bold** to Tables.

a_{23}, size dependence of, 88–89
Aβ. *See* amyloid β (Aβ)
acetamide, 167
acetone, 167
acidic fibroblast growth factor (aFGF), 130, **144**, 155, 172
AcP. *See* acylphosphatase (AcP)
active pharmaceutical ingredient (API), 440
acute infusion reactions, 405, 421
acylphosphatase (AcP), 158
adalimumab, 404
adaptive immune system, 409–410
ADAs. *See* antidrug antibodies (ADAs)
additives, 143–154, **144**
  amino acids, **275**
  chaotropes, **275**, 277–278
  kosmotropes, **275**, 276–277
  polyhydric alcohols, **275**, 276, 278
  polymers, **275**, 279
  protein aggregation and, **122**
  to reduce aggregation, **275**, 275–280
  small charged, 144, 145–147
  small neutral, 143–145, **144**
  sugars, **275**, 276, 278
  surfactants, **275**, 278–279
  use during refolding, 288–290
ADH, 145, 160
administration route, immunogenicity and, 408
administrative protocol, verification of, **335**
adsorption-mediated aggregation path, 80
aFGF. *See* acidic fibroblast growth factor (aFGF)
A53T mutant α-synuclein, 147
aggregate nature of protein, immunogenicity and, 425
aggregates. *See* protein aggregates
aggregation. *See* protein aggregation
aggregation-prone regions (APRs)
  identification and impact of, 103–115
  molecular simulations to identify potential, 113–114
  packing density and, 113
  structural content and, 113
  in therapeutic monoclonal antibodies, **112**
aggregation-prone sequences and sequence patterns, experimentally studied, 109–110
aggregation-prone species, 302
AGGRESCAN, 48–49, 108
agitation
  assessing protein stability after, 267–268
  IgG1 antibody and, 394–395, *395*
  protein aggregation and, 169–170
  silicone oil and, 309
air-water interfaces, aggregation and, 169–170, 305–306
alanine (Ala), 17, 19, 39
albumin, freeze-thaw-induced aggregation of, 347
Albutropin, 151, 170, 346
alcohols, protein stabilization and, **289**, 382
aldimine linkages, 43
α_2, 23
α amino acids, 4
α_1-antitrypsin, 148, 166, **339**, 353
α-β-crystallin, 145
α-chymotrypsin (CT), 130, 336, 343, 353
α-chymotrypsinogen, 160
α-chymotrypsinogen A, 79, *79*, 130, 134
α-crystallin, 150
α-cyclodextrin, 150, 151

α helix, 4, 16, 18
  domain swapping and, 26
α-lactalbumin (α-La), 128, 142, 155
  coaggregates, 160
α-spectrin, 126–127, 164
α-synuclein, 2, 8, 10, 108, 110
  aggregation propensity, 46, 47
  agitation and, 170
  amphoteric polymers and, 149
  coaggregates, 160
  covalent dimerization of, 43
  dissolving aggregates, 159
  self-aggregation of, 130, 131
α-synuclein A53T, 145, 150
α-zein, 44
aluminum, protein aggregation and, 157, 158, **158**
aluminum(III), protein aggregation and, 342
Alzheimer's disease, 1
amino acids, 4
  propensity to form alpha and beta structures, 17
  protein aggregation and, 145–146
  protein stability and, 21–22, 175, **275**, 288–289, **289**
amino acid sequence
  amyloid fibril formation and, 109–110
  protein folding and, 11–12
amino acid side chains, 4
  interaction parameter values for cosolutes and, 90–91, *91*
ammonium sulfate, 167, 277, 307
amorphous aggregates/aggregation, 5–6, *9*, 131
amorphous state, 176
amphoteric polymers/proteins, 149–150
amylin (IAPP), 45, 110
amylin-induced β-cell death, 1–2
amylin 20–29 peptide, 44
amyloid angiopathy, 26
amyloid β (Aβ), 8, 10, 108, 110
  aggregation propensity, 46, 47
  fibril polymerization of, 22
  proline isomerization in, 45
amyloid β (Aβ) peptides, 1, 36, 41
  $A\beta_{1-42}$, 113–114
amyloid cross-beta structure, 26
amyloid fibrils
  amino acid sequence and formation of, 109–110
  aromatic stacking and, 14
  cross-beta structure, *7*, 7–8, *8*
amyloid-like fibrils
  seeding and propagation of, 114
  structural content and formation of, 113
amyloidosis, oxidation and, 41
amyloid plaques, 6–8
  oxidative stress and, 41
amyloids, 10
AMYLPRED, 108–109
analytical techniques, 173
  for subvisible particles, 237–240
  for visible particles, 240–241
analytical tools for aggregate characterization, 212–246, **213–220**
  analytical ultracentrifugation, **214**, 228–231
  asymmetric flow field flow fractionation, **215**, 231–233
  capillary electrophoresis-sodium dodecyl sulfate, **215**, 234–235
  circular dichroism spectropolarimeter, **220**, 245–246
  Coulter counter, **218**, 237, 240
  dynamic light scattering, **214**, 226–228
  electrospray differential mobility analyzer, **218**, 243–244
  flow microscopy imaging, **218**
  fluorescence spectrometer, **219**, 241–243
  Fourier transform infrared (FTIR) spectrometer, **220**, 245–246
  light obscuration, **217**, 237, 238, **239**
  microfluidics chip-based electrophoresis, **216**
  microscopy, **218**, 237, 238–239, **239**
  polarization-induced differential scattering, **219**, 244
  Raman spectrometer, **220**, 245–246
  scanning electron microscope/transmission electron microscope, **219**, 244–245
  size-exclusion chromatography, **213**, 221–223

sodium dodecyl sulfate-polyacrylamide gel electrophoresis, **215**, 233–234
static light scattering, **213**, 223–226
turbidimetry/nephelometry, **217**, 235–237, **237**
analytical ultracentrifugation (AUC), **214**, 228–231, 442–443, 444
anaphylactic reactions, 333, 404
anergy, induction of, 415
8-anilinonaphthalene-1-sulfonate, 134
1-anilinonaphthalene 8-sulfonate (ANS) fluorescence, 241–242, 260–261, 344
animal models for predicting immunogenicity, 427
ANS. *See* 1-anilinonaphthalene 8-sulfonate (ANS) fluorescence
antibodies
cross-reacting, 426
to decrease aggregate formation, 291–292
freeze-thaw-induced aggregation of, 347
sugar protection for in solid state, 175
antibody assay-related factors in protein immunogenicity, *406*, 409
antibody induction, by protein aggregates, 420–423, *421*, *422*
antibody-vinca conjugate aggregation, solid-state pH and, 174
antidrug antibodies (ADAs), 403, 409, 424
antigen-binding fragment. *See* Fab (antigen-binding fragment)
antigen-presenting cells (APCs), 412, *413*, *414*, 420–421, *422*, 424
anti-IL8 mAb, 145
antioxidants, effect on aggregation, 153
antiparallel arrangements, 17–18
prediction of, 47
antiparallel sheets, 6
antiparallel strands, 4
anti-streptavidin, 276
APCs. *See* antigen-presenting cells (APCs)
API. *See* active pharmaceutical ingredient (API)
A30P mutant, 160
apoferritin, 142
apolarity, stability of protein folds and, 13
apolar surface area, solubility and, 20
apomyoglobin, 106, 130, 137, 155, 159, **337**
apomyoglobin analogue, 143
APRs. *See* aggregation-prone regions (APRs)
arc repressor, 67
L-argininamide, 167
L-arginine, 288
arginine (Arg), 15
aggregation and protective effect of, 145–146
aggregation during refolding and, 167
in interfaces, 40
overrepresentation in interfaces, 39
in thermophiles, 19
arginine hydrochloride, effect on structural perturbation and aggregation, 307
arginine kinase, 143, 161
aromatic residues, protein stability and, 14–15
aromatic stacking, 14
Arrhenius kinetics, 343
Arrhenius plots, 379–380
artificial chaperones, 167
Artocarpus hirsuta lectin, 134, 138, 160
asparaginase-2, 134
asparagine (Asn), 17, 19, 44
asparagine (Asn) side chains, deamidation of, 80
asparginase (Asn), 407, 410
aspartic acid (Asp), 15, 19, 44
aspartic acid (Asp) 76, 20
assay matrix, 443
association-limited aggregation, 127
association-limited kinetic pathway, **76**
asymmetric flow field flow fractionation (AF4), **215**, 231–233, 232–233
AUC. *See* analytical ultracentrifugation (AUC)
Avonex, 408

b_{23}, size dependence of, 88–89
Bacillus thermocatenulatus, 146, 152, 157
backbone
coarse-grained models and, 36–37

backbone (*cont'd*)
interaction parameter values for cosolutes and, 90–91, *91*
interactions in fibrillar aggregates, 7, *8*
planarity of peptide bond and, 15–16
torsion angles, 17
bacteria-derived proteins, 268–269, 285–292
BAFF (B-cell activating factor belonging to the TNF family), 414
B-cell activation mechanisms, 411–415, *413*, *414*
T cell dependent, 410, 411–412, *413*
T cell independent, 412–415, *414*, 421–422, *422*, 423–425
B-cell immune tolerance, breaking, 415
B-cell proliferation and differentiation (P/D), 421–422, *422*
B-cell receptors (BCRs), 411, 412, *413*, *414*, 417–418, 421, 422, 424–425
B cells, 410, 411
B-cell tolerance, 411
BCP. *See* blue copper protein (BCP)
BCRs. *See* B-cell receptors (BCRs)
beads on a string-like structure, 10
Beckman LS230 particle sizer, 393
Beckman Z1 coulter counter, 393
benzyl alcohol, 13, **144**, 153, 305
βA3 crystallins, deamidation of, 129
β-amyloid peptide, 130
beta barrels, 8, *9*, 10
β-elimination, 129
β-galactosidase, 131
freeze-drying and, 171
lyophilized, 176
reconstitution of, 350
sugar protection for in solid state, 174
β_H-crystallin, 150, 153
β-lactamase, 165, 166
β-lactoglobulin A
coaggregates, 160
pH and aggregation of, 138
β-lactoglobulin aggregation
amyloid aggregates and, 113
BSA and, 149
calcium binding and, 157, **158**
calcium chloride and, 342
characterizing self-association of using SLS, 225
chemical modification and, 81
excipients and, **144**, 145
glycosylation and, 175–176
ionic strength and, 141, **339**
metal ions and, 158–159
pH and, 139
protein concentration and, 155, 156
temperature and, 134
Tween 20 and, 151
unfolding and, 127
β-lactoglobulin B (β-Lg) aggregation, 128, 134, 141, 170
β-lactoglobulin-enriched whey protein isolate, 339–340
β_L-crystallin, 150, 153
β-mercaptoethanol, 283
β2-microglobulin (β2m), 10
β sheet-containing proteins, proline isomerization and, 45
β-sheet propensity, 17–18, 107, 108
β sheets, 4, *9*, 10
hydrophobic residues and, 16–17
information theory and, 39
prediction of aggregation-prone regions in, 107, 108
structure of fibrillar aggregates, 6
β strands, 16, 18
aggregation-prone regions in, 103, 111, 113, 115
contiguity of, 108
domain swapping and, 26–28
β structure, 16
predicting, 49
propensity to form, 46
β-synuclein, 149
β-trefoil motif, 369
Betaferon, 408
bFGF, 131
BGG. *See* bovine γ immunoglobulin (BGG)
biological activity, immunogenicity and, 407
biologics, 436
Biologics License Application (BLA), 436
biopharmaceuticals

processing steps during manufacturing, 310–315, **311**, *312*, *313*, *315*
production of (*see* bulk production, controlling protein aggregation during)
bioreactor conditions, 270–271
biosimilars, assessment of, 446–447
biotherapeutics, 103–104
predicting aggregation-prone regions in, 106–114
processing steps, 104
BiP, 269
4,4′-bis-1-anilinonaphthalene 8-sulfonate (bis-ANS), 241–242
bis-ANS. *See* 4,4′-bis-1-anilinonaphthalene 8-sulfonate (bis-ANS)
BLA. *See* Biologics License Application (BLA)
blue copper protein (BCP), 157
Boltzmann constant, 94, 132
Boltzmann distribution function, information theory and, 38
botulinum neurotoxin aggregation, 155
freeze-drying and, 171, 172
hydroxyethyl starch and, 342
pH and, 137, **337**
solid-state pH and, 174
sugar protection in solid state and, 174
bovine adenosine deaminase, 406–407
bovine β-lactoglobulin aggregation, 131
glycosylation and, 353
ionic strength and, 142–143
moisture content and, 177
bovine carbonic anhydrase, 130, 167
bovine carbonic anhydrase B (CAB), 165
bovine carbonic anhydrase II (BCA II), 139
bovine core histones, pH and aggregation of, 138
bovine fibrinogen, 155
bovine GCSF, 130
bovine γ immunoglobulin (BGG), 417
bovine insulin, 131, 134, 137, 404
bovine pancreatic chymotrypsin, 130, 135, 138, 149
bovine pancreatic ribonuclease A (RNase A), 137
bovine pancreatic ribonuclease (RNase), 177
bovine ribonuclease A (RNase A), 171
bovine serum albumin (BSA) aggregation, 139–140
dextran sulfate and, 139, 149
excipients and, 143, **144**, 149, 175, 340, 342
ionic strength and, 142, 143
metal ions and, 157, **158**
moisture content and, 177
multiplicity in, 131
pI and, 138
protein concentration and, 155, 156
silicone and, 163
solid-state pH and, 174
spray-freeze drying and, 173
sugar protection for in solid state, 174
sugars and, 143
temperature and, 134
Tween 80 and, 151
unfolding intermediates and, 127
urea and, 159
bovine somatotropin (bST), 148
Brij 35, 170, 346
bromide, 277
Bruton's tyrosine kinases, 412
BSA. *See* bovine serum albumin (BSA)
bST, reducing agents and, 160
buffer cosolvents, aggregation reduction and, 276
buffers
cooling and freezing and, 312
particulation and, 392
protein aggregation and, **121**, **125**, 140–141
bulk drug substance (DS) stability and storage, 292–293
bulk production, controlling protein aggregation during, 257–295
bacteria-derived protein, 268–269
candidate selection, 257–269
cell culture and protein aggregation, 269–271
mammalian cell culture-derived protein, 268
purification and protein aggregation, 271–293, *272*, *273*, **275**, *281*, *284*, **289**
scale-up, 293–295

bulk production, controlling protein aggregation during (*cont'd*)
sequence analysis, 258–259
stability to formulation and long-term storage conditions, 265–268, *267*
stability to process conditions, 259–265, *262*, *264*, *265*

CA. *See* cellulose acetate (CA)
CAB. *See* carbonic anhydrase B (CAB)
CAC. *See* critical association concentration (CAC)
Cα-L, 163
calcitonin, 344
calcium, protein aggregation and, 157, **158**, 158–159
calcium chloride, protein aggregation and, 342
calf lens proteins, 153
CamP method, 47
cancer patients, immunogenicity and, 408
candidate selection, 257–269
bacteria-derived protein, 268–269
mammalian cell culture-derived protein, 268
sequence analysis, 258–259
stability to formulation and long-term storage conditions, 265–268, *267*
stability to process conditions, 259–265, *262*, *264*, *265*
capillary electrophoresis-sodium dodecyl sulfate (CE-SDS), **215**, 234–235
carbohydrate
lyophilized state and, 389–390
protein bioactivity and, 268
as protein stabilizer, 386
carbonic anhydrase, 165
carbonic anhydrase B (CAB) aggregation, 127, 151
denaturants and, 166
protein concentration and, 166
carboxylic acids, inhibition of aggregation and, 175
carboxymethyl cellulose, protein aggregation and, 342
carnosine, effect on high-molecular-weight aggregation, 270
case studies, 367–396
aggregate variability, 394–396, *395*
aggregation in frozen state, 381–385
aggregation in liquid state, 368–381, *370*, *371*, *372*, *373*, *374*, *376*, *377*, *380*
aggregation in lyophilized state, 385–391
protein particulation due to nucleation by foreign material, 391–393
catalysts, refolding, 167
cationic surfactant cetyltrimethyl-ammonium bromide (CTAB), 152
cavitation, 293
cavity-formation thermodynamics, 38
CD. *See* circular dichroism (CD)
CD40, *413*, 414
CD80, *413*
CD86, *413*
CD40L, *413*, 414
CDR. *See* complementarity-determining region (CDR)
cell culture, protein aggregation and, 269–271
cellulose acetate (CA), 314
ceruloplasmin aggregation, 173, 351
CE-SDS. *See* capillary electrophoresis-sodium dodecyl sulfate (CE-SDS)
CFR. *See* Code of Federal Regulations (CFR)
CGMP. *See* current good manufacturing practice (CGMP)
chaotropes, **275**, 277–278, 288, **289**
chaperones, 150
aggregates and, 2, 3, 292
artificial, 167
protein folding and, 17–18, 269
charged residues, 15
solubility and, 18–19
charge neutralization, 22
chelating agent, in protein formulation, 342
chemical aggregation, ionic strength and, 142
chemical alteration, influences on aggregation, 40–46
chemical degradation, indirect aggregation through, 129
chemical equilibrium, classical models of protein folding based on, 29–31
chemical linkages, direct aggregation through, 127–129

chemically controlled kinetic pathway, **76**
Chemistry, Manufacturing, and Controls (CMC), 437
chemokines, 20
chicken egg white lysozyme, 141
Chinese hamster ovary (CHO) cells, 270
Chiti-Dobson equation, 46–47
chloramine-T, 173
CHO. *See* Chinese hamster ovary (CHO) cells
Chou-Fasman secondary structure propensity values, 108
chromatography
 hydrophobic interaction, 282–283
 hydroxyapatite, 282
 ion exchange, 282, 283, *284*
 to remove aggregates, 282–283
cIFN. *See* consensus interferon-alpha mutant (cIFN)
circular dichroism (CD), 245, 260, 261, *262*, 369
circular dichroism (CD) spectropolarimeter, **220**, 245–246
circulating pumping, 313
citrate buffer, aggregation and, 140, 141
citrate synthase aggregation
 quench freezing-induced, 347
 sugars and, 143
classical immune response, 410
classical models of protein folding, 29–31
classification, of aggregates, 206–212, **207–208**, **211**
clinical evidence of aggregates as risk factor for immunogenicity, 419–420
closed β-barrel formation, *9*, 10
closed interface, 26
closures, aggregation and, 162–163, **441**, 443
clotting factors, 104
CMC. *See* Chemistry, Manufacturing, and Controls (CMC); critical micelle concentration (CMC)
CNTF. *See* recombinant human ciliary neurotrophic factor (CNTF)
coarse-grained models, 35–37, 49
Code of Federal Regulations (CFR), 436, 445
collagen matrices, 43
colloidal stability
 role in reducing aggregation rates, 303, 305
 self-association and, 128
complementarity-determining region (CDR), aggregation-prone regions in, 103, 111, 115
computational analysis, 31–33
computational models of protein folding, 33–35
computational tools to predict aggregation-prone regions in proteins and biotherapeutics, 106–109
computer-assisted design of formulations, 344–345
concanavalin A, 163
concentration, industrial protein refolding and, 287–288
condensation particle counter (CPC), 243
condensed state, 31, 34
configurational entropy, 274
conformation, of protein aggregates, 206, **207**, 208
conformational selection, 5
conformational stability, role in reducing aggregation rates, 302–304
conformation of aggregate protein constituents, immunogenicity and, 423–424
Congo red (CR), 6, 105, 241, 242
consensus interferon-alpha mutant (cIFN), 164
container closure, aggregation and, 162–163, **441**, 443
containers, aggregation and, 162–163
contamination sources, in parenteral products, 323–324, *324*
controlled protein delivery systems, preparation of, 350–351
Coomassie Brilliant Blue R-250 dye, 234
copper binding, 45
copper ion, protein aggregation and, 157, **158**
copper sulfate, 164, 270
cosolute-cosolute interactions, 87

cosolvents/cosolutes, 82–94
additive approach to $\Delta\mu^{tr/2}$, $d\mu^{ex/2}/dm_3$, and $\tau^{(u3)/32}$, 89–93, *91*, *92*
interactions with solvents, 21
organic solvents as, 161
predicting effects on unfolding, association and aggregation, 93–94
protein stabilization and, 369, 372
relating $\tau^{(u3)/32}$ to molecular interactions, 85–89
transfer free energies and preferential interaction parameters, 82–85
Coulter counter, **218**, 237, 240, 317
particle detection using, 391, 393
covalent aggregates, **207**, 208
covalent bonds, 4
covalent cross-linking, 42–43
CPC. *See* condensation particle counter (CPC)
CR. *See* Congo red (CR)
creatine kinase aggregation, 135, 150, 156, 161
Creutzfeldt-Jakob's disease, 1
critical association concentration (CAC), 141
critical micelle concentration (CMC), 279
critical opalescence, 236
Crohn's disease, 404
Cro repressor protein mutant, high temperature and aggregation of, 133
cross-beta amyloid structure, 7, *7*, *8*
cross-β steric zipper motif, 107
cross-β structure, 11, 108
cross-reacting antibodies, 426
cryoprotectants, 381–382
temperature and, 375, *376*
cryostabilization, 382
cryptic epitope, *421*
crystalline state, 176
crystallins, 150, 153
crystallization, excipient, 175–176, 381–385
CS, aggregation of, 150, 171
CTAB. *See* cationic surfactant cetyltrimethylammonium bromide (CTAB)
CT aggregation, 133, 146, 150, 155, 161
cupric-silver staining, 234
current good manufacturing practice (CGMP), 437
cyclic polymers, 150–151
cyclodextrins, 150, 167, 290
cyclohexane, 161
cystatin C, 9
cysteine (Cys), 43
cysteine residues, 4, 42
cytochrome c, 133
cytokines, 103

darbepoetin α, spray-freeze drying and, 173
DC. *See* diffusion-collision (DC) model
DCs. *See* dendritic cells (DCs)
deamidation, 40, 44, 80, 129
deglycosylated *P. lycii* phytase (dgPhy), 130, 134, 142
denaturants, 159–160
as stabilizers, 277, 290
unfolding/refolding and, 166
dendritic cells (DCs), 412, 420
deoxy hemoglobin, 137, 170
design of experiments (DOEs), 345
detergents, 167
dextran, 279, 342, 353
dextran sulfate, 139, 148, 149
DF. *See* diafiltration (DF)
diabetes, 2
diafiltration (DF), 335
diaphragm pumps, 315
6-diazo-5-oxonorleucine (DON), 165
dielectric constants, protein folding/unfolding and, 13–14, *14*, 16
differential mobility analyzer (DMA), 243
differential scanning calorimetry (DSC), 389
in candidate analysis, 260, 261, 263, *264*
to probe thermal transitions of antibody domains, 378
diffusion coefficient, temperature and, 132
diffusion-collision (DC) model, 25–26
dihedral angles, 15–16
dimerization, as rate limiting, 82
dimers, domain swapping among, 26–28, *27*

dimethyl sulfoxide, aggregation inhibition during refolding and, 167
disaccharides, 342, 348
discrete molecular dynamics (DMD), 34, 49
DISEMBL, 49
disordered regions, of protein, 4
dissociable aggregates, **207**, 208
dissociation, 283–285
dissociation equilibrium constant, in self-association of native proteins, 71, *72*
disulfide agents, refolding and, 167
disulfide bonds, 4, 5
  covalent cross-linking and, 42–43
  protein folding intermediates and, *28*, 30
  thiols and formation of, 389
  UV irradiation and, 164
disulfide linkages, protein aggregation and, 128–129
disulfide polymerization mechanism, 81
dithiothreitol (DTT), 160, 234, 283
dithiothreitol (DTT)-induced aggregation, 145
dityrosine, 43, 129
DLS. *See* dynamic light scattering (DLS)
DMA. *See* differential mobility analyzer (DMA)
DMD. *See* discrete molecular dynamics (DMD)
$d\mu^{ex/2}/dm_3$, additive approach to, 89–93
DMSO, 161–162
  effect on shelf life, 370, 372, *372*
  protein destabilization and, 382, 384
  protein stabilization during freezing and, 384
  temperature-dependent effects of, 375
DOEs. *See* design of experiments (DOEs)
domain architecture, 4–5
domain swapping, 26–28, *27*, 49
  runaway, 28
DON. *See* 6-diazo-5-oxonorleucine (DON)
Donnan equilibrium, 82
dopamine, aggregation and, 146–147
dose, immunogenicity and, 408
DOWEX media, 290
DPPC, 160
drug development, 436–437
drug substance (DS)
  bulk stability and storage, 292–293
  processing, 274
  storage, 274, 292–293
drying. *See also* lyophilization
  protection of proteins in, 348–350, **349**
  protein aggregation and, **125**, 171–173
    freeze-drying, 171–172
    spray-drying, 172
    spray-freeze drying (SFD), 172–173
    supercritical fluid (SCF) drying, 173
    vacuum-drying, 171
  secondary, 386
DS. *See under* drug substance (DS)
DSC. *See* differential scanning calorimetry (DSC)
DSPC, 160
DTT. *See* dithiothreitol (DTT)
dynamic light scattering (DLS), **214**, 226–228
  assessing protein stability using, 267
  in candidate analysis, 261, 263
  detection of particulates using, 377–378

efficacy, antidrug antibodies and, 405
EGF, 149
Eisenberg model of amylin fiber, 7–8
electron microscopy
  scanning, 244–245
  transmission, 244–245
electrophoresis, 233–234
  CE-SDS, **215**, 234–235
  SDS-PAGE, **215**, 233–234
electrospray-differential mobility analysis (ES-DMA), **218**, 243–244
electrostatic forces, aggregation and, 127
Ellman's reagent, 389
energy, entropy *vs.*, *32*
energy landscapes of proteins, 105–106, *106*
enthalpy, protein folding and, 13
entropy
  energy *vs.*, *32*
  information theory and, 38
  protein folding and, 30–31, 32, *32*

environment, protein response to, 105–106, *106*
enzymes, 104
epitope presentation, 420–421, *421*, 422–423, 424
epitope spreading, 426
EPO. *See* erythropoietin (EPO)
Eprex bulk drug substance, 168
equilibrium constant, for monomer-dimer formation, 73
equilibrium dissociation constants, of multimers, 67–68
equlibrium state diagram, *272*
erythropoietin (EPO), 405–406, 407, 416
Escherichia coli
  expression of N-terminal Fc fusion protein in, 382
  immunogenicity of proteins made in, 407
  proteins derived from, 268
  recombinant KGF in, 369
Escherichia coli OsmB protein, 130
ES-DMA. *See* electrospray-differential mobility analysis (ES-DMA)
ESI-MS, 243
ethanol, 161
European Pharmacopoeia (Ph. Eur.)
  analytical techniques for subvisible particles, 238, **239**
  analytical techniques for visible particles, 240–241
  regulatory requirements on particulate contamination and control of subvisible and visible particles in parenteral products, 317, 320–321, *321*
  on visual clarity, 236–237, **237**
  visual inspection methods, 321, *321*
European Union, regulatory requirements on particulate contamination and control of subvisible and visible particles in parenteral products, 317
excipient crystallization, 176
  aggregation in frozen state and, 381–385
excipients
  aggregation in lyophilized state and, 385–391
  drying-induced aggregation and, 348, **349**
  effect on structural perturbation and aggregation at surfaces, 307–308
  preparation of final formulation with, *313*, 313–314
  protein aggregation and, **125**, 143–153, **144**
  protein stabilization and, 340
  solid-state condition and, 174–176
  temperature-dependent effects of, 373–375, *374*
excluded volume effects, 372–373, *373*
expression, aggregation and, 164–165
expression strategies, 269–270
extrinsic aggregate factors, immunogenicity and, 423
extrinsic material, 323

Fab (antigen-binding fragment), 368
F-actin, 150
factor VIII, 446
factor VIII antibodies, 407
factor IX, 347
factor XIII, 347
FAT domain proteins, 49
Fc conjugates, 285–286, 286
Fc (fragment crystallizable), 368
Fc fusion protein aggregation, **154**, 159, 368
  in frozen state, 382–385
  pH and, 137
  during purification, 168
Fc fusion protein candidates, analysis of, 260–261
FDA. *See* Food and Drug Administration (FDA)
Fenton chemistry, 43
fermentation, aggregation and, 164–165, 440
FFF. *See* field flow fractionation (FFF)
FGF. *See* fibroblast growth factor (FGF) family
fibril formation, 131
fibrillar aggregates, 5, 6–11, *7*, *8*, *9*
fibroblast growth factor (FGF) family, 137, 147, 369
Ficoll-70, 167, 175
Ficoll-70000, 147

field flow fractionation (FFF), 75, 173, 230, 231–233
fill/finish manufacturing, particulation due to nucleation by foreign materials in, 391–393
filling, 314–315, *315*, **441**
filling pump, 392
filtration, 314
firefly luciferase, 130, 134, 137
Flexboy bags, 163
flow field flow fractionation (F4), 232
flow microscopy imaging, **218**
fluorescence spectroscopy, **219**, 241–243, 260–261
  using extrinsically fluorescent dyes, 241–243
folded proteins, self-association of, 70–73, *72*
foldons, 25
Food, Drug, and Cosmetic (FD&C) Act, 436
Food and Drug Administration (FDA)
  biosimilars and, 447
  regulatory documents, 445, **445**
Food and Drug Administration Modernization Act (1997), 436
forced degradation, 442
formaldehyde-mediated cross-linking, 129
formazin turbidity unit (FTU), 237
formulation
  aggregation and, 440, **441**, 443
  immunogenicity and, 407
  preparation of final, with excipients, *313*, 313–314
formulation and stability evaluation, **335**, 336–343
formulation development, 334–345
  computer-assisted design, 344–345
  high-throughput, 344
  protein stability and, 265–268
  traditional, 334–343
fouled membranes, 245
four-bead models, 36–37
Fourier transform infrared (FTIR) spectroscopy, **220**, 245–246, 260, 324
  examining secondary structure using, 388, 389, 390, 391
  spectrum of IgG1 antibody after stresses, 394–396, *395*
fragment crystallizable. *See* Fc (fragment crystallizable)
free energy
  cosolvents and change in, 375, *377*
  of native folded state, 13
  thermodynamic stability of protein and, 132
freeze-dried solid products, 334
freeze-drying, **124**, 171–172
freeze-thaw, 294
  assessing protein stability after storage and, 266
  protection of proteins during, 346–347
  protein aggregation and, **123**, 168–169, **441**
freezing operations, 310–312, **311**, *312*
fructose, as protein stabilizer, 278
FTIR. *See* Fourier transform infrared (FTIR) spectroscopy
FTU. *See* formazin turbidity unit (FTU)
fusion constructs, to decrease aggregate formation, 291–292

GAGs. *See* glycosaminoglycans (GAGs)
γ-crystallin aggregation, 150, 153, 163–164
gas-phase electrophoretic mobility molecular analysis (GEMMA), 243
GCSF. *See* granulocyte colony-stimulating factor (GCSF)
gel filtration chromatography, 221–223
GEMMA. *See* gas-phase electrophoretic mobility molecular analysis (GEMMA)
Genentech, 388
GH. *See* growth hormone (GH)
Gibbs free energy
  information theory and, 38
  protein folding and, 12
  in self-association of native proteins, 70
globular domain folds, 4–5
globular proteins
  classical models of protein folding and, 29
  folding and unfolding processes of, 23
  thermodynamic stabilizing of, 19

glucose
  aggregation inhibition during refolding and, 167
  ε-amine lysyl moiety and, 43
  freeze-thaw-induced aggregation and, 347
  as protein stabilizer, 278, 290
glucose oxidase aggregation, moisture content and, 177
glutamic acid (Glu), 15, 19, 20
glutamic acid (Glu) residues, 20
glutamine (Gln), 19, 44
glutamine (Gln) side chains, deamidation of, 80
glutamyl endopeptidase, 160
glutathione, 42
  oxidized, 167
glyceraldehyde 3-phosphate dehydrogenase, **154**, 157
glycerol
  aggregation reduction and, 167, 276
  protein aggregation and, 161
  as protein stabilizer, 278, 290
  refolding and, 167
glyceryl monooleate, 346
glycine (Gly), 39, 146, 168, 407
glycine (Gly)37, 41
glycine (Gly)38, 41
glycodelin A (GdA), pH and aggregation of, 138
glycogen phosphorylase b, **154**
glycoprotein *P. lycii* phytase (Phy), 134
glycoproteins, deglycosylation of, 407
glycosaminoglycans (GAGs), 22
glycosylated interferon-α2, 407
glycosylated interleukin-2 (IL-2), 407
glycosylation, 142, 353
  aggregate-related immunogenicity and, 425
  aggregation control and, 165
GMP. *See* good manufacturing practice (GMP)
Go models, 34–35
good manufacturing practice (GMP), 321, 322, 446
grand canonical ensemble, 95
granulocyte colony-stimulating factor (GCSF), 13
  aggregation of, 127, 128, 212
  denaturants and aggregation of, 159
  detection of aggregates using SEC, 222
  PEGylation of, 352
  pH and aggregation of, 137
  protein concentration and aggregation of, **154**, 155, 156
granulocyte colony-stimulating factor (GCSF) folding, 291
GroEL-GroES, 167
growth factor-β aggregation, 344
growth factors, 103
growth hormone (GH), 42. *See also* human growth hormone (hGH) aggregation
  aggregation of, 147
  high pressure and disaggregation of, 171
  immunogenicity and, 404
  polyanions and, 148
guanidine, 159–160, 277, 290
guanidine hydrochloride, 5, 379
guanidinium, 147, 277

HA, 342
hard spheres (HSs), 83, 95
harvest, 440
HCC. *See* human cystatin C (HCC)
heat shock proteins (HSPs), 150
heat stress, IgG1 antibody and, 394, *395*, 395–396
3_{10}-helix, 4
helper T-lymphocytes, 410
hemoglobin, 20
  freeze-thaw-induced aggregation and, 168
  irradiation and, 173
  L-tyrosine and, 351
  protein concentration and aggregation of, **154**
  unfolding of, 134
hen egg white lysozyme (HEWL), 131, 135, 143, 145, 159
heparin, 135, **144**, 148–149
heterodimerization, 20
heterogeneity, in IgG2 aggregation during long-term storage, 376–381, *377*, *380*
HEWL. *See* hen egg white lysozyme (HEWL)

HGG. *See* human γ globulin (HGG)
hGH. *See* human growth hormone (hGH)
HIC. *See* hydrophobic interaction chromatography (HIC)
high-molecular weight (HMW) aggregates, 68–69, 143, 155
high-molecular weight (HMW) proteins, refolding problems, 286
high-molecular weight (HMW) soluble aggregate, 69
high-molecular-weight species (HMWS), 263, 269
high-pressure refolding of aggregates, 290–291
high-throughput formulation development (HTFD), 344
hinge regions, 368
histidine (His), 15, 19, 43, 145, 146, 175, 387
histidine (His) buffer, aggregation and, 140
histidine (His) tags, 292
HIV-1 envelope glycoproteins, 414
HMW. *See under* high-molecular weight (HMW)
HMWS. *See* high-molecular-weight species (HMWS)
Hofmeister ions, 277
holo-α-lactalbumin aggregation, 134, 141, **154**, 163, 342
homocomplexes, 39, 40
hormones, 103
horse heart apomyoglobin, 106
horse heart myoglobin, 131, 143
horse liver alcohol dehydrogenase (ADH), 134
HP-β-CD, 150, 342
HP (hydrophobic polar) model, 33–34
H-predictor, 49
HS. *See* hard spheres (HSs)
HSA. *See* human serum albumin (HSA)
HSPs. *See* heat shock proteins (HSPs)
HS repulsion, 86
HTFD. *See* high-throughput formulation development (HTFD)
human amylin (hIAPP), 113
human α-synuclein, 143
human cystatin C (HCC), 26–28, *27*
human factor XIII, 172
human γ globulin (HGG), 417, *418*, 419
human growth hormone (hGH)
  aggregates, 210, 212
  immunogenicity and, 419
human growth hormone (hGH) aggregation
  agitation and, 170
  binding antibodies and, 405
  denaturants and, 159
  drying-induced, **349**
  excipient inhibition of, 175
  freeze-drying and, 171
  freezing and, 312, *312*
  moisture content and, 177
  polyanions and, 148
  during shaking, 346
  spray-drying and, 172
  sugar protection for in solid state, 174
human hemoglobin A_0, 155
human interferon-α (hIFN-α), 418–419, 424
human interferon-γ aggregation, 134
humanized monoclonal antibody, 345
human lysozyme variants, 127
human muscle AcP aggregation, trifluoroethanol and, 161
human muscle acylphosphatase, **158**
human recombinant interleukin-1 receptor antagonist (IL-1ra), 130–131
human relaxin, 137
human serum albumin (HSA), 134, 407, 419, 438, 442
human transthyretin, 131, 143
Humicola lanuginosa lipase, sugar protection for in solid state, 175
Huntington's disease, 1
HX. *See* hydrogen-deuterium exchange (HX) experiments
hydration
  preferential, 373, *374*
  thermodynamics of, 38
hydration shell, 21
hydrogen bonding, 4
  coarse-grained models and, 36–37
  excipients and, 348
  PASTA and, 48

hydrogen bonding (*cont'd*)
  protein stability and, 20
  secondary structure and, 16–17
hydrogen-deuterium exchange (HX) experiments, 30
hydrophobic effect, 38
hydrophobic interaction chromatography (HIC), 282–283, 292
hydrophobicity
  aggregation and, 39, 41
  formation of condensed state and, 33–34
  immunogenicity and aggregate, 425
  stability of protein folds and, 13–15
hydrophobic protein interactions, kosmotropes and, 276–277
hydrophobic residues, 16–17
  aggregation and, 112–113
  in interfaces, 39–40
hydrostatic pressure, aggregation and high, 170–171
hydroxyapatite chromatography, 282
hydroxyethyl cellulose, inhibiting aggregation during freeze-drying and, 172
hydroxyethyl (heta) starch, 147, 172, 279, 342
hydroxyl-containing Ser/Thr residues, 110
2-hydroxylpropyl-β-cyclodextrin, 350, 351
hydroxylysine, 43
hydroxyproline, 146

ice formation, initiation of, 311–312
ice-water interface, freeze-thaw-induced aggregation and, 169
ICH. *See* International Conference on Harmonization (ICH)
ideal gas law, describing protein folding and, 30–31
IEX. *See* ion exchange chromatography (IEX)
IEXα-La, 163
IFN. *See under* interferon (IFN)
iFold server, 49
IgA, 412, 414
IgD, 412
IgD receptors, 412, *413*
IgE, 412
IGF-1, 291
IgG, 412, 414, 425
IgG aggregation, 368
  containers and, 162
  excipient inhibition of, 175
  freeze-drying and, 171
  freeze-thaw-induced, 347
  protein concentration and, 166
  during shaking, 346
  spray-drying and, 172
  temperature and, 134
IgG1 aggregation, 155, 368
  agitation and, 170
  drying-induced, **349**
  spray-drying and, 172
  sugar protection for in solid state, 174
IgG1 antibody
  FTIR spectra, 394–395, *395*
  solid-state pH and, 174
IgG1-X myeloma protein, 272
IgG2 aggregation, 131, 368
  hetereogenity and non-linearity in during long-term storage, 376–381, *377*, *380*
IgG2-A, 379–380, *380*
IgG2-B, 379–380, *380*
IgG2 disulfide-bonded isoforms, 379–380, *380*
IgG2 isoforms, 394
IgG2 mAb, 155
IgG3, 368
IgG4 aggregation, 170, 368
IgG light chain, 130
IgG monoclonal antibody, particulation due to nucleation of, 391–393
IgM, 168, 412, 414, 415
IgM cryoglobulin, 136
IgM receptors, 412, *413*
IL. *See under* interleukin (IL)
Ile, 16
imidazole, 15, 146
immobilization, 292
immune mechanisms, protein immunogenicity and, 409–415
  B-cell activation mechanisms, 411–415, *413*, *414*
  breaking B-cell immune tolerance, 415

breaking immune tolerance, 410–411
classical immune response, 410
immune tolerance, 411
immune response, therapeutic proteins with aggregates and, 2
immune system, 409
immune tolerance, 411
breaking, 410–411
immunogenicity, protein aggregates and, 333, 438–440
immunogenicity of therapeutic proteins, 403–428
aggregates and, 415–427, *418*, *421*, *422*
antibody assay-related, *407*, 409
consequences of for patients, 405–406
immune mechanisms, 409–415, *413*, *414*
incidence of immunogenicity during therapy, 404
product-related factors in, 406–407, *407*
treatment-related factors, *407*, 408
immunoglobulin aggregates, 133, 333. *See also under* Ig
immunoglobulin light chain, 10
immunon, 424
impurities, protein aggregation and, 160
incidence of immunogenicity during therapy, 404
inclusion bodies, 3, 5–6, 164, 286, 290
IND. *See* investigational new drug (IND)
indirect aggregation, through chemical degradation, 129
induced fit model, 5
infliximab, 404
information theory (IT), 37–40
infusion reactions, 405
inositol hexaphosphate, 135, 147, 149
in-process, 274
in silico models for predicting immunogenicity, 427
insoluble aggregates, 301, 302
pathways leading to, 66, *67*, 68
insulin, 42, 100
insulin aggregation, 210. *See also* bovine insulin
additives and, 145
agitation and, 170
containers and, 163
excipients and, 343
glycosylation of, 353
immunogenicity and, 404
ionic strength and, 141–142
moisture content and, 177
multiplicity of, 131
neutral polymers and fibrillation of, 147
polymers and, 148
protein concentration and, 155
reducing agents and, 160
during shaking, 346
solid-state pH and, 174
sugars in solid state and, 175
temperature and, 133, 134
unfolding intermediates and, 127
interfaces
effects on protein aggregation, 305–310
air-water, 305–306
effects of excipients, 307–308
interactions of proteins with microparticle surfaces, 309
shear effects, 309–310
silicone oil/water, 309
solid-liquid, 306–307
protein, 39–40
interferon-α (IFN-α), 222, 418, 420
interferon-β (IFN-β) products, immunogenicity and, 408
interferon-γ (IFN-γ)
detection of aggregates using SEC, 222
jet nebulization-induced aggregation, 351
reconstitution of lyophilized, 350
spray-freeze drying and, 173
interferon-γ (IFN-γ) dimer, 157
interferon (IFN) aggregates, 212
interferons, 42
interferon-τ (IFN-τ), 127, 130, 140, 143, 145
interleukin aggregates, 212
interleukin-1β (IL-1β), 129, 137, 151, 155, 336
interleukin-1ra (IL-1ra) aggregation, 145, 339, **339**
excipient inhibition of, 175

interleukin-1ra (IL-1ra) aggregation (*cont'd*)
  high hydrostatic pressure and, 170
  solid-state pH and, 174
  sugar protection for in solid state, 174
interleukin-1ra (IL-1ra) mutein, 135
interleukin-2 (IL-2), 175, 343, 438, 446
interleukin-2 (IL-2) mutein, 137
  aggregation during shaking/agitation, 170, 346
  glycosylation of, 353
  polysorbate 80 and, 341, *341*, *342*
interleukin-6 (IL-6), 172, 175
interleukin-10 (IL-10), 414
interleukin-11 (IL-11), 175, 342
interleukin-13 (IL-13), 269
interleukin-18 (IL-18), 258–259
intermediates, folding, 29–30
intermolecular disulfide bond formation/exchange, 128
International Conference on Harmonization (ICH), 442, 445, **445**
intersheet associations, 37
interstrand associations, 37
intracellular aggregation, 42–43
intramuscular administration, immunogenicity and, 408
intravenous administration, immunogenicity and, 408
intravenous immunoglobulins (IVIGs), hypersensitivity reactions to, 438
intrinsic aggregate features, immunogenicity and, 423
intrinsic disorder, predicting, 49
intrinsic particulates, 323
investigational new drug (IND), 436, 437
in vitro models for predicting immunogenicity, 427
in vivo models for predicting immunogenicity, 427
iodine, 277
ion exchange chromatography (IEX), 282, 283, *284*, 292
ionic strength, aggregation and, **121**, 141–143, 338–340, **339**
  colloidal stability and, 305
  pH and, 339–340
  during processing, 279–280
ionic surfactants, 152
ion-pairings, 19
ions, protein stability and, 21–22
IPOD site, 2
irradiation, 173, 351
irreversible aggregates, **207**, 209, 211, 287
islet amyloid protein, 10
isoaspartic acid (IsoAsp), 44
isoelectric point (pI), 15
  protein aggregation *vs.*, 138–139
isoelectric precipitation, 282
isoleucine, 40
isomerization, proline, 45–46
IT. *See* information theory (IT)
IVIGs. *See* intravenous immunoglobulins (IVIGs)

Japanese Pharmacopoiea (JP)
  analytical techniques for subvisible particles, 238, **239**
  analytical techniques for visible particles, 240–241
  regulatory requirements on particulate contamination and control of subvisible and visible particles in parenteral products, 317, 320
  visual inspection methods, 322
jet nebulization, 351
JP. *See* Japanese Pharmacopoiea (JP)
JUNQ site, 2

κ-casein (κ-CN), 128, 134
Karl Fischer titration method, 388
keratocyte growth factor (KGF), osmolytes and stabilization of, 368–376, *370*, *371*, *372*, *373*, *374*, *376*
KGF-2, 134
K97I IL-1β, 130, 166
kinetic pathways, distinguishing, 73–80
kinetics
  aggregation, 69, 75–80
    chemical modifications affecting, 80–82
    metal ions and, 158–159
  differences between conformational states, 32
  of fibril formation, 37
  linear Arrhenius, 377–378
  non-Arrhenius, 376, 378–380
  non-linear Arrhenius, 394
  of protein folding, 11, 17–18, 29

Kirkwood-Buff fluctuation theory, 83
Kirkwood-Buff integrals, 84
  for water-peptide, cosolute-peptide and osmolytes, *92*, 92–93
kosmotropes, **275**, 276–277, 282, 288

lactate dehydrogenase, 347
  freeze-thaw-induced aggregation and, 168
  protein denaturation in frozen state and, 384
lactose, as stabilizer, 172, 290
λ-cro, 20
Lamm equation, 229
laser-induced fluorescence (LIF), 235
late embryogenesis abundant (LEA) proteins, 171, 347
late infusion reactions, 405
lattice method, 34
L chirality, 4
LEA. *See* late embryogenesis abundant (LEA) proteins
lentil seedling amine oxidase, 161
LESs. *See* local elementary structures (LESs)
leucine, in protein interfaces, 40
LIF. *See* laser-induced fluorescence (LIF)
ligand, conformational stability and, 304
light exposure
  assessing protein stability after, 268
  protein aggregation and, **121**, 163–164
light obscuration, **217**, 237, 238, **239**
  subvisible particle counting by, *317*, 317–318, **319**, *320*
light obscuration particle count test, 238, **239**
light scattering
  dynamic, **214**, 226–228
  multi-angle laser, 224
  static, **213**, 223–226
light scattering system (SEC-LS), 388
linear Arrhenius kinetics, 377–378
line noise, 38
linkage, of protein aggregates, **207**, 208
linoleic acid, 147
lipids, protein aggregation and, 142, 160
liquid drug products, 334
  manufacturing processes for, 347–348
liquid state, aggregation in
  heterogeneity and non-linearity in IgG2 aggregation, 376–381, *377*, *380*
  role of osmolytes in stabilizing KGF, 368–376, *370*, *371*, *372*, *373*, *374*, *376*
lithium, 276
LMW-UK. *See* low-molecular-weight urokinase (LMW-UK)
local elementary structures (LESs), 35
lock-and-key binding, 5
long-term storage stability, **335**, 343
lot release, 442, 444
lot-to-lot consistency, 435
low-molecular-weight heparin, 148
low-molecular-weight urokinase (LMW-UK), 147–148, 149, 155, 277, 280
low temperature, protein aggregation and, 136
L68Q mutation, 26, *27*
lymphocytes, 410. *See also* B cells; T cells
lyophilization, 342
  steps in, 385–386
lyophilized state, aggregation in, 385–391
lysine hydrochloride, effect on structural perturbation and aggregation, 307
lysine (Lys), 15, 19, 20, 39, 145
lysine residues, covalent linkages with, 43
lysozyme aggregation, 160
  denaturants and, 166
  freeze-drying and, 171
  high pressure and disaggregation of, 171
  hydroxyproline and, 146
  ionic strength and, 141, **144**
  pH and, 336, **337**
  protein concentration and, 155, 166
  during refolding, 167, 168
  silicone and, 163
  spray-freeze drying and, 173
  supercritical fluid drying and, 173
  temperature and, 133, 134
lysozymes, 21

mAAT. *See* mitochondrial aspartate aminotransferase (mAAT)
mAb. *See under* monoclonal antibody (mAb)

macromolecular crowding, 154
macroscopic surfaces, influence of adsorption to, 80
magnesium, 276
magnesium chloride, 277
magnetic stirrer, 313
Maillard reaction, 129
major histocompatibility complex class II (MHC II), *413*, 421
maltosyl-β-cyclodextrin, 150
MALZ, 278
mammalian cell culture-derived protein, 268, 285
manganese (II), protein aggregation and, **158**
mannitol
  inhibiting aggregation and, 148
    during freeze-drying, 172
    during spray-drying, 172
  as protein stabilizer, 278
  sugar protection in solid state and, 175
manufacturing biopharmaceuticals. *See also* bulk production, controlling protein aggregation during
  fill/finish, 391–393
  liquid drug products, 347–348
  processing steps, 310–315, **311**, *312*, *313*, *315*
  protein aggregates and, 440, **441**
marketing, of new drug, 438
mass action view of folding, 30
MD. *See* molecular dynamics (MD)
MDC. *See* monomer-directed conversion (MDC) model
mechanical effectors, 293
megakaryocyte growth and differentiation factor (MGDF), 407
melting temperature
  protein aggregation and, 135
  shelf life and, 370–372, *372*
membrane permeability, aggregate structure and, 10
2-mercaptoethanol, 167, 234
mesophilic proteins, 19
Met, 43
metal ions, protein aggregation and, **122**, 157–159, **158**, 175, 341–342
Met4, 170
Met4 residues, 130
Met-hGH aggregation, 175
  drying-induced, **349**
  freeze-drying and, 171
  spray-drying and, 172
methionine oxidation, 41–42
methionyl pGH, 151
methoxy polyethylene glycol (mPEG), 352
metmyoglobin, 133
MFI. *See* micro-flow imaging (MFI)
MGDF. *See* megakaryocyte growth and differentiation factor (MGDF)
MHC II. *See* major histocompatibility complex class II (MHC II)
microdomains, 25
micro-flow imaging (MFI), 239, 318
microfluidics chip-based electrophoresis, **216**
microglobulin, 110
microparticle surfaces, proteins and, 309
microplates, HTFD, 344
microscopy, **218**, 237, 238–239, **239**
  subvisible particle counting by, 318, **319**
misfolding diseases, 1–2
mitochondrial aspartate aminotransferase (mAAT), 134
mixed micelles, 290
mixing, 294
  of bulk drug solution after thawing or pooling, **311**, 313
Miyazawa-Jernigan (MJ) amino acid interaction matrix, 34
moisture
  protein aggregation and, **126**
  role in aggregation in lyophilized state, 385–391
moisture content, 177
molecular dynamics (MD), 34, 105
molecular interactions, $\tau^{(u3)/32}$ and, 85–89
molecular sieve chromatography, 221–223
molecular simulations, to identify potential aggregation-prone regions and seeds, 113–114
molten globules, 120, 126–127. *See also* unfolding intermediates
molten globule states, 13
monoclonal antibody (mAb), 42

to bind hydrophobic patches, 291
as focus of drug development, 367
precipitation of, *313*, 313–314
monoclonal antibody (mAb) aggregation, 210, 212
aggregation-prone regions, 103–115, **112**
prediction of, 110–113, *111*
detection of aggregates using SEC, 222
excipient inhibition of, 175
moisture content and, 177
pH and, 137, 285, **337**
solid-liquid interface and, 306
spray-drying and, 172
monoclonal antibody (mAb) candidates
analysis of, 261–265, *262*, *264*, *265*
assessing stability after storage and freeze-thaw, 266
monoclonal antibody (mAb) IgG2, freeze-thaw-induced aggregation and, 169
monomer concentration, aggregation formation and, 77–78
monomer conversion, 23
monomer-dimer formation, 73–74
monomer-directed conversion (MDC) model, 23
monomers
kinetics of loss of, 77
in self-associated oligomers, 68
Monte Carlo simulations, 34, 114
mouse monoclonal antibody (mAb) aggregation
excipients and, 343
pH and, 137
multi-angle laser light scattering (MALLS), 224–225
multimeric folding, 67–68
multiplicity, in protein aggregation, 131
mutagenesis, 351–352
mutant superoxide dismutase-containing aggregates, 150
mutant transthyretin (ΔN50), 130, 134
mutations
aggregation rate and, 109
effect on protein folding, 32
protein stability and, 45
mutual information, 38
myristic acid, 147
N-acetyl glucosamine (NAG), aggregation inhibition during refolding and, 167
National Instrument Filmatic pump, 391
native aggregates, 206, **207**, 208, 210
native aggregation pathways, 66–69, *67*
native-like conformation, immunogenicity and, 424
native monomers, interaction with aggregates, 79, *79*
native proteins, self-association of, 70–73, *72*
native state, 11
N′-butyl derivatives, 167
NCC. *See* nucleation conformational and conversion (NCC) mechanism
NDSB. *See* 3-(1-pyridinio)-1-propanesulfonate (NDSB)
nebulization processes, 173
neo-epitopes, 420–421, *421*
nephelometric turbidity unit (NTU), 237
nephelometry, **217**, 235–237, **237**
N′-ethyl derivatives, 167
neurofilaments (NFs), 38–39
neutral crowders, 147
neutralizing antibodies, 426
neutral polymers, 147–148
NF. *See* U.S. National Formulary (NF)
NFs. *See* neurofilaments (NFs)
Nile red, 241, 242
NNQQNY, 109
nomenclature, aggregation, **64–66**, 209–212
non-Arrhenius kinetics, 376, 378–380
nonclinical evidence of aggregates as risk factor for immunogenicity, 417–419, *418*
non-covalent aggregates, 104, **207**, 208, 210
noncovalent bonds/association, 4, 5
non-dissociable aggregates, **207**, 208
non-Hodgkin's lymphoma, 408
nonionic surfactants, 151–152
non-linear Arrhenius kinetics, 394
non-linearity, in IgG2 aggregation during long-term storage, 376–381, *377*, *380*

nonnative aggregates, **207**, 208
nonnative aggregation kinetics, chemical modifications and, 80–82
nonnative aggregation pathways, *67*, 68–69
nonnative proteins, aggregation and, 63
nonnative self-association, 73–75
non-steric interactions, 86, 95
N-O pairings, 19
NP. *See* nucleation-polymerization (NP)
N′-substituted N-methylimidazolium chlorides, 167
N-terminal Fc fusion protein, 382
NTU. *See* nephelometric turbidity unit (NTU)
nuclear magnetic resonance spectroscopy, 245
nucleation, 68, **76**
  acceleration of, 22
  aggregation and, 130–131
  in protein folding, 25–26
  protein particulation due to, 391–393
  as rate-limiting step in aggregation, 24, 81–82
nucleation conformational and conversion (NCC) mechanism, 24–25
nucleation-dependent aggregation, 130–131
nucleation-growth mechanism, 37
nucleation hypothesis, 30–31
nucleation-polymerization (NP), 24, 130
nucleus, 68
Nyquist, Harry, 37

occupancy probability distribution (OPD), 39
octanoic acid, 152
octasulfate, 135
OD. *See* optical density (OD)
OFAAT. *See* one factor at a time (OFAAT)
off-lattice models, 34
O-fucosyltransferase 1, glycosylation of, 353
oligomeric intermediates, 24–25
oligomers, native structure, 19
O-linked N-acetylglucosamine (O-GlcNAc), 165, 353
one-bead models, 36
one factor at a time (OFAAT), 344
opalescence, 236, **237**
OPD. *See* occupancy probability distribution (OPD)
open β-barrels, *9*, 10
optical density (OD), 235
organic solvents, 161–162
  aggregation inhibition during refolding and, 167
  effects as function of temperature, 375, *376*
  protein aggregation and, **122**, 161–162
  protein destabilization and, 382
ornitine, 168
orthogonal β-sheets, *9*, 10
orthogonal methods, 442–443
osmolytes
  preferential interaction parameters of, 92–93
  role in stabilizing keratocyte growth factor, 368–376, *370*, *371*, *372*, *373*, *374*, *376*
  as stabilizers, 288, **289**
osmotic pressure, stabilization and, 21
osmotic second virial coefficient, 71–72, *72*
ovalbumin aggregation, 128, 131
  high temperature and, 133
  moisture content and, 177
  pH and, 137, **337**
  protein concentration and, 156
oxidation
  aggregation and, 129
  methionine, 41–42
oxidative modification, covalent cross-linking and, 42–43
oxidized glutathione, 167

p39A cellular retinoic acid-binding protein, 146, 164
packaging, primary, **311**
packing arrangements, 18–20
packing density, aggregation-prone regions and, 113
paddle mixer, 313
PAGE, 107, 110
paired helical filaments (PHFs), 130
PAMPs. *See* pathogen-associated molecular patterns (PAMPs)

pancreatic trypsin inhibitor, 133
parallel association, 4
parallel β-sheets, *9*, 10
parallel conformation, 17
  prediction of, 47
parallel sheets, in fibrillar aggregates, 6–7, *7*
parenteral products
  contamination sources, 323–324, *324*
  particles in, 316–324
Parkinson's disease, 2
particles/particulates. *See also* protein aggregates; protein aggregation
  defined, 320
  due to nucleation by foreign material in fill/finish manufacturing, 391–393
  formation of, 301–302
  irregular, 338, *338*
  limits for, 446
  in parenteral products, 316–324
    causes of particulate contamination, 323–324, *324*
    history of particle contamination control, 316–317
    regulatory requirements on, 317
    subvisible particle acceptance criteria, 318–320, *319*, *320*
    subvisible particle measurements, *317*, 317–318
    visual, meaning of, 322–323
    visual inspection, 316–322, *321*
  size distribution of, 393
particulate matter in injections test, **239**
particulate matter in ophthalmic solutions test, **239**
PASTA. *See* prediction of amyloid structure aggregation (PASTA)
pasteurized plasma solutions, 419
pathogen-associated molecular patterns (PAMPs), 409
patients
  consequences of protein immunogenicity for, 405–406
  immunogenicity and specific features of, 408
pattern recognition receptors (PRRs), 409
Pauling, Linus, 15
PBS, 128
  in candidate analysis, 261, 263, *265*, 266, *267*
PCS. *See* photon correlation spectroscopy (PCS)
PDB. *See* Protein Data Bank (PDB)
PDI. *See* protein disulfide isomerase (PDI)
PEG. *See* polyethylene glycol (PEG)
PEGylation, 80, 352, 407, 425
Peniophora lycii phytase (Phy), 130, 141, 142, 353
pentosan polysulfate, 148
peptibody A, 272–273, *273*
peptide bonds, 4
peptide interaction matrix analyzer (PIMA), 48
peptides, rationally designed, 150
peptyl-prolyl isomerases, 45
peripheral tolerance, 411
peristaltic pumps, 315
PES. *See* polyethersulfone (PES)
pH
  assessing protein stability upon transitioning into neutral, 266–267
  conformational stability and, 303–304
  freeze-thaw-induced aggregation and change in, 169
  increase in aggregation with changes in, 304, 305
  ionic strength and, 339–340
  mAb candidate analysis and, 261–262
  polycations/polyanions inhibition of aggregation and, 149
  during processing, 279–280
  protein aggregation and, **121**, 137–140, 336–337, **337**
    aggregate morphology and, 139–140
    aggregation mechanisms *vs.*, 140
    extreme pH values, 138
    indirect effect of, 139
    neutral and moderate acidic/basic conditions, 137–138
    protein aggregation *vs.* pI, 138–139
  scale-up and, 295
  self-aggregation and, 285
  self-association and, 272–273, *273*, 274
  solid-state, 174, 343
  stability of protein folds and, 13
  stability toward soluble aggregation and, 378–379

Ph. Eur. *See* European Pharmacopoeia (Ph. Eur.)
phage P22 coat protein mutants, 165
phage P22 wild-type coat protein, 127
pharmacokinetic/pharmacodynamic (PK/PD) behavior, 405
phase III trial, 437
phase II trial, 437, 438
phase I studies, 437, 438
phase I trial, 438
phase IV clinical trial, 437
phenylalanine (Phe), 91
phenyl methane sulfonyl fluoride (PMSF), 142
PHFs. *See* paired helical filaments (PHFs)
phosphate, 276
phosphate buffers, 140, 141, 148
photon correlation spectroscopy (PCS), 226–228
physical aggregates, **207**, 208
physical aggregation through formation of unfolding intermediates, 120, 126–127
physical state of solid, aggregation and, 176
physicochemical properties of proteins, 12–18
phytic acid, 147
pI. *See* isoelectric point (pI)
Pichia pastoris fermentation, 164
PIDS. *See* polarization intensity differential scattering (PIDS)
PIMA. *See* peptide interaction matrix analyzer (PIMA)
π-helix, 4
pi-pi interactions, 14
pi-stacking, 14
piston pumps, 314–315, *315*
pKa
  oxidation and, 42
  protein stability and, 20
planarity of peptide bond, backbone conformation and, 15–16
plasma-derived clotting factors, 404
Pluronic F68, 163, 170
Pluronic F127, 351
Pluronics, 346
PMSF. *See* phenyl methane sulfonyl fluoride (PMSF)
POE, 151
polarization intensity differential scattering (PIDS), **219**, 244
polar zipper sequences, 110
Poloxamers, 151
poly(acrylic acid), 148
polyalanine peptides, 37
polyalanine sequences, 16
polyanions, 135, 148–149
polycations, 148–149
polyclonal B-cell activation, 412
polyethersulfone (PES), 314
polyethylene glycol (PEG)
  aggregation inhibition and, 148, 167, 172, 279, 289, 342, 372–373
  excluded volume effects and, 372–373, *373*
  precipitation of proteins and, 282
polyethylene glycol (PEG)-300, 148
polyethylene glycol (PEG)-3500, 148
polyethylene glycol (PEG)-4000, 147, 148, 279
polyethylene glycol (PEG)-GCSF, **144**, 157, 341, 352
polyethylene terephthalate, 162
polyglutamine sequences, 41, 114
polyglutamine tracts, 1
polyhydric alcohols
  aggregation reduction and, **275**, 276, 278
  as stabilizers, **289**, 290
polyionic fusion constructs, 292
poly-L-glutamic acid, 148
polymer precipitation, 282
polymers
  aggregation reduction and, **144**, 147–151, **275**, 279
  amphoteric, 149–150
  BCRs and rigidity, 425
  cyclic, 150–151
  drying-induced aggregation and, 348–350
  neutral, 147–148
  polycations/polyanions, 148–149
  as stabilizers, 288–289, **289**
poly(methacrylic acid), 148
polyols
  aggregation and, 143–145
  effect on shelf life, 370, 371, *372*
  protein stabilization and, 369, 386

polypeptide chain, aggregation and length of, 41
polyphosphoric acid, 148
polypropylene containers, 163
polysorbate 20, 279, 284, 341, 346, 350, 351, 387. *See also* Tween 20
polysorbate 80, 279, 341, *341*, *342*, 347, 407. *See also* Tween 80
polysorbates, 308, 346
polyvinylidene difluoride (PVDF), 314
PONDR, 49
pooling, mixing of bulk drug solution after, **311**, 313
porcine insulin, immunogenicity and, 404
porcine somatotropin (pST), 148
positive displacement piston pumps, 314–315
potassium, 277
PRCA. *See* pure red cell aplasia (PRCA)
precipitation, of mAb, *313*, 313–314
predicting aggregation, 46–49
prediction of amyloid structure aggregation (PASTA), 48, 108, 109
preferential exclusion, 147, 373, 382, 384–385, 394
preferential hydration, 373, *374*
preferential interaction, 373–374, *374*
preferential interaction parameters, 82–85
  for backbone of amino acid ($\tau^{bb/32}$), 90
  of cosolute with protein ($\tau^{(u3)/32}$)
    additive approach to, 89–93
    molecular interactions and, 85–89
    simple perturbation theory for, 86–88
    size dependence of b_{23} and a_{23}, 88–89
  for side chain of amino acid *i* ($\tau^{sc,\ i/32}$), *90*, *91*, *93*
preformulation, **335**, 335–336
prenucleation, 68
preservatives, protein aggregation and, **122**, **144**, 152–153
pressure-induced aggregation, **123**, 170–171
prion protein (PrP), 1
  coaggregates, 160
prion protein (PrP) aggregation, 8, 25, 41, 110, 157, **158**, 159
prions, 23
PRL-1, 20
process conditions, stability to, 259–265, *262*, *264*, *265*
process control, 435
processing factors in protein aggregation, **123–124**
processing steps, 310–315, **311**
  drying, 348–350, **349**
  filling, **311**, 314–315, *315*
  filtration, **311**, 314
  freezing and thawing operations, 310–312, **311**, *312*, 346–347
  for liquid drug product, 347–348
  mixing of bulk drug solution after thawing or pooling, **311**, 313
  preparation of controlled protein delivery systems, 350–351
  preparation of final formulation with excipients, **311**, *313*, 313–314
  protection of proteins in, 345–351
  protein aggregation and, 164–173
    agitation, 169–170
    analytical methodologies, 173
    drying, 171–173
    fermentation/expression, 164–165
    freeze-thaw, 168–169
    high hydrostatic pressure, 170–171
    miscellaneous processes, 173
    protein delivery system preparation, 173
    purification, 168
    removal of aggregates, 281–285
    unfolding/refolding, 165–168
  reconstitution, 350
  shaking, 346
product aggregates, 271
product definition, **335**
product development, managing protein aggregation in, 333–354, *354*
  formulation development, 334–345, **335**, **337**, *338*, **339**, *341*, *342*, *354*
  protection of protein in processing steps, 345–351
    drying, 348–350, **349**
    freezing/thawing, 346–347

product development, managing protein aggregation in (*cont'd*)
manufacturing processes for liquid drug product, 347–348
preparation of controlled protein delivery systems, 350–351
reconstitution, 350
shaking, 346
structural modifications and, 351–353, *354*
production of biological/protein pharmaceuticals, controlling protein aggregation. *See* bulk production, controlling protein aggregation during
product modification, immunogenicity and, 407
product quality attribute, aggregates as, 435–447
product quality control, 436–437
product-related factors in protein immunogenicity, *406*, 406–407
product stability, 435
product storage
aggregation and, 440, **441**, 442
bulk DS stability and, 292–293
formulation development and, 334
heterogeneity in IgG2 aggregation during, 376–381, *377*, *380*
long-term, 265–268, **335**
verification of long-term stability, 343
proline
effect on shelf life, 370, *372*
protein stabilization and, 146, 288–289
proline isomerization, 45–46
PROSITE format, 110
protein
aggregation and function of, 104
aggregation-prone regions, 103–115
predicting, 106–114
energy landscapes, 105–106, *106*
microparticle surface interactions, 309
packing arrangements, 18–20
physicochemical properties, 12–18, *14*
solvent interactions and, 21–22
sources of, **123**, 163
structural features, 4–5
surface properties, 18–20
protein A affinity chromatography, 168
protein aggregates
amorphous, 5–6
assessment methods, 440–442
chromatography to remove, 282–283, *284*
classification of, 301–302, 416
detection and characterization of, 205–247
aggregate classification, 206–212, **207–208**, **211**
analytical tools, 212–246, **213–220**, **237**, **239**
dissociation of, 283–285
factors contributing to formation of, 440, **441**
fibrillar, 5, 6–11, *7*, *8*, *9*
forms of, 338, *338*
immunogenicity and, 407, 415–427
evidence of aggregates as risk factor, 417–420, *418*
factors responsible for, 423–425
mechanisms of antibody induction by protein aggregates, 420–423, *421*, *422*
predictive models, 426–427
risk analysis of aggregates, 426
irreversible, 287
pH and morphology of, 139–140
precipitation of, 282
as product quality attribute, 435–447
removal during processing, 281–285
reversible, 287
safety and biologic, 438–440
soluble, 75
stable, 10
submicron-sized, 446
therapeutic proteins and, 2
validation of test methods to detect and measure, 442–444
protein aggregation, 1–11, 63. *See also* particles/particulates
biotherapeutic production and, 104
case studies (*see* case studies)
cell culture and, 269–271
covalent cross-linking of strands, 42–43
deamidation, 44
domain swapping, 26–28, *27*
effects of interfaces on, 305–310
external factors, 119–178, *120*, **121–126**

agitation, **123**, 169–170
analytical methodologies, 173
buffering agents, **121**, **125**, 140–141
containers/closures, 162–163
denaturant and reducing agents, 159–160
drying, **123–124**, 171–173
excipients/additives, **122**, **125**, 143–153, **144**, 174–176
fermentation/expression, 164–165
freeze-thaw, **123**, 168–169
high hydrostatic pressure, **123**, 170–171
impurities, 160
ionic strength, **121**, 141–143
light, **121**, 163–164
metal ions, **122**, 157–159, **158**
organic solvents, 161–162
processing steps, 164–173
protein aggregation pathways, 120–131, *126*
protein concentration, **122**, **154**, 154–157
protein delivery system preparation, 173
purification, 168
solid-state condition and composition, **125–126**, 174–177
solution conditions and composition, 136–164
solution pH, **121**, 137–140
sources of protein, **123**, 163
temperature, **121**, 132–136, *133*
unfolding/refolding, **123**, 165–168
in frozen state, 381–385
influence of chemical alteration on, 40–46
length of polypeptide chain, 41
in liquid state, 368–381, *370*, *371*, *372*, *373*, *374*, *376*, *377*, *380*
in lyophilized state, 385–391
methionine oxidation, 41–42
models of, 22–28
monomer conversion, 23
mutations and rate of, 109
as natural consequence to changes in protein itself, 104, *106*
nomenclature, **64–66**, 209–212
non-covalent, 104
nucleation in protein folding, 25–26
oligomeric intermediates, 24–25
predicting, 46–49
preventing with solvent conditions, 274–281, **275**
processing steps and, 310–315, **311**
filling, 314–315, *315*
filtration, 314
freezing and thawing operations and, 310–312, *312*
mixing of bulk drug solution after thawing or pooling, 313
preparation of final formulation with excipients, *313*, 313–314
product development and (*see* product development, managing protein aggregation in)
proline isomerization, 45–46
protein folding and, 11–12, 105–106
protein function and, 104
purification and, 271–281
roles of conformational and colloidal stability in reducing rates of, 302–305
strategies to control during protein refolding, 286–292, **289**
structural features of aggregates, 5–11, *7*, *8*, *9*
in vivo, 2, 3
protein aggregation kinetics, 69, 75–80
chemical modifications affecting, 80–82
unfolded-limited aggregation and, 127
protein aggregation mechanisms, pH *vs.*, 140
protein aggregation nucleation, 130–131
protein aggregation pathways, 120–131, *126*
aggregation nucleation, 130–131
chemical modifications, 80–82
derivation of τ_{32} for van der Waals (vdW) mixture, 94–97
direct aggregation through self-association or chemical linkages, 127–129
distinguishing, 78–79
effects of cosolvents or cosolutes, 82–94

protein aggregation pathways (*cont'd*)
indirect aggregation through chemical degradation, 129
multiplicity in protein aggregation, 131
native and nonnative, 66–69, *67*
nomenclature, **64–66**
physical aggregation through formation of unfolding intermediates, 120, 126–127
protein aggregation from unfolded state, 129–130
thermodynamics of reversible self-association, 69–75, *72*
protein A resin, 168
protein-based vaccines, 104
protein CheY, 165
protein concentration
during processing, 280–281, *281*
protein aggregation and, **122**, **154**, 154–157
drying-induced, **349**, 350
enhanced protein aggregation at high concentrations, 155–156
protein behavior at high concentrations, 154–155
solubility-controlled protein aggregation in folded states, 157
unaltered or reduced protein aggregation at high concentrations, 156–157
unfolding/refolding and, 166
protein-cosolute interactions, 84–86, 87
Protein Data Bank (PDB), 19
protein delivery systems, 173
protein disulfide isomerase (PDI), 160
protein engineering, 258–259
protein folding, 3, 11–12, 66, *67*
aggregation and, 105–106
classical models based on chemical equilibrium, 29–31, *28*
coarse-grained models, 35–37
computational models, 33–35
information theory, 37–40
models, 29–40
multimeric, 67–68
nucleation in, 25–26
statistical mechanical models, 31–33, *32*
protein fusion, 351
protein-glutaminase, 44
protein interfaces, 39–40
protein L aggregation, trifluoroethanol and, 161
protein modifications, **441**
protein oligomerization, 20
protein products, regulatory oversight of. *See* regulatory oversight, of protein products
protein-protein interaction, 20
protein refolding, protein aggregation and, **123**, 165–168
catalysts, 167
denaturants and, 166
high-pressure, 290–291
protein concentration and, 166
strategies to control, 286–292, **289**
temperature and, 165–166
thiol/disulfide agents and, 167
use of additives/stabilizers and, 288–290
protein stability, 3
alcohols and, **289**, 382
amino acids and, 21–22, 175, **275**, 288–289, **289**
aromatic residues and, 14–15
assessing after agitation, 267–268
cosolvents and, 369, 372
excipients and, 340
formulation development and, 265–268
ionic strength and, 142–143
salt bridges and, 19–20
sugars and, **289**, 290, 369
temperature and, 132–133, *133*
protein tyrosine phosphatase (PTPase), 42
protein unfolding
denaturants and, 166
pH and, 272–273
process conditions and, 259
protein aggregation and, 66, *67*, 165–168
protein concentration and, 166
temperature and, 165–166
thermal, 132–133
of keratocyte growth factor, 369–370, *370*
protein vaccination, 416

protein-water interactions, 84–86
proteoglycans, 22
proteosomal proteins, aggregates and, 2
PRRs. *See* pattern recognition receptors (PRRs)
P22 tailspike, 67
P22 tailspike polypeptide, 127
Public Health Service Act (1944), 436
pure red cell aplasia (PRCA), 405–406
purification, 168, 259, 271–281, 440
putrescine, 146
PVDF. *See* polyvinylidene difluoride (PVDF)
3-(1-pyridinio)-1-propanesulfonate (NDSB), 152
pyrophosphate buffer, 141

QbD. *See* quality by design (QbD)
QELS. *See* quasi-elastic light scattering (QELS)
quality by design (QbD), 258
quasi-elastic light scattering (QELS), 226–228
quaternary structure, 5
quench freeze-thaw, 347

rabbit IgG, 155
radial distribution function (RDF), 84
radial piston pump, 173
radial spatial distribution (RDF), 39
raffinose, as protein stabilizer, 278
Ramachandran plot, of globular proteins, 15–16
Raman spectrometer, **220**, 245–246, 260
  assessment of contamination in parenteral products, 324, *324*
rate-limiting steps, in aggregation process, **76**, 78–79
rationally designed peptides, aggregation and, 150
Rayleigh's ratio of protein solution, 224
Rc. *See* rusticyanin (Rc)
RDF. *See* radial distribution function (RDF)
receptors, 104
recombinant consensus α-interferon (rConIFN), 351
recombinant factor VIIa (rFVIIa), 41, 134, **144**, 336, 340
recombinant factor VIII (rFVIII), 119, *120*, 134, 155, 156
recombinant factor XIII (rFXIII), 138, 170, 346
recombinant human albumin (rHA) aggregation, 174, 177
recombinant human ciliary neurotrophic factor (CNTF), 341
recombinant human DNase (rhDNase) aggregation
  excipient inhibition of, 175
  protein concentration and, 166
recombinant human factor VIII (rhFVIII) aggregates, 162, 212
  analysis of using bis-ANS, 241
  detection of using SEC, 222
  metal ions and, 157, 158
recombinant human granulocyte colony-stimulating factor (rhGCSF) aggregation
  colloidal stability and, 128, 305
  excipients and, **144**, 153
  ionic strength and, **339**
  pH and, 137, 139, **337**
  surfactants and, 164
  unfolding intermediates and, 127
recombinant human granulocyte-macrophage colony-stimulating factor (GM-CSF), 352
recombinant human growth hormone (rhGH), 127, 151, 152, 166, 222, 351
recombinant human interleukin-1 receptor antagonist (rhIL-1ra), 13
  buffers and, 140, 141
  drying-induced aggregation and, **349**, 350
  excipients and, 153
  freeze-drying and, 171
  lyophilization of, 172
  temperature and, 134
recombinant human interleukin-2 (rhIL-2), 137
  immunogenicity and, 420
recombinant humanized monoclonal antibody (rhuMAb) anti-CD20, 146
recombinant human keratinocyte growth factor (rhKGF), 148, 150

recombinant human megakaryocyte growth and development factor (rhMGDF), 133
recombinant human monoclonal antibody, aggregation in lyophilized, 388–389
recombinant human platelet-activating factor acetylhydrolase (rhPAF-AH), 134, 162, 163
recombinant human serum albumin (rHSA) aggregation, 175
recombinant interleukin-2 (rIL-2) aggregation, 306, 310
recombinant PEGylated staphylokinase SY161, 345
recombinant porcine growth hormone (rpGH), 133, 166
recombinant protein, production of, 2–3
reconstitution, **124**, 350
recovery, aggregate detection assays with low, 443
reducing agents, 159–160
regulatory oversight, of protein products, 435–447
  assessment of aggregates, 440–445, **441**
  biosimilars, 446–447
  control of subvisible particulates, 446
  regulatory documents, 436–437, 445, **445**
  safety concerns, 438–440
regulatory process, 436–438
  drug development to licensure, 436–437
  pre- and post-marketing comparability, 438
repetitive epitope, *421*
replica exchange protocols, 34
residues, aggregation and, 41
resin matrix, 290
reversibility, of protein aggregates, **207**, 208–209
reversible aggregates, **207**, 208–209, 210, 287
reversible self-association, thermodynamics of, 69–75
rF. *See under* recombinant factor
rh. *See under* recombinant human
rheumatoid arthritis, 404
rhGCSF. *See* recombinant human granulocyte colony-stimulating factor (rhGCSF) aggregation
rhGH. *See* recombinant human growth hormone (rhGH)
rhKGF. *See* recombinant humankeratinocyte growth factor (rhKGF)
rhMGDF. *See* recombinant human megakaryocyte growth and development factor (rhMGDF)
rhodanses, 17–18
rhPAF-AH. *See* recombinant human platelet-activating factor acetylhydrolase (rhPAF-AH)
rhuMAb, 175
rhuMab VEGF, detection of aggregates using SEC, 222, 223
ribonuclease A (RNase A), 49, 133, 137, 160, 163, 291
ribonuclease (RNase), 11, 136, 149, 174
ribonuclease (RNase) T1, 20
risk analysis, of protein aggregates, 426
rituximab, 408
RosettaDesign, 109
runaway domain swapping, 28
rusticyanin (Rc), 157, 159

saccharides, drying-induced aggregation and, 348
safety, product aggregates and, 438–440
salmon calcitonin, 407
SALSA, 108, 109
salt
  drying-induced aggregation and, **349**
  inhibition of aggregation and, 175
  protein aggregation and, **122**, 142
  protein concentration, self-aggregation, and, 280, *281*
  during purification process, 285
  self-association and, 272–273, *273*, 274
  as stabilizers, **289**
salt bridges, protein stability and, 19–20
"salting out" proteins, 282
SAP. *See* spatial aggregation propensity (SAP)
sarcosine, aggregation inhibition during refolding and, 167
SASA. *See* solvent-accessible surface area (SASA)

scale-up, 293–295
freeze/thaw, 294
mechanical effectors, 293
mixing, 294
pH, 295
temperature, 294–295
scanning electron microscopy (SEM), **219**, 244–245
S-carboxymethyl-α-lactalbumin, 147, 148, 149
Scatchard notation, 82
SCF. *See* supercritical fluid (SCF)
scrapie amyloid prion protein (PrP27–30), 127
SDS. *See* sodium dodecyl sulfate (SDS)
SDS-PAGE. *See* sodium dodecyl sulfate-polyacrylamide gel electrophoresis (SDS-PAGE)
SE-AUC. *See* sedimentation equilibrium analytical ultracentrifugation (SE-AUC)
SEC. *See* size-exclusion chromatography (SEC)
secondary drying, 386
secondary structure, 16, 17–18, 260
second osmotic virial coefficient, size dependence of, 88–89
second virial coefficient (B_{22}), 128
SEDFIT software, 229
sedimentation equilibrium analytical ultracentrifugation (SE-AUC), 228, 229–231
sedimentation velocity analytical ultracentrifugation (SV-AUC), 228–229, 231
sediment coefficients, 230
seeding, mechanistic features of, 114
seeds, molecular simulations to identify, 113–114
SE-HPLC. *See* size-exclusion high-performance liquid chromatography (SE-HPLC)
self-antigen organization, 425
self-association, 63, 66, *67*, 67–68, 271–272
cosolvents or cosolutes and, 82
direct aggregation through, 127–129
of folded/native proteins, 70–73, *72*
nonnative, 73–75
thermodynamics of reversible, 69–75
SEM. *See* scanning electron microscopy (SEM)
sequence analysis, 258–259
serine (Ser), 19, 39, 40
serpentine model of amylin fiber, 7
serum albumin, 136
serum cryoglobulins, 136
SFD. *See* spray-freeze drying (SFD)
shadow casting, 245
shaker tables, 336
shaking, 313
protection of proteins during, 346
protein aggregation and, **123**, 169–170
shearing, protein aggregation and, **123**, 169–170, 309–310
shelf life
of keratocyte growth factor, 370–372, *372*
of protein product, 334
sialylation, 351
sickle cell anemia, 1
side-chain pKa, 20
Sigmaα-La, 163
silicone, aggregation and, 163
silicone oil/water interfaces, aggregation and, 309
simple perturbation theory for $\tau^{(u3)/32}$, 86–88
simulated annealing, 34
site-specific mutagenesis, 351–352
size, aggregate, **207–208**, 209, 446
immune responses and, 424
size-exclusion chromatography (SEC), 75, **213**, 221–223
aggregation monitoring by, 378
assessing protein stability using, 266, 267
combined with static light scattering, 225–226
size-exclusion high-performance liquid chromatography (SE-HPLC), 267, 318, 440–441, 444
Sjogren's syndrome, 408
SLS. *See* static light scattering (SLS)
sodium chloride, aggregation during refolding and, 168
sodium cholate, 152

sodium dodecyl sulfate-polyacrylamide gel electrophoresis (SDS-PAGE), **215**, 233–234, 266, 389
sodium dodecyl sulfate (SDS), 5, 152, 159
sodium phosphate buffers, cooling and freezing and, 312
solid, physical state of, 176
solid-liquid interfaces, effects on protein aggregation, 306–307
solid-state condition and composition, effect on aggregation, 174–177
  excipients and excipient levels, 174–176
  moisture content, 177
  pH, 174
  physical state of the solid, 176
solid-state factors, protein aggregation and, **125**
solid-state pH, protein aggregation and, **125**, 343
solubility
  apolar surface area and, 20
  charged residues on surface and, 18–19
  immune response and aggregate, 424
solubility-controlled protein aggregation, 157
soluble aggregates, 68, 75, **207–208**, 209, 210, 211, 301
  pathways leading to, 66, *67*
soluble interleukin-13 receptor (sIL-13R), 269
solution conditions and composition, protein aggregation and, 136–164
  buffering agents, 140–141
  conformational and colloidal stability, 303
  containers/closures, 162–163
  denaturant and reducing agents, 159–160
  excipients/additives, 143–153, **144**
  impurities, 160
  ionic strength, 141–143
  light, 163–164
  metal ions, 157–159, **158**
  organic solvents, 161–162
  pH, 137–140
  protein concentration, **154**, 154–157
  protein sources, 163
solution NMR, 31
solvent-accessible surface area (SASA), 165, 353
solvent-averaged protein-cosolute interactions, 87
solvent conditions
  additives, 275–280
  preventing aggregation with, 274–281, **275**
  protein concentration during processing, 280–281, *281*
solvent interactions, 21–22. *See also* organic solvents
solvophobic attractions, 85
sorbitol
  aggregation and, 143
  as cryoprotectant, 381, 382–385
  inhibiting aggregation during spray-drying and, 172
  preferential interaction parameters of, 92
  as protein stabilizer, 278, 340
sorbitol crystallization, 136, 382
SOS, 147, 149
soybean trypsin inhibitor (STI), 129
spatial aggregation propensity (SAP), 112
specification, defined, 444
spectroscopy-based techniques, 245–246
spermidine, 146
spray-drying, **124**, 172
spray-freeze drying (SFD), **124**, 172–173
stability analysis techniques, 259–260, 444
stabilizers, using during refolding, 288–290
stable aggregates, 10
staphylococcal protein A, 34
staphylococcal protein A ligand, 291
static light scattering (SLS), **213**, 223–226
statistical mechanical models of protein folding, 31–33
steric hindrance of protein-protein interactions, 147
steric interactions, 86–87, 95
STI. *See* soybean trypsin inhibitor (STI)
stirring, protein aggregation and, **123**, 169–170

Stokes-Einstein equation, 132, 226–227
storage. *See* product storage
streptokinase, 406, 410
stressed conditions, 442
structural content, aggregation-prone regions and, 113
structural features
of protein aggregates, 5–11, *7*, *8*, *9*
of proteins, 4–5
structural modifications, aggregation control by, 351–353
subcutaneous administration, immunogenicity and, 408
"substantially free," 316
subtilisin, protein engineering and, 259
subvisible aggregates/particles, **208**, 209, 211, 301, 302
acceptance criteria, 318–320, *319*, **319**, *320*
analytical techniques for, 237–240
control of, 317, 446
immunogenicity and, 426
measurement of, *317*, 317–318
sucrose
aggregation and, 143
aggregation inhibition during refolding and, 167
aggregation reduction and, 276
as cryoprotectant, 381, 382–383
drying-induced aggregation and, 348
effect on structural perturbation and aggregation, 307
freeze-drying aggregation and, 172
freeze-thaw-induced aggregation and, 347
melting temperature and, 135
preferential interaction parameters of, 92, 93
as protein stabilizer, 278, 290, 304, 340
in lyophilized state, 386, 387, 388, 389–391
in solid state, 174–175
thermal unfolding and, 370, *371*
sucrose-water solutions, 85
sugars
aggregation reduction and, **125**, 143–145, **275**, 276, 278
drying-induced aggregation and, **349**
effect on shelf life, 371
effect on structural perturbation and aggregation, 308
as protein stabilizers, **289**, 290, 369
stabilizing against aggregation in solid state and, 174–175
thermal unfolding and, 370–371, *371*
sulfate, 276
sulfobetaine, 152
Sup35, 24
supercritical fluid (SCF) drying, 173
superoxide dismutase (SOD1), 42
surface, protein aggregation and, **122**
surface activity, 147
surface properties, 18–20
surface skin, 338, *338*
surface tension
excipients and, 307–308
stabilization and, 21
surfactants, 151–152
aggregation during fermentation and, 164
aggregation reduction and, **275**, 278–279, 346
inhibiting aggregation during spray-drying and, 172
ionic, 152
nonionic, 151–152
protein adsorption and, 162
protein aggregation and, **122**, **144**
shaking/shearing-induced aggregation and, 170
as stabilizers, **289**, 290
structural perturbation and aggregation and, 307–308
SV-AUC. *See* sedimentation velocity analytical ultracentrifugation (SV-AUC)
Svedberg equation, 228
symbols, **64–66**
syringes, aggregates in, 440
systemic lupus erythematosus, 408

TA. *See* templated assembly (TA)
TANGO, 47–48, 107, 110
tau protein, 10, 46, 108, 130
T cell dependent (Td) activation, 411–412, *413*
T cell independent (Ti) activation, 412–415, *414*, 421–422, *422*, 423–425

T-cell receptors (TCRs), 411, 412, *413*
T cells, 411
  helper, 410
T-cell tolerance, breaking, 423
TCRs. *See* T-cell receptors (TCRs)
Teflon containers, 163
TEM. *See* transmission electron microscopy (TEM)
temperature
  bioreactor conditions and culture, 270
  melting, 135, 370–372, *372*
  protein aggregation and, **121**
    fermentation and, 164
    high temperature and, 133–135
    low temperature and, 136
    melting temperature and, 135
    pressure-induced, 171
  protein degradation and, 379–381, *380*
    effects of excipients, 373–375, *374*
  protein folding and, 13–14, *14*
  protein refolding and, 165–166
  protein stability and, 132–133, *133*
  protein unfolding and, 10–11, 13–14, *14*, 165–166
  scale-up, 294–295
temperature-leap tactic, 288
templated assembly (TA), 23, 24
tertiary structure, 4
  conditions causing perturbations in, 302, 304
  in lyophilized antibodies, 388
tetanus toxoid, sugar protection for in solid state, 175
tetramer formation, 43
TFE. *See* trifluoroethanol (TFE)
thawing, 310–312
  aggregation and, **441**
  mixing of bulk drug solution after, **311**, 313
  protection of proteins during, 346–347
therapeutic monoclonal antibodies, aggregation-prone regions in, **112**
  prediction of, 110–113, *111*
therapeutic proteins, aggregates and, 2
  immunogenicity of, 403–428
thermal unfolding, 132–133
  of keratocyte growth factor, 369–370, *370*
thermodynamic hypothesis, 11, 31, 271
thermodynamics
  of hydration, 38
  of reversible self-association, 69–75
thermophilic proteins, 19
thioflavin T (ThT), 6, 105, 241, 242
thiols, 167, 389
three-D profile methods, 109
threonine (Thr), 19, 40
thrombin, 133
ThT. *See* thioflavin T (ThT)
thyroid-stimulating hormone (TSH), 165
time-pressure pumps, 315
titration scaling, risks of, 295
TLRs. *See* toll-like receptors (TLRs)
tobacco mosaic virus coat protein, 152
toll-like receptors (TLRs), 409, *414*
topical administration, immunogenicity and, 408
toxicity, protein aggregates and, 333
traditional formulation development, 334–343
  formulation and stability evaluation, 336–343, **337**, *338*, **339**, *341*, *342*
  preformulation, 335–336
  stages and tasks, **335**
transfer free energy, 82–85
  for amino acids in osmolyte concentration, 90
  for protein from water to water-cosolute mixture ($\Delta\mu^{tr/2}$), additive approach to, 89–93
transgenic animal models, 427
transmission electron microscopy (TEM), 79, *79*, **219**, 244–245
transparency, of parenteral products, 324
transthyretin (TTR), 10, 67
treatment-related factors in protein immunogenicity, *406*, 408
trehalose
  inhibiting aggregation during freeze-drying and, 172
  inhibiting aggregation during spray-drying and, 172
  preferential interaction parameters of, 93
  protein aggregation and, 143–144
    drying-induced, 348
    freeze-thaw-induced, 347
    solid state and, 174–175

as protein stabilizer, 278, 290
in lyophilized state, 386, 388, 390–391
thermal unfolding and, 370, *371*
vacuum-drying and, 171
trifluoroethanol (TFE), 16
trifluoroethanol (TFE)-induced aggregation, 131, 146, 161, 343
trimethylamine-N-oxide (TMAO), 10, 16, 146
Tris buffer, aggregation and, 140
trisubstituted p-succinamidophenyl-glucopyranoside derivatives, 353
Triton X-100, 151–152, 159
trypsin (Trp), 91, 133
tryptophan, 40, 260
T7 endonuclease I (T7EI), 28, *28*
TSH. *See* thyroid-stimulating hormone (TSH)
TTR. *See* transthyretin (TTR)
tubulin, 134
tumor necrosis factor (TFN)-alpha, 404
tumor necrosis factor (TNF) aggregation
excipients and, 343
sugar protection for in solid state, 174
tungsten oxides, aggregation and exposure to, 440
tunicamycin, 165
turbidimetry, **217**, 235–237, **237**
turbidity, 338, *338*
assessing, 322
turns, 4
Tween 20, 151–152, 164, 172, 175, 279. *See also* polysorbate 20
Tween 80, 151, 162, 164, 172, 279. *See also* polysorbate 80
Tweens, 170
drying-induced aggregation and, **349**
two-bead models, 36
two-domain γD-crystallin proteins, 44
two-moment model, 38
two-state cooperative folding model, 32, *32*
Tycko model of amylin fiber, 7
Tyndall effect, 322
type 1 activation, 412, *414*
type 2 activation, 412–413, *414*, 424–425
L-tyrosine, 351
tyrosine (Tyr), 15, 40, 260

UF. *See* ultrafiltration (UF)
UHPH. *See* ultra-high-pressure homogenization (UHPH)
ultrafiltration (UF), 335
ultra-high-pressure homogenization (UHPH), 170
unfolded-limited aggregation, aggregation kinetics and, 127
unfolded state. *See also* protein unfolding
protein aggregation from, 129–130
unfolding intermediates, 120, 126–127
unfolding-limited kinetic pathway, **76**
U.S. National Formulary (NF), 316
United States Pharmacopoeia (USP)
analytical techniques for subvisible particles, 238, **239**
analytical techniques for visible particles, 240–241
limits for particulates, 446
potential causes of particulate contamination, 323
regulatory requirements on particulate contamination and control of subvisible and visible particles in parenteral products, 317, 320
"substantially free," 316
on visual clarity, 236–237
visual inspection methods, 321
urea
backbone exposed to solvent, 91
effect on structural perturbation and aggregation, 307
preferential interaction parameters of, 93
protein aggregation and, 147, 159
as protein stabilizer, 277, 290
self-association and, 283, *284*
solubilization of proteins and, 5, 277
Ure2p (Ure2p$_{1-94}$), 113–114
USP. *See* United States Pharmacopoeia (USP)
UV irradiation, aggregation and, 163–164
UV light, assessing protein stability after exposure to, 268

vaccination, protein, 416
vaccine development, 438

vaccines, protein-based, 104
vacuum-drying, **123**, 171
valine (Val), 16, 40
valosin-containing protein (VCP), 149
van der Waals contacts, 14
van der Waals forces, 127
van der Waals (vdW) fluid, 86
van der Waals (vdW) mixture, derivation of τ_{32} for, 94–97
variable regions, 368
VCP. *See* valosin-containing protein (VCP); viral coat protein (VCP)
viral coat protein (VCP), 425
virial coefficient values, 72
virus-like particles (VLPs), 415
visible aggregates/particles, **208**, 209, 210, 211, 301, 302
  analytical techniques for, 240–241
  control of, 317
"visible," defining, 322–323
visible/visual inspection
  history of, 316–317
  methods, *321*, 321–322
  requirements, 320–321
VLPs. *See* virus-like particles (VLPs)
volume effects, 87
water, cosolutes and, 84–85
water binding sites, 387–388
water-cosolvent mixture, 92
water molecules, hydration shell of, 21
water substitution hypothesis, 390–391
weak binding, 147
"weakly" self-associated oligomers, 68
weight-average molecular weight, 75
whey protein, 160, 177
whey protein concentration solutions, 158–159
whey protein isolate, 170
white-light inspection, 322
Wyman linkage function, 304

xylitol, as protein stabilizer, 278

zeins, 161
Zimm plot, 224
zinc
  insulin and, 210
  protein aggregation and, 157, **158**, 342
Zyggregator, 47, 108

Printed in the USA/Agawam, MA
September 16, 2026

910661.029